Student's Solution

for use with

Intermediate Algebra
A Real-World Approach

Second Edition

Ignacio Bello
Hillsborough Community College

Fran Hopf
Hillsborough Community College

Prepared by
Joe Kemble
Lamar University

and

Jon Weerts
Triton College

 **Higher Education**

Boston   Burr Ridge, IL   Dubuque, IA   Madison, WI   New York   San Francisco   St. Louis
Bangkok   Bogotá   Caracas   Kuala Lumpur   Lisbon   London   Madrid   Mexico City
Milan   Montreal   New Delhi   Santiago   Seoul   Singapore   Sydney   Taipei   Toronto

The **McGraw·Hill** Companies

Student's Solutions Manual for use with
INTERMEDIATE ALGEBRA: A REAL-WORLD APPROACH, SECOND EDITION
IGNACIO BELLO AND FRAN HOPF

Published by McGraw-Hill Higher Education, an imprint of The McGraw-Hill Companies, Inc., 1221 Avenue of the Americas,
New York, NY 10020. Copyright © 2006 by The McGraw-Hill Companies, Inc. All rights reserved.

No part of this publication may be reproduced or distributed in any form or by any means, or stored in a database or retrieval system,
without the prior written consent of The McGraw-Hill Companies, Inc., including, but not limited to, network or other electronic
storage or transmission, or broadcast for distance learning.

 This book is printed on recycled, acid-free paper containing 10% postconsumer waste.

3 4 5 6 7 8 9 0 QSR/QSR 0 9 8 7

ISBN 978-0-07-294558-4
MHID 0-07-294558-3

www.mhhe.com

Contents

Chapter 4   Solving Systems of Linear Equations and Inequalities

Chapter 5   Polynomials

Chapter 6   Rational Expressions

Chapter 7   Rational Exponents and Radicals

Chapter 8   Quadratic Equations and Inequalities

Chapter 9   Quadratic Functions and the Conic Sections

Chapter 10   Functions—Inverse, Exponential, and Logarithmic

Appendix A: Sequences and Series

Preface

The <u>Student's Solutions Manual</u> to accompany <u>Intermediate Algebra: A Real-World Approach,</u> <u>2nd Edition</u> by Ignacio Bello and Fran Hopf contains detailed solutions of the margin exercises, the odd numbered exercises, the review exercises, and the cumulative review exercises. I have attempted to provide solutions consistent with the procedures introduced in the textbook. Every attempt has been made to make this manual as error free as possible. It is my desire that this manual will assist and enrich student's understanding of algebra.

A number of people need to be recognized for their contributions in the preparation of this manual. Thanks go to David Dietz of McGraw-Hill for inviting me to author this manual, to Cindy Trimble and Colleen Fitzpatrick for their assistance in trouble shooting the project, and to Jon Weerts of Triton College for the many hours of guidance, error correcting, editing and assembling of this manual. Finally, thanks to my colleagues at Lamar University and my family for their support of this project.

Joe Kemble
Lamar University
P.O. Box 10060
Beaumont, TX 77710
kemblejd@hal.lamar.edu

Chapter 1   The Real Numbers

1.1   Numbers and Their Properties

Problems 1.1

1.   **a.** $\{4, 5, 6, 7\}$ **b.** $\{0, 1, 2, 3\}$ **c.** $\{-1, -2, -3, -4\}$ **d.** $\{0\}$

2.   **a.** $\frac{3}{8} = 0.375$ **b.** $\frac{5}{11} = 0.454545\ldots$ or $0.\overline{45}$ **c.** $\frac{95}{60} = 1.583333\ldots$ or $1.58\overline{3}$

3.

Set	$\frac{8}{5}$	π	6	$\sqrt{3}$	-11
Natural numbers			✓		
Whole numbers			✓		
Integers			✓		✓
Rational numbers	✓		✓		✓
Irrational numbers		✓		✓	
Real numbers	✓	✓	✓	✓	✓

4.   **a.** -3 **b.** $-(-2.5) = 2.5$ **c.** $-\frac{3}{5}$

5.   **a.** $|-19| = 19$ **b.** $\left|\frac{1}{6}\right| = \frac{1}{6}$ **c.** $|-0| = 0$
     d. $|3.1| = 3.1$ **e.** $\left|0.\overline{6}\right| = 0.\overline{6}$ **f.** $-\left|-\frac{5}{7}\right| = -\frac{5}{7}$

6.   **a.** $-8 < -4$ **b.** $-7 < -6$ **c.** $0 > -\frac{1}{3}$
     d. $\frac{1}{3} < \frac{1}{2}$ **e.** $-2.1 > -2.3$

Exercises 1.1

1. $\{1, 2\}$ **3.** $\{5, 6, 7\}$

5. $\{-1, -2, -3\}$ **7.** $\{0, 1, 2, 3\}$

9. $\{1, 2, 3, 4, \ldots\}$ **11.** $\{x \mid x \text{ is an integer between } 0 \text{ and } 4\}$

13. $\{x \mid x \text{ is an integer between } -3 \text{ and } 3\}$ **15.** $\{x \mid x \text{ is an even number between } 19 \text{ and } 78\}$

17.   False; 0 is not a natural number   **19.** True; -0.3 is a rational number.

21.   True   **23.** True; 8.112233 is a rational number.

25.   True; fractions are rational numbers.   **27.** $\frac{2}{3} = 0.\overline{6}$

29. $\frac{7}{8} = 0.875$ **31.** $\frac{5}{2} = 2.5$

1

33. $\frac{7}{6} = 1.1\overline{6}$

35.

	N	W	I	Q	H	R
$\frac{-3}{8}$				✓		✓
37. $\sqrt{8}$					✓	✓
39. $0.\overline{3}$				✓		✓
41. 0.9				✓		✓
43. 3.1416				✓		✓

45. True; all natural numbers are rational numbers.

47. False; R cannot be a subset of the set W. For example: $\frac{1}{2} \notin W$. A true statement is $W \subseteq R$.

49. -8

51. 7

53. $-\frac{3}{4}$

55. $\frac{1}{5}$

57. -0.5

59. $-0.\overline{2}$

61. $1.\overline{36}$

63. $\pi\pi$

65. $\left|10\right| = 10$

67. $\left|-17\right| = 17$

69. $\left|\frac{3}{5}\right| = \frac{3}{5}$

71. $\left|0.\overline{5}\right| = 0.\overline{5}$

73. $\left|-3.\overline{61}\right| = 3.\overline{61}$

75. $-\left|\sqrt{2}\right| = -\sqrt{2}$

77. $\left|\pi\pi\right| = \pi$

79. $-5 < 2$

81. $-6 > -8$

83. $\frac{1}{2} > \frac{1}{4}$

85. $-\frac{3}{5} < -\frac{1}{4}$

87. $-3.5 < -3.4$

89. $\frac{11}{100} = 0.11$; Burger King has the larger percentage of fat since $0.11009 > 0.11$.

91. $\frac{2896}{38210} \approx 0.0758$; The Advanced Degree had the better increase in average earnings from 1977 to 1997 since $0.25 > 0.0758$.

93. The second tallest is Monjane at 8 feet 0.75 inch.

95. False; because 5 is a number, not a set. A true statement would be $\{5\} \subseteq N$ or $5 \in N$.

97. Sample answer: A rational number is: $\left\{ \frac{a}{b} \mid a \text{ and } b \text{ are integers and } b \neq 0 \right\}$. An irrational number is not rational.

99. Sample answer: Rational numbers include integers, fractions, and decimals, while integers do not include fractions (ex. $\frac{1}{2}$), or decimals.

101. $\frac{1}{8} = 0.125$ **103.** -8

105. $\left| -(-2) \right| = |2| = 2$

107. $\{ x \mid x \text{ is 3 or an integer obtained when 3, 4, 5, } \ldots \text{ is added in succession} \}$

109. $\frac{1}{3} > 0.331332333334\ldots$

1.2   Operations and Properties of Real Numbers

Problems 1.2

1.
  a. $-3 + (-16) = -19$ **b.** $0.7 + (-0.2) = 0.5$ **c.** $-0.5 + 0.3 = -0.2$
  d. $\frac{5}{7} + \left(-\frac{3}{7}\right) = \frac{2}{7}$ **e.** $-\frac{5}{6} + \frac{2}{3} = -\frac{5}{6} + \frac{4}{6} = -\frac{1}{6}$ **f.** $-0.4 + 0.8 = 0.4$

2.
  a. $-15 - 3 = -18$ **b.** $-0.7 - (-0.3) = -0.7 + 0.3 = -0.4$
  c. $-\frac{1}{5} - \left(-\frac{1}{5}\right) = -\frac{1}{5} + \frac{1}{5} = 0$

3.
  a. $9-(-7) = -63$ **b.** $(\cdot\, 4.5)\cdot 2 = \cdot\, 9.0 \text{ or } \cdot\, 9$

4.
  a. $\left(\cdot\, \frac{2}{5} \right) \cdot \frac{5}{7} = \cdot\, \frac{2 \cdot \cancel{5}}{\cancel{5} \cdot 7} = \cdot\, \frac{2}{7}$ **b.** $\left(\frac{3}{14} \right) \cdot \left(\frac{7}{6} \right) = \frac{\cancel{3} \cdot \cancel{7}}{2 \cdot \cancel{7} \cdot 2 \cdot \cancel{3}} = \frac{1}{4}$

5.
  a. $60 \div 10 = 6$ **b.** $\frac{48}{-3} = -16$ **c.** $\frac{-18}{-2} = 9$
  d. $-14 \div 2 = -7$ **e.** $-4 \div 0$ is undefined   **f.** $4.8 \div 1.6 = 3$
  g. $\frac{4.2}{-2.1} = -2$ **h.** $\frac{-3.8}{-1.9} = 2$ **i.** $0 \div 9.2 = 0$

6.
  a. $\frac{9}{5}$ **b.** $-\frac{8}{7}$ **c.** 2

7.
  a. $\frac{3}{5} \div \left(\cdot\, \frac{4}{7} \right) = \frac{3}{5} \cdot \left(\cdot\, \frac{7}{4} \right) = \cdot\, \frac{21}{20}$ **b.** $\left(\cdot\, \frac{6}{7} \right) \div \left(\cdot\, \frac{3}{5} \right) = \left(\cdot\, \frac{2\cancel{6}}{7} \right) \cdot \left(\cdot\, \frac{5}{\cancel{3}} \right) = \frac{10}{7}$
  c. $\left(\cdot\, \frac{4}{5} \right) \div \frac{8}{5} = \left(\cdot\, \frac{\cancel{4}}{\cancel{5}} \right) \cdot \frac{\cancel{5}}{\cancel{8}2} = \cdot\, \frac{1}{2}$

8.
  a. Commutative property of multiplication   **b.** Commutative property of multiplication
  c. Associative property of multiplication   **d.** Commutative property of multiplication

Exercises 1.2

1. $\frac{3}{5}+\left(-\frac{1}{5}\right)=\frac{2}{5}$

3. $-0.3+0.2=-0.1$

5. $-4+6=2$

7. $-0.5+(-0.3)=-0.8$

9. $-\frac{1}{5}+\frac{1}{4}+\frac{3}{20}=-\frac{4}{20}+\frac{5}{20}+\frac{3}{20}=\frac{4}{20}=\frac{1}{5}$

11. $6-13=-7$

13. $0.6-0.9=-0.3$

15. $\frac{1}{7}-\frac{3}{8}=\frac{8}{56}-\frac{21}{56}=-\frac{13}{56}$

17. $-8-4-2=-14$

19. $-0.4-0.2=-0.6$

21. $-\frac{3}{7}-\frac{2}{9}=-\frac{27}{63}-\frac{14}{63}=-\frac{41}{63}$

23. $-6-(-5)=-6+5=-1$

25. $-8-(-4)=-8+4=-4$

27. $-0.7-(-0.6)=-0.7+0.6=-0.1$

29. $-\frac{2}{7}-\left(-\frac{4}{3}\right)=-\frac{2}{7}+\frac{4}{3}=-\frac{6}{21}+\frac{28}{21}=\frac{22}{21}$

31. $-5(8)=-40$

33. $4(-3)=-12$

35. $-10(-5)=50$

37. $-3(4)(-5)=-12(-5)=60$

39. $-4(-2)(5)=8(5)=40$

41. $-3(5)(-2)=-15(-2)=30$

43. $4(-5)(2)=-20(2)=-40$

45. $-2.2(3.3)=-7.26$

47. $-1.3(-2.2)=2.86$

49. $\frac{5}{6}\left(-\frac{5}{7}\right)=-\frac{25}{42}$

51. $\cdot\ \frac{3}{5}\left(\cdot\ \frac{5}{12}\right)=\cdot\ \frac{\cancel{5}}{\cancel{5}}\left(\cdot\ \frac{\cancel{5}}{\cancel{5}\cdot4}\right)=\frac{1}{4}$

53. $\cdot\ \frac{6}{7}\left(\frac{35}{8}\right)=\cdot\ \frac{\cancel{2}\cdot3}{\cancel{7}}\left(\frac{\cancel{7}\cdot5}{\cancel{2}\cdot4}\right)=\cdot\ \frac{15}{4}$

55. $\frac{-18}{9}=-2$

57. $\frac{20}{-5}=-4$

59. $\frac{-14}{-7}=2$

61. $\frac{0}{-3}=0$

63. $\frac{4}{0}$ is undefined

65. $-\left(\frac{-4}{-2}\right)=-2$

67. $-\left(\frac{-27}{3}\right)=-(-9)=9$

69. $-\left(\frac{15}{-5}\right)=-(-3)=3$

71. $\frac{-3}{-3}=1$

73. $\frac{-16}{4}=-4$

75. $\frac{-56}{8}=-7$

77. $\frac{3}{5}\div\left(\cdot\ \frac{4}{7}\right)=\frac{3}{5}\cdot\left(\cdot\ \frac{7}{4}\right)=\cdot\ \frac{21}{20}$

79. $\cdot\ \frac{2}{3}\div\left(\cdot\ \frac{7}{6}\right)=\cdot\ \frac{2}{3}\cdot\left(\cdot\ \frac{6}{7}\right)=\frac{2}{\cancel{3}}\cdot\frac{2\cdot\cancel{3}}{7}=\frac{4}{7}$

81. $\cdot \frac{5}{8} \div \frac{7}{8} = \cdot \frac{5}{\cancel{8}} \cdot \frac{\cancel{8}}{7} = \cdot \frac{5}{7}$

83. $\frac{-3.1}{6.2} = -\frac{1}{2}$

85. $\frac{-1.6}{-9.6} = \frac{1}{6}$

87. Commutative property of addition

89. Commutative property of multiplication

91. Commutative property of multiplication

93. Multiplicative identity

95. $A = a(b + c) = ab + ac$

97. $5000°C - 1500°C = 3500°C$

99. $47 + 1 + 2 - 1 - 2 - 1 = \$46$

101. $15°C + 2°C + 1°C - 1°C - 3°C = 14°C$

	Number	Additive Inverse	Reciprocal
103.	7	-7	$\frac{1}{7}$
105.	0	0	undefined

107. Slightly wicked $= 0.54(-2.5) = -1.35$

109. Extremely disgusting $= 1.45(-2.1) = -3.045$

111. Very good $= 1.25(3.1) = 3.875$

113. Sample answer: No. Consider $3 + (5 \cdot 4)$. If addition is distributive over multiplication, the result would be: $3 + (5 \cdot 4) = (3 + 5) \cdot (3 + 4) = 8 \cdot 7 = 56$. By order of operations, we know that the solution is 23.

115. $\cdot \frac{2}{5} \div \left(\cdot \frac{5}{8}\right) = \cdot \frac{2}{5} \cdot \left(\cdot \frac{8}{5}\right) = \frac{16}{25}$

117. $\cdot \left(\frac{3}{5}\right)\left(\cdot \frac{10}{3}\right) = \left(\frac{\cancel{3}}{\cancel{5}}\right)\left(\frac{2 \cdot \cancel{5}}{\cancel{3}}\right) = \frac{2}{1} = 2$

119. $-\frac{5}{8} + \left(-\frac{3}{7}\right) = -\frac{35}{56} + \left(-\frac{24}{56}\right) = -\frac{59}{56}$

121. $3.2(-4) = -12.8$

123. $(-3 - 5)(-2) + 8(3 - 7 + 4) = -8(-2) + 8(0) = 16 + 0 = 16$

125. Inverse property of addition

127. Associative property of multiplication

129. Commutative property of multiplication

1.3   Properties of Exponents

Problems 1.3

1. a. $(-5)^2 = (-5)(-5) = 25$

b. $\cdot 5^2 = \cdot 1 \cdot 5^2 = \cdot 1 \cdot (5 \cdot 5) = \cdot 25$

c. $(-6)^3 = (-6)(-6)(-6) = -216$

d. $\cdot 6^3 = \cdot 1 \cdot 6^3 = \cdot 1 \cdot (6 \cdot 6 \cdot 6) = \cdot 216$

2. a. $5^{-2} = \dfrac{1}{5^2} = \dfrac{1}{25}$

b. $(-3)^{-3} = \dfrac{1}{(-3)^3} = \dfrac{1}{-27} = -\dfrac{1}{27}$

c. $x^{-5} = \dfrac{1}{x^5}$

d. $-b^{-5} = \dfrac{-1}{b^5} = -\dfrac{1}{b^5}$

e. $5x^{-3} = 5 \cdot x^{-3} = \dfrac{5}{x^3}$

3. a. $(-3x^{-3})(5x^5) = -3 \cdot 5(x^{-3+5}) = -15x^2$

b. $(3x^2)(-2x^3) = 3(-2)(x^{2+3}) = -6x^5$

c. $(5x^3y^4)(-2x^{-7}y) = 5(-2)(x^{3+(-7)})(y^{4+1}) = -10x^{-4}y^5 = -10 \cdot \dfrac{1}{x^4} \cdot y^5 = -\dfrac{10y^5}{x^4}$

4. a. $\dfrac{6x^9}{12x^3} = \dfrac{6}{12} \cdot \dfrac{x^9}{x^3} = \dfrac{1}{2}x^{9-3} = \dfrac{x^6}{2}$

b. $\dfrac{-18x^3}{-9x^8} = \dfrac{-18}{-9} \cdot \dfrac{x^3}{x^8} = 2x^{3-8} = 2x^{-5} = \dfrac{2}{x^5}$

c. $\dfrac{10x^{-6}}{-20x^{-8}} = \dfrac{10}{-20} \cdot \dfrac{x^{-6}}{x^{-8}} = \dfrac{-1}{2} \cdot x^{-6-(-8)} = \dfrac{-1}{2}x^2 = -\dfrac{x^2}{2}$

d. $\dfrac{45x^4}{-15x^{-6}} = \dfrac{45}{-15} \cdot \dfrac{x^4}{x^{-6}} = -3x^{4-(-6)} = -3x^{10}$

5. a. $(4x^3y^4)^{-2} = 4^{-2}(x^3)^{-2}(y^4)^{-2} = \dfrac{1}{4^2} \cdot x^{-6} \cdot y^{-8} = \dfrac{1}{4^2} \cdot \dfrac{1}{x^6} \cdot \dfrac{1}{y^8} = \dfrac{1}{16x^6y^8}$

b. $(-3x^2y^{-5})^3 = (-3)^3(x^2)^3(y^{-5})^3 = -27x^6y^{-15} = -27x^6 \cdot \dfrac{1}{y^{15}} = -\dfrac{27x^6}{y^{15}}$

6. a. $\left(\dfrac{x^5 \cdot {}^{-4}}{y^{-3}}\right) = \dfrac{(x^5)^{-4}}{(y^{-3})^{-4}} = \dfrac{x^{-20}}{y^{12}} = \dfrac{1}{x^{20}} \cdot \dfrac{1}{y^{12}} = \dfrac{1}{x^{20}y^{12}}$

b. $\left(\dfrac{-2x^{-3}y^3}{3y^3}\right)^2 = \left(\dfrac{-2}{3} \cdot x^{-3} \cdot y^{3-3}\right)^2 = \left(\dfrac{-2}{3}x^{-3}y^0\right)^2 = \dfrac{2^2}{3^2}(x^{-3})^2(y^0)^2 = \dfrac{4}{9}x^{-6}y^0 = \dfrac{4}{9} \cdot \dfrac{1}{x^6} \cdot 1 = \dfrac{4}{9x^6}$

7. a. $350{,}000 = 3.5 \times 10^5$

b. $0.000000378 = 3.78 - 10^{-7}$

8. a. $3.5 \times 10^5 = 350{,}000$

b. $8.2 - 10^{-3} = 0.0082$

9. a. $(3 - 10^3)(8 - 10^{-7}) = (3 - 8) - (10^3 - 10^{-7}) = 24 - 10^{3-7} = 2.4 - 10^1 - 10^{-4} = 2.4 - 10^{-3}$

b. $\dfrac{4 - 10^6}{2 - 10^{-5}} = \dfrac{4}{2} - \dfrac{10^6}{10^{-5}} = 2 - 10^{6-(-5)} = 2 - 10^{11}$

Exercises 1.3

1. $-4^2 = -(4 \cdot 4) = -16$

3. $(-5)^2 = (-5)(-5) = 25$

5. $-5^3 = -(5 \cdot 5 \cdot 5) = -125$

7. $(-6)^4 = (-6)(-6)(-6)(-6) = 1296$

9. $-2^5 = -(2 \cdot 2 \cdot 2 \cdot 2 \cdot 2) = -32$

11. $4^{-2} = \dfrac{1}{4^2} = \dfrac{1}{16}$

13. $5^{-3} = \dfrac{1}{5^3} = \dfrac{1}{125}$

15. $3^{-4} = \dfrac{1}{3^4} = \dfrac{1}{81}$

17. $x^{-6} = \dfrac{1}{x^6}$

19. $a^{-8} = \dfrac{1}{a^8}$

21. $2^{-4} \cdot 2^{-2} = 2^{-4+(-2)} = 2^{-6} = \dfrac{1}{2^6} = \dfrac{1}{64}$

23. $(3x^6)(-4x^{-4}) = (3-4)(x^6 \cdot x^{-4}) = -12x^{6+(-4)} = -12x^2$

25. $(-3y^{-3}) \cdot (5y^5) = (-3 \cdot 5)(y^{-3} \cdot y^5) = -15(y^{-3+5}) = -15y^2$

27. $(-4a^3) \cdot (-5a^{-8}) = -4(-5)(a^3 \cdot a^{-8}) = 20a^{3+(-8)} = 20a^{-5} = 20 \cdot \dfrac{1}{a^5} = \dfrac{20}{a^5}$

29. $(3x^{-5}) \cdot (5x^2y)(-2xy^2) = (3 \cdot 5 \cdot -2)(x^{-5} \cdot x^2 \cdot x)(y \cdot y^2) = -30x^{-5+2+1}y^{1+2} = -30x^{-2}y^3 = -\dfrac{30y^3}{x^2}$

31. $(-2x^{-3}y^2)(3x^{-2}y^3)(4xy) = (-2 \cdot 3 \cdot 4)(x^{-3} \cdot x^{-2} \cdot x)(y^2 \cdot y^3 \cdot y) = -24x^{-3+(-2)+1}y^{2+3+1} = -24x^{-4}y^6$
$$= -24 \cdot \dfrac{1}{x^4} \cdot y^6 = -\dfrac{24y^6}{x^4}$$

33. $(4a^{-2}b^{-3})(5a^{-1}b^{-1})(-2ab) = (4 \cdot 5 \cdot -2)(a^{-2} \cdot a^{-1} \cdot a)(b^{-3} \cdot b^{-1} \cdot b) = -40a^{-2+(-1)+1}b^{-3+(-1)+1}$
$$= -40a^{-2}b^{-3} = -40 \cdot \dfrac{1}{a^2} \cdot \dfrac{1}{b^3} = -\dfrac{40}{a^2b^3}$$

35. $(6a^{-3}b^3)(5a^2b^2)(-ab^{-5}) = (6 \cdot 5 \cdot -1)(a^{-3} \cdot a^2 \cdot a)(b^3 \cdot b^2 \cdot b^{-5}) = -30a^{-3+2+1}b^{3+2+(-5)} = -30a^0b^0$
$$= -30 \cdot 1 \cdot 1 = -30$$

37. $\dfrac{8x^7}{4x^3} = \dfrac{8}{4} \cdot \dfrac{x^7}{x^3} = 2x^{7-3} = 2x^4$

39. $\dfrac{-8a^4}{-16a^2} = \dfrac{-8}{-16} \cdot \dfrac{a^4}{a^2} = \dfrac{1}{2}a^{4-2} = \dfrac{a^2}{2}$

41. $\dfrac{12x^5y^3}{-6x^2y} = \dfrac{12}{-6} \cdot \dfrac{x^5}{x^2} \cdot \dfrac{y^3}{y} = -2x^{5-2}y^{3-1} = -2x^3y^2$

43. $\dfrac{\cdot\,6x^{\cdot 4}}{12x^{\cdot 5}} = \dfrac{\cdot\,6}{12} \cdot \dfrac{x^{\cdot 4}}{x^{\cdot 5}} = \dfrac{\cdot\,1}{2} x^{\cdot 4 \cdot (\cdot 5)} = \dfrac{\cdot\,1}{2} x^1 = \cdot\,\dfrac{x}{2}$

45. $\dfrac{\cdot\,14a^{\cdot 5}}{\cdot\,21a^{\cdot 2}} = \dfrac{\cdot\,14}{\cdot\,21} \cdot \dfrac{a^{\cdot 5}}{a^{\cdot 2}} = \dfrac{2}{3} a^{\cdot 5 \cdot (\cdot 2)} = \dfrac{2}{3} a^{\cdot 3} = \dfrac{2}{3} \cdot \dfrac{1}{a^3} = \dfrac{2}{3a^3}$

47. $\dfrac{\cdot\,27a^{\cdot 4}}{\cdot\,36a^{\cdot 4}} = \dfrac{\cdot\,27}{\cdot\,36} \cdot \dfrac{a^{\cdot 4}}{a^{\cdot 4}} = \dfrac{3}{4} a^{\cdot 4 \cdot (\cdot 4)} = \dfrac{3}{4} a^0 = \dfrac{3}{4} \cdot 1 = \dfrac{3}{4}$

49. $\dfrac{3a^{\cdot 2}b^5}{2a^4b^2} = \dfrac{3}{2} \cdot \dfrac{a^{\cdot 2}}{a^4} \cdot \dfrac{b^5}{b^2} = \dfrac{3}{2} a^{\cdot 2 \cdot 4} b^{5 \cdot 2} = \dfrac{3}{2} a^{\cdot 6} b^3 = \dfrac{3}{2} \cdot \dfrac{1}{a^6} \cdot b^3 = \dfrac{3b^3}{2a^6}$

51. $\left(2x^3y^{\cdot 2}\right)^3 = 2^3\left(x^3\right)^3\left(y^{\cdot 2}\right)^3 = 8x^9y^{\cdot 6} = 8x^9 \cdot \dfrac{1}{y^6} = \dfrac{8x^9}{y^6}$

53. $\left(2x^{\cdot 2}y^3\right)^2 = 2^2\left(x^{\cdot 2}\right)^2\left(y^3\right)^2 = 4x^{\cdot 4}y^6 = 4 \cdot \dfrac{1}{x^4} \cdot y^6 = \dfrac{4y^6}{x^4}$

55. $\left(\cdot\,3x^3y^2\right)^{\cdot 3} = \left(\cdot\,3\right)^{\cdot 3}\left(x^3\right)^{\cdot 3}\left(y^2\right)^{\cdot 3} = \left(\cdot\,3\right)^{\cdot 3} x^{\cdot 9} y^{\cdot 6} = \dfrac{1}{\left(\cdot\,3\right)^3} \cdot \dfrac{1}{x^9} \cdot \dfrac{1}{y^6} = \dfrac{1}{\cdot\,27x^9y^6} = \cdot\,\dfrac{1}{27x^9y^6}$

57. $\left(x^{\cdot 6}y^{\cdot 3}\right)^2 = \left(x^{\cdot 6}\right)^2\left(y^{\cdot 3}\right)^2 = x^{\cdot 12}y^{\cdot 6} = \dfrac{1}{x^{12}} \cdot \dfrac{1}{y^6} = \dfrac{1}{x^{12}y^6}$

59. $\left(x^{-4}y^{-4}\right)^{-3} = \left(x^{-4}\right)^{-3}\left(y^{-4}\right)^{-3} = x^{12}y^{12}$

61. $\left|\dfrac{a}{b^3}\right|^2 = \dfrac{a^2}{\left(b^3\right)^2} = \dfrac{a^2}{b^6}$

63. $\left|\dfrac{\mid 3a}{2b^2}\right|^{\mid 3} = \left|\dfrac{2b^2}{\mid 3a}\right|^{\mid 3} = \dfrac{2^3\left(b^2\right)^3}{\left(\mid 3\right)^3 a^3} = \dfrac{8\,b^6}{\mid 27a^3} = \mid\dfrac{8\,b^6}{27a^3}$

65. $\dfrac{\cdot\,a^{\cdot 4}}{\vdots\,b^2\vdots}^{\cdot 2} = \dfrac{\cdot\,b^2}{\cdot\,a^{\cdot 4}}^{\cdot 2} = \dfrac{\left(b^2\right)^2}{\left(a^{\cdot 4}\right)^2} = \dfrac{b^4}{a^{\cdot 8}} = \dfrac{1}{a^{\cdot 8}} \cdot b^4 = a^8 b^4$

67. $\dfrac{\cdot\,x^5}{\vdots\,y^{\cdot 2}\vdots}^{\cdot 3} = \dfrac{\left(x^5\right)^{\cdot 3}}{\left(y^{\cdot 2}\right)^{\cdot 3}} = \dfrac{x^{\cdot 15}}{y^6} = \dfrac{1}{x^{15}} \cdot \dfrac{1}{y^6} = \dfrac{1}{x^{15}y^6}$

69. $\left|\dfrac{x^{\mid 4}y^3}{x^5y^5}\right|^{\mid 3} = \left(x^{\mid 4\mid 5}y^{3\mid 5}\right)^{\mid 3} = \left(x^{\mid 9}y^{\mid 2}\right)^{\mid 3} = \left(x^{\mid 9}\right)^{\mid 3}\left(y^{\mid 2}\right)^{\mid 3} = x^{27}y^6$

71. $268{,}000{,}000 = 2.68 \times 10^8$ **73.** $0.00024 = 2.4 - 10^{-4}$

75. $8 \times 10^6 = 8,000,000$

77. $2.3 - 10^{-1} = 0.23$

79. $\dfrac{2.8 - 10^8}{1.4 - 10^5} = \dfrac{2.8}{1.4} - \dfrac{10^8}{10^5} = 2 - 10^{8-5} = 2 - 10^3$ hr

81. $\dfrac{1.47 - 10^{11}}{490}$ m / sec $= \dfrac{1.47 - 10^{11}}{4.90 - 10^2} = \dfrac{1.47}{4.90} - \dfrac{10^{11}}{10^2} = 0.3 - 10^9 = 3 - 10^{-1} - 10^9 = 3 - 10^8$ m / sec

83. $\dfrac{6.28 - 10^{11}}{2.0 - 10^{10}} = \dfrac{6.28}{2.0} - \dfrac{10^{11}}{10^{10}} - 3.1 - 10^{11-10} - 3.1 - 10^1 - 31$ yr

85. **a.** $(-3.2)(-1.4)(-2.2) = -9.856$ **b.** $(-1.1)(-1.2)(-2.1) = -2.772$

87. 3.34×10^5

89. **a.** 7.3   10 **b.** 1.23 -07

91. Sample answer: Zero raised to any natural number power (except zero) is equal to zero since zero times any number of zero factors is equal to zero.

93. Sample answer: $x^m x^n = x^{m+n}$; for example:
$$x^2 x^3 = x^{2+3}$$
$$xx \cdot xxx = x^5$$
$$x^5 = x^5$$

95. Sample answer: $\left(x^m\right)^n = x^{m \cdot n}$; for example:
$$\left(x^2\right)^3 = x^{2 \cdot 3}$$
$$x^2 \cdot x^2 \cdot x^2 = x^6$$
$$x^{2+2+2} = x^6$$
$$x^6 = x^6$$

97. $\left(\cdot 2x^{\cdot 5}y^{\cdot 3}\right)\left(\cdot 4xy\right) = \cdot 2(\cdot 4)\left(x^{\cdot 5} \cdot x\right)\left(y^{\cdot 3} \cdot y\right) = 8x^{\cdot 5+1}y^{\cdot 3+1} = 8x^{\cdot 4}y^{\cdot 2} = 8 \cdot \dfrac{1}{x^4} \cdot \dfrac{1}{y^2} = \dfrac{8}{x^4 y^2}$

99. $\dfrac{\cdot 6x^{\cdot 8}}{30x^6} = \dfrac{\cdot 6}{30} \cdot \dfrac{x^{\cdot 8}}{x^6} = \dfrac{\cdot 1}{5}x^{\cdot 8 \cdot 6} = \dfrac{\cdot 1}{5}x^{\cdot 14} = \dfrac{\cdot 1}{5} \cdot \dfrac{1}{x^{14}} = \cdot \dfrac{1}{5x^{14}}$

101. $-2^6 = -(2)(2)(2)(2)(2)(2) = -64$

103. $\left(\cdot 3x^{\cdot 4}y^5\right)^{\cdot 2} = \left(\cdot 3\right)^{\cdot 2}\left(x^{\cdot 4}\right)^{\cdot 2}\left(y^5\right)^{\cdot 2} = \dfrac{1}{\left(\cdot 3\right)^2} \cdot x^8 \cdot y^{\cdot 10} = \dfrac{1}{9} \cdot x^8 \cdot \dfrac{1}{y^{10}} = \dfrac{x^8}{9y^{10}}$

105. $(-x)^{-5} = \dfrac{1}{(-x)^5} = \dfrac{1}{-x^5} = -\dfrac{1}{x^5}$

107. $\left.\vdots\,\dfrac{5x^{\cdot 5}y^{7\cdot}}{7y^3}\,\vdots\,\right.^{\cdot 2} = \vdots\,\dfrac{5}{7}\cdot x^{\cdot 5}\cdot y^{7\cdot 3}\,\vdots^{\cdot 2} = \vdots\,\dfrac{5}{7}x^{\cdot 5}y^{4\cdot}\,\vdots^{\cdot 2} = \vdots\,\dfrac{5y^{4\cdot}}{7x^5}\,\vdots^{\cdot 2} = \vdots\,\dfrac{7x^5\cdot}{5y^4}\,\vdots^{2} = \vdots\,\dfrac{7\cdot}{5\cdot}^2\cdot\dfrac{(x^5)^2}{(y^4)^2}$

$$= \dfrac{49}{25}\cdot\dfrac{x^{10}}{y^8} = \dfrac{49x^{10}}{25y^8}$$

109. $\left(4 - 10^{-5}\right) - \left(6 - 10^2\right) = (4-6) - \left(10^{-5} - 10^2\right) = 24 - 10^{-3} = 2.4 - 10^1 - 10^{-3} = 2.4 - 10^{-2}$

1.4 Algebraic Expressions and the Order of Operations

Problems 1.4

1. **a.** $(\cdot\,6\cdot4) + 7 = (\cdot\,24) + 7 = \cdot\,17$ **b.** $\cdot\,6\cdot(4+7) = \cdot\,6\cdot11 = \cdot\,66$

   c. $\cdot\,36 \div (3\cdot4) = \cdot\,36 \div 12 = \cdot\,3$ **d.** $(\cdot\,36 \div 3)\cdot4 = (\cdot\,12)\cdot4 = \cdot\,48$

2. **a.** $[\cdot\,8\cdot(7+2)] + 8 = [\cdot\,8\cdot9] + 8 = \cdot\,72 + 8 = \cdot\,64$

   b. $[\cdot\,10\cdot(9\cdot\,6)]\cdot\,9 = [\cdot\,10\cdot3]\cdot\,9 = \cdot\,30\cdot\,9 = \cdot\,39$

3. $\dfrac{5\text{-}(149 - 32)}{9} = \dfrac{5\text{-}117}{9} = \dfrac{585}{9} = 65$

4. $-7^2 + \dfrac{(2-8)}{2} + 20 \div 5 = -7^2 + \dfrac{-6}{2} + 20 \div 5 = -49 + \dfrac{-6}{2} + 20 \div 5 = -49 - 3 + 4 = -48$

5. **a.** $\dfrac{5(50-32)}{9} = \dfrac{5(18)}{9} = \dfrac{90}{9} = 10°C$

   b. $3(-1)^2 - (7+11) + 10 \div 2 = 3(-1)^2 - 18 + 10 \div 2 = 3(1) - 18 + 10 \div 2 = 3 - 18 + 5 = -10$

6. **a.** $-3(a+b) = -3a - 3b$

   b. $0.2(6-b) = 0.2(6) - 0.2b = 1.2 - 0.2b$

7. **a.** $\cdot\,(y\cdot\,6) = \cdot\,1\cdot(y\cdot\,6) = \cdot\,1\cdot y + (\cdot\,1)(\cdot\,6) = \cdot\,y + 6$

   b. $\cdot\,(xy+7) = \cdot\,1\cdot(xy+7) = \cdot\,1\cdot xy + (\cdot\,1)(7) = \cdot\,xy\cdot\,7$

8. **a.** $5(a - 3b + 4) = 5(a) + 5(-3b) + 5(4) = 5a - 15b + 20$

   b. $-3(4x + y - z) = -3(4x) - 3(y) - 3(-z) = -12x - 3y + 3z$

   c. $0.5(-2x + 3y - 6z - 4) = 0.5(-2x) + 0.5(3y) + 0.5(-6z) + 0.5(-4) = -x + 1.5y - 3z - 2$

   d. $0.3x(y + 2z - 4) = 0.3x(y) + 0.3x(2z) + 0.3x(-4) = 0.3xy + 0.6xz - 1.2x$

9. **a.** $8y + 3(y - 4) = 8y + 3y - 12 = 11y - 12$
   b. $-4(y + 2) - 3y = -4y - 8 - 3y = -7y - 8$
   c. $6y - 3(y + 2) + (y + 8) = 6y - 3y - 6 + y + 8 = 4y + 2$

10. $\left[(2x^2 - 3) + (3x + 1)\right] - \left[(x + 1) + (x^2 - 2)\right] = \left[2x^2 - 3 + 3x + 1\right] - \left[x + 1 + x^2 - 2\right]$
    $$= (2x^2 + 3x - 2) - (x^2 + x - 1) = 2x^2 + 3x - 2 - x^2 - x + 1 = x^2 + 2x - 1$$

Exercises 1.4

1. **a.** $(\cdot\, 10 \cdot 3) + 4 = \cdot\, 30 + 4 = \cdot\, 26$ **b.** $\cdot\, 10 \cdot (3 + 4) = \cdot\, 10 \cdot 7 = \cdot\, 70$

3. **a.** $(36 \div 4) \cdot 3 = 9 \cdot 3 = 27$ **b.** $36 \div (4 \cdot 3) = 36 \div 12 = 3$

5. $\left[\cdot\, 5 \cdot (8 + 2)\right] + 3 = \left[\cdot\, 5 \cdot 10\right] + 3 = \cdot\, 50 + 3 = \cdot\, 47$

7. $\cdot\, 7 + \left[3 \cdot (4 + 5)\right] = \cdot\, 7 + \left[3 \cdot 9\right] = \cdot\, 7 + 27 = 20$

9. $\left[\cdot\, 6 \cdot (4 \cdot 2)\right] \cdot 3 = \left[\cdot\, 6 \cdot 2\right] \cdot 3 = \cdot\, 12 \cdot 3 = \cdot\, 15$

11. $3 \cdot \left[8 \cdot (5 \cdot 3)\right] = 3 \cdot \left[8 \cdot 2\right] = 3 \cdot 16 = \cdot\, 13$

13. $\rfloor 8 \rfloor 3 \rfloor 2(4 + 1) \rfloor + 1 = \rfloor 8 \rfloor 3 \rfloor 2(5) \rfloor + 1 = \rfloor 8 [3 \rfloor 10] + 1 = \rfloor 8 [\rfloor 7] + 1 = 56 + 1 = 57$

15. $48 \div \left\{4(8 - 2[3 - 1])\right\} = 48 \div \left\{4(8 - 2[2])\right\} = 48 \div \left\{4(8 - 4)\right\} = 48 \div \left\{4(4)\right\} = 48 \div \{16\} = 3$

17. $\left|\dfrac{9 \mid (\mid 3)}{8 \mid 6}\right| \left|\dfrac{3 + (\mid 8)}{7 \mid 2}\right| = \left|\dfrac{12}{2}\right| \left|\dfrac{\mid 5}{5}\right| = [6][\mid 1] = \mid 6$

19. $\dfrac{3 \mid 5 \left|\dfrac{4 + 2}{2 + 1}\right| \mid 2}{\mid 4 + 3 \left|\dfrac{4 \mid 2}{4 \mid 6}\right| \mid 2} = \dfrac{3 \mid 5 \left|\dfrac{6}{3}\right| \mid 2}{\mid 4 + 3 \left|\dfrac{2}{\mid 2}\right| \mid 2} = \dfrac{3 \mid 5(2) \mid 2}{\mid 4 + 3(\mid 1) \mid 2} = \dfrac{3 \mid 10 \mid 2}{\mid 4 \mid 3 \mid 2} = \dfrac{\mid 9}{\mid 9} = 1$

21. $\cdot\, 5 \cdot 6 \cdot 6 = \cdot\, 30 \cdot 6 = \cdot\, 36$

23. $\cdot\, 7 \cdot 3 \div 3 \cdot 3 = \cdot\, 21 \div 3 \cdot 3 = \cdot\, 7 \cdot 3 = \cdot\, 10$

25. $(-20 - 5 + 3 \div 3) \div 6 = (-20 - 5 + 1) \div 6 = -24 \div 6 = -4$

27. $\dfrac{8 + (-3)}{5} - 1 = \dfrac{5}{5} - 1 = 1 - 1 = 0$

29. $\dfrac{4-(6-2)}{-8}-\dfrac{6}{-2}=\dfrac{4-4}{-8}-\dfrac{6}{-2}=\dfrac{16}{-8}-\dfrac{6}{-2}=-2+3=1$

31. $4\div 2+3-5^2=4\div 2+3-25=2+3-25=-20$

33. $4+6-4\div 2-2^3=4+6-4\div 2-8=4+24\div 2-8=4+12-8=8$

35. $-5^2+\dfrac{2-10}{4}+12\div 4=-5^2+\dfrac{-8}{4}+12\div 4=-25+\dfrac{-8}{4}+12\div 4=-25-2+3=-24$

37. $\cdot\ 3^3+4\cdot\ 6\cdot 8\div 4\cdot\ \dfrac{8\cdot 2}{\cdot\ 3}=\cdot\ 3^3+4\cdot\ 6\cdot 8\div 4\cdot\ \dfrac{6}{\cdot\ 3}=\cdot\ 27+4\cdot\ 6\cdot 8\div 4\cdot\ \dfrac{6}{\cdot\ 3}$

$=-27+4-48\div 4-\dfrac{6}{-3}=-27+4-12+2=-33$

39. $4-9\div 3-10^3-2-10^2=4-9\div 3-1000-2-100=36\div 3-1000-2-100=12-1000-2-100$

$=12,000-200=11,800$

41. $2(l+w);\, l=12\tfrac{1}{2},\, w=6;\quad =2\left(12\tfrac{1}{2}+6\right)=2\left(18\tfrac{1}{2}\right)=37$

43. $P(1+r);\, P=\$1{,}000,\, r=5\%;\quad =\$1{,}000(1+0.05)=\$1{,}000(1.05)=\$1{,}050$

45. $2x-y;\, x=4,\, y=-5;\quad =2(4)-(-5)=8+5=13$

47. $t^2-5t+8;\, t=-3;\quad =(-3)^2-5(-3)+8=9-5(-3)+8=9+15+8=32$

49. $5n^2+2(n-m);\, n=-4,\, m=6;\quad =5(-4)^2+2(-4-6)=5(-4)^2+2(-10)$

$=5(16)+2(-10)=80-20=60$

51. $4(x-y)=4(x)+4(-y)=4x-4y$

53. $-9(a-b)=-9(a)-9(-b)=-9a+9b$

55. $0.3(4x-2)=0.3(4x)+0.3(-2)=1.2x-0.6$

57. $\Big|\dfrac{3a}{2}\Big|\dfrac{6}{7}\Big|=\Big|1\Big|\dfrac{3a}{2}\Big|\Big|1\Big|\Big|\dfrac{6}{7}\Big|=\Big|\dfrac{3a}{2}+\dfrac{6}{7}$ or $\dfrac{|21a+12}{14}$

59. $-(2x-6y)=-1(2x)-1(-6y)=-2x+6y$ **61.** $-(2.1+3y)=-1(2.1)-1(3y)=-2.1-3y$

63. $-4(a+5)=-4(a)-4(5)=-4a-20$ **65.** $-x(6+y)=-x(6)-x(y)=-6x-xy$

67. $-8(x-y)=-8(x)-8(-y)=-8x+8y$

69. $-3(2a - 7b) = -3(2a) - 3(-7b) = -6a + 21b$

71. $0.5(x + y - 2) = 0.5(x) + 0.5(y) + 0.5(-2) = 0.5x + 0.5y - 1$

73. $-\dfrac{6}{5}(a - b + 5) = -\dfrac{6}{5}(a) - \dfrac{6}{5}(-b) - \dfrac{6}{5}(5) = -\dfrac{6a}{5} + \dfrac{6b}{5} - 6$

75. $-2(x - y + 3z + 5) = -2(x) - 2(-y) - 2(3z) - 2(5) = -2x + 2y - 6z - 10$

77. $-0.3(x + y - 2z - 6) = -0.3(x) - 0.3(y) - 0.3(-2z) - 0.3(-6) = -0.3x - 0.3y + 0.6z + 1.8$

79. $-\dfrac{5}{2}(a - 2b + c + 2d - 2) = -\dfrac{5}{2}(a) - \dfrac{5}{2}(-2b) - \dfrac{5}{2}(c) - \dfrac{5}{2}(2d) - \dfrac{5}{2}(-2)$

$\qquad\qquad = -\dfrac{5a}{2} + 5b - \dfrac{5c}{2} - 5d + 5$

81. $6x + 3(x - 2) = 6x + 3x - 6 = 9x - 6$

83. $-4(x + 2) - 5x = -4x - 8 - 5x = -9x - 8$

85. $(5L - 3W) - (W - 6L) = 5L - 3W - W + 6L = 11L - 4W$

87. $5x - (8x + 1) + (x + 1) = 5x - 8x - 1 + x + 1 = -2x$

89. $\dfrac{2x}{9} - \left(\dfrac{x}{9} - 2\right) = \dfrac{2x}{9} - \dfrac{x}{9} + 2 = \dfrac{x}{9} + 2$

91. $4a - (a + b) + 3(b + a) = 4a - a - b + 3b + 3a = 6a + 2b$

93. $7x - 3(x + y) - (x + y) = 7x - 3x - 3y - x - y = 3x - 4y$

95. $-(x + y - 2) + 3(x - y + 6) - (x + y - 16) = -x - y + 2 + 3x - 3y + 18 - x - y + 16 = x - 5y + 36$

97. $\left(x^2 + 7 - x\right) + \left[-2x^3 + \left(8x^2 - 2x\right) + 5\right] = \left(x^2 + 7 - x\right) + \left[-2x^3 + 8x^2 - 2x + 5\right]$

$\qquad\qquad = x^2 + 7 - x - 2x^3 + 8x^2 - 2x + 5 = -2x^3 + 9x^2 - 3x + 12$

99. $\left[\left(\dfrac{5}{7}x^2 + \dfrac{1}{5}x\right) - \dfrac{1}{8}\right] - \left[\left(\dfrac{3}{7}x^2 - \dfrac{3}{5}x\right) + \dfrac{5}{8}\right] = \left(\dfrac{5}{7}x^2 + \dfrac{1}{5}x - \dfrac{1}{8}\right) - \left(\dfrac{3}{7}x^2 - \dfrac{3}{5}x + \dfrac{5}{8}\right)$

$\qquad\qquad = \dfrac{5}{7}x^2 + \dfrac{1}{5}x - \dfrac{1}{8} - \dfrac{3}{7}x^2 + \dfrac{3}{5}x - \dfrac{5}{8} = \dfrac{2x^2}{7} + \dfrac{4x}{5} - \dfrac{6}{8} = \dfrac{2x^2}{7} + \dfrac{4x}{5} - \dfrac{3}{4}$

101. $\left[3(2a - 4) + 5\right] - \left[2(a - 1) + 6\right] = \left[6a - 12 + 5\right] - \left[2a - 2 + 6\right] = \left[6a - 7\right] - \left[2a + 4\right]$

$\qquad\qquad = 6a - 7 - 2a - 4 = 4a - 11$

103. $[4a - (3 + 2b)] - [6(a - 2b) + 5a] = [4a - 3 - 2b] - [6a - 12b + 5a] = [4a - 3 - 2b] - [11a - 12b]$
$$= 4a - 3 - 2b - 11a + 12b = -7a + 10b - 3$$

105. $-[-(0.2x + y) + 3(x - y)] - [2(x + 0.3y) - 5] = -[-0.2x - y + 3x - 3y] - [2x + 0.6y - 5]$
$$= -[2.8x - 4y] - [2x + 0.6y - 5] = -2.8x + 4y - 2x - 0.6y + 5 = -4.8x + 3.4y + 5$$

107. $[(-2)(-2)][(-2)(-2)] = [4][4] = 16$ **109.** $\dfrac{(-2)(-2)(-2)}{(-2)(-2)} = \dfrac{-8}{4} = -2$

111. $v_a = \dfrac{1}{2}(v_1 + v_2) = \dfrac{1}{2}v_1 + \dfrac{1}{2}v_2$ **113.** $KE = \dfrac{1}{2}m\left(v_1^2 + v_2^2\right) = \dfrac{1}{2}mv_1^2 + \dfrac{1}{2}mv_2^2$

115. Sample answer: For $(32 \div 4) \cdot 2$, the solution is: $(8) \cdot 2 = 16$. For $32 \div (4 \cdot 2)$, the solution is:
$32 \div 8 = 4$.

117. $\left| 3^2 + \left| \dfrac{4 \mid 8}{2} \right| + 48 \div 6 \right| = \left| 3^2 + \left| \dfrac{\mid 4}{2} \right| + 48 \div 6 \right| = \left| 3^2 \mid 2 + 48 \div 6 \right| = \mid 9 \mid 2 + 48 \div 6$
$$= -9 - 2 + 8 = -3$$

119. $-2(x - 3y + 2z - 4) = -2(x) - 2(-3y) - 2(2z) - 2(-4) = -2x + 6y - 4z + 8$

121. $\dfrac{3}{8}x \mid \left| \dfrac{x}{8} \right| 5 \mid = \dfrac{3}{8}x \mid \dfrac{x}{8} + 5 = \dfrac{2}{8}x + 5 = \dfrac{1}{4}x + 5$

123. $[(a^2 - 5) + (2a^3 - 3)] - [(4a^3 + a) - (a^2 - 9)] = [a^2 - 5 + 2a^3 - 3] - [4a^3 + a - a^2 + 9]$
$$= a^2 + 2a^3 - 8 - 4a^3 - a + a^2 - 9 = -2a^3 + 2a^2 - a - 17$$

125. $\left| \dfrac{8 \mid (\mid 4)}{8 \mid 10} \right| \left| \dfrac{5 + (\mid 9)}{7 \mid 3} \right| = \left| \dfrac{12}{\mid 2} \right| \left| \dfrac{\mid 4}{4} \right| = [\mid 6][\mid 1] = 6$

Review Exercises

1.   a. $\{4, 5, 6, 7, 8\}$ **b.** $\{5, 6, 7\}$

2.   a. $\frac{1}{5} = 0.2$ **b.** $\frac{2}{5} = 0.4$

3.   a. $\frac{1}{9} = 0.\overline{1}$ **b.** $\frac{2}{9} = 0.\overline{2}$

4.

Set	0.3	0	$\frac{-3}{4}$	-5	$\sqrt{3}$
Natural numbers					
Whole numbers		✓			
Integers		✓		✓	
Rational numbers	✓	✓	✓	✓	
Irrational numbers					✓
Real numbers	✓	✓	✓	✓	✓

5. a. 3.5 **b.** $-\dfrac{3}{4}$

6. a. $\left|-9\right| = 9$ **b.** $\left|4.2\right| = 4.2$

7. a. $\left|-\dfrac{1}{8}\right| = \dfrac{1}{8}$ **b.** $\left|0.\overline{4}\right| = 0.\overline{4}$

8. a. $-8 < -7$ **b.** $-4 < -3$

9. a. $\dfrac{1}{4} > \dfrac{1}{5}$ **b.** $\dfrac{3}{4} = 0.75$

10. a. $\dfrac{1}{5} < 0.25$ **b.** $0.\overline{6} = \dfrac{2}{3}$

11. a. $-3 + (-8) = -11$ **b.** $-5 + 2 = -3$

12. a. $\dfrac{1}{7} - \dfrac{3}{7} = -\dfrac{2}{7}$ **b.** $-0.2 - 0.4 = -0.6$

13. a. $8 - (-4) = 8 + 4 = 12$ **b.** $-3 - (-7) = -3 + 7 = 4$

14. a. $\dfrac{3}{4} \mid \left| \dfrac{1}{5} \right| = \dfrac{3}{4} + \dfrac{1}{5} = \dfrac{15}{20} + \dfrac{4}{20} = \dfrac{19}{20}$ **b.** $\dfrac{5}{6} \mid \left| \dfrac{1}{4} \right| = \dfrac{5}{6} + \dfrac{1}{4} = \dfrac{10}{12} + \dfrac{3}{12} = \dfrac{13}{12}$

15. a. $9 - (-4) = -36$ **b.** $\cdot\, 2.4 \cdot 6 = \cdot\, 14.4$

16. a. $\therefore \dfrac{3}{4} \cdot \dfrac{7}{8} = \cdot\, \dfrac{21}{32}$ **b.** $\therefore \dfrac{5}{6} \cdots \dfrac{2}{7} = \dfrac{5}{\cancel{2} \cdot 3} \cdot \dfrac{\cancel{2}}{7} = \dfrac{5}{21}$

17. a. $\dfrac{0}{7} = 0$ **b.** $\dfrac{8}{0}$ is undefined

18. a. $-\dfrac{5}{3}$ **b.** $\dfrac{1}{0.3}$ or $\dfrac{10}{3}$

19. a. $\cdot \dfrac{3}{5} \div \dfrac{4}{15} = \cdot \dfrac{3}{5} \cdot \dfrac{15}{4} = \cdot \dfrac{3}{\cancel{5}} \cdot \dfrac{\cancel{5} \cdot 3}{4} = \cdot \dfrac{9}{4}$ **b.** $\dfrac{3.6}{-1.2} = -3$

20. a. Commutative property of addition   **b.** Associative property of addition

21. a. $(-3)^4 = (-3)(-3)(-3)(-3) = 81$ **b.** $-3^4 = -1(3)(3)(3)(3) = -81$

22. a. $9^0 = 1$ **b.** $\left|\dfrac{1}{7}\right|^0 = 1$

23. a. $(-8)^{-3} = \dfrac{1}{(-8)^3} = \dfrac{1}{(-8)(-8)(-8)} = -\dfrac{1}{512}$ **b.** $x^{-10} = \dfrac{1}{x^{10}}$

24. a. $\left(3x^{\cdot 4}y\right)\left(\cdot 5x^{\cdot 8}y^9\right) = 3(\cdot 5)\left(x^{\cdot 4 + (\cdot 8)}\right)\left(y^{1+9}\right) = \cdot 15x^{\cdot 4}y^{10} = \cdot 15 \cdot \dfrac{1}{x^4} \cdot y^{10} = \cdot \dfrac{15y^{10}}{x^4}$

b. $\left(4x^{\cdot 3}y^{\cdot 1}\right)\left(\cdot 6x^{\cdot 8}y^{\cdot 7}\right) = 4(\cdot 6)\left(x^{\cdot 3 + (\cdot 8)}\right)\left(y^{\cdot 1 + (\cdot 7)}\right) = \cdot 24x^{\cdot 11}y^{\cdot 8} = \cdot 24 \cdot \dfrac{1}{x^{11}} \cdot \dfrac{1}{y^8} = \cdot \dfrac{24}{x^{11}y^8}$

25. a. $\dfrac{48x^4}{16x^6} = \dfrac{48}{16} \cdot \dfrac{x^4}{x^6} = 3 \cdot x^{4-6} = 3x^{-2} = 3 \cdot \dfrac{1}{x^2} = \dfrac{3}{x^2}$

b. $\dfrac{8x^5}{\cdot 2x^{\cdot 6}} = \dfrac{8}{\cdot 2} \cdot \dfrac{x^5}{x^{\cdot 6}} = \cdot 4 \cdot x^{5 \cdot (\cdot 6)} = \cdot 4x^{11}$

26. a. $\dfrac{\cdot 5x^{\cdot 3}}{15x^{\cdot 4}} = \dfrac{\cdot 5}{15} \cdot \dfrac{x^{\cdot 3}}{x^{\cdot 4}} = \dfrac{\cdot 1}{3} \cdot x^{\cdot 3 \cdot (\cdot 4)} = \dfrac{\cdot 1}{3} \cdot x^1 = \cdot \dfrac{x}{3}$

b. $\dfrac{8x^{\cdot 4}}{\cdot 4x^7} = \dfrac{8}{\cdot 4} \cdot \dfrac{x^{\cdot 4}}{x^7} = \cdot 2 \cdot x^{\cdot 4 \cdot 7} = \cdot 2x^{\cdot 11} = \cdot 2 \cdot \dfrac{1}{x^{11}} = \cdot \dfrac{2}{x^{11}}$

27. a. $\left(\cdot 2x^7y^{\cdot 6}\right)^3 = (\cdot 2)^3\left(x^7\right)^3\left(y^{\cdot 6}\right)^3 = \cdot 8x^{21}y^{\cdot 18} = \cdot 8x^{21} \cdot \dfrac{1}{y^{18}} = \cdot \dfrac{8x^{21}}{y^{18}}$

b. $\left(\cdot 2x^{\cdot 6}y^{\cdot 6}\right)^4 = (\cdot 2)^4\left(x^{\cdot 6}\right)^4\left(y^{\cdot 6}\right)^4 = 16x^{\cdot 24}y^{\cdot 24} = 16 \cdot \dfrac{1}{x^{24}} \cdot \dfrac{1}{y^{24}} = \dfrac{16}{x^{24}y^{24}}$

28. a. $\left(\dfrac{\cdot x^6}{\cdot y^{\cdot 3}}\right)^{\cdot 4} = \dfrac{\left(x^6\right)^{\cdot 4}}{\left(y^{\cdot 3}\right)^{\cdot 4}} = \dfrac{x^{\cdot 24}}{y^{12}} = \dfrac{1}{x^{24}} \cdot \dfrac{1}{y^{12}} = \dfrac{1}{x^{24}y^{12}}$

b. $\left|\dfrac{x^{|5}}{y^3}\right|^{|5} = \dfrac{\left(x^{|5}\right)^{|5}}{\left(y^3\right)^{|5}} = \dfrac{x^{25}}{y^{|15}} = x^{25}y^{15}$

29. a. $340{,}000 = 3.4 \times 10^5$ **b.** $0.000047 = 4.7 - 10^{-5}$

30. a. $3.7 \times 10^4 = 37{,}000$ **b.** $7.8 - 10^{-3} = 0.0078$

31. **a.** $\left[\cdot \ 8\cdot(9+2)\right]+13=\left[\cdot \ 8\cdot 11\right]+13=\cdot \ 88+13=\cdot \ 75$

  b. $\left[-7(3-8)\right]+15=\left[-7(-5)\right]+15=35+15=50$

32. **a.** $6^2\div 3\cdot \ 9\cdot 2\div 3+3=36\div 3\cdot \ 9\cdot 2\div 3+3=12\cdot \ 9\cdot 2\div 3+3=12\cdot \ 18\div 3+3$
$$=12-6+3=9$$

  b. $\dfrac{5-(68-32)}{9}=\dfrac{5-36}{9}=\dfrac{180}{9}=20$

33. **a.** $-3^2+\dfrac{4-10}{2}+15\div 3=-3^2+\dfrac{-6}{2}+15\div 3=-9+\dfrac{-6}{2}+15\div 3=-9-3+15\div 3$
$$=-9-3+5=-7$$

  b. $-4^3+\dfrac{2-10}{2}-25\div 5=-4^3+\dfrac{-8}{2}-25\div 5=-64+\dfrac{-8}{2}-25\div 5=-64-4-25\div 5$
$$=-64-4-5=-73$$

34. **a.** $2(lw+lh+wh);\ l=12\,\text{in},\ w=8\,\text{in},\ h=4\,\text{in}$
$$=2(12\,\text{in}\cdot 8\,\text{in}+12\,\text{in}\cdot 4\,\text{in}+8\,\text{in}\cdot 4\,\text{in})=2\left(96\,\text{in}^2+48\,\text{in}^2+32\,\text{in}^2\right)=2\left(176\,\text{in}^2\right)=352\,\text{in}^2$$

  b. $-2x^2-(4-y)5;\ x=-2,\ y=-1$
$$=-2(-2)^2-(4-(-1))5=-2(-2)^2-(4+1)5=-2(-2)^2-(5)5=-2(4)-(5)5$$
$$=-8-25=-33$$

35. **a.** $-3(x-7)=-3(x)-3(-7)=-3x+21$

  b. $3(x+8)-(x+7)=3(x)+3(8)-1(x)-1(7)=3x+24-x-7=2x+17$

36. **a.** $3(x+2y-2)-2(2x-2y+5)=3x+6y-6-4x+4y-10=-x+10y-16$

  b. $\left\lfloor(5x^2\,\rfloor\,3)+(4x+5)\rfloor\,\rfloor\,\rfloor(x\,\rfloor\,4)+(2x^2\,\rfloor\,2)\rfloor=\rfloor 5x^2\,\rfloor\,3+4x+5\rfloor\,\rfloor\,\rfloor x\,\rfloor\,4+2x^2\,\rfloor\,2\rfloor$
$$=\left[5x^2+4x+2\right]-\left[2x^2+x-6\right]=5x^2+4x+2-2x^2-x+6=3x^2+3x+8$$

Chapter 2   Linear Equations and Inequalities

2.1   Linear Equations in One Variable

Problems 2.1

1. **a.** Yes; substituting 6 for x in $x - 2 = 4$, we have $6 - 2 = 4$, which is a true statement.
   b. Yes; substituting 7 for y in $4 = 11 - y$, we have $4 = 11 - 7$, which is a true statement.
   c. No; substituting 4 for z in $\frac{1}{2}z - 2 = z - 2$, we have $\frac{1}{2}(4) - 2 = 4 - 2$, which simplifies to $0 = 2$, and is a false statement.

2. **a.**
$$3x - 8 = 4$$
$$3x - 8 + 8 = 4 + 8$$
$$3x = 12$$
$$\frac{1}{3} \cdot 3x = \frac{1}{3} \cdot 12$$
$$x = 4$$

   b.
$$\frac{3}{4}x - 5 = 10$$
$$\frac{3}{4}x - 5 + 5 = 10 + 5$$
$$\frac{3}{4}x = 15$$
$$\frac{4}{3} \cdot \frac{3}{4}x = \frac{4}{3} \cdot 15$$
$$x = 20$$

3.
$$5a - 8 = a + 5$$
$$5a + (-a) - 8 = a + (-a) + 5$$
$$4a - 8 = 5$$
$$4a - 8 + 8 = 5 + 8$$
$$4a = 13$$
$$\frac{1}{4} \cdot 4a = \frac{1}{4} \cdot 13$$
$$a = \frac{13}{4}$$

4.
$$3 + x = 2(2x - 3)$$
$$3 + x = 4x - 6$$
$$3 + x + (-x) = 4x + (-x) - 6$$
$$3 = 3x - 6$$
$$3 + 6 = 3x - 6 + 6$$
$$9 = 3x$$
$$\frac{1}{3} \cdot 9 = \frac{1}{3} \cdot 3x$$
$$3 = x$$

5. **a.**
$$\frac{x + 2}{4} + \frac{x - 1}{5} = 3$$
$$\text{LCD} = 20$$
$$20 \cdot \left[\frac{x + 2}{4} + \frac{x - 1}{5} \right] = 20 \cdot 3$$
$$5(x + 2) + 4(x - 1) = 60$$
$$5x + 10 + 4x - 4 = 60$$
$$9x + 6 = 60$$
$$9x = 54$$
$$x = 6$$

   b.
$$\frac{x + 3}{2} - \frac{x - 2}{3} = 5$$
$$\text{LCD} = 6$$
$$6 \cdot \left[\frac{x + 3}{2} - \frac{x - 2}{3} \right] = 6 \cdot 5$$
$$3(x + 3) - 2(x - 2) = 30$$
$$3x + 9 - 2x + 4 = 30$$
$$x + 13 = 30$$
$$x = 17$$

6. a.
$$\frac{7}{12} = \frac{x}{4} + \frac{1}{3}$$
$$12\left[\frac{7}{12}\right] = 12\left[\frac{x}{4} + \frac{1}{3}\right]$$
$$12\left[\frac{7}{12}\right] = 12\left[\frac{x}{4}\right] + 12\left[\frac{1}{3}\right]$$
$$7 = 3x + 4$$
$$3 = 3x$$
$$1 = x$$

b.
$$\frac{1}{3}\left[\frac{x}{5}\right] = \frac{8(x+2)}{15}$$
$$15\left[\frac{1}{3}\left[\frac{x}{5}\right]\right] = 15\left[\frac{8(x+2)}{15}\right]$$
$$5\mid 3x = 8(x+2)$$
$$5\mid 3x = 8x + 16$$
$$5 = 11x + 16$$
$$\mid 11 = 11x$$
$$\mid 1 = x$$

7. a. $4(x+1) + 3 = 4x + 7$
   $4x + 4 + 3 = 4x + 7$
   $4x + 7 = 4x + 7$
   Identity. Solution set is the set of
   all real numbers

b. $5(x+2) + 1 = 3 + 5x$
   $5x + 10 + 1 = 3 + 5x$
   $5x + 11 = 5x + 3$
   Contradiction. No solution - the solution set
   is the empty set.

8.
$$12.5 \cdot 1.25x = 6.5$$
$$100 \cdot (12.5 \cdot 1.25x) = 100 \cdot (6.5)$$
$$1250 \cdot 125x = 650$$
$$1250 \cdot 1250 \cdot 125x = 650 \cdot 1250$$
$$\cdot 125x = \cdot 600$$
$$\frac{\cdot 125x}{\cdot 125} = \frac{\cdot 600}{\cdot 125}$$
$$x = \frac{600}{125}$$
$$x = \frac{24}{5} \text{ or } 4.8$$

9.
$$3.30d + 1.30(30 - d) = 42.50$$
$$100(3.30d) + 100(1.30)(30 - d) = 100(42.50)$$
$$330d + 130(30 - d) = 4250$$
$$330d + 3900 - 130d = 4250$$
$$200d + 3900 = 4250$$
$$200d + 3900 - 3900 = 4250 - 3900$$
$$200d = 350$$
$$\frac{200d}{200} = \frac{350}{200}$$
$$d = \frac{7}{4} \text{ or } 1.75$$

Exercises 2.1

1. Yes; substituting 3 for x in $2x + 8 = 14$, we have $2(3) + 8 = 14$, which simplifies to $14 = 14$, and is a true statement.

3. Yes; substituting -1 for x in $-2x + 1 = 3$, we have $-2(-1) + 1 = 3$ which simplifies to $3 = 3$, and is a true statement.

5. Yes; substituting 3 for y in $2y - 5 = y - 2$, we have $2(3) - 5 = 3 - 2$ which simplifies to $1 = 1$, and is a true statement.

7. No; substituting $-\frac{1}{5}$ for t in $\frac{4}{5}t - 1 = 5t$, we have $\frac{4}{5}\left(-\frac{1}{5}\right) - 1 = 5\left(-\frac{1}{5}\right)$, which simplifies to $-\frac{29}{25} = -1$, and is a false statement.

9. No; substituting 3 for x in $\frac{1}{2}x + 5 = 5 - \frac{1}{3}x$, we have $\frac{1}{2}(3) + 5 = 5 - \frac{1}{3}(3)$, which simplifies to $\frac{13}{2} = 4$, and is a false statement.

11.
$$3x - 4 = 8$$
$$3x - 4 + 4 = 8 + 4$$
$$3x = 12$$
$$\frac{1}{3} \cdot 3x = \frac{1}{3} \cdot 12$$
$$x = 4$$

13.
$$2y + 8 = 10$$
$$2y + 8 - 8 = 10 - 8$$
$$2y = 2$$
$$\frac{1}{2} \cdot 2y = \frac{1}{2} \cdot 2$$
$$y = 1$$

15.
$$-3z - 6 = -12$$
$$-3z - 6 + 6 = -12 + 6$$
$$-3z = -6$$
$$-\frac{1}{3} \cdot (-3z) = -\frac{1}{3} \cdot (-6)$$
$$z = 2$$

17.
$$-5y + 2 = -8$$
$$-5y + 2 - 2 = -8 - 2$$
$$-5y = -10$$
$$-\frac{1}{5} \cdot (-5y) = -\frac{1}{5} \cdot (-10)$$
$$y = 2$$

19.
$$3x + 5 = x + 19$$
$$3x + (-x) + 5 = x + (-x) + 19$$
$$2x + 5 = 19$$
$$2x + 5 + (-5) = 19 + (-5)$$
$$2x = 14$$
$$\frac{1}{2} \cdot 2x = \frac{1}{2} \cdot 14$$
$$x = 7$$

21.
$$7(x - 1) - 3 + 5x = 3(4x - 3) + x$$
$$7x - 7 - 3 + 5x = 12x - 9 + x$$
$$12x - 10 = 13x - 9$$
$$12x + (-12x) - 10 = 13x + (-12x) - 9$$
$$-10 = x - 9$$
$$-10 + 9 = x - 9 + 9$$
$$-1 = x$$

23.
$$6v - 8 = 8v + 8$$
$$6v + (-6v) - 8 = 8v + (-6v) + 8$$
$$-8 = 2v + 8$$
$$-8 + (-8) = 2v + 8 + (-8)$$
$$-16 = 2v$$
$$\frac{1}{2} \cdot (-16) = \frac{1}{2} \cdot 2v$$
$$-8 = v$$

25.
$$7m - 4m + 12 = 0$$
$$3m + 12 = 0$$
$$3m + 12 + (-12) = 0 + (-12)$$
$$3m = -12$$
$$\frac{1}{3} \cdot (3m) = \frac{1}{3} \cdot (-12)$$
$$m = -4$$

27.
$$4(2 \cdot z) + 8 = 8(2 \cdot z)$$
$$8 \cdot 4z + 8 = 16 \cdot 8z$$
$$\cdot \, 4z + 16 = 16 \cdot 8z$$
$$\cdot \, 4z + 8z + 16 = 16 \cdot 8z + 8z$$
$$4z + 16 = 16$$
$$4z + 16 + (\cdot \, 16) = 16 + (\cdot \, 16)$$
$$4z = 0$$
$$\frac{1}{4} \cdot 4z = \frac{1}{4} \cdot 0$$
$$z = 0$$

29.
$$5(x + 3) = 3(x + 3) + 6$$
$$5x + 15 = 3x + 9 + 6$$
$$5x + 15 = 3x + 15$$
$$5x + (\cdot \, 3x) + 15 = 3x + (\cdot \, 3x) + 15$$
$$2x + 15 = 15$$
$$2x + 15 + (\cdot \, 15) = 15 + (\cdot \, 15)$$
$$2x = 0$$
$$\frac{1}{2} \cdot 2x = \frac{1}{2} \cdot 0$$
$$x = 0$$

31.
$$5(4 \cdot 3a) = 7(3 \cdot 4a)$$
$$20 \cdot 15a = 21 \cdot 28a$$
$$20 \cdot 15a + 28a = 21 \cdot 28a + 28a$$
$$20 + 13a = 21$$
$$20 + (\cdot \, 20) + 13a = 21 + (\cdot \, 20)$$
$$13a = 1$$
$$\frac{1}{13} \cdot 13a = \frac{1}{13} \cdot 1$$
$$a = \frac{1}{13}$$

33.
$$\cdot \, \frac{7}{8}c + 5 = \cdot \, \frac{5}{8}c + 3$$
$$8 \cdot \cdot \, \frac{7}{8}c + 5 \cdot = 8 \cdot \cdot \, \frac{5}{8}c + 3 \cdot$$
$$\cdot \, 7c + 40 = \cdot \, 5c + 24$$
$$\cdot \, 7c + 7c + 40 = \cdot \, 5c + 7c + 24$$
$$40 = 2c + 24$$
$$40 + (\cdot \, 24) = 2c + 24 + (\cdot \, 24)$$
$$16 = 2c$$
$$\frac{1}{2} \cdot 16 = \frac{1}{2} \cdot 2c$$
$$8 = c$$

35.
$$\cdot \, 2x + \frac{1}{4} = 2x + \frac{4}{5}$$
$$20 \cdot \cdot \, 2x + \frac{1}{4} \cdot = 20 \cdot 2x + \frac{4}{5} \cdot$$
$$\cdot \, 40x + 5 = 40x + 16$$
$$\cdot \, 40x + 40x + 5 = 40x + 40x + 16$$
$$5 = 80x + 16$$
$$5 + (\cdot \, 16) = 80x + 16 + (\cdot \, 16)$$
$$\cdot \, 11 = 80x$$
$$\frac{1}{80} \cdot (\cdot \, 11) = \frac{1}{80} \cdot 80x$$
$$\cdot \, \frac{11}{80} = x$$

37.
$$\frac{t}{6} + \frac{t}{8} = 7$$
$$24 \cdot \frac{t}{6} + \frac{t}{8} \cdot = 24 \cdot 7$$
$$24 \cdot \frac{t}{6} + 24 \cdot \frac{t}{8} = 24 \cdot 7$$
$$4t + 3t = 168$$
$$7t = 168$$
$$\frac{7t}{7} = \frac{168}{7}$$
$$t = 24$$

39.
$$\frac{x}{2} + \frac{x}{5} = \frac{7}{10}$$
$$10 \cdot \frac{x}{2} + \frac{x}{5} \cdot = 10 \cdot \frac{7}{10}$$
$$10 \cdot \frac{x}{2} + 10 \cdot \frac{x}{5} = 10 \cdot \frac{7}{10}$$
$$5x + 2x = 7$$
$$7x = 7$$
$$\frac{7x}{7} = \frac{7}{7}$$
$$x = 1$$

41.
$$\frac{c}{3} \cdot \frac{c}{5} = 2$$
$$15 \cdot \frac{c}{3} \cdot \frac{c}{5} \cdot = 15 \cdot 2$$
$$15 \cdot \frac{c}{3} \cdot 15 \cdot \frac{c}{5} = 15 \cdot 2$$
$$5c \cdot 3c = 30$$
$$2c = 30$$
$$\frac{2c}{2} = \frac{30}{2}$$
$$c = 15$$

43.
$$\frac{W}{6} \cdot \frac{W}{8} = \frac{5}{12}$$
$$24 \cdot \frac{W}{6} \cdot \frac{W}{8} \cdot = 24 \cdot \frac{5}{12}$$
$$24 \cdot \frac{W}{6} \cdot 24 \cdot \frac{W}{8} = 24 \cdot \frac{5}{12}$$
$$4W \cdot 3W = 10$$
$$W = 10$$

45.
$$\frac{x}{5} \cdot \frac{3}{10} = \frac{1}{2}$$
$$10 \cdot \frac{x}{5} \cdot \frac{3}{10} \cdot = 10 \cdot \frac{1}{2}$$
$$10 \cdot \frac{x}{5} \cdot 10 \cdot \frac{3}{10} = 10 \cdot \frac{1}{2}$$
$$2x \cdot 3 = 5$$
$$2x \cdot 3 + 3 = 5 + 3$$
$$2x = 8$$
$$\frac{2x}{2} = \frac{8}{2}$$
$$x = 4$$

47.
$$\frac{x+4}{4} \cdot \frac{x+2}{3} = \cdot \frac{1}{2}$$
$$12 \cdot \frac{x+4}{4} \cdot \frac{x+2}{3} \cdot = 12 \cdot \cdot \frac{1}{2} \cdot$$
$$12 \cdot \frac{x+4}{4} \cdot \cdot 12 \cdot \frac{x+2}{3} \cdot = 12 \cdot \cdot \frac{1}{2} \cdot$$
$$3(x+4) \cdot 4(x+2) = \cdot 6$$
$$3x + 12 \cdot 4x \cdot 8 = \cdot 6$$
$$\cdot x + 4 = \cdot 6$$
$$\cdot x + 4 \cdot 4 = \cdot 6 \cdot 4$$
$$\cdot x = \cdot 10$$
$$\frac{\cdot x}{\cdot 1} = \frac{\cdot 10}{\cdot 1}$$
$$x = 10$$

49.
$$\frac{x+1}{4} \cdot \frac{2x \cdot 2}{3} = 3$$
$$12 \cdot \frac{x+1}{4} \cdot \frac{2x \cdot 2}{3} \cdot = 12 \cdot 3$$
$$12 \cdot \frac{x+1}{4} \cdot \cdot 12 \cdot \frac{2x \cdot 2}{3} \cdot = 12 \cdot 3$$
$$3(x+1) \cdot 4(2x \cdot 2) = 36$$
$$3x + 3 \cdot 8x + 8 = 36$$
$$\cdot 5x + 11 = 36$$
$$\cdot 5x + 11 \cdot 11 = 36 \cdot 11$$
$$\cdot 5x = 25$$
$$\frac{\cdot 5x}{\cdot 5} = \frac{25}{\cdot 5}$$
$$x = \cdot 5$$

51.
$$\frac{2h+1}{3} = \frac{h+4}{12}$$
$$12\left(\frac{2h+1}{3}\right) = 12\left(\frac{h+4}{12}\right)$$
$$4(2h+1) = h+4$$
$$8h+4 = h+4$$
$$8h - h + 4 = h - h + 4$$
$$7h+4 = 4$$
$$7h+4-4 = 4-4$$
$$7h = 0$$
$$\frac{7h}{7} = \frac{0}{7}$$
$$h = 0$$

53.
$$\frac{2w+3}{2} - \frac{3w+1}{4} = 1$$
$$4\left(\frac{2w+3}{2} - \frac{3w+1}{4}\right) = 4\cdot 1$$
$$4\cdot\frac{2w+3}{2} - 4\cdot\frac{3w+1}{4} = 4$$
$$2(2w+3) - (3w+1) = 4$$
$$4w+6 - 3w-1 = 4$$
$$w+5 = 4$$
$$w+5-5 = 4-5$$
$$w = -1$$

55.
$$\frac{7r+2}{6} + \frac{1}{2} = \frac{r}{4}$$
$$12\left(\frac{7r+2}{6} + \frac{1}{2}\right) = 12\left(\frac{r}{4}\right)$$
$$12\cdot\frac{7r+2}{6} + 12\cdot\frac{1}{2} = 12\cdot\frac{r}{4}$$
$$2(7r+2) + 6(1) = 3r$$
$$14r+4+6 = 3r$$
$$14r+10 = 3r$$
$$14r - 14r + 10 = 3r - 14r$$
$$10 = -11r$$
$$\frac{10}{-11} = \frac{-11r}{-11}$$
$$-\frac{10}{11} = r$$

57.
$$\frac{x-5}{2} - \frac{x-4}{3} = \frac{x-3}{2} - (x-2)$$
$$6\left(\frac{x-5}{2} - \frac{x-4}{3}\right) = 6\left(\frac{x-3}{2} - (x-2)\right)$$
$$6\cdot\frac{x-5}{2} - 6\cdot\frac{x-4}{3} = 6\cdot\frac{x-3}{2} - 6(x-2)$$
$$3(x-5) - 2(x-4) = 3(x-3) - 6(x-2)$$
$$3x-15 - 2x+8 = 3x-9 - 6x+12$$
$$x-7 = -3x+3$$
$$3x+x-7 = -3x+3x+3$$
$$4x-7 = 3$$
$$4x-7+7 = 3+7$$
$$4x = 10$$
$$\frac{4x}{4} = \frac{10}{4}$$
$$x = \frac{5}{2}$$

59.
$$4(x-2)+4 = 4x-4$$
$$4x-8+4 = 4x-4$$
$$4x-4 = 4x-4$$
Identity. Solution set is the set of all real numbers.

61.
$$6.3x - 8.4 = 16.8$$
$$10(6.3x - 8.4) = 10(16.8)$$
$$10(6.3x) - 10(8.4) = 10(16.8)$$
$$63x - 84 = 168$$
$$63x - 84 + 84 = 168 + 84$$
$$63x = 252$$
$$\frac{63x}{63} = \frac{252}{63}$$
$$x = 4$$

63.
$$-12.6y - 25.2 = 50.4$$
$$10(-12.6y - 25.2) = 10(50.4)$$
$$-126y - 252 = 504$$
$$-126y - 252 + 252 = 504 + 252$$
$$-126y = 756$$
$$\frac{-126y}{-126} = \frac{756}{-126}$$
$$y = -6$$

65.
$$2.1y + 3.5 = 0.7y + 83.3$$
$$10(2.1y + 3.5) = 10(0.7y + 83.3)$$
$$21y + 35 = 7y + 833$$
$$21y - 7y + 35 = 7y - 7y + 833$$
$$14y + 35 = 833$$
$$14y + 35 - 35 = 833 - 35$$
$$14y = 798$$
$$\frac{14y}{14} = \frac{798}{14}$$
$$y = 57$$

67.
$$3.5(x + 3) = 2.1(x + 3) + 4.2$$
$$10[3.5(x + 3)] = 10[2.1(x + 3) + 4.2]$$
$$35(x + 3) = 21(x + 3) + 42$$
$$35x + 105 = 21x + 63 + 42$$
$$35x + 105 = 21x + 105$$
$$35x - 21x + 105 = 21x - 21x + 105$$
$$14x + 105 = 105$$
$$14x + 105 - 105 = 105 - 105$$
$$14x = 0$$
$$\frac{14x}{14} = \frac{0}{14}$$
$$x = 0$$

69.
$$0.40y + 0.20(32 - y) = 9.60$$
$$100[0.40y + 0.20(32 - y)] = 100(9.60)$$
$$40y + 20(32 - y) = 960$$
$$40y + 640 - 20y = 960$$
$$20y + 640 = 960$$
$$20y + 640 - 640 = 960 - 640$$
$$20y = 320$$
$$\frac{20y}{20} = \frac{320}{20}$$
$$y = 16$$

71.
$$0.65x + 0.40(50 - x) = 25.375$$
$$1000[0.65x + 0.40(50 - x)] = 1000(25.375)$$
$$650x + 400(50 - x) = 25{,}375$$
$$650x + 20{,}000 - 400x = 25{,}375$$
$$250x + 20{,}000 = 25{,}375$$
$$250x + 20{,}000 - 20{,}000 = 25375 - 20000$$
$$250x = 5{,}375$$
$$\frac{250x}{250} = \frac{5{,}375}{250}$$
$$x = 21.5$$

73.
$$0.06P + 0.08(2000 - P) = 130$$
$$100[0.06P + 0.08(2000 - P)] = 100(130)$$
$$6P + 8(2000 - P) = 13{,}000$$
$$6P + 16{,}000 - 8P = 13{,}000$$
$$-2P + 16{,}000 = 13{,}000$$
$$-2P = -3000$$
$$\frac{-2P}{-2} = \frac{-3000}{-2}$$
$$P = 1500$$

75.
$$0.30y + 1.80 = 0.20(y + 12)$$
$$100[0.30y + 1.80] = 100[0.20(y + 12)]$$
$$30y + 180 = 20(y + 12)$$
$$30y + 180 = 20y + 240$$
$$30y - 20y + 180 = 20y - 20y + 240$$
$$10y + 180 = 240$$
$$10y + 180 - 180 = 240 - 180$$
$$10y = 60$$
$$\frac{10y}{10} = \frac{60}{10}$$
$$y = 6$$

77.
$$3x - (4x + 7) - 2x = 3x - 4x - 7 - 2x$$
$$= -3x - 7$$

79.
$$6x - 2(3x - 1) = 6x - 6x + 2$$
$$= 2$$

81.
$$\frac{400 \cdot 2 \cdot 150}{2} = \frac{400 \cdot 300}{2} = \frac{100}{2} = 50$$

83.
$$\frac{5}{9}(F - 32) \text{ when } F = 41$$
$$= \frac{5}{9}(41 - 32) = \frac{5}{9}(9) = 5$$

85.
$$S = 3L - 22; \quad S = 11$$
$$11 = 3L - 22$$
$$11 + 22 = 3L - 22 + 22$$
$$33 = 3L$$
$$\frac{33}{3} = \frac{3L}{3}$$
$$11\text{in} = L$$

87.
$$S = 3L - 22; \quad S = 7$$
$$7 = 3L - 22$$
$$7 + 22 = 3L - 22 + 22$$
$$29 = 3L$$
$$\frac{29}{3} = \frac{3L}{3}$$
$$\frac{29}{3} = L$$

$$S = 3L \mid 21; \quad L = \frac{29}{3}$$
$$S = 3 \left| \frac{29}{3} \right| \mid 21$$
$$S = 29 \mid 21$$
$$S = 8$$

Sue wears a size 8 shoe.

89.
$$S = 3L - 22; \quad S = 14$$
$$14 = 3L - 22$$
$$14 + 22 = 3L - 22 + 22$$
$$36 = 3L$$
$$\frac{36}{3} = \frac{3L}{3}$$
$$12\text{in} = L$$

91. **a.**
$$S = 3L - 22; \quad L = 23$$
$$S = 3(23) - 22$$

b.
$$S = 69 - 22$$
$$S = 47$$

$$S = 3L - 21; \quad L = 23$$
$$S = 3(23) - 21$$
$$S = 69 - 21$$
$$S = 48$$

93. Sample answer: A contradictory equation is an equation that yields a false statement such as $1 = 0$.

95.

$$\frac{x+4}{8} \mid \frac{x+2}{6} = \mid \frac{1}{4}$$
$$24\left|\frac{x+4}{8} \mid \frac{x+2}{6}\right| = 24\left|\mid \frac{1}{4}\right|$$
$$24\left|\frac{x+4}{8}\right| \mid 24\left|\frac{x+2}{6}\right| = 24\left|\mid \frac{1}{4}\right|$$
$$3(x+4) \mid 4(x+2) = \mid 6$$
$$3x + 12 \mid 4x \mid 8 = \mid 6$$
$$\mid x + 4 = \mid 6$$
$$\mid x + 4 \mid 4 = \mid 6 \mid 4$$
$$\mid x = \mid 10$$
$$\frac{\mid x}{\mid 1} = \frac{\mid 10}{\mid 1}$$
$$x = 10$$

97.

$$0.8t + 0.4(32 - t) = 19.2$$
$$10[0.8t + 0.4(32 - t)] = 10(19.2)$$
$$8t + 4(32 - t) = 192$$
$$8t + 128 - 4t = 192$$
$$4t + 128 = 192$$
$$4t + 128 - 128 = 192 - 128$$
$$4t = 64$$
$$\frac{4t}{4} = \frac{64}{4}$$
$$t = 16$$

99.

$$7(x-1) - x = 3 - 5x + 3(4x - 3)$$
$$7x - 7 - x = 3 - 5x + 12x - 9$$
$$6x - 7 = 7x - 6$$
$$6x - 6x - 7 = 7x - 6x - 6$$
$$-7 = x - 6$$
$$-7 + 6 = x - 6 + 6$$
$$-1 = x$$

101. Substitute -5 for x in:

$$-\frac{1}{5}x + 2 = -x + 4$$
$$-\frac{1}{5}(-5) + 2 = -(-5) + 4$$
$$1 + 2 = 5 + 4$$
$$3 = 9$$

False, therefore $x - -5$.

2.2 Formulas, Geometry, and Problem Solving

Problems 2.2

1. **a.**

$$H = 2.75\,h + 71.48$$
$$H - 71.48 = 2.75\,h + 71.48 - 71.48$$
$$H - 71.48 = 2.75\,h$$
$$\frac{H - 71.48}{2.75} = h$$

b.

$$h = \frac{H - 71.48}{2.75};\quad H = 126.48$$
$$h = \frac{126.48 - 71.48}{2.75}$$
$$h = \frac{55}{2.75}$$
$$h = 20\text{ cm}$$

2.   a.

$$A = \frac{1}{2}h(a+b); \quad \text{solve for } a$$

$$2 \cdot A = 2 \cdot \frac{1}{2}h(a+b)$$

$$2A = h(a+b)$$

$$2A = ha + hb$$

$$2A - hb = ha$$

$$a = \frac{2A - bh}{h} \quad \text{or} \quad a = \frac{2A}{h} - b$$

b.

$$a = \frac{2A - bh}{h}; \quad A = 40, h = 10, b = 5$$

$$a = \frac{2(40) - 5(10)}{10}$$

$$a = \frac{80 - 50}{10}$$

$$a = \frac{30}{10}$$

$$a = 3$$

3.   a.

$$P = 2L + 2W; \quad \text{solve for } L$$

$$P - 2W = 2L + 2W - 2W$$

$$P - 2W = 2L$$

$$\frac{P - 2W}{2} = L$$

b.

$$L = \frac{P - 2W}{2}; \quad P = 102.8, W = 20.9$$

$$L = \frac{102.8 - 2(20.9)}{2}$$

$$L = \frac{102.8 - 41.8}{2}$$

$$L = \frac{61}{2} = 30.5 \text{ in}$$

4.   a.

$$2A = ah + bh; \quad \text{solve for } h$$

$$2A = h(a+b)$$

$$\frac{2A}{a+b} = \frac{h(a+b)}{a+b}$$

$$\frac{2A}{a+b} = h$$

b.

$$h = \frac{2A}{a+b}; \quad A = 150, a = 10, b = 20$$

$$h = \frac{2(150)}{10 + 20}$$

$$h = \frac{300}{30}$$

$$h = 10$$

5.   a.

$$A = \frac{n(P_1 + P_n)}{2}; \quad \text{solve for } n$$

$$2 \cdot A = 2 \cdot \frac{n(P_1 + P_n)}{2}$$

$$2A = n(P_1 + P_n)$$

$$\frac{2A}{P_1 + P_n} = \frac{n(P_1 + P_n)}{P_1 + P_n}$$

$$\frac{2A}{P_1 + P_n} = n$$

b.

$$n = \frac{2A}{P_1 + P_n}; \quad A = 50, P_1 = 32, P_n = 18$$

$$n = \frac{2(50)}{32 + 18}$$

$$n = \frac{100}{50}$$

$$n = 2$$

6. a. $C = 0.25 + 0.10\,(t \ge 1);\ \ t \ge 1$

   c. $t = \dfrac{C - 0.15}{0.1};\ \ C = \$2.75$

b.

Solve for t; $C = 0.25 + 0.10(t - 1)$

$$C = 0.25 + 0.10t - 0.10$$

$$C = 0.15 + 0.10t$$

$$C - 0.15 = 0.10t$$

$$\frac{C - 0.15}{0.1} = t$$

$$t = \frac{2.75 - 0.15}{0.1}$$

$$t = \frac{2.6}{0.1} = 26 \text{ min}$$

7.

$$8x + 70 = 2x + 100$$

$$8x - 2x + 70 = 2x - 2x + 100$$

$$6x + 70 = 100$$

$$6x + 70 - 70 = 100 - 70$$

$$6x = 30$$

$$x = 5$$

measure of one angle:

$$8x + 70 = 8(5) + 70$$

$$= 40 + 70$$

$$= 110°$$

measure of the other angle:

$$2x + 100 = 2(5) + 100$$

$$= 10 + 100$$

$$= 110°$$

Exercises 2.2

1.

$$V = \pi r^2 h$$

$$\frac{V}{\pi r^2} = \frac{\pi r^2 h}{\pi r^2}$$

$$\frac{V}{\pi r^2} = h$$

3.

$$V = LWH$$

$$\frac{V}{LH} = \frac{LWH}{LH}$$

$$\frac{V}{LH} = W$$

5.

$$P = s_1 + s_2 + b$$

$$P - s_1 - s_2 = s_1 - s_1 + s_2 - s_2 + b$$

$$P - s_1 - s_2 = b$$

7.

$$A = -\left(r^2 + rs\right)$$

$$A = -r^2 + -rs$$

$$A - -r^2 = -r^2 - -r^2 + -rs$$

$$A - -r^2 = -rs$$

$$\frac{A - -r^2}{-r} = \frac{-rs}{-r}$$

$$\frac{A - -r^2}{-r} = s \text{ or } s = \frac{A}{-r} - \frac{-r^2}{-r} = \frac{A}{-r} - r$$

9.

$$\frac{V_2}{V_1} = \frac{P_1}{P_2}$$

$$V_1 \cdot \frac{V_2}{V_1} = V_1 \cdot \frac{P_1}{P_2}$$

$$V_2 = \frac{P_1 V_1}{P_2}$$

11. $2x + 3y = 12$

$$3y = 12 - 2x$$

$$y = \frac{12 - 2x}{3} \text{ or } y = 4 - \frac{2}{3}x$$

13. **a.** $D = RT$

$$\frac{D}{R} = \frac{RT}{R}$$

$$\frac{D}{R} = T$$

b. $T = \dfrac{D}{R}; \quad D = 220, R = 55$

$$T = \frac{220}{55}$$

$$T = 4\,\text{hr}$$

15. **a.**
$$H = 17 - \frac{A}{2}$$

$$2H = 2 \cdot 17 - 2\left|\frac{A}{2}\right|$$

$$2H = 34 - A$$

$$2H - 34 = -A$$

$$-1(2H - 34) = -1(-A)$$

$$34 - 2H = A$$

b. $A = 34 - 2H; \quad H = 8$

$A = 34 - 2(8)$

$A = 34 - 16$

$A = 18\ \text{yr}$

17. **a.**
$$O = \frac{C + E}{N}$$

$$N \cdot O = N \cdot \frac{C + E}{N}$$

$$N \cdot O = C + E$$

$$N \cdot O - E = C$$

b. $C = N \cdot O - E;$

$E = 18{,}500,\ O = 0.96,\ N = 50{,}000$

$C = 50{,}000 \cdot 0.96 - 18{,}500$

$C = 48{,}000 - 18{,}500$

$C = \$29{,}500$

19. **a.**
$$A + B + C = 180$$

$$A - A + B + C - C = 180 - A - C$$

$$B = 180 - A - C$$

b. $B = 180 - A - C; \quad A = 47, C = 119$

$B = 180 - 47 - 119$

$B = 14^\circ$

21. **a.**
$$H = 1.88L + 32$$

$$H - 32 = 1.88L$$

$$\frac{H - 32}{1.88} = L$$

b. Substitute $H = 72,\ L = 20$

$$\frac{72 - 32}{1.88} = 20$$

$$\frac{40}{1.88} = 20$$

$$21.3 = 20$$

The answer is no.

c. Substitute $H = 69.6,\ L = 20$

$$\frac{69.6 - 32}{1.88} = 20$$

$$\frac{37.6}{1.88} = 20$$

$$20 = 20$$

The answer is yes.

d.
$$H = 1.95L + 29$$

$$H - 29 = 1.95L$$

$$\frac{H - 29}{1.95} = L$$

e. Substitute $H = 68$, $L = 20$

$$\frac{68 - 29}{1.95} = 20$$

$$\frac{39}{1.95} = 20$$

$$20 = 20$$

The answer is yes.

23. **a.** $C = 2000 + 309m$

b. Substitute $m = 39$

$$C = 2000 + 309(39)$$

$$C = 2000 + 12,051$$

$$C = \$14,051$$

25. **a.** $C = 0.36 + 0.08(t \geq 1);\ t \geq 1$

b. Substitute $C = 4.04$

$$4.04 = 0.36 + 0.08(t - 1)$$

$$4.04 - 0.36 = 0.36 - 0.36 + 0.08(t - 1)$$

$$3.68 = 0.08(t - 1)$$

$$\frac{3.68}{0.08} = \frac{0.08(t - 1)}{0.08}$$

$$46 = t - 1$$

$$46 + 1 = t - 1 + 1$$

$$47 \text{ min} = t$$

27. **a.** $F = 20 + 10m$

b. Substitute $F = 140$

$$140 = 20 + 10m$$

$$140 - 20 = 20 - 20 + 10m$$

$$120 = 10m$$

$$\frac{120}{10} = \frac{10m}{10}$$

$$12 \text{ months} = m$$

29.
$$15 - 4x = 25 - 2x$$

$$15 - 4x + 4x = 25 - 2x + 4x$$

$$15 = 25 + 2x$$

$$15 - 25 = 25 - 25 + 2x$$

$$-10 = 2x$$

$$\frac{-10}{2} = \frac{2x}{2}$$

$$-5 = x$$

$$15 - 4x = 15 - 4(-5) = 15 + 20 = 35°$$

$$25 - 2x = 25 - 2(-5) = 25 + 10 = 35°$$

31.
$$80 + 3x = 40 + 5x$$

$$80 + 3x - 5x = 40 + 5x - 5x$$

$$80 - 2x = 40$$

$$80 - 80 - 2x = 40 - 80$$

$$-2x = -40$$

$$\frac{-2x}{-2} = \frac{-40}{-2}$$

$$x = 20$$

$$80 + 3x = 80 + 3(20) = 80 + 60 = 140°$$

$$40 + 5x = 40 + 5(20) = 40 + 100 = 140°$$

33.
$$14x + 8 = 16x + 2$$

$$14x - 16x + 8 = 16x - 16x + 2$$

$$-2x + 8 = 2$$

$$-2x + 8 - 8 = 2 - 8$$

$$-2x = -6$$

$$\frac{-2x}{-2} = \frac{-6}{-2}$$

$$x = 3$$

$$14x + 8 = 14(3) + 8 = 42 + 8 = 50°$$

$$16x + 2 = 16(3) + 2 = 48 + 2 = 50°$$

35.
$$50 - 10x + 42 - 6x = 180$$

$$-16x + 92 = 180$$

$$-16x + 92 - 92 = 180 - 92$$

$$-16x = 88$$

$$\frac{-16x}{-16} = \frac{88}{-16}$$

$$x = -5.5$$

$$50 - 10x = 50 - 10(-5.5) = 50 + 55 = 105°$$

$$42 - 6x = 42 - 6(-5.5) = 42 + 33 = 75°$$

37. $n + (n + 2) + (n + 4) = n + n + 2 + n + 4 = 3n + 6$

39.
$$p + (p + 60{,}000) = 110{,}000$$
$$2p + 60{,}000 = 110{,}000$$
$$20 + 60{,}000 - 60{,}000 = 110{,}000 - 60{,}000$$
$$2p = 50{,}000$$
$$\frac{2p}{2} = \frac{50{,}000}{2}$$
$$p = 25{,}000$$

41.
$$0.25m + 20 = 37.50$$
$$100(0.25m + 20) = 100(37.50)$$
$$25m + 2000 = 3750$$
$$25m + 2000 - 2000 = 3750 - 2000$$
$$25m = 1750$$
$$\frac{25m}{25} = \frac{1750}{25}$$
$$m = 70$$

43.
$$(90 - x) + x + 5x = 180$$
$$90 - x + x + 5x = 180$$
$$5x + 90 = 180$$
$$5x + 90 - 90 = 180 - 90$$
$$5x = 90$$
$$\frac{5x}{5} = \frac{90}{5}$$
$$x = 18$$

45.
$$t = \frac{C - 14.84}{0.08}; \quad C = 20$$
$$t = \frac{20 - 14.84}{0.08} = \frac{5.16}{0.08} = 64.5$$

$$1960 + 64.5 = 2024.5 \approx 2025$$

47. Substitute $2000 - 1960 = 40$ for t.
$$C = 14.84 + 0.08(40)$$
$$C = 14.84 + 3.2$$
$$C = 18.04°\text{C}$$

Between 1962 and 2000, the average global temperature increased $18.04° - 15° = 3.04°$ which means that sea level increased 3.04 ft.

49. Sample answer: To solve for x, first subtract b from both sides of the equation. Then, divide both sides of the equation by a.

51.

$$90 - 4x = 50 - 6x$$
$$90 - 4x + 6x = 50 - 6x + 6x$$
$$90 + 2x = 50$$
$$90 - 90 + 2x = 50 - 90$$
$$2x = -40$$
$$\frac{2x}{2} = \frac{-40}{2}$$
$$x = -20$$

$$90 - 4x = 90 - 4(-20) = 90 + 80 = 170°$$
$$50 - 6x = 50 - 6(-20) = 50 + 120 = 170°$$

53. **a.** $C = 0.37 + 0.23(w \geq 1); \quad w \geq 1$

b.
$$C = 0.37 + 0.23w - 0.23$$
$$C = 0.14 + 0.23w$$
$$C - 0.14 = 0.14 - 0.14 + 0.23w$$
$$C - 0.14 = 0.23w$$
$$\frac{C - 0.14}{0.23} = \frac{0.23w}{0.23}$$
$$\frac{C - 0.14}{0.23} = w$$

c. Substitute $w = 10$
$$C = 0.37 + 0.23(10 - 1) = 0.37 + 0.23(9)$$
$$= 0.37 + 2.07 = \$2.44$$

55. **a.**

$$S = \frac{1}{6}(C \cdot 4)$$
$$6 \cdot S = 6 \cdot \frac{1}{6}(C \cdot 4)$$
$$6S = C \cdot 4$$
$$6S + 4 = C \cdot 4 + 4$$
$$6S + 4 = C$$

b. Substitute $S = 2$;
$$C = 6(2) + 4 = 12 + 4 = 16°C$$

2.3   Problem Solving: Integers and Geometry

Problems 2.3

1. **a.** $x + 5 = 8$

b. $\dfrac{x}{x - 4} = 16$

c. $2x - 7 = \dfrac{14}{x}$

d. $9 - 6x = 21$

e. $n - (n + 2) = -2$

2. Let m = number of miles
$$25 + 0.20m = 50$$
$$25 - 25 + 0.20m = 50 - 25$$
$$0.20m = 25$$
$$\frac{0.20m}{0.20} = \frac{25}{0.20}$$
$$m = 125 \text{ miles}$$

3.   Let $x =$ first odd integer
$x + 2 =$ second consecutive odd integer
$x + 4 =$ third consecutive odd integer

$$(x) + (x + 2) + (x + 4) = 249$$
$$3x + 6 = 249$$
$$3x + 6 - 6 = 249 - 6$$
$$3x = 243$$
$$\frac{3x}{3} = \frac{243}{3}$$
$$x = 81$$
$$x + 2 = 83$$
$$x + 4 = 85$$

4. $$L = 100 + W$$
$$P = 2L + 2W; \quad P = 360$$
$$360 = 2(100 + W) + 2W$$
$$360 = 200 + 2W + 2W$$
$$360 = 4W + 200$$
$$160 = 4W$$
$$40\,\text{in} = W$$
$$L = 100 + W = 100 + 40 = 140\,\text{in}$$

5.   Let $x =$ first angle
$2x =$ second angle
$x + 20 =$ third angle

$$(x) + (2x) + (x + 20) = 180$$
$$x + 2x + x + 20 = 180$$
$$4x + 20 = 180$$
$$4x = 160$$
$$x = 40° \text{ (first angle)}$$
$$\text{second angle} = 2(40) = 80°$$
$$\text{third angle} = 20 + 40 = 60°$$

Exercises 2.3

1. $4m = m + 18$

3. $\dfrac{1}{2}x - 12 = \dfrac{2x}{3}$

5. $4x + 5 = 29$
$4x = 24$
$x = 6$

7. $3x + 8 = 35$
$3x = 27$
$x = 9$

9. $3x - 2 = 16$
$3x = 18$
$x = 6$

11. $5x = 12 + 2x$
$3x = 12$
$x = 4$

13.

$$\frac{1}{3}x \cdot 2 = 10$$

$$3\left(\frac{1}{3}x \cdot 2\right) = 3 \cdot 10$$

$$x \cdot 6 = 30$$

$$x = 36$$

15.   Let $x =$ votes received by other four

$3x =$ votes received by golf

$$x + 3x = 100$$

$$4x = 100$$

$$x = 25$$

$$3x = 75$$

Golf received 75% of the votes.

17.   Let $x =$ number of years from now

$$12 + x = 2(2 + x)$$

$$12 + x = 4 + 2x$$

$$12 = 4 + x$$

$$8 = x$$

Alex will be twice as old as Ramie 8 years from now.

19.   Let $m =$ number of miles driven

$$44 = 18 + 0.20\,m$$

$$26 = 0.20\,m$$

$$130 = m$$

Margie traveled 130 miles.

21.   Let $x =$ first even integer

$x + 2 =$ second consecutive even integer

$x + 4 =$ third consecutive even integer

$$(x) + (x + 2) + (x + 4) = 138$$

$$x + x + 2 + x + 4 = 138$$

$$3x + 6 = 138$$

$$3x = 132$$

$$x = 44$$

$$x + 2 = 46$$

$$x + 4 = 48$$

The even integers are 44, 46, and 48.

23.   Let $x =$ first odd integer

$x + 2 =$ second consecutive odd integer

$x + 4 =$ third consecutive odd integer

$$(x) + (x + 2) + (x + 4) = -27$$

$$x + x + 2 + x + 4 = -27$$

$$3x + 6 = -27$$

$$3x = -33$$

$$x = -11$$

$$x + 2 = -9$$

$$x + 4 = -7$$

The odd integers are –11, –9, and –7.

25.   Let $x =$ first number

$x + 5 =$ second number

$$(x) + (x + 5) = 179$$

$$x + x + 5 = 179$$

$$2x + 5 = 179$$

$$2x = 174$$

$$x = 87$$

$$x + 5 = 92$$

The numbers are 87 and 92.

27. a.   Let $x =$ F - 0 wind speed

$3x + 100 =$ F - 5 wind speed

$$(x) + (3x + 100) = 388$$

$$x + 3x + 100 = 388$$

$$4x + 100 = 388$$

$$4x = 288$$

$$x = 72$$

The F-0 wind rate is 72 mi/hr.

b.

$$3x + 100 = 3(72) + 100$$

$$= 216 + 100 = 316$$

The F-5 wind rate is 316 mi/hr.

29.   Let $M =$ weight on moon

$E =$ weight on earth

$M = \dfrac{E}{6};\quad M = 35.5$

$35.5 = \dfrac{E}{6}$

$213 = E$

The rocks weigh 213 lb on the earth.

31.   Let $x =$ McCartney awards

$x + 20 =$ Beatles awards

$(x) + (x + 20) = 74$

$x + x + 20 = 74$

$2x + 20 = 74$

$2x = 54$

$x = 27$

$x + 20 = 47$

The Beatles have 47 awards.

33.   Let $w =$ width of painting

$w + 4988 =$ length of painting

$2l + 2w = P$

$2(w + 4988) + 2w = 10024$

$2w + 9976 + 2w = 10024$

$4w + 9976 = 10024$

$4w = 48$

$w = 12$

$w + 4988 = 5000$

The dimensions of the painting are 12 ft by 5000 ft.

35.   Let $l =$ length of the building

$l - 198 =$ width of building

$2l + 2w = P$

$2l + 2(l - 198) = 2468$

$2l + 2l - 396 = 2468$

$4l - 396 = 2468$

$4l = 2864$

$l = 716$

$l - 198 = 518$

The dimensions of the building are 518 ft by 716 ft.

37.   Let $x =$ measure of angle

$90 - x =$ measure of complement

$x = 4(90 - x)$

$x = 360 - 4x$

$5x = 360$

$x = 72°$

The measure of the angle is 72°.

39.   Let $x =$ measure of angle

$180 - x =$ measure of supplement

$x = 3(180 - x)$

$x = 540 - 3x$

$4x = 540$

$x = 135$

The measure of the angle is 135°.

41. $3x + 10 + 2x - 15 = 90$

$5x - 5 = 90$

$5x = 95$

$x = 19$

$3x + 10 = 3(19) + 10 = 57 + 10 = 67°$

$2x - 15 = 2(19) - 15 = 38 - 15 = 23°$

43. $3x + 10 + 2x - 25 = 180$

$5x - 15 = 180$

$5x = 195$

$x = 39$

$3x + 10 = 3(39) + 10 = 117 + 10 = 127°$

$2x - 25 = 2(39) - 25 = 78 - 25 = 53°$

45. Let $x=$ measure of the first angle

$90-x=$ measure of complement

$3x=$ measure of the third angle

$$x+(90-x)+3x=180$$
$$90+3x=180$$
$$3x=90$$
$$x=30°$$
$$90-x=60°$$
$$3x=90°$$

47.
$$n+0.30n=26$$
$$1.30n=26$$
$$n=20$$

49.
$$220=2(W+70)+2W$$
$$220=2W+140+2W$$
$$220=4W+140$$
$$80=4W$$
$$20=W$$

51.
$$0.05x+0.10(800-x)=20$$
$$100[0.05x+0.10(800-x)]=100(20)$$
$$5x+10(800-x)=2000$$
$$5x+8000-10x=2000$$
$$-5x+8000=2000$$
$$-5x=-6000$$
$$x=1200$$

53.
$$110T=80(T+3)$$
$$110T=80T+240$$
$$30T=240$$
$$T=8$$

55.
$$\frac{1+x}{4+x}=\frac{2}{3}$$
$$3(4+x)\left|\frac{1+x}{4+x}\right|=3(4+x)\frac{2}{3}$$
$$3(1+x)=(4+x)2$$
$$3+3x=8+2x$$
$$3+x=8$$
$$x=5$$

57. Sample answer: Determine what is to be found.

59. Answers will vary.

61. Sample answer: Make a table or chart, use guess and check, or work backwards.

63. Let $x=$ first odd integer

$x+2=$ second consecutive odd integer

$x+4=$ third consecutive odd integer

$$(x)+(x+2)+(x+4)=99$$
$$3x+6=99$$
$$3x=93$$
$$x=31$$
$$x+2=33 \qquad x+4=35$$

The odd integers are 31, 33, and 35.

65. $x-12=7x$

67.
$$3x + 4 + 4x - 13 = 180$$
$$7x - 9 = 180$$
$$7x = 189$$
$$x = 27$$
$$3x + 4 = 3(27) + 4 = 81 + 4 = 85°$$
$$4x - 13 = 4(27) - 13 = 108 - 13 = 95°$$

69.   Let C = students in college (in millions)

$C + 2$ = students in secondary school (millions)

$3C$ = students in elementary school (millions)

$$C + (C + 2) + 3C = 65$$
$$5C + 2 = 65$$
$$5C = 63$$
$$C = 12.60$$

There were about 12.60 million students enrolled in college in 1995.

2.4   Problem Solving: Percent, Investment, Motion, and Mixture Problems

Problems 2.4

1. **a.** Let x = amount of the down payment.

Percent·(cost of house) = down payment
$$(0.10)($122,852) = x$$
$$$12,285 = x$$

b. Cost of house – down payment = amt of mortgage
$$122,852 - 12,285 = $110,567$$

2. Let x = number of passengers in 2001

number in 2001 – decrease = number in 2002
$$x - (0.013)(x) = 67,000,000$$
$$0.987x = 67,000,000$$
$$x \approx 67,882,000$$

The number of passengers in 2001 was approximately 67,882,000.

3.

Investment	Principal	Rate	Interest
First	x	0.07	$0.07x$
Second	$8000-x$	0.10	$0.10(8000-x)$

$$0.07x + 0.10(8,000 - x) = 710$$
$$0.07x + 800 - 0.10x = 710$$
$$800 - 0.03x = 710$$
$$-0.03x = -90$$
$$\frac{-0.03x}{-0.03} = \frac{-90}{-0.03}$$
$$x = $3,000 \text{ at } 7\%$$
$$8000 - x = $5,000 \text{ at } 10\%$$

$3,000 is invested at 7% and $5,000 at 10%.

4.

	Rate	Time	Distance
Bus	50	$T+1$	$50(T+1)$
Car	60	T	$60T$

Distance of bus = Distance of car

$$50(T+1) = 60T$$
$$50T + 50 = 60T$$
$$50 = 10T$$
$$5\,\text{hrs} = T$$

Since $D = RT = 60(5) = 300\,\text{miles}$

The car catches the bus 300 miles from Los Angeles.

5.

Percent of Salt	Amount to be mixed	Amount of Pure Salt
10% or 0.10	x	$0.10x$
20% or 0.20	15	$0.20(15)$
18% or 0.18	$x + 15$	$0.18(x + 15)$

$$0.10x + 0.20(15) = 0.18(x+15)$$
$$0.10x + 3 = 0.18x + 2.7$$
$$0.3 = 0.08x$$
$$\frac{0.3}{0.08} = \frac{0.08x}{0.08}$$
$$3.75\,\text{gal} = x$$

3.75 gallons of 10% salt solution must be added.

Exercises 2.4

1. a. $150(0.43) = 64.5\,\text{lb}$

b. $150(0.065) = 9.75\,\text{lb}$

3. $196.50(0.055) = 10.81$

The sales tax on the bicycle is $10.81.

5. Let x = number of homes with televisions in major cities

$$0.417x = 41.2\,\text{million}$$
$$\frac{0.417x}{0.417} = \frac{41.2\,\text{million}}{0.417}$$
$$x = 98.8\,\text{million}$$

There are approximately 98.8 million homes with televisions in major cities.

7. Selling price = cost + markup

$$\text{Selling price} = 18.50 + 0.25(18.50)$$
$$= 18.50 + 4.63$$
$$= 23.13$$

The selling price is $23.13.

9. Selling price = cost + markup

$$54 = 30 + m$$
$$24 = m$$
$$\frac{24}{54} = 0.44444$$

The markup is $24 and the percent of markup on selling price is approximately 44.4%.

11. $25 + n(25) = 35$

$$25n = 10$$
$$n = \frac{10}{25} = 0.40$$

The percent of increase is 40%.

13. $n = 40 + 0.25(40) = 40 + 10 = 50$

15.
$$25,000 + n(25,000) = 30,000$$
$$25,000n = 5,000$$
$$n = \frac{5,000}{25,000} = 0.20$$

Luisa's percent of increase is 20%.

17. $m = 1.20(25) = 30$

Tran's mileage would increase to 30 mi/gal.

19. Let x = number of beverages introduced the first year.
$$x - 0.06x = 611$$
$$0.94x = 611$$
$$\frac{0.94x}{0.94} = \frac{611}{0.94}$$
$$x = 650$$

The number of beverages introduced during the first year was 650.

21. Let x = amount the average person spends on entertainment now
$$x + 0.33x = 1463$$
$$1.33x = 1463$$
$$\frac{1.33x}{1.33} = \frac{1463}{1.33}$$
$$x = 1100$$

The average person is currently spending $1100 on entertainment.

23.

Investment	Principal	Rate	Interest
First	x	0.06	0.06x
Second	20000–x	0.08	0.08(20000–x)

$$0.06x + 0.08(20,000 - x) = 1500$$
$$0.06x + 1600 - 0.08x = 1500$$
$$1600 - 0.02x = 1500$$
$$-0.02x = -100$$
$$\frac{-0.02x}{-0.02} = \frac{-100}{-0.02}$$
$$x = \$5,000 \text{ at } 6\%$$
$$20,000 - x = \$15,000 \text{ at } 8\%$$

$5,000 is invested at 6% and $15,000 at 8%.

25.

Investment	Principal	Rate	Interest
Sav acct	x	0.05	0.05x
CD	18000–x	0.07	0.07(18000–x)

$$0.05x + 0.07(18,000 - x) = 1100$$
$$0.05x + 1260 - 0.07x = 1100$$
$$-0.02x + 1260 = 1100$$
$$-0.02x = -160$$
$$\frac{-0.02x}{-0.02} = \frac{-160}{-0.02}$$
$$x = \$8,000$$

$8,000 is in the savings account.

27.

	Rate	Time	Distance
Car	96	$T+1$	96($T+1$)
Plane	144	T	144T

a. Distance of car = Distance of plane
$$96(T + 1) = 144T$$
$$96T + 96 = 144T$$
$$96 = 48T$$
$$2\,\text{hr} = T$$

The border patrol reaches the smugglers in 2 hours.

b. $D = 144T = 144(2) = 288\,\text{km}$

The border patrol overtakes the smugglers in 288 km from the border.

29.

	Rate	Time	Distance
Bus	60	$T+2$	60($T+2$)
Car	90	T	90T

Distance of bus = Distance of car
$$60(T + 2) = 90T$$
$$60T + 120 = 90T$$
$$120 = 30T$$
$$4\,\text{hrs} = T$$

Since $D = RT = 90(4) = 360\,\text{km}$

The car overtakes the bus 360 km from the station.

31.

	Rate	Time	Distance
Bicycle	15	$T+\frac{1}{2}$	$15\left(T+\frac{1}{2}\right)$
Car	60	T	$60T$

Distance of bicycle = Distance of car

$$15\left(T+\tfrac{1}{2}\right)=60T$$
$$15T+\tfrac{15}{2}=60T$$
$$2\left(15T+\tfrac{15}{2}\right)=2\left(60T\right)$$
$$30T+15=120T$$
$$15=90T$$
$$\tfrac{1}{6}\,\text{hr}=T$$

Since $D=RT=60\left(\tfrac{1}{6}\right)=10\text{ mi}$

It is 10 miles from the house to the school.

35.

	Price of one lb	Amount to be mixed	Value of tea
Oolong	19	x	$19x$
Regular	4	$50-x$	$4(50-x)$
Mixture	7	50	$7(50)$

$$19x+4\left(50-x\right)=7\left(50\right)$$
$$19x+200-4x=350$$
$$15x+200=350$$
$$15x=150$$
$$x=10\,\text{lbs}$$

10 pounds of Oolong tea and 40 pounds of regular tea should be mixed to make 50 pounds of the mixture.

33.

Percent of Acetic Acid	Amount to be mixed	Amount of Acetic Acid
99.5% or 0.995	x	$0.995x$
10% or 0.10	100	$0.10(100)$
28% or 0.28	$x+100$	$0.28(x+100)$

$$0.995x+0.10\left(100\right)=0.28\left(x+100\right)$$
$$0.995x+10=0.28x+28$$
$$0.715x=18$$
$$\frac{0.715x}{0.715}=\frac{18}{0.715}$$
$$x\approx25.2\,\text{parts}$$

Approximately 25.2 parts of 99.5% acetic acid must be added.

37. **a.** $-3<-1$ **b.** $-1.3>-1.4$ **c.** $\frac{1}{3}<\frac{1}{2}$

39. Difference $=1090.5\text{ billion}-1380.9\text{ billion}=-\$290.4\text{ billion}$

41. Personal taxes $=0.35\left(1090.5\text{ billion}\right)=\$381.675\text{ billion}$

43. Answers will vary.

45. Answers will vary.

47.

Investment	Principal	Rate	Interest
First	x	0.05	$0.05x$
Second	$8000 - x$	0.10	$0.10(8000 - x)$

$$0.05x + 0.10(8,000 - x) = 650$$
$$0.05x + 800 - 0.10x = 650$$
$$-0.05x + 800 = 650$$
$$-0.05x = -150$$
$$x = \$3,000 \text{ at } 5\%$$
$$8,000 - x = \$5,000 \text{ at } 10\%$$

$3000 is invested at 5% and $5000 at 10%.

49. Let x = number of passengers that used the Atlanta airport in 1999.

$$x + 0.30x = 31.2 \text{ million}$$
$$1.30x = 31.2 \text{ million}$$
$$\frac{1.30x}{1.30} = \frac{31.2 \text{ million}}{1.30}$$
$$x = 24 \text{ million}$$

24 million passengers used the airport in 1999.

51.
$$12 \text{ million} + n(12 \text{ million}) = 14.4 \text{ million}$$
$$n(12 \text{ million}) = 2.4 \text{ million}$$
$$n = \frac{2.4 \text{ million}}{12 \text{ million}} = 0.20$$

The percent of increase is 20%.

2.5 Linear and Compound Inequalities

Problems 2.5

1. **a.**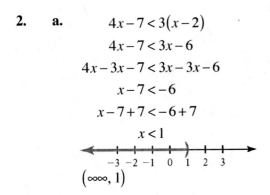

b.

2. **a.**
$$4x - 7 < 3(x - 2)$$
$$4x - 7 < 3x - 6$$
$$4x - 3x - 7 < 3x - 3x - 6$$
$$x - 7 < -6$$
$$x - 7 + 7 < -6 + 7$$
$$x < 1$$

$(\infty, 1)$

b.
$$3(x + 1) - 2x + 5$$
$$3x + 3 - 2x + 5$$
$$3x - 2x + 3 - 2x - 2x + 5$$
$$x + 3 - 5$$
$$x + 3 - 3 - 5 - 3$$
$$x - 2$$

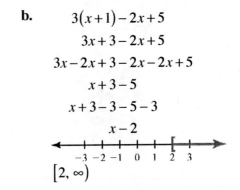

$[2, \infty)$

3. **a.**
$$3(x \geq 1) \geq 5x + 1$$
$$3x \geq 3 \geq 5x + 1$$
$$3x \geq 3x \geq 3 \geq 5x \geq 3x + 1$$
$$\geq 3 \geq 2x + 1$$
$$\geq 3 \geq 1 \geq 2x + 1 \geq 1$$
$$\geq 4 \geq 2x$$
$$\frac{\geq 4}{2} \geq \frac{2x}{2}$$
$$\geq 2 \geq x \text{ or } x \geq \geq 2$$

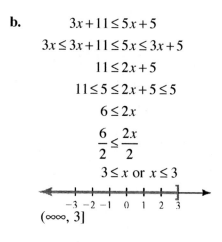

$$[\infty 2, \infty)$$

b.
$$3x + 11 \leq 5x + 5$$
$$3x \leq 3x + 11 \leq 5x \leq 3x + 5$$
$$11 \leq 2x + 5$$
$$11 \leq 5 \leq 2x + 5 \leq 5$$
$$6 \leq 2x$$
$$\frac{6}{2} \leq \frac{2x}{2}$$
$$3 \leq x \text{ or } x \leq 3$$

$$(\infty\infty, 3]$$

4.
$$\frac{x}{3} \mid \frac{x}{4} > \frac{x \mid 4}{4}$$
$$12 \left| \frac{x}{3} \mid \frac{x}{4} \right| > 12 \left| \frac{x \mid 4}{4} \right|$$
$$12 \left| \frac{x}{3} \right| \mid 12 \left| \frac{x}{4} \right| > 12 \left| \frac{x \mid 4}{4} \right|$$
$$4x \mid 3x > 3(x \mid 4)$$
$$x > 3x \mid 12$$
$$x \mid 3x > 3x \mid 3x \mid 12$$
$$\mid 2x > \mid 12$$
$$\frac{\mid 2x}{\mid 2} < \frac{\mid 12}{\mid 2}$$
$$x < 6$$

$$(\infty\infty, 6)$$

5.
$$(-\infty, 1] \cup (4, \infty)$$

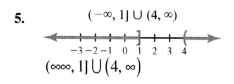

$$(\infty\infty, 1] \cup (4, \infty)$$

6.

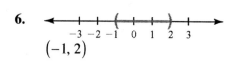

$$(-1, 2)$$

7. **a.**
$$1 < x - 1 \quad \text{and } x < 4$$
$$1 + 1 < x - 1 + 1 \text{ and } x < 4$$
$$2 < x \quad \text{and } x < 4$$
$$2 < x < 4$$

$$(2, 4)$$

b.
$$3 \leq \leq x \quad \text{and} \quad x + 6 \leq 5$$
$$\frac{3}{\leq 1} \leq \frac{\leq x}{\leq 1} \text{ and } x + 6 \leq 6 \leq 5 \leq 6$$
$$\leq 3 \leq x \quad \text{and} \quad x \leq \leq 1$$
$$\leq 3 \leq x \leq \leq 1$$

$$[-3, -1]$$

c. $x + 2 - 6$ and $-3x < 12$

$x + 2 - 2 - 6 - 2$ and $\dfrac{-3x}{-3} > \dfrac{12}{-3}$

$x - 4$ and $x > -4$

$-4 < x$ and $x - 4$

$-4 < x - 4$

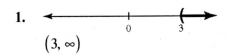

$(-4, 4]$

8. $\geq 4 \geq \geq 2x \geq 2 < 6$

$\geq 4 + 2 \geq \geq 2x \geq 2 + 2 \geq 6 + 2$

$\geq 2 \geq \geq 2x < 8$

$\dfrac{\geq 2}{\geq 2} \geq \dfrac{\geq 2x}{\geq 2} > \dfrac{8}{\geq 2}$

$1 \geq x > \geq 4$

$\geq 4 < x \geq 1$

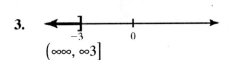

$(-4, 1]$

9. **a.** $x \leq 23$
  b. $y \leq 180$
  c. $z \leq 10$
  d. $p \geq 45$

Exercises 2.5

1.
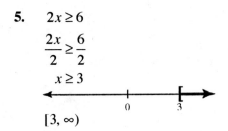

$(3, \infty)$

3.
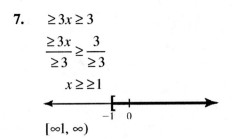

$(\infty\infty, \infty 3]$

5. $2x \geq 6$

$\dfrac{2x}{2} \geq \dfrac{6}{2}$

$x \geq 3$

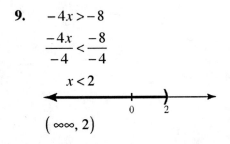

$[3, \infty)$

7. $\geq 3x \geq 3$

$\dfrac{\geq 3x}{\geq 3} \geq \dfrac{3}{\geq 3}$

$x \geq \geq 1$

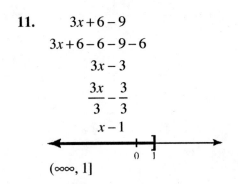

$[\infty 1, \infty)$

9. $-4x > -8$

$\dfrac{-4x}{-4} < \dfrac{-8}{-4}$

$x < 2$

$(\infty\infty, 2)$

11. $3x + 6 - 9$

$3x + 6 - 6 - 9 - 6$

$3x - 3$

$\dfrac{3x}{3} - \dfrac{3}{3}$

$x - 1$

$(\infty\infty, 1]$

13.
$$\leq 2y \leq 4 \leq \, \leq 10$$
$$\leq 2y \leq 4 + 4 \leq \, \leq 10 + 4$$
$$\leq 2y \leq \, \leq 6$$
$$\frac{\leq 2y}{\leq 2} \leq \frac{\leq 6}{\leq 2}$$
$$y \leq 3$$

$$(\infty\infty, \, 3]$$

15.
$$-3x + 1 < -14$$
$$-3x + 1 - 1 < -14 - 1$$
$$-3x < -15$$
$$\frac{-3x}{-3} > \frac{-15}{-3}$$
$$x > 5$$

$$(5, \, \infty)$$

17.
$$3a + 6 - a + 10$$
$$3a - a + 6 - a - a + 10$$
$$2a + 6 - 10$$
$$2a + 6 - 6 - 10 - 6$$
$$2a - 4$$
$$\frac{2a}{2} - \frac{4}{2}$$
$$a - 2$$

$$(\infty\infty, \, 2]$$

19.
$$7z - 12 > 8z - 8$$
$$7z - 7z - 12 > 8z - 7z - 8$$
$$-12 > z - 8$$
$$-12 + 8 > z - 8 + 8$$
$$-4 > z$$
$$z < -4$$

$$(\infty\infty, \, \infty 4)$$

21.
$$10 \leq 5x \leq 7 \leq 8x$$
$$10 \leq 5x + 8x \leq 7 \leq 8x + 8x$$
$$10 + 3x \leq 7$$
$$10 \leq 10 + 3x \leq 7 \leq 10$$
$$3x \leq \, \leq 3$$
$$\frac{3x}{3} \leq \frac{\leq 3}{3}$$
$$x \leq \, \leq 1$$

$$(\infty\infty, \, \infty 1]$$

23.
$$5(x + 2) - 3(x + 3) + 1$$
$$5x + 10 - 3x + 9 + 1$$
$$5x + 10 - 3x + 10$$
$$5x - 3x + 10 - 3x - 3x + 10$$
$$2x + 10 - 10$$
$$2x + 10 - 10 - 10 - 10$$
$$2x - 0$$
$$\frac{2x}{2} - \frac{0}{2}$$
$$x - 0$$

$$(\infty\infty, \, 0]$$

45

25.

$$4x+\frac{1}{2} > 4x+\frac{8}{5}$$

$$10\left|\,4x+\frac{1}{2}\,\right| > 10\left|4x+\frac{8}{5}\right|$$

$$10\left(\,4x\right)+10\left|\frac{1}{2}\right| > 10(4x)+10\left|\frac{8}{5}\right|$$

$$40x+5 > 40x+16$$

$$40x+40x+5 > 40x+40x+16$$

$$5 > 80x+16$$

$$5\ |16 > 80x+16\ |16$$

$$11 > 80x$$

$$\frac{11}{80} > \frac{80x}{80}$$

$$\frac{11}{80} > x$$

$$x < \frac{11}{80}$$

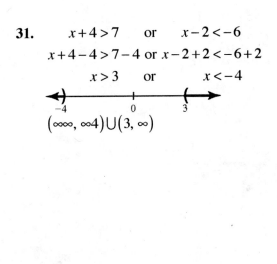

$$\left|\,|\,,\ \frac{11}{80}\,\right|$$

27.

$$\frac{x}{5} \geq \frac{x}{4} \geq 1$$

$$20\underset{\geq}{\geq}\frac{x}{5}\geq\frac{x}{4}\underset{\geq}{\geq} 20(1)$$

$$20\underset{\geq}{\geq}\frac{x}{5}\underset{\geq}{\geq} 20\underset{\geq}{\geq}\frac{x}{4}\underset{\geq}{\geq} 20(1)$$

$$4x \geq 5x \geq 20$$

$$\geq x \geq 20$$

$$\frac{\geq x}{\geq 1} \geq \frac{20}{\geq 1}$$

$$x \geq \geq 20$$

$$[\infty 20,\ \infty)$$

29.

$$\frac{7x+2}{6}+\frac{1}{2} \leq \frac{3}{4}x$$

$$12\underset{\leq}{\leq}\frac{7x+2}{6}+\frac{1}{2}\underset{\leq}{\leq} 12\underset{\leq}{\leq}\frac{3}{4}x\underset{\leq}{\leq}$$

$$12\underset{\leq}{\leq}\frac{7x+2}{6}\underset{\leq}{\leq}+12\underset{\leq}{\leq}\frac{1}{2}\underset{\leq}{\leq} 12\underset{\leq}{\leq}\frac{3}{4}x\underset{\leq}{\leq}$$

$$2(7x+2)+6(1) \leq 3(3x)$$

$$14x+4+6 \leq 9x$$

$$14x+10 \leq 9x$$

$$14x \leq 14x+10 \leq 9x \leq 14x$$

$$10 \leq \leq 5x$$

$$\frac{10}{\leq 5} \leq \frac{\leq 5x}{\leq 5}$$

$$\leq 2 \leq x$$

$$x \leq \leq 2$$

$$[\infty 2,\ \infty)$$

31.

$$x+4 > 7 \quad \text{or} \quad x-2 < -6$$

$$x+4-4 > 7-4 \ \text{or}\ x-2+2 < -6+2$$

$$x > 3 \quad \text{or} \quad x < -4$$

$$\left(\infty\infty,\ \infty 4\right) \cup \left(3,\ \infty\right)$$

33. $\qquad 3x+4>10 \qquad$ or $\qquad 3x-1<2$

$\qquad 3x+4-4>10-4$ or $3x-1+1<2+1$

$\qquad\qquad 3x>6 \qquad$ or $\qquad 3x<3$

$\qquad\qquad \dfrac{3x}{3}>\dfrac{6}{3} \qquad$ or $\qquad \dfrac{3x}{3}<\dfrac{3}{3}$

$\qquad\qquad x>2 \qquad$ or $\qquad x<1$

$\left(\infty\infty, 1\right)\cup\left(2, \infty\right)$

35. $\qquad 2x-x+4 \qquad$ or $\qquad x-2>3$

$\qquad 2x-x-x+4$ or $x-2+2>3+2$

$\qquad\qquad x-4 \qquad$ or $\qquad x>5$

$\left(\infty\infty, 4\right]\cup\left(5, \infty\right)$

37. $\qquad -3x+2<-4 \qquad$ or $\qquad 2x-1<3$

$\qquad -3x+2-2>-4-2$ or $2x-1+1<3+1$

$\qquad\qquad -3x<-6 \qquad$ or $\qquad 2x<4$

$\qquad\qquad \dfrac{-3x}{-3}>\dfrac{-6}{-3} \qquad$ or $\qquad \dfrac{2x}{2}<\dfrac{4}{2}$

$\qquad\qquad x>2 \qquad$ or $\qquad x<2$

$\left(\infty\infty, 2\right)\cup\left(2, \infty\right)$

39. $\qquad \le 6x\le 2\le\le 14 \qquad$ or $\qquad \le 7x+2<\le 19$

$\qquad \le 6x\le 2+2\le\le 14+2$ or $\le 7x+2\le 2<\le 19\le 2$

$\qquad\qquad \le 6x\le\le 12 \qquad$ or $\qquad \le 7x<\le 21$

$\qquad\qquad \dfrac{\le 6x}{\le 6}\le\dfrac{\le 12}{\le 6} \qquad$ or $\qquad \dfrac{\le 7x}{\le 7}>\dfrac{\le 21}{\le 7}$

$\qquad\qquad x\le 2 \qquad$ or $\qquad x>3$

$\left(\infty\infty, 2\right]\cup\left(3, \infty\right)$

41. $\qquad x-4$ and $x--2$

$\qquad -2-x-4$

$\left[-2, 4\right]$

43. $\qquad x+1-7 \qquad$ and $x>2$

$\qquad x+1-1-7-1$ and $x>2$

$\qquad\qquad x-6 \qquad$ and $x>2$

$\qquad\qquad 2<x-6$

$(2, 6]$

45. $\qquad 2x-1>1 \qquad$ and $\qquad x+1<4$

$\qquad 2x-1+1>1+1$ and $x+1-1<4-1$

$\qquad\qquad 2x>2 \qquad$ and $\qquad x<3$

$\qquad\qquad x>1 \qquad$ and $\qquad x<3$

$\qquad\qquad 1<x<3$

$\left(1, 3\right)$

47. $x<\varnothing 5$ and $x>5$

No solution $\varnothing$

49. $\qquad x+1-2 \qquad$ and $x-4$

$\qquad x+1-1-2-1$ and $x-4$

$\qquad\qquad x-1 \qquad$ and $x-4$

$\qquad\qquad 1-x-4$

$\qquad\left[1, 4\right]$

51. $x<3$ and $-x<-2$

$\qquad x<3$ and $\dfrac{-x}{-1}>\dfrac{-2}{-1}$

$\qquad x<3$ and $\quad x>2$

$\qquad 2<x<3$

$\left(2, 3\right)$

53. $x+1<4$ and $-x<-1$

$x+1-1<4-1$ and $\dfrac{-x}{-1}>\dfrac{-1}{-1}$

$x<3$ and $x>1$

$1<x<3$

$(1,3)$

55. $x-2<3$ and $2>-x$

$x-2+2<3+2$ and $\dfrac{2}{-1}<\dfrac{-x}{-1}$

$x<5$ and $-2<x$

$-2<x<5$

$(-2,5)$

57. $x+2<3$ and $-4<x+1$

$x+2-2<3-2$ and $-4-1<x+1-1$

$x<1$ and $-5<x$

$-5<x<1$

$(-5,1)$

59. $x-1>2$ and $x+7<11$

$x-1+1>2+1$ and $x+7-7<11-7$

$x>3$ and $x<4$

$3<x<4$

$(3,4)$

61. $-3<x-1<3$

$-3+1<x-1+1<3+1$

$-2<x<4$

$(-2,4)$

63. $-8<2y+4<6$

$-8-4<2y+4-4<6-4$

$-12<2y<2$

$\dfrac{-12}{2}<\dfrac{2y}{2}<\dfrac{2}{2}$

$-6<y<1$

$(-6,1)$

65. $4-3y-8-10$

$4+8-3y-8+8-10+8$

$12-3y-18$

$\dfrac{12}{3}-\dfrac{3y}{3}-\dfrac{18}{3}$

$4-y-6$

$[4,6]$

67. $|1<\dfrac{x}{2}<2$

$2(|1)<2\left|\dfrac{x}{2}\right|<2(2)$

$|2<x<4$

$(-2,4)$

69.

$$2 < 4 + \frac{2}{3}a < 6$$

$$3(2) < 3 \cdot 4 + \frac{2}{3}a \cdot < 3(6)$$

$$6 < 12 + 2a < 18$$

$$6 - 12 < 12 - 12 + 2a < 18 - 12$$

$$-6 < 2a < 6$$

$$\frac{-6}{2} < \frac{2a}{2} < \frac{6}{2}$$

$$-3 < a < 3$$

$$(-3, 3)$$

71. $h \le 29,028$

73. $e \ge 2$

75. $n \times 4 \times 10^{25}$

77.

$$R > C$$

$$20n > 12n + 160,000$$

$$20n - 12n > 12n - 12n + 160,000$$

$$8n > 160,000$$

$$\frac{8n}{8} > \frac{160,000}{8}$$

$$n > 20,000$$

The minimum number of units that must be sold to make a profit is 20,001.

79.

$$1 + 0.75(h - 1) < 10$$

$$1 + 0.75h - 0.75 < 10$$

$$0.25 + 0.75h < 10$$

$$0.75h < 9.75$$

$$h < 13$$

The cost is less than $10 when you park for less than 13 hours.

81. $\frac{1}{3} < \frac{1}{2}$

83. $-0.234 < -0.233$

85. $\left| -\frac{1}{5} \right| = \frac{1}{5}$

87. $\left| \sqrt{2} \right| = \sqrt{2}$

89.

$$4200 - 70x < 1000$$

$$4200 - 4200 - 70x < 1000 - 4200$$

$$-70x < -3200$$

$$\frac{-70x}{-70} > \frac{-3200}{-70}$$

$$x > 45.71$$

$$1960 + 46 = 2006$$

The consumption of cigarettes will be less than 1000 cigarettes per person annually in 2006.

91. Sample answer: All real numbers greater than 1.

93. Sample answer: All real number greater than 4.

95. Sample answer: Since $a > b$, we know that $b - a < 0$. This means that in the second step both sides are being multiplied by a negative number and thus the inequality sign should be reversed. Thus in the last step, $b < a$.

97. $p \geq 45$

99.

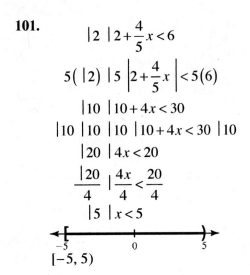

101.

$$2 \left| 2 + \frac{4}{5}x \right| < 6$$

$$5(2) \left| 5 \left| 2 + \frac{4}{5}x \right| \right| < 5(6)$$

$$10 \left| 10 + 4x < 30 \right.$$

$$10 \left| 10 \right| 10 \left| 10 + 4x < 30 \right| 10$$

$$20 \left| 4x < 20 \right.$$

$$\frac{20}{4} \left| \frac{4x}{4} < \frac{20}{4} \right.$$

$$5 \left| x < 5 \right.$$

$[-5, 5)$

103.

$$3x + 10 < 6x + 4$$
$$3x - 3x + 10 < 6x - 3x + 4$$
$$10 < 3x + 4$$
$$10 - 4 < 3x + 4 - 4$$
$$6 < 3x$$
$$\frac{6}{3} < \frac{3x}{3}$$
$$2 < x$$
$$x > 2$$

$(2, \infty)$

105.

$$2x + 3 > 1 \quad \text{or} \quad 5x - 3 < -13$$
$$2x + 3 - 3 > 1 - 3 \quad \text{or} \quad 5x - 3 + 3 < -13 + 3$$
$$2x > -2 \quad \text{or} \quad 5x < -10$$
$$\frac{2x}{2} > \frac{-2}{2} \quad \text{or} \quad \frac{5x}{5} < \frac{-10}{5}$$
$$x > -1 \quad \text{or} \quad x < -2$$

$(\infty, \infty 2) \cup (\infty 1, \infty)$

107.

$$2x + 3 < 5x - 3 \quad \text{and} \quad 3x - 2 < 3 + 2x$$
$$2x + 3 - 2x < 5x - 2x - 3 \quad \text{and} \quad 3x - 2x - 2 < 3 + 2x - 2x$$
$$3 < 3x - 3 \quad \text{and} \quad x - 2 < 3$$
$$3 + 3 < 3x - 3 + 3 \quad \text{and} \quad x - 2 + 2 < 3 + 2$$
$$6 < 3x \quad \text{and} \quad x < 5$$
$$\frac{6}{3} < \frac{3x}{3} \quad \text{and} \quad x < 5$$
$$2 < x \quad \text{and} \quad x < 5$$
$$2 < x < 5$$

$(2, 5)$

2.6   Absolute-Value Equations and Inequalities

Problems 2.6

1. **a.** $x = -7$ or $x = 7$ **b.** $y = -3.4$ or $y = 3.4$ **c.** No solution, $\varnothing$

2. **a.** $|3x+1| = 5$

$$3x+1 = -5 \text{ or } 3x+1 = 5$$
$$3x = -6 \text{ or } \quad 3x = 4$$
$$x = -2 \text{ or } \quad x = \frac{4}{3}$$
$$\left\{ 2, \frac{4}{3} \right\}$$

b. $\left| \frac{3}{2}x+5 \right| \, | 4 = 6$

$$\left| \frac{3}{2}x+5 \right| = 10$$
$$\frac{3}{2}x+5 = |10 \quad \text{or} \quad \frac{3}{2}x+5 = 10$$
$$2\left| \frac{3}{2}x+5 \right| = 2(|10) \text{ or } 2\left| \frac{3}{2}x+5 \right| = 2(10)$$
$$3x+10 = |20 \quad \text{or} \quad 3x+10 = 20$$
$$3x = |30 \quad \text{or} \quad 3x = 10$$
$$x = |10 \quad \text{or} \quad x = \frac{10}{3}$$
$$\left\{ 10, \frac{10}{3} \right\}$$

3. $|x-5| = |x-8|$

$$x-5 = x-8 \text{ or } x-5 = -(x-8)$$
$$-5 = -8 \quad \text{ or } x-5 = -x+8$$

Contradiction. $x = -x+13$

No solution $2x = 13$

for this case. $x = \dfrac{13}{2}$

$$\left\{ \frac{13}{2} \right\}$$

4. $|3x \le 2| \le 5$

$$\le 5 \le 3x \le 2 \le 5$$
$$\le 5+2 \le 3x \le 2+2 \le 5+2$$
$$\le 3 \le 3x \le 7$$
$$\frac{\le 3}{3} \le \frac{3x}{3} \le \frac{7}{3}$$
$$\le 1 \le x \le \frac{7}{3}$$

$$\left| 1, \frac{7}{3} \right|$$

5. $|-2x+1| > 5$

$$-2x+1 < -5 \quad \text{or} \quad -2x+1 > 5$$
$$-2x+1-1 < -5-1 \text{ or } -2x+1-1 > 5-1$$
$$-2x < -6 \quad \text{or} \quad -2x > 4$$
$$\frac{-2x}{-2} > \frac{-6}{-2} \quad \text{or} \quad \frac{-2x}{-2} < \frac{4}{-2}$$
$$x > 3 \quad \text{or} \quad x < -2$$

$$(\infty, \infty 2) \cup (3, \infty)$$

6.

$$5 \le 2|2x \le 6| \le 3$$
$$8 \le 2|2x \le 6|$$
$$4 \le |2x \le 6|$$
$$|2x \le 6| \le 4$$
$$\le 4 \le 2x \le 6 \le 4$$
$$2 \le 2x \le 10$$
$$1 \le x \le 5$$

$$[1, 5]$$

51

Exercises 2.6

1. $|x| = 13$

$\quad x = -13 \text{ or } x = 13$

3. $|y| - 2.3 = 0$

$\quad |y| = 2.3$

$\quad\quad y = -2.3 \text{ or } y = 2.3$

5. $|x| = 0$

$\quad x = 0$

7. $|z| = \varnothing 4$

$\quad$ No solution, $\varnothing$

9. $|x + 7| = 2$

$\quad x + 7 = -2 \text{ or } x + 7 = 2$

$\quad\quad x = -9 \text{ or } \quad x = -5$

$\{-9, -5\}$

11. $|2x - 4| = 8$

$\quad 2x - 4 = -8 \text{ or } 2x - 4 = 8$

$\quad\quad 2x = -4 \text{ or } \quad 2x = 12$

$\quad\quad\quad x = -2 \text{ or } \quad\quad x = 6$

$\{-2, 6\}$

13. $|5a - 2| - 8 = 0$

$\quad |5a - 2| = 8$

$\quad\quad 5a - 2 = -8 \text{ or } 5a - 2 = 8$

$\quad\quad\quad 5a = -6 \text{ or } \quad 5a = 10$

$\quad\quad\quad\quad a = -\dfrac{6}{5} \text{ or } \quad a = 2$

$\left\{ \dfrac{6}{5}, 2 \right\}$

15. $\left| \dfrac{1}{2}x + 4 \right| = 6$

$\quad \dfrac{1}{2}x + 4 = |6 \quad \text{ or } \quad \dfrac{1}{2}x + 4 = 6$

$2\left| \dfrac{1}{2}x \right| + 2(4) = 2(|6) \text{ or } 2\left| \dfrac{1}{2}x \right| + 2(4) = 2(6)$

$\quad\quad x + 8 = |12 \quad \text{ or } \quad\quad x + 8 = 12$

$\quad\quad\quad x = |20 \quad \text{ or } \quad\quad\quad x = 4$

$\{-20, 4\}$

17. $\left| \dfrac{2}{3}z \mid 3 \right| = 9$

$\quad \dfrac{2}{3}z \mid 3 = |9 \quad \text{ or } \quad \dfrac{2}{3}z \mid 3 = 9$

$3\left| \dfrac{2}{3}z \mid 3 \right| = 3(|9) \text{ or } 3\left| \dfrac{2}{3}z \mid 3 \right| = 3(9)$

$\quad 2z \mid 9 = |27 \quad \text{ or } \quad 2z \mid 9 = 27$

$\quad\quad 2z = |18 \quad \text{ or } \quad\quad 2z = 36$

$\quad\quad\quad z = |9 \quad \text{ or } \quad\quad\quad z = 18$

$\{-9, 18\}$

19. $|x + 2| = |x + 4|$

$\quad x + 2 = x + 4 \text{ or } x + 2 = -(x + 4)$

$\quad\quad 2 = 4 \quad \text{ or } x + 2 = -x - 4$

Contradiction. $\quad\quad x = -x - 6$

No solution $\quad\quad\quad 2x = -6$

for this case. $\quad\quad\quad x = -3$

$\{-3\}$

21. $|2y-4|=|4y+6|$

$2y-4=4y+6$ or $2y-4=-(4y+6)$

$-4=2y+6$ or $2y-4=-4y-6$

$-10=2y$ or $6y-4=-6$

$-5=y$ or $6y=-2$

$y=-5$ or $y=-\dfrac{2}{6}=-\dfrac{1}{3}$

$\left\{J\,5,\,J\,\dfrac{1}{3}\right\}$

23. $2|a+1|-3=9$

$2|a+1|=12$

$|a+1|=6$

$a+1=-6$ or $a+1=6$

$a=-7$ or $a=5$

$\{-7,5\}$

25. $3|2x+1|-4=-6$

$3|2x+1|=-2$

$|2x+1|=-\dfrac{2}{3}$

No solution because the value of an absolute value is never negative.

27. $|x-4|=|4-x|$

$x-4=4-x$ or $x-4=-(4-x)$

$2x-4=4$ or $x-4=-4+x$

$2x=8$ or $x-4=x-4$

$x=4$ or $x=x$ Identity.

Solution is all real numbers.

29. $|5x-10|=|10-5x|$

$5x-10=10-5x$ or $5x-10=-(10-5x)$

$10x-10=10$ or $5x-10=-10+5x$

$10x=20$ or $5x=5x$

$x=2$ or $x=x$ Identity.

Solution is all real numbers.

31. $|x|<4$

$-4<x<4$

$(-4,4)$

33. $|z|+4-6$

$|z|-2$

$-2-z-2$

$[-2,2]$

35. $|a|\le 2\le 2$

$|a|\le 4$

$\le 4\le a\le 4$

$[-4,4]$

37. $|x-1|<2$

$-2<x-1<2$

$-2+1<x-1+1<2+1$

$-1<x<3$

$(-1,3)$

39. $|x+3|<-2$

No solution because the value of an absolute value cannot be less than a negative number.

41. $|2x+3|-1$

$-1-2x+3-1$

$-1-3-2x+3-3-1-3$

$-4-2x--2$

$\dfrac{-4}{2}-\dfrac{2x}{2}-\dfrac{-2}{2}$

$-2-x--1$

$[-2,-1]$

43. $|4x+2|-4<2$

$|4x+2|<6$

$-6<4x+2<6$

$-8<4x<4$

$-2<x<1$

$(-2,1)$

45. $|x|>2$

$x<-2 \text{ or } x>2$

$(\infty\infty,\infty2)\cup(2,\infty)$

47. $|z|+5-6$

$|z|-1$

$z--1 \text{ or } z-1$

$(\infty\infty,\infty1]\cup[1,\infty)$

49. $|a|\le 1\le 2$

$|a|\le 3$

$a\le\le 3 \text{ or } a\le 3$

$(\infty\infty,\infty3]\cup[3,\infty)$

51. $|x-1|>1$

$x-1<-1 \text{ or } x-1>1$

$x<0 \quad \text{or} \quad x>2$

$(\infty\infty,0)\cup(2,\infty)$

53. $|x+3|>-1$

All real numbers because an absolute value is always greater than a negative number.

$(\infty\infty,\infty)$

55. $|2x+3|-1$

$2x+3--1 \text{ or } 2x+3-1$

$2x--4 \text{ or } \quad 2x--2$

$x--2 \text{ or } \quad\quad x--1$

$(\infty\infty,\infty2]\cup[\infty1,\infty)$

57. $|3-4x|>7$

$3-4x<-7 \quad \text{or } 3-4x>7$

$-4x<-10 \text{ or } \quad -4x>4$

$\dfrac{-4x}{-4}>\dfrac{-10}{-4} \text{ or } \quad \dfrac{-4x}{-4}<\dfrac{4}{-4}$

$x>\dfrac{5}{2} \quad \text{or} \quad x<-1$

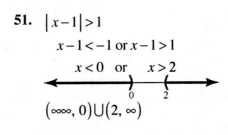

$(|1,|1)\cup\left|\dfrac{5}{2},|\right|$

59. $|4x+2|\le 4\le 2$

$|4x+2|\le 6$

$4x+2\le\le 6 \text{ or } 4x+2\le 6$

$4x\le\le 8 \text{ or } \quad 4x\le 4$

$x\le\le 2 \text{ or } \quad\quad x\le 1$

$(\infty\infty,\infty2]\cup[1,\infty)$

61.

$$\left|2-\tfrac{1}{2}a\right|>1$$

$$2-\tfrac{1}{2}a<-1 \quad \text{or} \quad 2-\tfrac{1}{2}a>1$$

$$-\tfrac{1}{2}a<-3 \quad \text{or} \quad -\tfrac{1}{2}a>-1$$

$$-2\left(-\tfrac{1}{2}a\right)>-2(-3) \quad \text{or} \quad -2\left(-\tfrac{1}{2}a\right)<-2(-1)$$

$$a>6 \quad \text{or} \quad a<2$$

$$(-\infty, 2)\cup(6, \infty)$$

63. $-2<\left|2x+4\right|$

All real numbers because an absolute value is always greater than a negative number.

$$(-\infty, \infty)$$

65.

$$-3\left|x-5\right|>6$$

$$\left|x-5\right|<-2$$

No solution because the value of an absolute value cannot be less than a negative number.

67.

$$9\le 5\left|3x-2\right|+4$$

$$5\le 5\left|3x-2\right|$$

$$1\le \left|3x-2\right|$$

$$\left|3x-2\right|\le 1$$

$$-1\le 3x-2\le 1$$

$$1\le 3x\le 3$$

$$\tfrac{1}{3}\le x\le 1$$

$$\left[\tfrac{1}{3}, 1\right]$$

69.

$$\left|500-a\right|\ge 50$$

$$-50\ge 500-a\ge 50$$

$$-550\ge -a\ge -450$$

$$550\ge a\ge 450$$

$$\$450\ge a\ge \$550$$

71.

$$\left|300-290\right|\le 0.05(300)$$

$$\left|10\right|\le 15$$

$$10\le 15$$

Yes, George was within his budget variance.

73. $-16t^2+10t-15; \ t=2$

$$-16(2)^2+10(2)-15=-16(4)+10(2)-15$$

$$=-64+20-15=-59$$

75. $\left(-6x^2+3x+5\right)+\left(7x^2+8x+2\right)=-6x^2+3x+5+7x^2+8x+2=x^2+11x+7$

77. $\left(3x-7x^2+4\right)-\left(3x^2+5x+1\right)=3x-7x^2+4-3x^2-5x-1=-10x^2-2x+3$

79. Sample answer: The absolute value of a number is the distance that number is from zero on the number line.

81. Sample answer: Set A(x) equal to positive a and negative a. Then solve the two linear equations. The solution set of this equation is empty when a is negative.

83. $|x-1| = |x-8|$

$\quad x-1 = x-8 \text{ or } \quad x-1 = -(x-8)$

$\quad -1 = -8 \quad \text{ or } \quad x-1 = -x+8$

Contradiction. $2x - 1 = 8$

There is no solution $2x = 9$

for this case. $x = \dfrac{9}{2}$

$\left\{ \dfrac{9}{2} \right\}$

85. $|3x+1| = 5$

$\quad 3x+1 = -5 \text{ or } 3x+1 = 5$

$\quad 3x = -6 \text{ or } \quad 3x = 4$

$\quad x = -2 \text{ or } \quad x = \dfrac{4}{3}$

$\left\{ 2, \dfrac{4}{3} \right\}$

87. $\left| \dfrac{2}{3}x+1 \right| + 3 = 9$

$\quad \left| \dfrac{2}{3}x+1 \right| = 6$

$\quad \dfrac{2}{3}x+1 = -6 \text{ or } \dfrac{2}{3}x+1 = 6$

$\quad \dfrac{2}{3}x = -7 \text{ or } \quad \dfrac{2}{3}x = 5$

$\quad 2x = -21 \text{ or } \quad 2x = 15$

$\quad x = -\dfrac{21}{2} \text{ or } \quad x = \dfrac{15}{2}$

$\left\{ \dfrac{21}{2}, \dfrac{15}{2} \right\}$

89. $|-2x+1| - 1 > 4$

$\quad |-2x+1| > 5$

$\quad -2x+1 < -5 \text{ or } -2x+1 > 5$

$\quad -2x < -6 \text{ or } \quad -2x > 4$

$\quad \dfrac{-2x}{-2} > \dfrac{-6}{-2} \text{ or } \quad \dfrac{-2x}{-2} < \dfrac{4}{-2}$

$\quad x > 3 \quad \text{ or } \quad x < -2$

$(\infty, \infty 2) \cup (3, \infty)$

91. $|\leq 2+3x| \leq 5$

$\quad \leq 2+3x \leq \leq 5 \text{ or } \leq 2+3x \leq 5$

$\quad 3x \leq \leq 3 \text{ or } \quad 3x \leq 7$

$\quad x \leq \leq 1 \text{ or } \quad x \leq \dfrac{7}{3}$

$\left(| |, |1] \cup \left| \dfrac{7}{3}, | \right| \right.$

Review Exercises

1. **a.** No; substituting -3 for x in $7 = 8 - x$, we have $7 = 8 - (-3)$, which simplifies to $7 = 11$, and is a false statement.

b. No; substituting -3 for x in $9 = 8 + x$, we have $9 = 8 + (-3)$, which simplifies to $9 = 5$, and is a false statement.

c. Yes; substituting -3 for x in $4 = 1 - x$, we have $4 = 1 - (-3)$, which simplifies to $4 = 4$, and is a true statement.

2. **a.**
$$\frac{2}{3}y \cdot 3 = 5$$
$$\frac{2}{3}y = 8$$
$$\frac{3}{2} \cdot \frac{2}{3}y = \frac{3}{2} \cdot 8$$
$$y = 12$$

b.
$$\frac{2}{3}y \cdot 5 = 5$$
$$\frac{2}{3}y = 10$$
$$\frac{3}{2} \cdot \frac{2}{3}y = \frac{3}{2} \cdot 10$$
$$y = 15$$

c.
$$\frac{2}{3}y \cdot 7 = 5$$
$$\frac{2}{3}y = 12$$
$$\frac{3}{2} \cdot \frac{2}{3}y = \frac{3}{2} \cdot 12$$
$$y = 18$$

3. **a.**
$$x + 2 = 2(2x - 2)$$
$$x + 2 = 4x - 4$$
$$2 = 3x - 4$$
$$6 = 3x$$
$$2 = x$$

b.
$$x + 3 = 3(2x - 4)$$
$$x + 3 = 6x - 12$$
$$3 = 5x - 12$$
$$15 = 5x$$
$$3 = x$$

c.
$$x + 4 = 4(2x - 6)$$
$$x + 4 = 8x - 24$$
$$4 = 7x - 24$$
$$28 = 7x$$
$$4 = x$$

4. **a.**
$$\frac{x+4}{3} - \frac{x-4}{5} = 4$$
$$15\left(\frac{x+4}{3} - \frac{x-4}{5}\right) = 15(4)$$
$$15 \cdot \frac{x+4}{3} - 15 \cdot \frac{x-4}{5} = 15(4)$$
$$5(x+4) - 3(x-4) = 60$$
$$5x + 20 - 3x + 12 = 60$$
$$2x + 32 = 60$$
$$2x = 28$$
$$x = 14$$

b.
$$\frac{x+9}{4} - \frac{x-12}{5} = 6$$
$$20\left(\frac{x+9}{4} - \frac{x-12}{5}\right) = 20(6)$$
$$20 \cdot \frac{x+9}{4} - 20 \cdot \frac{x-12}{5} = 20(6)$$
$$5(x+9) - 4(x-12) = 120$$
$$5x + 45 - 4x + 48 = 120$$
$$x + 93 = 120$$
$$x = 27$$

c.

$$\frac{x+8}{3} - \frac{x-8}{5} = 8$$

$$15\left(\frac{x+8}{3} - \frac{x-8}{5}\right) = 15(8)$$

$$15\left(\frac{x+8}{3}\right) - 15\left(\frac{x-8}{5}\right) = 15(8)$$

$$5(x+8) - 3(x-8) = 120$$

$$5x+40 - 3x+24 = 120$$

$$2x+64 = 120$$

$$2x = 56$$

$$x = 28$$

5.   a.

$$\frac{x}{4} - \frac{x}{3} = \frac{x-4}{4}$$

$$12\left(\frac{x}{4} - \frac{x}{3}\right) = 12\left(\frac{x-4}{4}\right)$$

$$12\left(\frac{x}{4}\right) - 12\left(\frac{x}{3}\right) = 12\left(\frac{x-4}{4}\right)$$

$$3x - 4x = 3(x-4)$$

$$-x = 3x - 12$$

$$-4x = -12$$

$$x = 3$$

b.

$$\frac{x}{5} - \frac{x}{4} = \frac{x-5}{5}$$

$$20\left(\frac{x}{5} - \frac{x}{4}\right) = 20\left(\frac{x-5}{5}\right)$$

$$20\left(\frac{x}{5}\right) - 20\left(\frac{x}{4}\right) = 20\left(\frac{x-5}{5}\right)$$

$$4x - 5x = 4(x-5)$$

$$-x = 4x - 20$$

$$-5x = -20$$

$$x = 4$$

c.

$$\frac{x}{7} - \frac{x}{3} = \frac{x-7}{7}$$

$$21\left(\frac{x}{7} - \frac{x}{3}\right) = 21\left(\frac{x-7}{7}\right)$$

$$21\left(\frac{x}{7}\right) - 21\left(\frac{x}{3}\right) = 21\left(\frac{x-7}{7}\right)$$

$$3x - 7x = 3(x-7)$$

$$-4x = 3x - 21$$

$$-7x = -21$$

$$x = 3$$

6.   a.

$$0.05P + 0.10(2000 - P) = 175$$

$$100[0.05P + 0.10(2000 - P)] = 100(175)$$

$$5P + 10(2000 - P) = 17,500$$

$$5P + 20,000 - 10P = 17,500$$

$$-5P + 20,000 = 17,500$$

$$-5P = -2,500$$

$$P = 500$$

b.

$$0.08P + 0.10(5000 - P) = 460$$

$$100[0.08P + 0.10(5000 - P)] = 100(460)$$

$$8P + 10(5000 - P) = 46,000$$

$$8P + 50,000 - 10P = 46,000$$

$$-2P + 50,000 = 46,000$$

$$-2P = -4,000$$

$$P = 2,000$$

c.

$$0.06P + 0.10(10{,}000 - P) = 840$$
$$100\left[0.06P + 0.10(10{,}000 - P)\right] = 100(840)$$
$$6P + 10(10{,}000 - P) = 84{,}000$$
$$6P + 100{,}000 - 10P = 84{,}000$$
$$-4P + 100{,}000 = 84{,}000$$
$$-4P = -16{,}000$$
$$P = 4{,}000$$

7. a.

$$H = 2.5h + 72.48$$
$$H - 72.48 = 2.5h$$
$$\frac{H - 72.48}{2.5} = h$$
$$h = \frac{82.48 - 72.48}{2.5} = \frac{10}{2.5} = 4$$

b.

$$H = 2.5h + 77.48$$
$$H - 77.48 = 2.5h$$
$$\frac{H - 77.48}{2.5} = h$$
$$h = \frac{82.48 - 77.48}{2.5} = \frac{5}{2.5} = 2$$

c.

$$H = 2.5h + 84.98$$
$$H - 84.98 = 2.5h$$
$$\frac{H - 84.98}{2.5} = h$$
$$h = \frac{82.48 - 84.98}{2.5} = \frac{-2.5}{2.5} = -1$$

8. a.

$$B = \frac{2}{7}(A - 7); \ \text{solve for } A$$
$$7 \cdot B = 7 \cdot \frac{2}{7}(A - 7)$$
$$7B = 2(A - 7)$$
$$7B = 2A - 14$$
$$7B + 14 = 2A$$
$$\frac{7B + 14}{2} = A \ \text{ or } \ \frac{7B}{2} + 7 = A$$

b.

$$B = \frac{3}{7}(A - 5); \ \text{solve for } A$$
$$7 \cdot B = 7 \cdot \frac{3}{7}(A - 5)$$
$$7B = 3(A - 5)$$
$$7B = 3A - 15$$
$$7B + 15 = 3A$$
$$\frac{7B + 15}{3} = A \ \text{ or } \ \frac{7B}{3} + 5 = A$$

c.

$$B = \frac{4}{5}(A - 7); \ \text{solve for A}$$
$$5 \cdot B = 5 \cdot \frac{4}{5}(A - 7)$$
$$5B = 4(A - 7)$$
$$5B = 4A - 28$$
$$5B + 28 = 4A$$
$$\frac{5B + 28}{4} = A \ \text{ or } \ \frac{5B}{4} + 7 = A$$

9.

$$P = 2L + 2W; \quad \text{solve for L}$$
$$P - 2W = 2L$$
$$\frac{P - 2W}{2} = \frac{2L}{2}$$
$$\frac{P - 2W}{2} = L$$

a.

$$P = 180; \quad L = W + 10$$
$$P = 2L + 2W$$
$$180 = 2(W + 10) + 2W$$
$$180 = 2W + 20 + 2W$$
$$180 = 4W + 20$$
$$160 = 4W$$
$$40 \text{ ft} = W$$
$$L = W + 10 = 40 + 10 = 50 \text{ ft}$$

b.

$$P = 220; \quad L = W + 10$$
$$P = 2L + 2W$$
$$220 = 2(W + 10) + 2W$$
$$220 = 2W + 20 + 2W$$
$$220 = 4W + 20$$
$$200 = 4W$$
$$50 \text{ ft} = W$$
$$L = W + 10 = 50 + 10 = 60 \text{ ft}$$

c.

$$P = 260; \quad L = W + 10$$
$$P = 2L + 2W$$
$$260 = 2(W + 10) + 2W$$
$$260 = 2W + 20 + 2W$$
$$260 = 4W + 20$$
$$240 = 4W$$
$$60 \text{ ft} = W$$
$$L = W + 10 = 60 + 10 = 70 \text{ ft}$$

10. a.

$$7x - 20 = 4x + 10$$
$$3x - 20 = 10$$
$$3x = 30$$
$$x = 10$$
$$7x - 20 = 7(10) - 20 = 70 - 20 = 50°$$
$$4x + 10 = 4(10) + 10 = 40 + 10 = 50°$$

b.

$$3x + 12 = 4x - 3$$
$$12 = x - 3$$
$$15 = x$$
$$3x + 12 = 3(15) + 12 = 45 + 12 = 57°$$
$$4x - 3 = 4(15) - 3 = 60 - 3 = 57°$$

c.

$$2x + 18 = 3x - 2$$
$$18 = x - 2$$
$$20 = x$$
$$2x + 18 = 2(20) + 18 = 40 + 18 = 58°$$
$$3x - 2 = 3(20) - 2 = 60 - 2 = 58°$$

11. a.

$$C = 30 + 0.15m$$
$$C = 45, \quad 1 \text{ day}$$
$$45 = 30 + 0.15m$$
$$15 = 0.15m$$
$$100 \text{ miles} = m$$

b.

$$C = 30 + 0.15m$$
$$C = 52.50, \quad 1 \text{ day}$$
$$52.50 = 30 + 0.15m$$
$$22.50 = 0.15m$$
$$150 \text{ miles} = m$$

c.

$$C = 30 + 0.15m$$
$$C = 60, \quad 1 \text{ day}$$
$$60 = 30 + 0.15m$$
$$30 = 0.15m$$
$$200 \text{ miles} = m$$

12. a. Let $x =$ first odd integer

$x + 2 =$ second consecutive odd integer

$x + 4 =$ third consecutive odd integer

$(x) + (x + 2) + (x + 4) = 153$

$x + x + 2 + x + 4 = 153$

$3x + 6 = 153$

$3x = 147$

$x = 49$

$x + 2 = 51$

$x + 4 = 53$

The odd integers are 49, 51, and 53.

b. Let $x =$ first odd integer

$x + 2 =$ second consecutive odd integer

$x + 4 =$ third consecutive odd integer

$(x) + (x + 2) + (x + 4) = 159$

$x + x + 2 + x + 4 = 159$

$3x + 6 = 159$

$3x = 153$

$x = 51$

$x + 2 = 53$

$x + 4 = 55$

The odd integers are 51, 53, and 55.

c. Let $x =$ first odd integer

$x + 2 =$ second consecutive odd integer

$x + 4 =$ third consecutive odd integer

$(x) + (x + 2) + (x + 4) = 207$

$x + x + 2 + x + 4 = 207$

$3x + 6 = 207$

$3x = 201$

$x = 67$

$x + 2 = 69$

$x + 4 = 71$

The odd integers are 67, 69, and 71.

13. a. Let $x =$ old salary

$x + (\% \text{ increase } \cdot x) = $ new salary

$x + 0.20x = \$24,000$

$1.20x = \$24,000$

$\dfrac{1.2x}{1.2} = \dfrac{\$24,000}{1.2}$

$x = \$20,000$

b. Let $x =$ old salary

$x + (\% \text{ increase } \cdot x) = $ new salary

$x + 0.20x = \$36,000$

$1.20x = \$36,000$

$\dfrac{1.2x}{1.2} = \dfrac{\$36,000}{1.2}$

$x = \$30,000$

c. Let $x =$ old salary

$x + (\% \text{ increase } \cdot x) = $ new salary

$x + 0.20x = \$18,000$

$1.20x = \$18,000$

$\dfrac{1.2x}{1.2} = \dfrac{\$18,000}{1.2}$

$x = \$15,000$

14. a.

Invest	Principal	Rate	Interest
Bonds	x	0.05	$0.05x$
CDs	20000–x	0.10	$0.10(20000–x)$

$$0.05x + 0.10(20,000 - x) = 1750$$
$$0.05x + 2000 - 0.10x = 1750$$
$$-0.05x + 2000 = 1750$$
$$-0.05x = -250$$
$$\frac{-0.05x}{-0.05} = \frac{-250}{-0.05}$$
$$x = \$5,000$$
$$20,000 - x = \$15,000$$
$5,000 is in bonds and $15,000 in CDs.

b.

Invest	Principal	Rate	Interest
Bonds	x	0.05	$0.05x$
CDs	20000–x	0.10	$0.10(20000–x)$

$$0.05x + 0.10(20,000 - x) = 1150$$
$$0.05x + 2000 - 0.10x = 1150$$
$$-0.05x + 2000 = 1150$$
$$-0.05x = -850$$
$$\frac{-0.05x}{-0.05} = \frac{-850}{-0.05}$$
$$x = \$17,000$$
$$20,000 - x = \$3,000$$
$17,000 is in bonds and $3,000 in CDs.

c.

Invest	Principal	Rate	Interest
Bonds	x	0.05	$0.05x$
CDs	20000–x	0.10	$0.10(20000–x)$

$$0.05x + 0.10(20,000 - x) = 1500$$
$$0.05x + 2000 - 0.10x = 1500$$
$$-0.05x + 2000 = 1500$$
$$-0.05x = -500$$
$$\frac{-0.05x}{-0.05} = \frac{-500}{-0.05}$$
$$x = \$10,000$$
$$20,000 - x = \$10,000$$
$10,000 is in bonds and $10,000 in CDs.

15. a.

	Rate	Time	Distance
Car 1	40	$T+1$	$40(T+1)$
Car 2	50	T	$50T$

Distance of car 1 = Distance of car 2
$$40(T +1) = 50T$$
$$40T + 40 = 50T$$
$$40 = 10T$$
$$4\,\text{hrs} = T$$
Since $D = RT = 50(4) = 200\,\text{mi}$

Car 2 overtakes car 1 200 miles from the town.

b.

	Rate	Time	Distance
Car 1	50	$T+1$	$50(T+1)$
Car 2	60	T	$60T$

Distance of car 1 = Distance of car 2
$$50(T +1) = 60T$$
$$50T + 50 = 60T$$
$$50 = 10T$$
$$5\,\text{hrs} = T$$
Since $D = RT = 60(5) = 300\,\text{mi}$

Car 2 overtakes car 1 300 miles from the town.

c.

	Rate	Time	Distance
Car 1	40	$T+1$	$40(T+1)$
Car 2	60	T	$60T$

Distance of car 1 = Distance of car 2
$$40(T +1) = 60T$$
$$40T + 40 = 60T$$
$$40 = 20T$$
$$2\,\text{hrs} = T$$
Since $D = RT = 60(2) = 120\,\text{mi}$

Car 2 overtakes car 1 120 miles from the town.

16. a.

Percent of Salt	Amount to be mixed	Amount of Pure Salt
40% or 0.40	x	$0.40x$
10% or 0.10	50	$0.10(50)$
30% or 0.30	$x + 50$	$0.30(x + 50)$

$$0.40x + 0.10(50) = 0.30(x + 50)$$
$$0.40x + 5 = 0.30x + 15$$
$$0.10x + 5 = 15$$
$$0.10x = 10$$
$$x = 100\,\text{L}$$

100 liters of 40% salt solution must be added.

b.

Percent of Salt	Amount to be mixed	Amount of Pure Salt
40% or 0.40	x	$0.40x$
10% or 0.10	50	$0.10(50)$
20% or 0.20	$x + 50$	$0.20(x + 50)$

$$0.40x + 0.10(50) = 0.20(x + 50)$$
$$0.40x + 5 = 0.20x + 10$$
$$0.20x + 5 = 10$$
$$0.20x = 5$$
$$x = 25\,\text{L}$$

25 liters of 40% salt solution must be added.

c.

Percent of Salt	Amount to be mixed	Amount of Pure Salt
40% or 0.40	x	$0.40x$
10% or 0.10	50	$0.10(50)$
10% or 0.10	$x + 50$	$0.10(x + 50)$

$$0.40x + 0.10(50) = 0.10(x + 50)$$
$$0.40x + 5 = 0.10x + 5$$
$$0.30x + 5 = 5$$
$$0.30x = 0$$
$$x = 0\,\text{L}$$

0 liters of 40% salt solution must be added.

17. a.

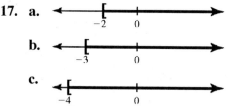

b.

c.

18. a. $2(x \geq 1) \geq 4x + 4$
$$2x \geq 2 \geq 4x + 4$$
$$\geq 2 \geq 2x + 4$$
$$\geq 6 \geq 2x$$
$$\geq 3 \geq x$$
$$x \geq \geq 3$$

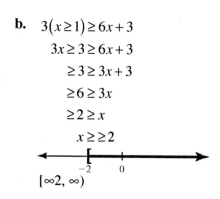

$[\infty 3, \infty)$

b. $3(x \geq 1) \geq 6x + 3$
$$3x \geq 3 \geq 6x + 3$$
$$\geq 3 \geq 3x + 3$$
$$\geq 6 \geq 3x$$
$$\geq 2 \geq x$$
$$x \geq \geq 2$$

$[\infty 2, \infty)$

c. $4(x \geq 1) \geq 8x + 4$

$\qquad 4x \geq 4 \geq 8x + 4$

$\qquad \geq 4 \geq 4x + 4$

$\qquad \geq 8 \geq 4x$

$\qquad \geq 2 \geq x$

$\qquad x \geq \geq 2$

$[\infty 2, \infty)$

b. $\dfrac{x}{5} \mid \dfrac{x}{3} < \dfrac{x \mid 5}{5}$

$15 \left| \dfrac{x}{5} \mid \dfrac{x}{3} \right| < 15 \left| \dfrac{x \mid 5}{5} \right|$

$\qquad 3x \mid 5x < 3(x \mid 5)$

$\qquad \mid 2x < 3x \mid 15$

$\qquad \mid 5x < \mid 15$

$\qquad \dfrac{\mid 5x}{\mid 5} > \dfrac{\mid 15}{\mid 5}$

$\qquad x > 3$

$(3, \infty)$

20. a. $-1 < x < 2$

$(-1, 2)$

c. $-3 < x < 4$

$(-3, 4)$

b.

$\left(\infty\infty, \infty 3\right) \cup [2, \infty)$

19. a. $\dfrac{x}{4} \mid \dfrac{x}{3} < \dfrac{x \mid 4}{4}$

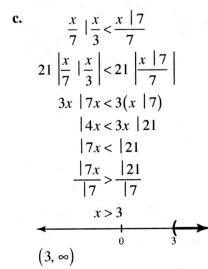

$12 \left| \dfrac{x}{4} \mid \dfrac{x}{3} \right| < 12 \left| \dfrac{x \mid 4}{4} \right|$

$\qquad 3x \mid 4x < 3(x \mid 4)$

$\qquad \mid x < 3x \mid 12$

$\qquad \mid 4x < \mid 12$

$\qquad \dfrac{\mid 4x}{\mid 4} > \dfrac{\mid 12}{\mid 4}$

$\qquad x > 3$

$(3, \infty)$

c. $\dfrac{x}{7} \mid \dfrac{x}{3} < \dfrac{x \mid 7}{7}$

$21 \left| \dfrac{x}{7} \mid \dfrac{x}{3} \right| < 21 \left| \dfrac{x \mid 7}{7} \right|$

$\qquad 3x \mid 7x < 3(x \mid 7)$

$\qquad \mid 4x < 3x \mid 21$

$\qquad \mid 7x < \mid 21$

$\qquad \dfrac{\mid 7x}{\mid 7} > \dfrac{\mid 21}{\mid 7}$

$\qquad x > 3$

$(3, \infty)$

b. $-2 < x < 3$

$(-2, 3)$

21. a.

$\left(\infty\infty, \infty 2\right) \cup [3, \infty)$

c.

$\left(\infty\infty, \infty 4\right) \cup [1, \infty)$

22. a. $x + 1 - 3$ and $-4x < 8$

$x - 2$ and $\dfrac{-4x}{-4} > \dfrac{8}{-4}$

$x - 2$ and $x > -2$

$-2 < x - 2$

$(-2, 2]$

b. $x + 1 - 4$ and $-3x < 9$

$x - 3$ and $\dfrac{-3x}{-3} > \dfrac{9}{-3}$

$x - 3$ and $x > -3$

$-3 < x - 3$

$(-3, 3]$

c. $x + 1 - 5$ and $-2x < 4$

$x - 4$ and $\dfrac{-2x}{-2} > \dfrac{4}{-2}$

$x - 4$ and $x > -2$

$-2 < x - 4$

$\left(-2, 4\right]$

23. a. $\geq 4 \geq \geq 2x \geq 6 < 4$

$2 \geq \geq 2x < 10$

$\dfrac{2}{\geq 2} \geq \dfrac{\geq 2x}{\geq 2} > \dfrac{10}{\geq 2}$

$\geq 1 \geq x > \geq 5$

$\geq 5 < x \geq \geq 1$

$(-5, -1]$

b. $\geq 3 \geq \geq 2x \geq 5 < 3$

$2 \geq \geq 2x < 8$

$\dfrac{2}{\geq 2} \geq \dfrac{\geq 2x}{\geq 2} > \dfrac{8}{\geq 2}$

$\geq 1 \geq x > \geq 4$

$\geq 4 < x \geq \geq 1$

$(-4, -1]$

c. $\geq 6 \geq \geq 2x \geq 4 < 2$

$\geq 2 \geq \geq 2x < 6$

$\dfrac{\geq 2}{\geq 2} \geq \dfrac{\geq 2x}{\geq 2} > \dfrac{6}{\geq 2}$

$1 \geq x > \geq 3$

$\geq 3 < x \geq 1$

$(-3, 1]$

24. a. $\left|\dfrac{2}{7}x + 2\right| + 5 = 9$

$\left|\dfrac{2}{7}x + 2\right| = 4$

$\dfrac{2}{7}x + 2 = \cdot\ 4 \quad$ or $\quad \dfrac{2}{7}x + 2 = 4$

$\dfrac{2}{7}x = \cdot\ 6 \quad$ or $\quad \dfrac{2}{7}x = 2$

$\dfrac{7}{2} \cdot \dfrac{2}{7}x = \dfrac{7}{2}(\cdot\ 6)$ or $\dfrac{7}{2} \cdot \dfrac{2}{7}x = \dfrac{7}{2} \cdot 2$

$x = \cdot\ 21 \quad$ or $\quad x = 7$

$\{-21, 7\}$

b. $\left|\dfrac{2}{7}x + 2\right| + 3 = 9$

$\left|\dfrac{2}{7}x + 2\right| = 6$

$\dfrac{2}{7}x + 2 = \cdot\ 6 \quad$ or $\quad \dfrac{2}{7}x + 2 = 6$

$\dfrac{2}{7}x = \cdot\ 8 \quad$ or $\quad \dfrac{2}{7}x = 4$

$\dfrac{7}{2} \cdot \dfrac{2}{7}x = \dfrac{7}{2}(\cdot\ 8)$ or $\dfrac{7}{2} \cdot \dfrac{2}{7}x = \dfrac{7}{2} \cdot 4$

$x = \cdot\ 28 \quad$ or $\quad x = 14$

$\{-28, 14\}$

c. $\left|\dfrac{2}{7}x+2\right|+1=9$

$\left|\dfrac{2}{7}x+2\right|=8$

$\dfrac{2}{7}x+2=\cdot\,8$ or $\dfrac{2}{7}x+2=8$

$\dfrac{2}{7}x=\cdot\,10$ or $\dfrac{2}{7}x=6$

$\dfrac{7}{2}\cdot\dfrac{2}{7}x=\dfrac{7}{2}(\cdot\,10)$ or $\dfrac{7}{2}\cdot\dfrac{2}{7}x=\dfrac{7}{2}\cdot6$

$x=\cdot\,35$ or $x=21$

$\{-35,\,21\}$

b. $|x-3|=|x-5|$

$x-3=x-5$ or $x-3=-(x-5)$

$-3=-5$ or $x-3=-x+5$

Contradiction. $2x-3=5$

There is no $2x=8$

solution for $x=4$

this case. $\{4\}$

25. a. $|x-1|=|x-3|$

$x-1=x-3$ or $x-1=-(x-3)$

$-1=-3$ or $x-1=-x+3$

Contradiction. $2x-1=3$

There is no $2x=4$

solution for $x=2$

this case. $\{2\}$

c. $|x-5|=|x-7|$

$x-5=x-7$ or $x-5=-(x-7)$

$-5=-7$ or $x-5=-x+7$

Contradiction. $2x-5=7$

There is no $2x=12$

solution for $x=6$

this case. $\{6\}$

26. a. $|3x\le1|\le2$

$\le2\le3x\le1\le2$

$\le1\le3x\le3$

$\dfrac{\le1}{3}\le\dfrac{3x}{3}\le\dfrac{3}{3}$

$\le\dfrac{1}{3}\le x\le1$

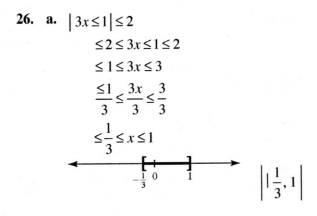

$\left|\,\middle|\dfrac{1}{3},1\middle|\,\right|$

b. $|4x\le1|\le3$

$\le3\le4x\le1\le3$

$\le2\le4x\le4$

$\dfrac{\le2}{4}\le\dfrac{4x}{4}\le\dfrac{4}{4}$

$\le\dfrac{1}{2}\le x\le1$

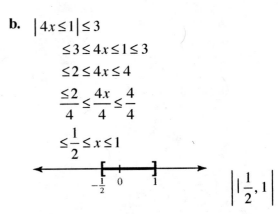

$\left|\,\middle|\dfrac{1}{2},1\middle|\,\right|$

c. $|5x\le1|\le4$

$\le4\le5x\le1\le4$

$\le3\le5x\le5$

$\dfrac{\le3}{5}\le\dfrac{5x}{5}\le\dfrac{5}{5}$

$\le\dfrac{3}{5}\le x\le1$

$\left|\,\middle|\dfrac{3}{5},1\middle|\,\right|$

27. a. $|3x\le1|\le2$

$3x\le1\le\le2$ or $3x\le1\le2$

$3x\le\le1$ or $3x\le3$

$\dfrac{3x}{3}\le\le\dfrac{1}{3}$ or $\dfrac{3x}{3}\le\dfrac{3}{3}$

$x\le\le\tfrac{1}{3}$ or $x\le1$

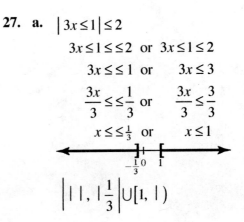

$\left|\,\middle|\,\middle|,\,\middle|\dfrac{1}{3}\middle|\cup[1,\,|)\right.$

b. $|4x \le 1| \le 3$

$\quad 4x \le 1 \le\le 3$ or $4x \le 1 \le 3$

$\quad\quad 4x \le\le 2$ or $\quad 4x \le 4$

$\quad\quad \dfrac{4x}{4} \le \dfrac{\le 2}{4}$ or $\dfrac{4x}{4} \le \dfrac{4}{4}$

$\quad\quad x \le\le \dfrac{1}{2}$ or $\quad x \le 1$

$\xleftarrow{\hspace{3cm}}\!\!\!\underset{-\frac{1}{2}\;\;0\;\;1}{\rule{0pt}{0pt}}\!\!\!\xrightarrow{\hspace{3cm}}$

$\left|\;|,\;|\dfrac{1}{2}\right| \cup [1,\;|)$

c. $|5x \le 1| \le 4$

$\quad 5x \le 1 \le\le 4$ or $5x \le 1 \le 4$

$\quad\quad 5x \le\le 3$ or $\quad 5x \le 5$

$\quad\quad \dfrac{5x}{5} \le \dfrac{\le 3}{5}$ or $\dfrac{5x}{5} \le \dfrac{5}{5}$

$\quad\quad x \le\le \dfrac{3}{5}$ or $\quad x \le 1$

$\xleftarrow{\hspace{3cm}}\!\!\!\underset{-\frac{3}{5}\;\;0\;\;1}{\rule{0pt}{0pt}}\!\!\!\xrightarrow{\hspace{3cm}}$

$\left|\;|,\;|\dfrac{3}{5}\right| \cup [1,\;|)$

Cumulative Review Chapters 1–2

1. $\{2,4,6\}$

2. $\dfrac{19}{100}$

3. $-5+(-6)=-11$

4. $5-(-4)=-20$

5. $\lfloor(3x^2 \rfloor 3)+(7x+1)\rfloor\rfloor \lfloor(x \rfloor 2)+(8x^2 \rfloor 5)\rfloor = \lfloor 3x^2 \rfloor 3+7x+1\rfloor\rfloor \lfloor x \rfloor 2+8x^2 \rfloor 5\rfloor$

$\quad\quad = \lfloor 3x^2+7x \rfloor 2\rfloor\rfloor \lfloor 8x^2+x \rfloor 7\rfloor = 3x^2+7x \rfloor 2 \rfloor 8x^2 \rfloor x+7 = \rfloor 5x^2+6x+5$

6. $\dfrac{80x^8}{16x^{\,6}} = \dfrac{80}{16}\cdot\dfrac{x^8}{x^{\,6}} = 5\cdot x^{8\cdot(\cdot 6)} = 5x^{14}$

7. $\rfloor\rfloor 2(4+4)\rfloor +7 = \rfloor\rfloor 2(8)\rfloor +7 = [\rfloor 16]+7 = \rfloor 9$

8. No; substituting -4 for x in $5=9-x$, we have $5=9-(-4)$, which simplifies to $5=13$, and is a false statement.

9. $x+5=3(2x-1)$
$x+5=6x-3$
$5=5x-3$
$8=5x$
$\dfrac{8}{5}=x$

10.
$\dfrac{x+2}{5} - \dfrac{x-2}{7} = 4$

$35\left|\dfrac{x+2}{5} - \dfrac{x-2}{7}\right| = 35(4)$

$7(x+2) - 5(x-2) = 140$

$7x+14 - 5x+10 = 140$

$2x+24 = 140$

$2x = 116$

$x = 58$

11.
$0.04P + 0.06(1700 - P) = 65$

$100\left[0.04P + 0.06(1700 - P)\right] = 100(65)$

$4P + 6(1700 - P) = 6500$

$4P + 10200 - 6P = 6500$

$-2P + 10200 = 6500$

$-2P = -3700$

$P = 1850$

12.
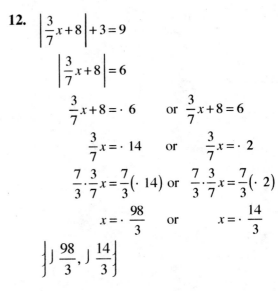

$\left|\dfrac{3}{7}x + 8\right| + 3 = 9$

$\left|\dfrac{3}{7}x + 8\right| = 6$

$\dfrac{3}{7}x + 8 = -6 \qquad \text{or} \quad \dfrac{3}{7}x + 8 = 6$

$\dfrac{3}{7}x = -14 \qquad \text{or} \qquad \dfrac{3}{7}x = -2$

$\dfrac{7}{3}\cdot\dfrac{3}{7}x = \dfrac{7}{3}(-14) \quad \text{or} \quad \dfrac{7}{3}\cdot\dfrac{3}{7}x = \dfrac{7}{3}(-2)$

$x = -\dfrac{98}{3} \qquad \text{or} \qquad x = -\dfrac{14}{3}$

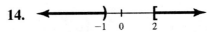

$\left\{-\dfrac{98}{3}, -\dfrac{14}{3}\right\}$

13. $3(x+1) \geq 4x \geq 3$
$3x+3 \geq 4x \geq 3$
$3 \geq x \geq 3$
$6 \geq x$
$x \geq 6$

(number line: 0, 6)

14.

(number line: −1, 0, 2)

15. $-4 \geq -4x \geq 8 < 4$
$4 \geq -4x < 12$
$\dfrac{4}{-4} \geq \dfrac{-4x}{-4} > \dfrac{12}{-4}$
$-1 \geq x > -3$
$-3 < x \geq -1$

(number line: −3, −1, 0)

16. $|2x+3| > 5$
$2x+3 < -5 \quad \text{or} \quad 2x+3 > 5$
$2x < -8 \quad \text{or} \qquad 2x > 2$
$x < -4 \quad \text{or} \qquad x > 1$

(number line: −4, 0, 1)

17.

$$B = \frac{6}{7}(A - 11); \text{ solve for } A$$

$$7(B) = 7 \cdot \frac{6}{7}(A - 11)$$

$$7B = 6(A - 11)$$

$$7B = 6A - 66$$

$$7B + 66 = 6A$$

$$A = \frac{7B + 66}{6} \quad \text{or} \quad \frac{7B}{6} + 11 = A$$

18.

$$P = 100; \quad L = W + 10$$

$$P = 2L + 2W$$

$$100 = 2(W + 10) + 2W$$

$$100 = 2W + 20 + 2W$$

$$100 = 4W + 20$$

$$80 = 4W$$

$$20 \text{ ft} = W$$

$$L = W + 10 = 20 + 10 = 30 \text{ ft}$$

19. Let x = old salary

$$x + (\% \text{ increase} \cdot x) = \text{new salary}$$

$$x + 0.30x = \$29{,}900$$

$$1.3x = \$29{,}900$$

$$x = \$23{,}000$$

20.

Percent of Salt	Amount to be mixed	Amount of Pure Salt
60% or 0.60	x	$0.60x$
21% or 0.21	60	$0.21(60)$
50% or 0.50	$x + 60$	$0.50(x + 60)$

$$0.60x + 0.21(60) = 0.50(x + 60)$$

$$0.60x + 12.6 = 0.50x + 30$$

$$0.10x + 12.6 = 30$$

$$0.10x = 17.4$$

$$x = 174 \text{ gallons}$$

174 gallons of 60% salt solution must be added.

Chapter 3   Graphs and Functions

3.1   Graphs

Problems 3.1

1.

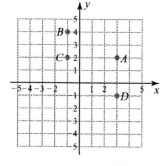

2.
$$A(-3, -1)$$
$$B(1, 2)$$
$$C(0, -3)$$
$$D(3, -2)$$

3.

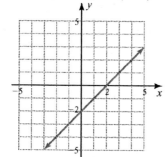

4.   a.   For $x = 0$, $y = x - 2$ becomes
$$y = 0 - 2 = -2.$$
The y-intercept is $(0, -2)$.
For $y = 0$, $y = x - 2$ becomes
$$0 = x - 2$$
$$2 = x$$
The x-intercept is $(2, 0)$.

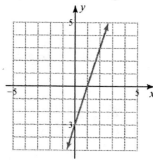

b.   For $x = 0$, $3x + y = 0$ becomes
$$3 \cdot 0 + y = 0 \text{ or } y = 0.$$
The y-intercept is $(0, 0)$.
For $y = 0$, $3x + y = 0$ becomes
$$3x + 0 = 0$$
$$3x = 0$$
$$x = 0$$
The x-intercept is $(0, 0)$.

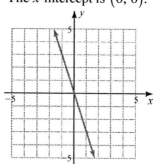

5. a. 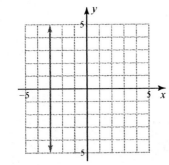   **b.**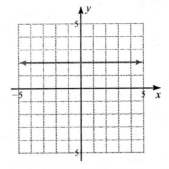

Exercises 3.1

1–8.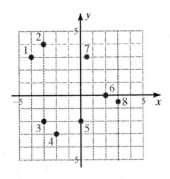

9.   QI, QII

11.   QIII, QIV

13. $C(3, 5)$ **15.** $E(-2, 3)$ **17.** $G(-4, 0)$ **19.** $I(0, -3)$

21.

$y = x + 3$	x	$x + 3$	(x, y)
	-2	$-2 + 3 = 1$	$(-2, 1)$
	-1	$-1 + 3 = 2$	$(-1, 2)$
	0	$0 + 3 = 3$	$(0, 3)$
	1	$1 + 3 = 4$	$(1, 4)$
	2	$2 + 3 = 5$	$(2, 5)$

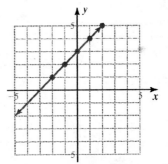

23. $x - y = 4$

For $x = -1$	For $x = 0$	For $x = 1$
$-1 - y = 4$	$0 - y = 4$	$1 - y = 4$
$-y = 5$	$-y = 4$	$-y = 3$
$y = -5$	$y = -4$	$y = -3$
$(-1, -5)$	$(0, -4)$	$(1, -3)$

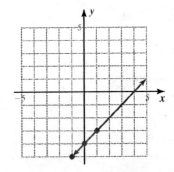

25.　　$2x - y - 3 = 0$

For $x = -1$　　For $x = 0$　　For $x = 1$

$2(-1) - y - 3 = 0$　$2 \cdot 0 - y - 3 = 0$　$2 \cdot 1 - y - 3 = 0$

$-2 - y - 3 = 0$　　$0 - y - 3 = 0$　　$2 - y - 3 = 0$

$-y - 5 = 0$　　　$-y - 3 = 0$　　　$-y - 1 = 0$

$-y = 5$　　　　$-y = 3$　　　　$-y = 1$

$y = -5$　　　　$y = -3$　　　　$y = -1$

$(-1, -5)$　　$(0, -3)$　　$(1, -1)$

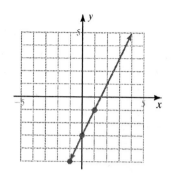

27.　　$y = x - 5$

For $y = 0$　　For $x = 0$

$0 = x - 5$　　$y = 0 - 5$

$5 = x$　　　$y = -5$

$(5, 0)$　　$(0, -5)$

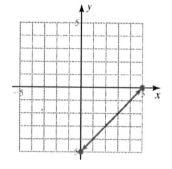

29.　　$2x + 3y = 6$

For $y = 0$　　　For $x = 0$

$2x + 3(0) = 6$　　$2(0) + 3y = 6$

$2x = 6$　　　　$3y = 6$

$x = 3$　　　　$y = 2$

$(3, 0)$　　　$(0, 2)$

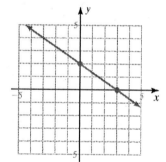

31.　　$2x - y = 4$

For $y = 0$　　　For $x = 0$

$2x - 0 = 4$　　$2(0) - y = 4$

$2x = 4$　　　　$-y = 4$

$x = 2$　　　　$y = -4$

$(2, 0)$　　　$(0, -4)$

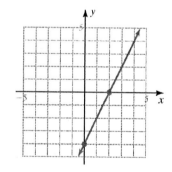

33. $2x + y - 4 = 0$

$$\text{For } y = 0 \qquad \text{For } x = 0$$

$$2x + 0 - 4 = 0 \qquad 2(0) + y - 4 = 0$$

$$2x = 4 \qquad\qquad y - 4 = 0$$

$$x = 2 \qquad\qquad y = 4$$

$$(2, 0) \qquad\qquad (0, 4)$$

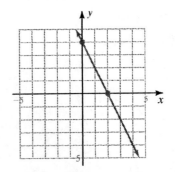

35. $y + 4x = 0$

$$\text{For } y = 0 \qquad \text{For } x = 0$$

$$0 + 4x = 0 \qquad y + 4(0) = 0$$

$$4x = 0 \qquad\qquad y + 0 = 0$$

$$x = 0 \qquad\qquad y = 0$$

$$(0, 0) \qquad\qquad (0, 0)$$

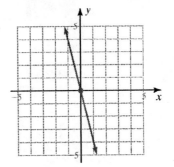

37. $2x - 5y = -10$

$$\text{For } y = 0 \qquad \text{For } x = 0$$

$$2x - 5(0) = -10 \qquad 2(0) - 5y = -10$$

$$2x = -10 \qquad\qquad -5y = -10$$

$$x = -5 \qquad\qquad y = 2$$

$$(-5, 0) \qquad\qquad (0, 2)$$

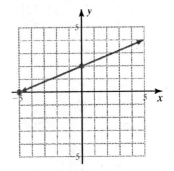

39. $x - y - 5 = 0$

$$\text{For } y = 0 \qquad \text{For } x = 0$$

$$x - 0 - 5 = 0 \qquad 0 - y - 5 = 0$$

$$x - 5 = 0 \qquad\qquad -y = 5$$

$$x = 5 \qquad\qquad y = -5$$

$$(5, 0) \qquad\qquad (0, -5)$$

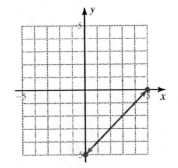

41. $\dfrac{\cdot 7}{2} x = 14 \cdot \quad x = \dfrac{2}{\cdot 7} \cdot 14 = \cdot 4$

vertical line: $x = \cdot 4$

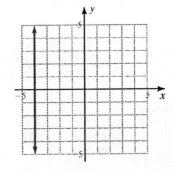

43. $\dfrac{3}{2}x = 6 \cdot$ $x = \dfrac{2}{3} \cdot 6 = 4$

vertical line: $x = 4$

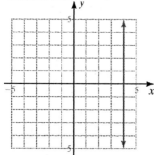

45. $\dfrac{\cdot\,3}{4}x = 3 \cdot$ $x = \dfrac{4}{\cdot\,3} \cdot 3 = \cdot\,4$

vertical line: $x = \cdot\,4$

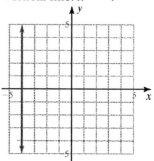

47. $\dfrac{-1}{3} + y = \dfrac{2}{3} \rightarrow y = \dfrac{3}{3} = 1$

horizontal line: $y = 1$

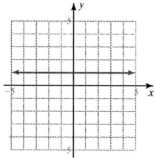

49. $\dfrac{2}{3} = x \dfrac{4}{3} \rightarrow x = \dfrac{6}{3} = 2$

vertical line: $x = 2$

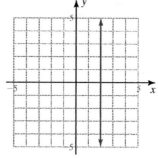

51. **a.**   Substituting 60 for T and 80 for N, we have $80 = 4(60 - 40) = 4(20) = 80$. The ordered pair (60, 80) satisfies the equation.

  b.   The cricket makes 80 chirps when the temperature is 60°F.

  c.   The cricket makes $N = 4(80 - 40) = 4(40) = 160$ chirps when the temperature is 80°F.

53. The lower limit pulse rate for a 20-year-old is 140. (20, 140)

55. The upper limit pulse rate for a 45-year-old is approximately 148. (45, 148)

57. $y = 3x + 6$; for $x = 2$
$y = 3(2) + 6 = 6 + 6 = 12$

59. $y = \dfrac{-2}{3}x + 4$; for $x = 3$
$y = \dfrac{-2}{3}(3) + 4 = -2 + 4 = 2$

61. $y = 3x + 6$; for $y = 0$
$0 = 3x + 6$
$-6 = 3x$
$-2 = x$

63. The safe zone for exercising allows the temperature to rise to 86°F when the humidity is 50%.

65. The safe zone for exercising at a temperature of 100°F requires the humidity to be less than 10%.

67. To reach the danger zone when the humidity is 60% and the temperature is currently 86°F, the temperature would have to rise to 97°F or rise 11°F from the current temperature.

69. Sample answer: Linear equations are named linear because their graphs are straight lines.

71. Sample answer: When $B = 0$ in the equation $Ax + By = C$ the equation becomes $Ax = C$ which is the equation of a vertical line.

73. Sample answer: The x-intercept for any line has the form $(a, 0)$, which is the result of setting $y = 0$ in a linear equation.

75. Sample answer: (a) Choose several values for x and evaluate the equation to find the corresponding values of y. This will determine the coordinates of points that are on the graph of the line. Plot these points and draw a line through them. (b) Set $x = 0$ to find the coordinate of the y-intercept and set $y = 0$ to find the coordinate of the x-intercept. Plot these points and draw a line through them.

77. $2y = -6 \rightarrow y = -3$

79. $-3x - y = -6$

For $x = 0$	For $x = 3$	For $x = 1$
$-3(0) - y = -6$	$-3(3) - y = -6$	$-3(1) - y = -6$
$-y = -6$	$-9 - y = -6$	$-3 - y = -6$
$y = 6$	$-y = 3$	$-y = -3$
$y = 6$	$y = -3$	$y = 3$
$(0, 6)$	$(3, -3)$	$(1, 3)$

81. $3x - 2y = -6$

For $y = 0$	For $x = 0$
$3x - 2(0) = -6$	$3(0) - 2y = -6$
$3x = -6$	$-2y = -6$
$x = -2$	$y = 3$
$(-2, 0)$	$(0, 3)$

83. $A\left(-2,3\right)$
$B\left(-3,0\right)$
$C\left(4,-2\right)$

3.2   Introduction to Functions

Problems 3.2

1. $D=\{2, 7, 9\}$ $R=\{1, 5, 8\}$

2.

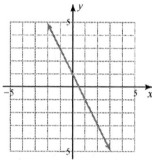

$D=\{\text{Real numbers}\}$ $R=\{\text{Real numbers}\}$

3.   **a.** $D=\{x|\leq 1\leq x\leq 1\}$ $R=\{y|\leq 3\leq y\leq 3\}$ **b.** $D=\{x|\leq 4\leq x\leq 0\}$ $R=\{y|0\leq y\leq 4\}$
c. $D=\{x|0-x-2\}$ $R=\{y|-2-y-2\}$

4.   **a.**   The relation is a function because no vertical line crosses the graph more than once.
b.   The relation is a function because no vertical line crosses the graph more than once.
c.   The relation is not a function because a vertical line can be drawn that crosses the graph more than once.

5.   **a.** $D=\{x|x-\text{Reals, where }x--1, 4\}$ -1 and 4 cause division by zero and must be excluded.
b. $D=\{x|x--2\}$ $x+2-0$ or $x--2$ because the square root of a negative number is not a real number.

6. $f\left(x\right)=2x-3$
a. $f\left(5\right)=2\left(5\right)-3=10-3=7$
b. $f\left(1\right)=2\left(1\right)-3=2-3=-1$
c. $f\left(1\right)-f\left(5\right)=-1-7=-8$
d. $f\left(x-1\right)=2\left(x-1\right)-3=2x-2-3=2x-5$

7. **a.** In the ordered pair $(2, 3)$, $x = 2$ and $y = 3$. Thus, $f(2) = 3$.

     b. In the ordered pair $(-1, 4)$, $x = -1$ and $y = 4$. Thus, $f(-1) = 4$.

     c. $f(2) + f(-1) = 3 + 4 = 7$

Exercises 3.2

1. $D = \{-3, -2, -1\}$ $R = \{0, 1, 2\}$ a function

3. $D = \{3, 4, 5\}$ $R = \{0\}$ a function

5. $D = \{1, 2\}$ $R = \{2, 3\}$ not a function; 1 and 2 are each paired with more than one number.

7. $D = \{1, 3, 5, 7\}$ $R = \{-1\}$ a function

9. $D = \{2\}$ $R = \{1, 0, -1, -2\}$ not a function; 2 is paired with more than one number.

11. $D = \{x \mid \leq 5 \leq x \leq 5\}$ $R = \{y \mid \leq 5 \leq y \leq 5\}$ The relation is not a function because a vertical line can be drawn that crosses the graph more than once.

13. $D = \{x \mid \leq 5 \leq x \leq 5\}$ $R = \{y \mid 0 \leq y \leq 5\}$ The relation is a function because no vertical line crosses the graph more than once.

15. $D = \{x \mid 0 - x - 5\}$ $R = \{y \mid -5 - y - 5\}$ The relation is not a function because a vertical line can be drawn that crosses the graph more than once.

17. $D = \{x \mid x \text{ is a real number}\}$ $R = \{y \mid y \geq 0\}$ The relation is a function because no vertical line crosses the graph more than once.

19. $D = \{x \mid x \text{ is a real number}\}$ $R = \{y \mid y \leq 0\}$ The relation is a function because no vertical line crosses the graph more than once.

21. $D = \{x \mid x \geq 0\}$ $R = \{y \mid y \text{ is a real number}\}$ The relation is not a function because a vertical line can be drawn that crosses the graph more than once.

23. $D = \{x \mid x \text{ is a real number}\}$ $R = \{y \mid y \text{ is a real number}\}$ The relation is a function because no vertical line crosses the graph more than once.

25. $D = \{x \mid x \text{ is a real number}\}$ $R = \{y \mid y \text{ is a real number}\}$ The relation is a function because no vertical line crosses the graph more than once.

27. $D = \left\{ x \mid \leq 3 \leq x \leq 3 \right\}$ $R = \left\{ y \mid \leq 2 \leq y \leq 2 \right\}$ The relation is not a function because a vertical line can be drawn that crosses the graph more than once.

29. $D = \left\{ x \mid x - 2 \text{ or } x - -2 \right\}$ $R = \left\{ y \mid y \text{ is a real number} \right\}$ The relation is not a function because a vertical line can be drawn that crosses the graph more than once.

31. The relation is a function because no vertical line crosses the graph more than once.

33. The relation is a function because no vertical line crosses the graph more than once.

35. The relation is a function because no vertical line crosses the graph more than once.

37. Find when the radicand is positive or zero.
$$x \geq 5 \geq 0$$
$$x \geq 5$$
$$D = \left\{ x \mid x \geq 5 \right\}$$

39. Find when the radicand is positive or zero.
$$4 \leq 2x \leq 0$$
$$\leq 2x \leq \leq 4$$
$$x \leq 2$$
$$D = \left\{ x \mid x \leq 2 \right\}$$

41. Find when the radicand is positive or zero.
$$x^2 + 1 - 0$$
$$x^2 - -1 \text{ is always true}$$
$$D = \left\{ x \mid x \text{ is a real number} \right\}$$

43. Exclude those values of x that make the denominator zero.
$$x \neq 5 = 0$$
$$x = 5$$
$$D = \left\{ x \mid x \text{ is a real number and } x \neq 5 \right\}$$

45. Exclude those values of x that make the denominator zero.
$$x + 5 = 0$$
$$x = \neq 5$$
$$D = \left\{ x \mid x \text{ is a real number and } x \neq \neq 5 \right\}$$

47. Exclude those values of x that make the denominator zero.
$$(x + 2)(x + 1) = 0$$
$$x + 2 = 0 \quad \text{or } x + 1 = 0$$
$$x = -2 \text{ or} \quad x = -1$$
$$D = \left\{ x \mid x \text{ is a real number and } x - -2 \text{ or } x - -1 \right\}$$

49. Exclude those values of x that make the denominator zero.
$$(x + 4)(x \neq 4) = 0$$
$$x + 4 = 0 \quad \text{or } x \neq 4 = 0$$
$$x = \neq 4 \text{ or} \quad x = 4$$
$$D = \left\{ x \mid x \text{ is a real number and } x \neq \neq 4 \text{ or } x \neq 4 \right\}$$

51. $f(x) = 3x + 1$

   a. $f(0) = 3(0) + 1 = 0 + 1 = 1$

   b. $f(2) = 3(2) + 1 = 6 + 1 = 7$

   c. $f(-2) = 3(-2) + 1 = -6 + 1 = -5$

53. $F(x) = \sqrt{x - 1}$

   a. $F(1) = \sqrt{1 - 1} = \sqrt{0} = 0$

   b. $F(5) = \sqrt{5 - 1} = \sqrt{4} = 2$

   c. $F(26) = \sqrt{26 - 1} = \sqrt{25} = 5$

55. $f(x) = \dfrac{1}{3x+1}$

    a. $f(1) = \dfrac{1}{3(1)+1} = \dfrac{1}{3+1} = \dfrac{1}{4}$

    b. $f(1) \cdot f(2) = \dfrac{1}{4} \cdot \dfrac{1}{3 \cdot 2 + 1} = \dfrac{1}{4} \cdot \dfrac{1}{7} = \dfrac{7}{28} \cdot \dfrac{4}{28} = \dfrac{3}{28}$

    c. $\dfrac{f(1) \cdot f(2)}{3} = \dfrac{\dfrac{1}{4} \cdot \dfrac{1}{3 \cdot 2 + 1}}{3} = \dfrac{\dfrac{1}{4} \cdot \dfrac{1}{7}}{3} = \dfrac{\dfrac{7}{28} \cdot \dfrac{4}{28}}{3} = \dfrac{\dfrac{3}{28}}{3} = \dfrac{3}{28} \cdot \dfrac{1}{3} = \dfrac{1}{28}$

57. $f(x) = 3x - 4 \quad g(x) = x^2 + 2x + 4$

    a. $f(3) = 3(3) - 4 = 9 - 4 = 5$

    b. $g(3) = 3^2 + 2(3) + 4 = 9 + 6 + 4 = 19$

    c. $f(3) + g(3) = 5 + 19 = 24$

59. $f(x) = 3x - 4 \quad g(x) = x^2 + 2x + 4$

    a. $f(-2) = 3(-2) - 4 = -6 - 4 = -10$

    b. $g(-3) = (-3)^2 + 2(-3) + 4 = 9 - 6 + 4 = 7$

    c. $f(\cdot 2) \cdot g(\cdot 3) = \cdot 10 \cdot 7 = \cdot 70$

61.   **a.** In the ordered pair $(1, 3)$, $x = 1$ and $y = 3$. Thus, $f(1) = 3$.

    b. In the ordered pair $(-2, 4)$, $x = -2$ and $y = 4$. Thus, $g(-2) = 4$.

    c. $f(1) + g(-2) = 3 + 4 = 7$

63.   **a.** In the ordered pair $(-3, 7)$, $x = -3$ and $y = 7$. Thus, $f(-3) = 7$.

    b. In the ordered pair $(2, 8)$, $x = 2$ and $y = 8$. Thus, $g(2) = 8$.

    c. $f(\cdot 3) \cdot g(2) = 7 \cdot 8 = 56$

65.   **a.** $P(x) = R(x) - C(x)$

$$= (30x - 0.0005x^2) - (100,000 + 6x)$$
$$= 30x - 0.0005x^2 - 100,000 - 6x$$
$$= -0.0005x^2 + 24x - 100,000$$

    b. $P(10,000) = -0.0005(10,000)^2 + 24(10,000) - 100,000 = 90,000$

    The profit on 10,000 books is $90,000.

67.   **a.** $U(50) = -50 + 190 = 140$ The highest safe pulse rate is 140.

    b. $U(60) = -60 + 190 = 130$ The highest safe pulse rate is 130.

69.   **a.** $w(70) = 5(70) - 190 = 350 - 190 = 160$ The ideal weight for a 70 in. man is 160 pounds.

    b. $200 = 5h - 190 \rightarrow 390 = 5h \rightarrow 78 = h$ A 200 pound man should be 78 in. tall.

71.   **a.** $P(10) = 63.9(10) = 639$ The pressure at a depth of 10 feet is 639 lb/ft^2.

    b. $P(100) = 63.9(100) = 6390$ The pressure at a depth of 100 feet is 6390 lb/ft^2.

73. **a.** The independent variable is *L*.

   b. The dependent variable is *S*.

   c. $S = m(11) = 3(11) - 22 = 33 - 22 = 11$ The man's shoe size is 11.

   d. $S = m(11) = 3(11) - 21 = 33 - 21 = 12$ The woman's shoe size is 12.

75. **a.** $R(20) = 1.85(20)^2 - 19.14(20) + 262 = 1.85(400) - 382.8 + 262 = 740 - 382.8 + 262 = 619.2$
   The number of robberies per 100,000 in 2000 was 619.2.

   b. $R(30) = 1.85(30)^2 - 19.14(30) + 262 = 1.85(900) - 574.2 + 262 = 1665 - 574.2 + 262 = 1352.8$
   The number of robberies per 100,000 in 2010 will be 1352.8.

77. **a.** The two graphs represent functions.

   b. $D = \{x \mid 1 \le x \le 6\}$ $R = \{y \mid 0 \le y \le 0.11\}$

   c. $D = \{x \mid 1 \le x \le 4\}$ $R = \{y \mid 0 \le y \le 0.10\}$

79. Sample answer: A function will have no repetitions of the first member of the ordered pairs, while a relation may have repetitions of the first member of the ordered pairs.

81. Sample answer: The domain of a function is the set of all values for *x* that will make it possible to find a value for *y* in the function.

83. False; some relations are functions, but not all relations are functions.

85. Reflexive – a triangle is similar to itself. In fact it is congruent to itself.
   Symmetric – If triangle A is similar to triangle B, then triangle B is certainly similar to triangle A.
   Transitive – If triangle A is similar to triangle B and triangle B is similar to triangle C, then triangle
      A is similar to triangle C. All similar triangles are similar.
   Yes, this is an equivalence relation.

87. Reflexive – if *x* is odd then *x* is odd; if *x* is even then *x* is even.
   Symmetric – if *x* is odd then *y* is odd; if *y* is odd then *x* is odd. Similarly, if *x* is even then *y* is even;
      if *y* is even then *x* is even.
   Transitive - if *x* is odd then *y* is odd and if *y* is odd then *z* is odd, then if *x* is odd then *z* is odd.
      Similarly, if *x* is even then *y* is even and if *y* is even then *z* is even, then if *x* is even then *z* is
      even.
   Yes, this is an equivalence relation.

89. Reflexive – *x* has the same remainder as *x* when *x* is divided by 3.
   Symmetric – if *x* and *y* have the same remainders when divided by 3, then *y* and *x* also have the same
      remainders when divided by 3.
   Transititve – if *x* and *y* have the same remainders when divided by 3 and if *y* and *z* have the same
      remainders when divided by 3, then *x* and *z* have the same remainders when divided by 3.
   Yes, this is an equivalence relation.

91. **a.** In the ordered pair $(4, 3)$, $x = 4$ and $y = 3$. Thus, $f(4) = 3$.

   b. In the ordered pair $(5, -1)$, $x = 5$ and $y = -1$. Thus, $f(5) = -1$.

   c. $f(4) - f(5) = 3 - (-1) = 3 + 1 = 4$

93. Exclude those values of x that make the denominator zero.

$$x \neq 2 = 0$$

$$x = 2$$

$$D = \{x \mid x \text{ is a real number and } x \neq 2\}$$

95. The relation is not a function because a vertical line can be drawn that crosses the graph more than once.

97. $D = \{x \mid \leq 3 \leq x \leq 3\} \quad R = \{y \mid \leq 3 \leq y \leq 3\}$ **99.** $D = \{x \mid \leq 3 \leq x \leq 3\} \quad R = \{y \mid \leq 3 \leq y \leq 0\}$

101. $D = \{7, 8, 9\} \quad R = \{8, 9, 10\}$

3.3 Using Slopes to Graph Lines

Problems 3.3

1. **a.** $m = \dfrac{5-3}{1-(-2)} = \dfrac{2}{3}$ **b.** $m = \dfrac{3-(-5)}{4-4} = \dfrac{8}{0}$ is undefined

     c. $m = \dfrac{-4-2}{4-2} = \dfrac{-6}{2} = -3$ **d.** $m = \dfrac{-1-(-1)}{7-6} = \dfrac{0}{1} = 0$

2. $m = \dfrac{290-260}{1990-1985} = \dfrac{30}{5} = 6$ $m = \dfrac{240-125}{1990-1985} = \dfrac{115}{5} = 23$

The spending per resident for public safety increased at the rate of $6 per year from 1985 to 1990.

The spending per resident for health and welfare increased at the rate of $23 per year from 1985 to 1990.

3. **a.** $m = \dfrac{9-6}{6-2} = \dfrac{3}{4}$

Parallel, since the slope of AB is $\frac{3}{4}$.

    b. $m = \dfrac{-5-(-1)}{-6-(-9)} = \dfrac{-5+1}{-6+9} = \dfrac{-4}{3}$

Perpendicular, since the slope of CD is $\frac{-4}{3}$.

4. $m = \dfrac{8-y}{-4-5} = \dfrac{8-y}{-9}$

The slope of the line through AB is $\frac{4}{3}$.

Solving for y:

$$\dfrac{8 \mid y}{\mid 9} = \dfrac{4}{3}$$

$$\left| 9 \left| \dfrac{8 \mid y}{\mid 9} \right| \right| = \left| 9 \left| \dfrac{4}{3} \right| \right|$$

$$8 \mid y = \mid 12$$

$$-y = -20$$

$$y = 20$$

5.

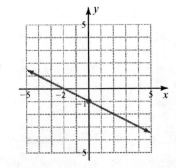

6. **a.** $y = -4x + 1$

slope $= -4$; y-intercept $= 1$

b. $-y = \frac{1}{2}x - 3$

$y = -\frac{1}{2}x + 3$

slope $= -\frac{1}{2}$; y-intercept $= 3$

c. $2x - 3y = 5$

$-3y = -2x + 5$

$-\frac{1}{3}(-3y) = -\frac{1}{3}(-2x + 5)$

$y = \frac{2}{3}x - \frac{5}{3}$

slope $= \frac{2}{3}$; y-intercept $= -\frac{5}{3}$

Exercises 3.3

1. $m = \dfrac{0 - 4}{-1 - 2} = \dfrac{-4}{-3} = \dfrac{4}{3}$

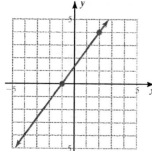

3. $m = \dfrac{3 - (-5)}{-1 - (-4)} = \dfrac{8}{3}$

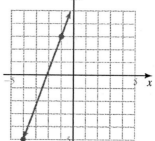

5. $m = \dfrac{-1 - 8}{1 - 4} = \dfrac{-9}{-3} = 3$

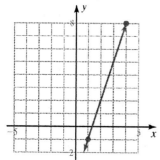

7. $m = \dfrac{-1 - (-1)}{-2 - 3} = \dfrac{0}{-5} = 0$

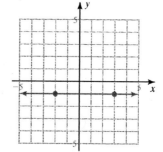

9. **a.** $m = \dfrac{54,000 - 50,000}{2004 - 1998} = \dfrac{4,000}{6} = \dfrac{2000}{3}$

b. The salary increased approximately $667 per year.

11. $m_{AB} = \dfrac{4 - 6}{-1 - 1} = \dfrac{-2}{-2} = 1$

$m_{CD} = \dfrac{-1 - (-2)}{2 - 1} = \dfrac{1}{1} = 1$

The lines are parallel since the slopes are the same.

83

13.
$$m_{AB} = \frac{5-0}{4-2} = \frac{5}{2}$$
$$m_{CD} = \frac{2-0}{3-8} = \frac{2}{-5} = -\frac{2}{5}$$

The lines are perpendicular since the slopes are negative reciprocals of each other.

15.
$$m_{AB} = \frac{2-1}{1-(-1)} = \frac{1}{2}$$
$$m_{CD} = \frac{-1-(-1)}{0-1} = \frac{0}{-1} = 0$$

The lines are neither parallel nor perpendicular.

17.
$$m_{AB} = \frac{\frac{-1}{2}-(-1)}{2-1} = \frac{\frac{1}{2}}{1} = \frac{1}{2}$$
$$m_{CD} = \frac{0-(-2)}{1-2} = \frac{2}{-1} = -2$$

The lines are perpendicular since the slopes are negative reciprocals of each other.

19.
$$m_{AB} = \frac{-1-1}{14-0} = \frac{-2}{14} = -\frac{1}{7}$$
$$m_{CD} = \frac{1-2}{7-0} = \frac{-1}{7} = -\frac{1}{7}$$

The lines are parallel since the slopes are the same.

21.
$$m = \frac{8-4}{6-x} = \frac{4}{6-x}$$

The slope of the parallel line is 1.
Solving for x:
$$\frac{4}{6-x} = 1$$
$$4 = 6-x$$
$$-2 = -x$$
$$2 = x$$

23.
$$m = \frac{6-2}{2-x} = \frac{4}{2-x}$$

The slope of the perpendicular line is $\frac{1}{2}$.
Solving for x:
$$\frac{4}{2-x} = -2$$
$$4 = -2(2-x)$$
$$4 = -4+2x$$
$$8 = 2x$$
$$4 = x$$

25.
$$m = \frac{-1-(-6)}{-2-x} = \frac{5}{-2-x}$$

The slope of the perpendicular line is $\frac{-2}{3}$.
Solving for x:
$$\frac{5}{-2-x} = \frac{3}{2}$$
$$5(2) = 3(-2-x)$$
$$10 = -6-3x$$
$$16 = -3x$$
$$-\frac{16}{3} = x$$

27.
$$m = \frac{-2-y}{1-3} = \frac{-2-y}{-2}$$

The slope of the parallel line is -3.
Solving for y:
$$\frac{-2-y}{-2} = -3$$
$$-2-y = -3(-2)$$
$$-2-y = 6$$
$$-y = 8$$
$$y = -8$$

29.
$$m = \frac{5-4}{3-x} = \frac{1}{3-x}$$

The slope of the horizontal line is 0.
The perpendicular line will be a vertical line whose equation is $x = 3$ which makes the slope undefined.

31.

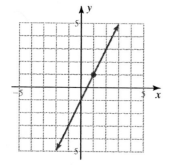

33.

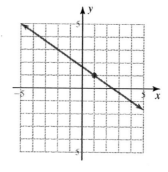

35.

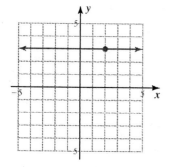

37.

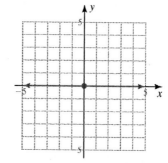

39.

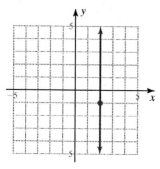

41.
$$x_m = \frac{3+7}{2} = \frac{10}{2} = 5$$
$$y_m = \frac{4+2}{2} = \frac{6}{2} = 3$$
midpoint is $(5, 3)$

43.
$$x_m = \frac{0+(-8)}{2} = \frac{-8}{2} = -4$$
$$y_m = \frac{-8+0}{2} = \frac{-8}{2} = -4$$
midpoint is $(-4, -4)$

45.
$$x_m = \frac{|5+(|6)}{2} = \frac{|11}{2}$$
$$y_m = \frac{|2+2}{2} = \frac{0}{2} = 0$$
midpoint is $\left| |\frac{11}{2}, 0 \right|$

47.　$y = x - 4$
slope $= 1$; y-intercept $= -4$

49.
$$-3y = x + 6$$
$$-\frac{1}{3}(-3y) = -\frac{1}{3}(x+6)$$
$$y = -\frac{1}{3}x - 2$$
slope $= -\frac{1}{3}$; y-intercept $= -2$

51.
$$4x + 2y = 8$$
$$2y = -4x + 8$$
$$\frac{1}{2}(2y) = \frac{1}{2}(-4x+8)$$
$$y = -2x + 4$$
slope $= -2$; y-intercept $= 4$

53. Find the midpoint:

$$\left|\frac{x_1 + x_2}{2}, \frac{y_1 + y_2}{2}\right| = \left|\frac{90 + 76}{2}, \frac{30 + 40}{2}\right| = \left|\frac{166}{2}, \frac{70}{2}\right| = (83, 35)$$

55. Find the slope of each side:

$$m_{AB} = \frac{5 - 2}{0 - 2} = \frac{3}{-2} = -\frac{3}{2}$$

$$m_{BC} = \frac{12 - 5}{-20 - 0} = \frac{7}{-20} = -\frac{7}{20}$$

$$m_{AC} = \frac{12 - 2}{-20 - 2} = \frac{10}{-22} = -\frac{5}{11}$$

Since none of the slopes are negative reciprocals of each other, there are no perpendicular sides and this is not a right triangle.

57. Find the slope of each side:

$$m_{AB} = \frac{5 - 2}{0 - 2} = \frac{3}{-2} = -\frac{3}{2}$$

$$m_{BC} = \frac{-12 - 5}{-19 - 0} = \frac{-17}{-19} = \frac{17}{19}$$

$$m_{AC} = \frac{-12 - 2}{-19 - 2} = \frac{-14}{-21} = \frac{2}{3}$$

Since two of the slopes are negative reciprocals of each other, there are perpendicular sides and this is a right triangle.

59. Find the slope of each side:

$$m_{AB} = \frac{-14 - 2}{-4 - 2} = \frac{-16}{-6} = \frac{8}{3}$$

$$m_{BC} = \frac{-8 - (-14)}{-20 - (-4)} = \frac{6}{-16} = -\frac{3}{8}$$

$$m_{AC} = \frac{-8 - 2}{-20 - 2} = \frac{-10}{-22} = \frac{5}{11}$$

Since two of the slopes are negative reciprocals of each other, there are perpendicular sides and this is a right triangle.

61. **a.** In 1990, $x = 30$
$y = 2.2(30) + 180 = 246$ million

b. In 2000, $x = 40$
$y = 2.2(40) + 180 = 268$ million

c. In 2010, $x = 50$
$y = 2.2(50) + 180 = 290$ million

d. $312 = 2.2x + 180$
$132 = 2.2x$
$60 = x$
The population reaches 312 million in 2020.

e.

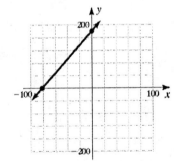

63. $6x + 3y = 12$

$3y = -6x + 12$

$\dfrac{1}{3}(3y) = \dfrac{1}{3}(-6x + 12)$

$y = -2x + 4$

65. $3y - 2x = 12$

$3y = 2x + 12$

$\dfrac{1}{3}(3y) = \dfrac{1}{3}(2x + 12)$

$y = \dfrac{2}{3}x + 4$

67. x-intercept: (2, 0); y-intercept: (0, –3)

$m = \dfrac{-3 - 0}{0 - 2} = \dfrac{-3}{-2} = \dfrac{3}{2}$

69. $m = \dfrac{3.6 - 1.3}{2 - 3} = \dfrac{2.3}{-1} = -2.3$

The parallel line has the same slope –2.3.

71. $m = \dfrac{0.\overline{3} - 0.\overline{6}}{2 - 3} = \dfrac{-0.\overline{3}}{-1} = 0.\overline{3} = \dfrac{1}{3}$

The slope of the perpendicular line is –3.

73. The slope of the parallel line is 2.

$m = \dfrac{3c - c}{2 - 1} = \dfrac{2c}{1} = 2c$

$2c = 2$

$c = 1$

75. Sample answer. The slope of a horizontal line is zero because the difference of the y-coordinates of two points on a horizontal line is zero.

77. Sample answer. Two parallel lines have the same slope because they must head in the same direction, otherwise they will intersect each other.

79. If the slope of a line is negative, the line will fall as the movement is from left to right.

81.
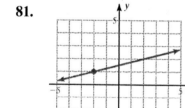

83. The slope of the parallel line is 1.

$\dfrac{4 - 2}{3 - x} = 1$

$\dfrac{2}{3 - x} = 1$

$2 = 3 - x$

$-1 = -x$

$1 = x$

85. **a.** $m = \dfrac{-3 - 2}{-2 - (-4)} = \dfrac{-5}{2} = -\dfrac{5}{2}$

b. $m = \dfrac{7 - 4}{-2 - (-2)} = \dfrac{3}{0}$ is undefined

c. $m = \dfrac{5 - 1}{4 - 2} = \dfrac{4}{2} = 2$

d. $m = \dfrac{3 - 3}{1 - 2} = \dfrac{0}{-1} = 0$

3.4   Equations of Lines

Problems 3.4

1.

$$m = \frac{3-1}{4-3} = \frac{2}{1} = 2$$
$$y - 1 = 2(x - 3)$$
$$y - 1 = 2x - 6$$
$$y = 2x - 5$$
$$-2x + y = -5$$
$$2x - y = 5$$

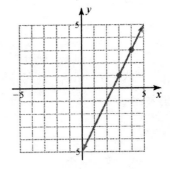

2.

$$y - 2 = -3(x - 1)$$
$$y - 2 = -3x + 3$$
$$y = -3x + 5$$
$$3x + y = 5$$

3. $m = 3; \quad b = 2$
$$y = 3x + 2$$

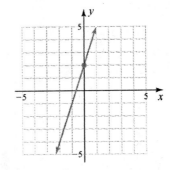

4.   **a.** $y = -2x + 7 \qquad m = -2$

The parallel line has the same slope.

$$y - 3 = -2(x - 4)$$
$$y - 3 = -2x + 8$$
$$y = -2x + 11$$

b. $y = -2x + 7 \qquad m = -2$

The perpendicular line has slope $\frac{1}{2}$.

$$y - 3 = \frac{1}{2}(x - 4)$$
$$y - 3 = \frac{1}{2}x - 2$$
$$y = \frac{1}{2}x + 1$$

5. **a.** Slope zero equation has the form $y = k$.

$y = -3$

b. Undefined slope equation has the form $x = h$.

$x = 1$

6.
$$m = \frac{76.9 - 68.2}{50 - 0} = \frac{8.7}{50} = 0.174$$
$$y - 68.2 = 0.174(t - 0)$$
$$y - 68.2 = 0.174t$$
$$y = 0.174t + 68.2$$

The equation is similar to the result of Example 6.

7. **a.**
$$m = \frac{1860 - 1710}{12,000 - 9,000} = \frac{150}{3,000} = 0.05$$
$$P - 1710 = 0.05(s - 9000)$$
$$P - 1710 = 0.05s - 450$$
$$P = 0.05s + 1260$$

b. The rate of change of the monthly gross pay is $0.05 = \frac{5}{100} = \frac{1}{20}$ or $1 gross pay per $20 in sales.

c.
$$P = 0.05(20,000) + 1260$$
$$= 1000 + 1260 = \$2260$$

Exercises 3.4

1.
$$m = \frac{2 - (-1)}{2 - 1} = \frac{3}{1} = 3$$
$$y - (-1) = 3(x - 1)$$
$$y + 1 = 3x - 3$$
$$y = 3x - 4$$
$$-3x + y = -4$$
$$3x - y = 4$$

3.
$$m = \frac{3 - 2}{2 - 3} = \frac{1}{-1} = -1$$
$$y - 2 = -1(x - 3)$$
$$y - 2 = -x + 3$$
$$y = -x + 5$$
$$x + y = 5$$

5. x-intercept: (2, 0)   y-intercept: (0, 4)
$$m = \frac{4 - 0}{0 - 2} = \frac{4}{-2} = -2$$
$$y - 0 = -2(x - 2)$$
$$y = -2x + 4$$
$$2x + y = 4$$

7.
$$y - 5 = 2(x - (-3))$$
$$y - 5 = 2x + 6$$
$$y = 2x + 11$$
$$-2x + y = 11$$
$$2x - y = -11$$

9.
$$y - (-2) = -3(x - (-1))$$
$$y + 2 = -3x - 3$$
$$y = -3x - 5$$
$$3x + y = -5$$

11. $m = 5; \quad b = 2$

$y = 5x + 2$

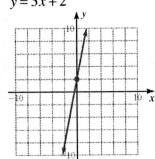

13.

$m = -\dfrac{1}{5}; \quad b = -\dfrac{1}{3}$

$y = -\dfrac{1}{5}x - \dfrac{1}{3}$

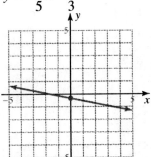

15.   The slope of the given line is 2. The parallel line has slope 2.

$$y - (-2) = 2(x - 1)$$
$$y + 2 = 2x - 2$$
$$y = 2x - 4$$

17. $2y + 6x = 8$

$$2y = -6x + 8$$
$$y = -3x + 4$$

The slope of the given line is –3. The parallel line has slope –3.

$$y - 3 = -3(x - (-5))$$
$$y - 3 = -3x - 15$$
$$y = -3x - 12$$

19. $2y = x + 6$

$$y = \dfrac{1}{2}x + 3$$

The slope of the given line is $\frac{1}{2}$. The perpendicular line has slope –2.

$$y - 1 = -2(x - 1)$$
$$y - 1 = -2x + 2$$
$$y = -2x + 3$$

21. $y - x = 3$

$$y = x + 3$$

The slope of the given line is 1. The perpendicular line has slope –1.

$$y - (-4) = -1(x - (-2))$$
$$y + 4 = -x - 2$$
$$y = -x - 6$$

23.   A slope zero equation has the form $y = k$.

$y = 4$

25.   An undefined slope equation has the form $x = h$.

$x = -2$

27.   A vertical line has an undefined slope. An undefined slope equation has the form $x = h$.

$x = -2$

29.   A horizontal line has slope zero. A slope zero equation has the form $y = k$.

$y = 2$

31. a.

x	y
0	24.50
1	25.40
2	23.90
3	23.10
4	22.90
8	?

b.

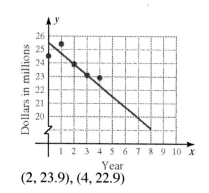

$(2, 23.9), (4, 22.9)$

c.
$$m = \frac{22.9 - 23.9}{4 - 2} = \frac{-1}{2} = -\frac{1}{2}$$

$$y - 23.9 = -\frac{1}{2}(x - 2)$$

$$y - 23.9 = -\frac{1}{2}x + 1$$

$$y = -\frac{1}{2}x + 24.9$$

d.
$$y = -\frac{1}{2}(8) + 24.9 = -4 + 24.9 = 20.9$$

$20.9 million

e.   Each year decreases by \$0.5 million.

33.   A horizontal line has slope zero. A slope zero equation has the form $y = k$.

$y = -4$

35.   Two points on the line are $(-3, 0)$ and $(0, 2)$.
$$m = \frac{2 - 0}{0 - (-3)} = \frac{2}{3} \qquad y\text{-intercept: } (0, 2)$$

$$y = \frac{2}{3}x + 2$$

$$3y = 2x + 6$$

$$2x - 3y = -6$$

37. a.
$$m = \frac{8 - 5.5}{10 - 0} = \frac{2.5}{10} = 0.25$$

$$N - 5.5 = 0.25(x - 0)$$

$$N - 5.5 = 0.25x$$

$$N = 0.25x + 5.5$$

b.   Each year there is a 0.25 million increase in the number of cases of diabetes.

c. $x = 2008 - 1983 = 25$

d. $N = 0.25(25) + 5.5 = 6.25 + 5.5 = 11.75$ million

39. a.
$$m = \frac{212,500 - 127,500}{2500 - 1500} = \frac{85000}{1000} = 85$$

$$C - 127,500 = 85(x - 1500)$$

$$C - 127,500 = 85x - 127,500$$

$$C = 85x$$

b.   The cost is \$85 per square foot.

c. $C = 85(1475) = \$125,375$

d. $138,000 = 85x$

$x = 1624$ sq. ft.

41. **a.** Use the ordered pairs $(2, 6)$ and $(6, 4)$.

$$m = \frac{4-6}{6-2} = \frac{-2}{4} = -\frac{1}{2}$$

$$p - 6 = -\frac{1}{2}(d - 2)$$

$$p - 6 = -\frac{1}{2}d + 1$$

$$2p - 12 = -d + 2$$

$$d + 2p = 14$$

b.
$$d + 2(2) = 14$$
$$d + 4 = 14$$
$$d = 10$$

43. **a.** Use ordered pairs $(50, 35)$ and $(20, 20)$.

$$m = \frac{20 - 35}{20 - 50} = \frac{-15}{-30} = \frac{1}{2}$$

$$p - 35 = \frac{1}{2}(s - 50)$$

$$p - 35 = \frac{1}{2}s - 25$$

$$2p - 70 = s - 50$$

$$-s + 2p = 20$$

$$s - 2p = -20$$

b.
$$0 - 2p = -20$$
$$-2p = -20$$
$$p = \$10$$

c.
$$s - 2(40) = -20$$
$$s - 80 = -20$$
$$s = 60$$

45.
$$3p - 6 = -2p + 14$$
$$5p - 6 = 14$$
$$5p = 20$$
$$p = \$4$$

47. For $x = 0$, $2x - 4y = 8$ becomes
$$2(0) - 4y = 8$$
$$-4y = 8$$
$$y = -2$$
The y-intercept is $(0, -2)$.
For $y = 0$, $2x - 4y = 8$ becomes
$$2x - 4(0) = 8$$
$$2x = 8$$
$$x = 4$$
The x-intercept is $(4, 0)$.

49. For $x = 0$, $y = 2x + 6$ becomes
$$y = 2(0) + 6$$
$$y = 6$$
The y-intercept is $(0, 6)$.
For $y = 0$, $y = 2x + 6$ becomes
$$0 = 2x + 6$$
$$-2x = 6$$
$$x = -3$$
The x-intercept is $(-3, 0)$.

51.
$$3(x - 1) = 6x + 3$$
$$3x - 3 = 6x + 3$$
$$-3x - 3 = 3$$
$$-3x = 6$$
$$x = -2$$

53. $y = 2x + 50$

55.   **a.**   The productions cost of each unit is $2.   **b.**   The fixed cost is $75.

57.   Sample answer. Plot the point and use the slope to determine the rise and run from the given point to find another point on the line. Draw a line through the points.

59.   Sample answer. Plot the x- and y-intercepts. Draw a line through the points.

61. $8x + 4y = 16$
    $\quad 4y = -8x + 16$
    $\qquad y = -2x + 4$
    $\qquad m = -2 \quad b = 4$

63. $m = 3 \quad b = 2$
    $y = 3x + 2$

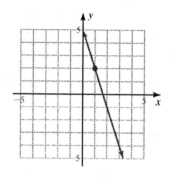

65. $y - 2 = -3(x - 1)$
    $y - 2 = -3x + 3$
    $\qquad y = -3x + 5$

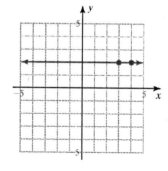

67. $m = \dfrac{2 - 2}{4 - 3} = \dfrac{0}{1} = 0$
    $y - 2 = 0(x - 3)$
    $y - 2 = 0$
    $\qquad y = 2$

3.5   Linear Inequalities in Two Variables

Problems 3.5

1. $3x - 2y < -6$

When $x = 0$, $-2y = -6$ and $y = 3$

When $y = 0$, $3x = -6$ and $x = -2$

The boundary is a dashed line.

Test point $(0,0)$: $3(0) - 2(0) < -6$, $0 < -6$

False, so shade the opposite side of the line.

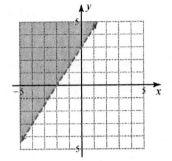

2. $y - -x - 3$

When $x = 0$, $y = -3$

When $y = 0$, $0 = -x - 3$ and $x = -3$

The boundary is a solid line.

Test point $(0,0)$: $0 - -0 - 3$, $0 - -3$

True, so shade this side of the line.

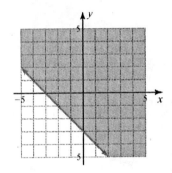

3. $x < 2$

Graph the boundary line $x = 2$.

The boundary is a dashed line.

Shade to the left of the line.

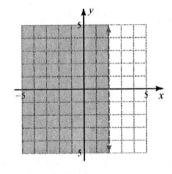

4. $|x \le 3| \le 4$

$\le 4 \le x \le 3 \le 4$

$\le 1 \le x \le 7$

Graph the boundary lines $x = \le 1$ and $x = 7$.

The boundaries are solid lines.

Shade between the two lines.

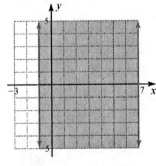

5. $|y| - 1$

$y - -1$ or $y - 1$

Graph the boundary lines $y = -1$ and $y = 1$.

The boundaries are solid lines.

Shade outside the two lines.

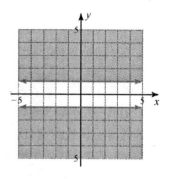

Exercises 3.5

1. $x + 2y > 4$

When $x = 0$, $2y = 4$ and $y = 2$

When $y = 0$, $x = 4$

The boundary is a dashed line.

Test point $(0,0)$: $0 + 2(0) > 4$, $0 > 4$

False, so shade the opposite side of the line.

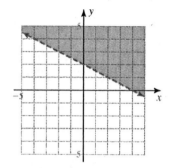

3. $\leq 2x \leq 5y \leq \leq 10$

When $x = 0$, $\leq 5y = \leq 10$ and $y = 2$

When $y = 0$, $\leq 2x = \leq 10$ and $x = 5$

The boundary is a solid line.

Test point $(0,0)$: $\leq 2(0) \leq 5(0) \leq \leq 10$, $0 \leq \leq 10$

False, so shade the opposite side of the line.

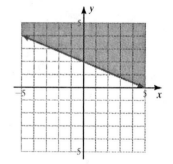

5. $y - 2x - 2$

When $x = 0$, $y = -2$

When $y = 0$, $0 = 2x - 2$, $2 = 2x$ and $x = 1$

The boundary is a solid line.

Test point $(0,0)$: $0 - 2(0) - 2$, $0 - -2$

True, so shade this side of the line.

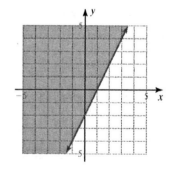

7. $6 < 3x - 2y$

When $x = 0$, $6 = -2y$ and $y = -3$

When $y = 0$, $6 = 3x$ and $x = 2$

The boundary is a dashed line.

Test point $(0,0)$: $6 < 3(0) - 2(0)$, $6 < 0$

False, so shade the opposite side of the line.

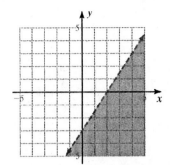

9. $4x + 3y - 12$

When $x = 0$, $3y = 12$ and $y = 4$

When $y = 0$, $4x = 12$ and $x = 3$

The boundary is a solid line.

Test point $(0,0)$: $4(0) - 3(0) - 12$, $0 - 12$

False, so shade the opposite side of the line.

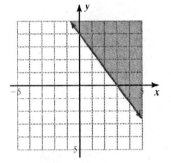

11. $10 < -5x + 2y$

When $x = 0$, $10 = 2y$ and $y = 5$

When $y = 0$, $10 = -5x$ and $x = -2$

The boundary is a dashed line.

Test point $(0,0)$: $10 < -5(0) + 2(0)$, $10 < 0$

False, so shade the opposite side of the line.

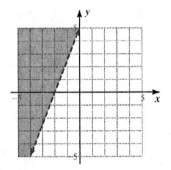

13. $2x - 2y - 4$

When $x = 0$, $0 = 2y - 4$, $-2y = -4$ and $y = 2$

When $y = 0$, $2x = -4$ and $x = -2$

The boundary is a solid line.

Test point $(0,0)$: $2(0) - 2(0) - 4$, $0 - -4$

True, so shade this side of the line.

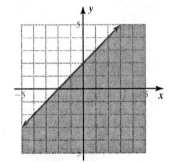

15. $2y < -4x + 8$

When $x = 0$, $2y = 8$ and $y = 4$

When $y = 0$, $0 = -4x + 8$, $4x = 8$ and $x = 2$

The boundary is a dashed line.

Test point $(0,0)$: $2(0) < -4(0) + 8$, $0 < 8$

True, so shade this side of the line.

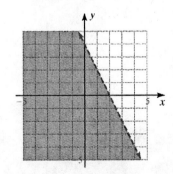

17. $x \ge -3$

Graph the border line $x = -3$.

The boundary is a solid line.

Shade to the right of the line.

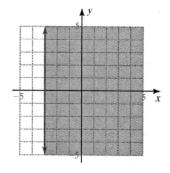

19. $y < 3$

Graph the border line $y = 3$.

The boundary is a dashed line.

Shade below the line.

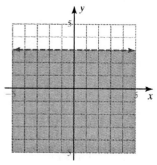

21. $|x| < 1$

$-1 < x < 1$

Graph the boundary lines $x = -1$ and $x = 1$.

The boundaries are dashed lines.

Shade between the two lines.

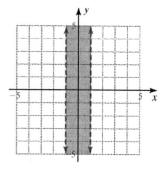

23. $|y| < 4$

$-4 < y < 4$

Graph the boundary lines $y = -4$ and $y = 4$.

The boundaries are dashed lines.

Shade between the two lines.

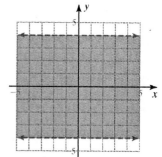

25. $|x| \ge 1$

$x \le -1$ or $x \ge 1$

Graph the boundary lines $x = -1$ and $x = 1$.

The boundaries are solid lines.

Shade outside the two lines.

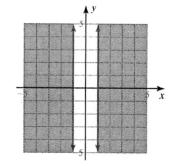

27. $|y| - 2$

$y - -2$ or $y - 2$

Graph the boundary lines $y = -2$ and $y = 2$.

The boundaries are solid lines.

Shade outside the two lines.

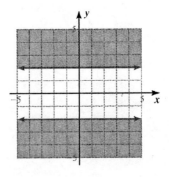

29. $|x + 2| < 1$

$-1 < x + 2 < 1$

$-3 < x < -1$

Graph the boundary lines $x = -3$ and $x = -1$.

The boundaries are dashed lines.

Shade between the two lines.

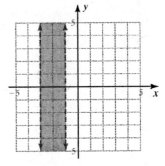

31. $|y + 2| < 1$

$-1 < y + 2 < 1$

$-3 < y < -1$

Graph the boundary lines $y = -3$ and $y = -1$.

The boundaries are dashed lines.

Shade between the two lines.

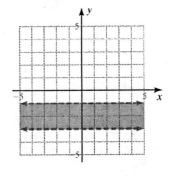

33. $|x + 1| - 3$

$x + 1 - -3$ or $x + 1 - 3$

$x - -4$ or $x - 2$

Graph the boundary lines $x = -4$ and $x = 2$.

The boundaries are solid lines.

Shade outside the two lines.

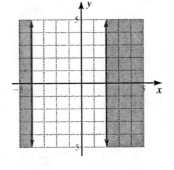

35. 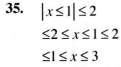 ≤ 2

$\leq 2 \leq x \leq 1 \leq 2$

$\leq 1 \leq x \leq 3$

Graph the boundary lines $x = \leq 1$ and $x = 3$.

The boundaries are solid lines.

Shade between the two lines.

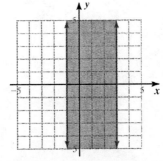

37. $|y-2|<1$

$-1<y-2<1$

$1<y<3$

Graph the boundary lines $y=1$ and $y=3$.

The boundaries are dashed lines.

Shade between the two lines.

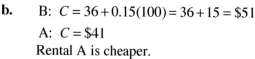

39. $56 = k \cdot 14$

$4 = k$

41. $60 = \dfrac{k}{3}$

$180 = k$

43. **a.** $C = 36 + 0.15m$

$41 = 36 + 0.15m$

$5 = 0.15m$

$\dfrac{5}{0.15} = m$

$m \approx 33$

b. B: $C = 36 + 0.15(100) = 36 + 15 = \$51$

A: $C = \$41$

Rental A is cheaper.

45. If you plan to drive more than 33 miles, Rental A will be less expensive.

47. Sample answer. The solution includes all points to the right of the vertical line $x = k$ and the line $x = k$.

49. Sample answer. A solid line is used as the boundary when the inequality is less than or equal or greater than or equal. The points on the boundary are included in the solution of the inequality.

51. $|x-1|<4$

$-4<x-1<4$

$-3<x<5$

Graph the boundary lines $x = -3$ and $x = 5$.

The boundaries are dashed lines.

Shade between the two lines.

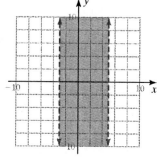

53. $|y|>3$

$y<-3$ or $y>3$

Graph the boundary lines $y = -3$ and $y = 3$.

The boundaries are dashed lines.

Shade outside the two lines.

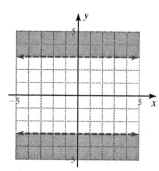

55. $x < 3$

Graph the border line $x = 3$.

The boundary is a dashed line.

Shade to the left of the line.

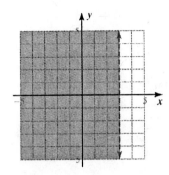

57. $\geq 2x \geq 3y \geq \geq 6$

When $x = 0$, $\geq 3y = \geq 6$ and $y = 2$

When $y = 0$, $\geq 2x = \geq 6$ and $x = 3$

The boundary is a solid line.

Test point $(0,0)$: $\geq 2(0) \geq 3(0) \geq \geq 6$, $0 \geq \geq 6$

True, so shade this side of the line.

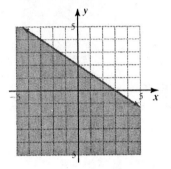

Review Exercises

1.   **a.**

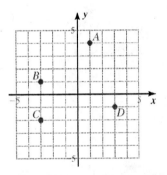

b.

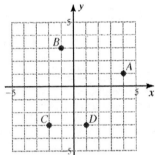

2.   A $(2, 0)$; B $(-1, 1)$; C $(-3, -3)$; D $(4, -4)$

3.   A $(2, 1)$; B $(0, 3)$; C $(-3, -1)$; D $(3, -3)$

4.   **a.** $x + 2y = 4$

When $x = 0$, $2y = 4$ and $y = 2$

When $y = 0$, $x = 4$

Graph $(0, 2)$ and $(4, 0)$

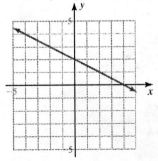

b. $2x - y = 2$

When $x = 0$, $-y = 2$ and $y = -2$

When $y = 0$, $2x = 2$ and $x = 1$

Graph $(0, -2)$ and $(1, 0)$

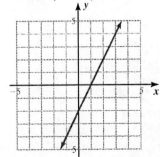

5. **a.** $y = 3x + 3$

If $x = 0$, $y = 3$

If $y = 0$, $0 = 3x + 3$, $-3x = 3$ and $x = -1$

y-intercept $(0, 3)$ and x-intercept $(-1, 0)$

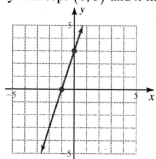

b. $y = 2x - 4$

If $x = 0$, $y = -4$

If $y = 0$, $0 = 2x - 4$, $-2x = -4$ and $x = 2$

y-intercept $(0, -4)$ and x-intercept $(2, 0)$

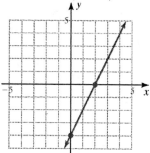

6. **a.** $2x = 6 \rightarrow x = 3$

vertical line: $x = 3$

b. $3y = 6 \rightarrow y = 2$

horizontal line: $y = 2$

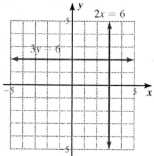

7. **a.** $2x = -6 \rightarrow x = -3$

vertical line: $x = -3$

b. $3y = -6 \rightarrow y = -2$

horizontal line: $y = -2$

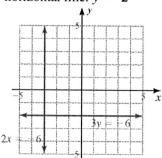

8. **a.** $D: \{0, 2, 3, 5\}$ $R: \{5, 8, 9, 10\}$

b. $D: \{0, 2, 3, 5\}$ $R: \{6, 9, 10, 11\}$

9. **a.** $D = \{x \mid x \text{ is a real number}\}$

$R = \{y \mid y \text{ is a real number}\}$

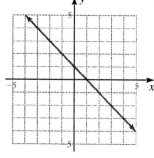

b. $D = \{x \mid x \text{ is a real number}\}$

$R = \{y \mid y \text{ is a real number}\}$

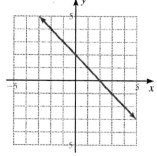

10. **a.** $D = \{x \mid \le 2 \le x \le 2\}$ $R = \{y \mid \le 2 \le y \le 2\}$

b. $D = \{x \mid \le 3 \le x \le 3\}$ $R = \{y \mid \le 3 \le y \le 3\}$

11. a. $D = \{x \mid \le 2 \le x \le 2\}$ $R = \{y \mid 0 \le y \le 2\}$ **b.** $D = \{x \mid \le 3 \le x \le 3\}$ $R = \{y \mid 0 \le y \le 3\}$

12. a. $D = \{x \mid \le 2 \le x \le 2\}$ $R = \{y \mid \le 2 \le y \le 0\}$ **b.** $D = \{x \mid \le 3 \le x \le 3\}$ $R = \{y \mid \le 3 \le y \le 0\}$

13. a. The relation is a function because no vertical line crosses the graph more than once.
   b. The relation is not a function because a vertical line can be drawn that crosses the graph more than once.

14. a. Exclude those values of x that make the denominator zero.
   $x \ne 1 = 0$
   $x = 1$
   $D = \{x \mid x \text{ is a real number and } x \ne 1\}$
   b. Exclude those values of x that make the denominator zero.
   $x \ne 2 = 0$
   $x = 2$
   $D = \{x \mid x \text{ is a real number and } x \ne 2\}$

15. a. Find when the radicand is positive or zero.
   $x \ge 3 \ge 0$
   $x \ge 3$
   $D = \{x \mid x \ge 3\}$
   b. Find when the radicand is positive or zero.
   $x \ge 4 \ge 0$
   $x \ge 4$
   $D = \{x \mid x \ge 4\}$

16. a. $f(2) = 2 - 4 = -2$ **b.** $f(1) = 1 - 4 = -3$ **c.** $f(2) - f(1) = -2 - (-3) = 1$

17. a. $f(2) = 2^2 - 3 = 4 - 3 = 1$ **b.** $f(1) = 1^2 - 3 = 1 - 3 = -2$ **c.** $f(2) - f(1) = 1 - (-2) = 3$

18. a. $f(2) = 0$ **b.** $f(1) = -1$ **c.** $f(2) - f(1) = 0 - (-1) = 1$

19. a. $f(2) = 1$ **b.** $f(1) = 0$ **c.** $f(2) - f(1) = 1 - 0 = 1$

20. a. $m = \dfrac{0-2}{1-(-3)} = \dfrac{-2}{4} = -\dfrac{1}{2}$ **b.** $m = \dfrac{-7-(-2)}{4-4} = \dfrac{-5}{0}$ is undefined

21. a. $m = \dfrac{3-4}{4-3} = \dfrac{-1}{1} = -1$ **b.** $m = \dfrac{5-(-1)}{-3-1} = \dfrac{6}{-4} = -\dfrac{3}{2}$

22. a. $m = \dfrac{7-3}{-2-1} = \dfrac{4}{-3} = -\dfrac{4}{3}$ Perpendicular, since the slope of AB is $\frac{-4}{3}$. **b.** $m = \dfrac{6-3}{5-1} = \dfrac{3}{4}$ Parallel, since the slope of AB is $\frac{3}{4}$.

23. a. $m = \dfrac{1-(-1)}{1-2} = \dfrac{2}{-1} = -2$ Parallel, since the slope of AB is -2. **b.** $m = \dfrac{-3-(-1)}{2-3} = \dfrac{-2}{-1} = 2$ The lines are neither parallel nor perpendicular.

24. **a.**
$$m = \frac{y-4}{5-2} = \frac{y-4}{3}$$
The slope of the perpendicular line is $-\frac{2}{3}$.
Solving for y:
$$\frac{y-4}{3} = -\frac{2}{3}$$
$$y - 4 = -2$$
$$y = 2$$

b.
$$m = \frac{y-4}{5-2} = \frac{y-4}{3}$$
The slope of the perpendicular line is $\frac{1}{2}$.
Solving for y:
$$\frac{y \cdot 4}{3} = \frac{1}{2}$$
$$2(y \cdot 4) = 3 \cdot 1$$
$$2y \cdot 8 = 3$$
$$2y = 11$$
$$y = \frac{11}{2} \text{ or } 5\frac{1}{2}$$

25. **a.**
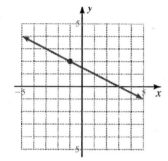

b.
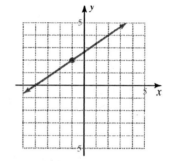

26. **a.** $6x - y = 2$
$$-y = -6x + 2$$
$$y = 6x - 2$$
$$m = 6 \quad \text{y-intercept} = -2$$

b. $4y = 2x - 8$
$$y = \frac{1}{2}x - 2$$
$$m = \frac{1}{2} \quad \text{y-intercept} = -2$$

27. **a.**
$$m = \frac{2-5}{-1-2} = \frac{-3}{-3} = 1$$
$$y - 5 = 1(x - 2)$$
$$y - 5 = x - 2$$
$$-x + y = 3$$
$$x - y = -3$$

b.
$$m = \frac{-2-3}{-2-(-4)} = \frac{-5}{2} = -\frac{5}{2}$$
$$y - 3 = -\frac{5}{2}\left(x - (-4)\right)$$
$$2(y - 3) = -5(x + 4)$$
$$2y - 6 = -5x - 20$$
$$5x + 2y = -14$$

28. **a.**
$$y - (-3) = 2\left(x - (-1)\right)$$
$$y + 3 = 2x + 2$$
$$-2x + y = -1$$
$$2x - y = 1$$

b.
$$y - 0 = 2(x - 5)$$
$$y = 2x - 10$$
$$-2x + y = -10$$
$$2x - y = 10$$

29. **a.** $y = 3x + 2$

b. $y = -3x + 4$

30. **a.** $2x + y = 7$

$y = -2x + 7$

The slope of the given line is –2. The parallel line has slope –2.

$y - 1 = -2(x - 2)$

$y - 1 = -2x + 4$

$y = -2x + 5$

b. $3x - y = 4$

$-y = -3x + 4$

$y = 3x - 4$

The slope of the given line is 3. The parallel line has slope 3.

$y - 1 = 3(x - 2)$

$y - 1 = 3x - 6$

$y = 3x - 5$

31. **a.** $2x + 3y = 7$

$3y = -2x + 7$

$y = -\dfrac{2}{3}x + \dfrac{7}{3}$

The slope of the given line is $-\frac{2}{3}$. The perpendicular line has slope $\frac{3}{2}$.

$y - 1 = \dfrac{3}{2}(x - 2)$

$2(y - 1) = 3(x - 2)$

$2y - 2 = 3x - 6$

$2y = 3x - 4$

$y = \dfrac{3}{2}x - 2$

b. $3x - 2y = 4$

$-2y = -3x + 4$

$y = \dfrac{3}{2}x - 2$

The slope of the given line is $\frac{3}{2}$. The perpendicular line has slope $-\frac{2}{3}$.

$y - 1 = -\dfrac{2}{3}(x - 2)$

$3(y - 1) = -2(x - 2)$

$3y - 3 = -2x + 4$

$3y = -2x + 7$

$y = -\dfrac{2}{3}x + \dfrac{7}{3}$

32. **a.** Since the slope of the line is 0, the equation has the form $y = k$.

$y = 7$

b. Since the slope of the line is undefined, the equation has the form $x = h$.

$x = -3$

33.

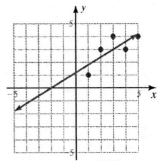

Answers for the equation may vary.

$m = \dfrac{4 - 1}{5 - 1} = \dfrac{3}{4} = 0.75$

$y - 1 = 0.75(x - 1)$

$y - 1 = 0.75x - 0.75$

$y = 0.75x + 0.25$

34. y-intercept: (0, 1)

rise = –1; run = 3

slope: $m = -\dfrac{1}{3}$

$y = -\dfrac{1}{3}x + 1$

35. **a.** $2x - y < 1$

When $x = 0$, $-y = 1$ and $y = -1$

When $y = 0$, $2x = 1$ and $x = \dfrac{1}{2}$

The boundary is a dashed line.

Test point $(0,0)$: $2(0) - 0 < 1$, $0 < 1$

True, so shade this side of the line.

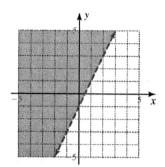

b. $y - x - 3$

When $x = 0$, $y = -3$

When $y = 0$, $0 = x - 3$, $-x = -3$ and $x = 3$

The boundary is a solid line.

Test point $(0,0)$: $0 - 0 - 3$, $0 - -3$

True, so shade this side of the line.

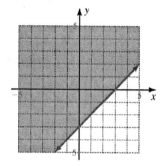

c. $x \geq 4$

Graph the border line $x = 4$.

The boundary is a solid line.

Shade to the right of the line.

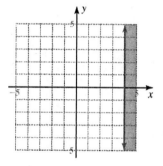

d. $2y < -6$

$y < -3$

Graph the border line $y = -3$.

The boundary is a dashed line.

Shade below the line.

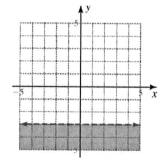

36. **a.** $|y| > 2$

$y < -2$ or $y > 2$

Graph the boundary lines $y = -2$ and $y = 2$.

The boundaries are dashed lines.

Shade outside the two lines.

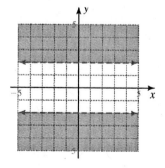

b. $|x+2|-3$

$-3-x+2-3$

$-5-x-1$

Graph the boundary lines $x=-5$ and $x=1$.

The boundaries are solid lines.

Shade between the two lines.

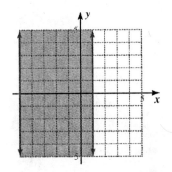

Cumulative Review Chapters 1-3

1. $\sqrt{13}$ is an irrational number and a real number.

2.

3. $\cdot\dfrac{1}{6}\div\cdot\dfrac{7}{12}\cdot=\cdot\dfrac{1}{\cancel{6}}\cdot\cdot\dfrac{\cancel{6}\cdot2\cdot}{7}\cdot=\dfrac{2}{7}$

4. $\left(2x^4y^{-4}\right)^3=2^3\left(x^4\right)^3\left(y^{-4}\right)^3=8x^{12}y^{-12}=\dfrac{8x^{12}}{y^{12}}$

5. $-7^3+\dfrac{14-10}{2}+35\div7=-7^3+\dfrac{4}{2}+35\div7$

$=-343+\dfrac{4}{2}+35\div7=-343+2+5=-336$

6. $\dfrac{3}{7}y\cdot\,4=2$

$\dfrac{3}{7}y=6$

$\dfrac{7}{3}\cdot\dfrac{3}{7}y=\dfrac{7}{3}\cdot6$

$y=14$

7. $|x-5|=|x-9|$

$x-5=x-9$ or $x-5=-(x-9)$

$-5=-9$ or $x-5=-x+9$

Contradiction $2x=14$

$x=7$

8. $x\geq1$

9. $\{x|x>-4$ and $x<4\}$

10. $H=2.45h+72.98;\quad H=136.68$

$136.68=2.45h+72.98$

$63.7=2.45h$

$26=h$

106

11.

$x = $ first odd number

$x + 2 = $ second consecutive odd number

$x + 4 = $ third consecutive odd number

$x + x + 2 + x + 4 = 75$

$$3x + 6 = 75$$
$$3x = 69$$
$$x = 23$$
$$x + 2 = 25$$
$$x + 4 = 27$$

The three consecutive odd numbers are 23, 25, and 27.

12.

	Rate	Time	Distance
Freight	30	$T+2$	$30(T+2)$
Passenger	40	T	$40T$

Distance of freight = Distance of passenger

$$30(T + 2) = 40T$$
$$30T + 60 = 40T$$
$$60 = 10T$$
$$6\,\text{hrs} = T$$

Since $D = RT = 40(6) = 240\,\text{miles}$

The passenger train overtakes the freight train 240 miles from the station.

13. $\{-3, -2, -4\}$

14. $D = \{x \mid x \text{ is a real number}\}$

$R = \{y \mid y \text{ is a real number}\}$

15. Find when the radicand is positive or zero.

$x + 25 - 0$

$x - -25$

$D = \{x \mid x - -25\}$

16. $f(-3) - f(-2) = 2 - 1 = 1$

17.

$$L(24) = -\frac{2}{3}(24) + 150 = -16 + 150 = 134$$

18. $x - y = 2$

When $x = 0$, $-y = 2$ and $y = -2$

When $y = 0$, $x = 2$

Graph $(0, -2)$ and $(2, 0)$

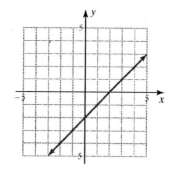

19. $y = |5x + 2$

If $x = 0$, $y = 2$

If $y = 0$, $0 = |5x + 2$, $5x = 2$ and $x = \dfrac{2}{5}$

y-intercept $(0, 2)$ and x-intercept $\left| \dfrac{2}{5}, 0 \right|$

20. $2x = -2 \rightarrow x = -1$

vertical line: $x = -1$

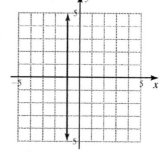

21.

$$m = \frac{3-6}{7-4} = \frac{-3}{3} = -1$$

Perpendicular, since the slope of the line is -1.

22.

$$y - 1 = -2\big(x - (-2)\big)$$
$$y - 1 = -2x - 4$$
$$2x + y = -3$$

23.

$$6x - 3y = 36$$
$$-3y = -6x + 36$$
$$y = 2x - 12$$
$$m = 2 \quad \text{y-intercept} = -12$$

24.

$$4x - 3y = -1$$
$$-3y = -4x - 1$$
$$y = \frac{4}{3}x + \frac{1}{3}$$

The slope of the given line is $\frac{4}{3}$. The parallel line has slope $\frac{4}{3}$.

$$y - 6 = \frac{4}{3}\big(x - (-2)\big)$$
$$3(y - 6) = 4(x + 2)$$
$$3y - 18 = 4x + 8$$
$$-4x + 3y = 26$$
$$4x - 3y = -26$$

25.

$$y - 6x - 6$$

When $x = 0$, $y = -6$

When $y = 0$, $0 = 6x - 6$, $-6x = -6$ and $x = 1$

The boundary is a solid line.

Test point $(0, 0)$: $0 - 6(0) - 6$, $0 - -6$

False, so shade the opposite side of the line.

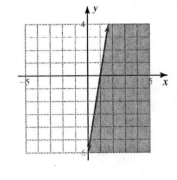

Chapter 4
Solving Systems of Linear Equations and Inequalities

4.1　Systems with Two Variables

Problems 4.1

1.　Graph each equation.
$$x - y = -1$$
$$y = -x - 1$$

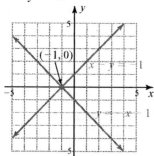

The solution is (–1, 0).

2.　Graph each equation.
$$y - 3x = 3$$
$$2y - 6x = 12$$

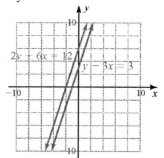

Since the lines are parallel, there is no solution. The system is inconsistent.

3.　Graph each equation.
$$x + \frac{1}{2}y = -2$$
$$y = -2x - 4$$

Rewriting the first equation:
$$x + \frac{1}{2}y = -2$$
$$2x + y = -4$$
$$y = -2x - 4$$

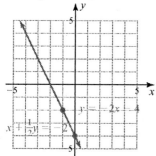

Since the two equations are equivalent, there are infinitely many solutions. The system is dependent.

4.　Solve the second equation for x, substitute into the first equation, and solve:
$$x + 2y = 3$$
$$x - 3y = 0 \quad | \quad x = 3y$$
$$3y + 2y = 3$$
$$5y = 3$$
$$y = \frac{3}{5}$$
$$x = 3\left(\frac{3}{5}\right) = \frac{9}{5}$$

Solution: $\left(\frac{9}{5}, \frac{3}{5}\right)$

5. $x - y = 2$

$y = x$

Substitute the second equation into the first and solve:

$x - x = 2$

$0 = 2$

This is a contradiction. There is no solution and the system is inconsistent. The solution set is $\varnothing$.

6.   Solve the first equation for x and substitute into the second equation:

$x - 3y = 4 \rightarrow x = 3y + 4$

$2x = 8 + 6y$

$2(3y + 4) = 8 + 6y$

$6y + 8 = 8 + 6y$

This is an identity so there are infinitely many solutions. The system is dependent. The solution is: $\left\{ (x, y) \mid y = \frac{1}{3}x - \frac{4}{3} \right\}$.

7.   To eliminate y, multiply both sides of the first equation by 10, both sides of the second equation by -3, add the results, and solve:

$x + 0.2y = 4 \quad - \quad 10x + 2y = 40$

$4x + \dfrac{2}{3}y = 2 \quad - \quad -12x - 2y = -6$

$-2x = 34$

$x = -17$

$-17 + 0.2y = 4$

$0.2y = 21$

$y = 105$

Solution: $(-17, 105)$

8.   To eliminate x, multiply both sides of the first equation by 5, both sides of the second equation by 2, add the results, and solve:

$-2x + 6y = 1 \rightarrow -10x + 30y = 5$

$5x - 15y = 2 \rightarrow 10x - 30y = 4$

$0 = 9$

This is a contradiction. There is no solution and the system is inconsistent. The solution set is $\varnothing$.

9.   To eliminate y, multiply both sides of the first equation by 60, both sides of the second equation by -15, add the results, and solve:

$\dfrac{x}{20} + \dfrac{y}{12} = \dfrac{1}{2} \quad - \quad 3x + 5y = 30$

$\dfrac{x}{5} + \dfrac{y}{3} = 2 \quad - \quad -3x - 5y = -30$

$0 = 0$

This is an identity so there are infinitely many solutions. The system is dependent. The solution is: $\left\{ (x, y) \mid y = -\frac{3}{5}x + 6 \right\}$.

10.   To find the equilibrium point, equate the supply and demand functions:

$S(p) = 0.4p + 3$

$D(p) = -0.2p + 27$

$0.4p + 3 = -0.2p + 27$

$0.6p + 3 = 27$

$0.6p = 24$

$p = 40$

$S(40) = 0.4(40) + 3 = 16 + 3 = 19$ items

The equilibrium point occurs when the price is $40 and the number of items is 19.

11.   Let $x =$ the first angle

Let $y =$ the second angle

$x + y = 180 \qquad$ angles are supplementary

$y = x - 40 \qquad$ one angle 40° less than the other

Substitute the second equation into the first equation and solve:

$x + x - 40 = 180$

$2x - 40 = 180$

$2x = 220$

$x = 110°$

$y = 110 - 40 = 70°$

The angles are 110° and 70°.

Exercises 4.1

1. Graph each equation.

$x - 2y = 6$

$y = 2x$

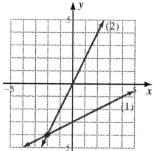

The solution is $(-2, -4)$. The system is consistent.

3. Graph each equation.

$y = x - 3$

$y = 2x - 4$

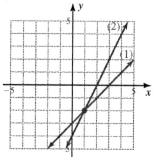

The solution is $(1, -2)$. The system is consistent.

5. Graph each equation.

$2y = -x + 4$

$y = -2x + 4$

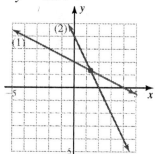

The solution is $\left(\frac{4}{3}, \frac{4}{3}\right)$ or $\left(1\frac{1}{3}, 1\frac{1}{3}\right)$. The system is consistent.

7. Graph each equation.

$2x - y = -2$

$y = 2x + 4$

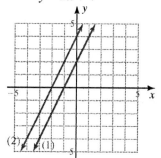

Since the lines are parallel, there is no solution. The system is inconsistent.

9. Graph each equation.

$3x + 4y = 12$

$8y = 24 - 6x$

Rewriting the first equation:

$3x + 4y = 12 - \ 4y = 12 - 3x - \ 8y = 24 - 6x$

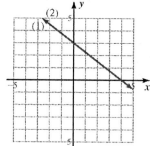

Since the two equations are equivalent, there are infinitely many solutions. The system is dependent.

11. $y = 2x - 4$

$-2x = y - 4$

Substitute the first equation into the second and solve:

$-2x = 2x - 4 - 4$

$-4x = -8$

$x = 2$

$y = 2(2) - 4 = 4 - 4 = 0$

Solution: $(2, 0)$

The system is consistent.

13. Solve the first equation for y, substitute into the second equation, and solve:

$$x + y = \frac{5}{2} \quad | \quad y = \frac{5}{2} \mid x$$

$$3x + y = \frac{9}{2}$$

$$3x + \frac{5}{2} \mid x = \frac{9}{2}$$

$$2x = 2$$

$$x = 1$$

$$y = \frac{5}{2} \mid 1 = \frac{3}{2}$$

Solution: $\left| 1, \frac{3}{2} \right|$

The system is consistent.

15. $y - 4 = 2x$

$$y = 2x + 2$$

Substitute the second equation into the first and solve:

$$2x + 2 - 4 = 2x$$

$$-2 = 0$$

This is a contradiction. There is no solution and the system is inconsistent. The solution set is $\varnothing$.

17. $x = 8 - 2y$

$$x + 2y = 4$$

Substitute the first equation into the second and solve:

$$8 - 2y + 2y = 4$$

$$8 = 4$$

This is a contradiction. There is no solution and the system is inconsistent. The solution set is $\varnothing$.

19. $x + 2y = 4$

$$x = -2y + 4$$

Substitute the second equation into the first and solve:

$$-2y + 4 + 2y = 4$$

$$4 = 4$$

This is an identity so there are infinitely many solutions. The system is dependent.

21. $x = 2y + 1$

$$y = 2x + 1$$

Substitute the first equation into the second and solve:

$$y = 2(2y + 1) + 1$$

$$y = 4y + 2 + 1$$

$$-3y = 3$$

$$y = -1$$

$$x = 2(-1) + 1 = -2 + 1 = -1$$

Solution: $(-1, -1)$

The system is consistent.

23. Solve the first equation for y and substitute into the second equation:

$$2x - y = -4 \;\rightarrow\; -y = -2x - 4 \;\rightarrow\; y = 2x + 4$$

$$4x = 4 + 2y$$

$$4x = 4 + 2(2x + 4)$$

$$4x = 4 + 4x + 8$$

$$0 = 12$$

This is a contradiction. There is no solution and the system is inconsistent.

25.
$x = 5 - y$
$0 = x - 4y$
Substitute the first equation into the second and solve:
$0 = 5 - y - 4y$
$-5 = -5y$
$1 = y$
$x = 5 - 1 = 4$
Solution: (4, 1). The system is consistent.

27. To eliminate y, add the two equations, and solve:
$x + y = 8$
$x - y = 2$
$2x = 10$
$x = 5$
$5 + y = 8 \rightarrow y = 3$
Solution: (5, 3). The system is consistent.

29. To eliminate y, add the two equations, and solve:
$x + 4y = 2$
$x - 4y = -2$
$2x = 0$
$x = 0$
$0 + 4y = 2 \rightarrow y = \dfrac{1}{2}$
Solution: $\left(0, \frac{1}{2}\right)$. The system is consistent.

31. To eliminate x, add the two equations, and solve:
$-x - 2y = -2$
$x - 2y = -2$
$-4y = -4$
$y = 1$
$x - 2(1) = -2 \rightarrow x - 2 = -2 \rightarrow x = 0$
Solution: (0, 1). The system is consistent.

33. To eliminate y, multiply the first equation by 2, add the result to the second equation, and solve:
$2x + y = 7 - \quad 4x + 2y = 14$
$3x - 2y = 0 - \quad 3x - 2y = \ 0$
$7x = 14$
$x = 2$
$2(2) + y = 7 \rightarrow 4 + y = 7 \rightarrow y = 3$
Solution: (2, 3). The system is consistent.

35. To eliminate y, multiply the second equation by 2, add the result to the first equation, and solve:
$2x - 2y = 6 \rightarrow 2x - 2y = 6$
$x + y = 2 \rightarrow 2x + 2y = 4$
$4x = 10$
$x = \dfrac{5}{2}$
$\dfrac{5}{2} + y = 2 - \quad y = -\dfrac{1}{2}$
Solution: $\left(\frac{5}{2}, -\frac{1}{2}\right)$. The system is consistent.

37. To eliminate y, multiply the first equation by 2, add the result to the second equation, and solve:
$3x + 5y = 1 - \quad 6x + 10y = 2$
$-6x - 10y = 2 - \quad -6x - 10y = 2$
$0 = 4$
This is a contradiction. There is no solution and the system is inconsistent.

39. To eliminate y, add the two equations, and solve:
$2x + y = 8$
$3x - y = 7$
$5x = 15$
$x = 3$
$2(3) + y = 8 \rightarrow 6 + y = 8 \rightarrow y = 2$
Solution: (3, 2). The system is consistent.

41. To eliminate y, multiply the first equation by -2, add the result to the second equation, and solve:

$$2x - 5y = \ 9 \ \rightarrow \ -4x + 10y = -18$$
$$4x - 10y = 18 \ \rightarrow \ \ \ 4x - 10y = \ 18$$
$$0 = 0$$

This is an identity so there are infinitely many solutions. The system is dependent.

43. To eliminate y, multiply both sides of the first equation by 4, both sides of the second equation by 5, add the results, and solve:

$$6x + 5y = 12 \ - \ 24x + 20y = \ \ 48$$
$$9x - 4y = -5 \ - \ 45x - 20y = -25$$
$$69x = 23$$
$$x = \frac{1}{3}$$
$$6\left|\frac{1}{3}\right| + 5y = 12$$
$$2 + 5y = 12$$
$$5y = 10$$
$$y = 2$$

Solution: $\left(\frac{1}{3}, \ 2\right)$. The system is consistent.

45. To eliminate x, multiply the second equation by -2, add the result to the first equation, and solve:

$$2x - 3y = 16 \ \rightarrow \ \ \ 2x - 3y = \ 16$$
$$x - y = \ 7 \ \rightarrow \ -2x + 2y = -14$$
$$-y = 2$$
$$y = -2$$
$$x - (-2) = 7 \ \rightarrow \ x + 2 = 7 \ \rightarrow \ x = 5$$

Solution: $(5, -2)$. The system is consistent.

47. To eliminate y, multiply both sides of the first equation by 4, both sides of the second equation by -5, add the results, and solve:

$$18x - 15y = 1 \ \rightarrow \ \ \ 72x - 60y = \ \ \ 4$$
$$10x - 12y = 3 \ \rightarrow \ -50x + 60y = -15$$
$$22x = -11$$
$$x = -\frac{1}{2}$$
$$10\left|-\frac{1}{2}\right| - 12y = 3$$
$$-5 - 12y = 3$$
$$-12y = 8$$
$$y = -\frac{2}{3}$$

Solution: $\left(-\frac{1}{2}, -\frac{2}{3}\right)$. The system is consistent.

49. Multiply each equation by its LCD. To eliminate x, multiply both sides of the first equation by -4, add to the second equation, and solve:

$$\frac{x}{3} + \frac{y}{6} = \frac{2}{3} \ - \ 2x + \ y = 4 \ - \ -8x - 4y = -16$$
$$\frac{2}{5}x + \frac{y}{4} = \frac{1}{5} \ - \ 8x + 5y = 4 \ - \ \ \ 8x + 5y = \ \ \ 4$$
$$y = -12$$
$$2x - 12 = 4$$
$$2x = 16$$
$$x = 8$$

Solution: $(8, -12)$. The system is consistent.

51. To eliminate y, multiply both sides of the first equation by 12, both sides of the second equation by 24, add the results, and solve:

$$\frac{5}{6}x + \frac{y}{4} = 7 \ - \ 10x + 3y = 84$$
$$\frac{2}{3}x - \frac{y}{8} = 3 \ - \ 16x - 3y = 72$$
$$26x = 156$$
$$x = 6$$
$$10(6) + 3y = 84$$
$$60 + 3y = 84$$
$$3y = 24$$
$$y = 8$$

Solution: $(6, 8)$. The system is consistent.

53. To eliminate y, multiply both sides of the first equation by 2, both sides of the second equation by 3, add the results, and solve:

$$\frac{2}{x}+\frac{3}{y}=\frac{1}{2} \rightarrow \frac{4}{x}+\frac{6}{y}=-1$$

$$\frac{3}{x}-\frac{2}{y}=\frac{17}{12} \rightarrow \frac{9}{x}-\frac{6}{y}=\frac{17}{4}$$

$$\frac{13}{x}=\frac{13}{4}$$

$$4x\cdot\frac{13}{x}=4x\cdot\frac{13}{4}$$

$$52=13x$$

$$4=x$$

$$\frac{2}{4}+\frac{3}{y}=-\frac{1}{2}$$

$$\frac{3}{y}=-1$$

$$3=-y$$

$$y=-3$$

Solution: $(4,-3)$. The system is consistent.

55. To eliminate y, multiply the first equation by 5, add the result to the second equation, and solve:

$$\frac{2}{x}-\frac{1}{y}=0 \rightarrow \frac{10}{x}-\frac{5}{y}=0$$

$$\frac{3}{x}+\frac{5}{y}=\frac{13}{4} \rightarrow \frac{3}{x}+\frac{5}{y}=\frac{13}{4}$$

$$\frac{13}{x}=\frac{13}{4}$$

$$4x\cdot\frac{13}{x}=4x\cdot\frac{13}{4}$$

$$52=13x$$

$$4=x$$

$$\frac{2}{4}-\frac{1}{y}=0$$

$$\frac{2}{4}=\frac{1}{y}$$

$$2y=4$$

$$y=2$$

Solution: $(4,2)$. The system is consistent.

57. Equate the supply and demand functions, and solve:

$$S(p)=D(p)$$
$$20p+200=620-10p$$
$$30p=420$$
$$p=14$$

59. Equate the supply and demand functions, and solve:

$$S(p)=D(p)$$
$$300+p=1500-5p$$
$$6p=1200$$
$$p=200$$

61. Equate the supply and demand functions, and solve:

$$S(p)=D(p)$$
$$2p+170=450-5p$$
$$7p=280$$
$$p=40$$

63. Equate the supply and demand functions, and solve:

$$S(p)=D(p)$$
$$14p+120=960-16p$$
$$30p=840$$
$$p=28$$
$$D(28)=960-16(28)=960-448=512$$

65. Equate the supply and demand functions, and solve:

$$S(p)=D(p)$$
$$25p+60=480-3p$$
$$28p=420$$
$$p=15$$

$$D(15)=480-3(15)=480-45=435$$
$$S(15)=25(15)+60=375+60=435$$

67. Let x = the first angle
Let y = the second angle
$x + y = 90$ angles are complementary
$y = x + 15$ one angle 15° more than other
Substitute the second equation into the first equation and solve:
$x + x + 15 = 90$
$2x + 15 = 90$
$2x = 75$
$x = 37.5°$
$y = 37.5 + 15 = 52.5°$
The angles are 37.5° and 52.5°.

69. Let x = the first angle
Let y = the second angle
$x + y = 180$ angles are supplementary
$y = 4x$ one angle is four times other
Substitute the second equation into the first equation and solve:
$x + 4x = 180$
$5x = 180$
$x = 36°$
$y = 4(36) = 144°$
The angles are 36° and 144°.

71. Let x = the weight of first tusk
Let y = the weight of second tusk
$x + y = 465$ weight of pair of tusks
$x = y + 15$ one is 15 more than other
Substitute the second equation into the first equation and solve:
$y + 15 + y = 465$
$2y + 15 = 465$
$2y = 450$
$y = 225$ lb
$x = 225 + 15 = 240$ lb
The tusks weigh 225 lb and 240 lb.

73. Let x = cigarettes per person in U.S.
Let y = cigarettes per person in Japan
$x + y = 4637$ total cigarettes per person
$y = x + 437$ Japan 437 more than U.S.
Substitute the second equation into the first equation and solve:
$x + x + 437 = 4637$
$2x + 437 = 4637$
$2x = 4200$
$x = 2100$
$y = 2100 + 437 = 2537$
Annual cigarette consumption per person:
U.S.: 2100; Japan: 2537

75. Let x = weight of first ring
Let y = weight of second ring

$\dfrac{x}{7} + \dfrac{y}{11} = 1$ sum of the parts is 1

$\dfrac{6x}{7} = \dfrac{10y}{11}$ remaining parts equal

Multiply each equation by the LCD, solve the first equation for y, substitute into the second equation and solve:

$\dfrac{x}{7} + \dfrac{y}{11} = 1 \;\Big|\; 11x + 7y = 77$

$\dfrac{6x}{7} = \dfrac{10y}{11} \;\Big|\; 66x = 70y$

$7y = 77 \;\Big|\; 11x \;\Big|\; y = 11 \;\Big|\; \dfrac{11}{7}x$

$66x = 70 \;\Big|\; 11 \;\Big|\; \dfrac{11}{7}x \;\Big|$

$66x = 770 \cdot 110x$

$176x = 770$

$x = \dfrac{770}{176} = \dfrac{35}{8}$

$y = 11 \cdot \dfrac{11}{7} \cdot \dfrac{35}{8} = 11 \cdot \dfrac{55}{8} = \dfrac{33}{8}$

The total weight of the rings originally is $\frac{68}{8} = 8.5$ shekels.

77. Let x = age of first cat
Let y = age of second cat
$x + y = 70$ sum of ages
$x - y = 2$ difference of ages
To eliminate y, add the equations and solve:
$2x = 72$
$x = 36$
$36 + y = 70 \rightarrow y = 34$
The ages are 36 and 34 years.

79. Let x = height of building
Let y = height of antenna
$x + y = 1472$ sum of heights
$x - y = 1028$ difference of heights
To eliminate y, add the equations and solve:
$2x = 2500$
$x = 1250$
$1250 + y = 1472 \rightarrow y = 222$
The height of the building is 1250 ft and the height of the antenna is 222 ft.

81. $2x + y = 6$
When $y = 0$, $2x + 0 = 6$ and $x = 3$
When $x = 0$, $2(0) + y = 6$ and $y = 6$
x-intercept: $(3, 0)$; y-intercept: $(0, 6)$

83. $-2x + 3y = 9$
When $y = 0$, $-2x + 3(0) = 9$ and $x = -4.5$
When $x = 0$, $-2(0) + 3y = 9$ and $y = 3$
x-intercept: $(-4.5, 0)$; y-intercept: $(0, 3)$

85. $-2x - 3y = 7$

When $y = 0$, $-2x - 3(0) = 7$ and $x = -\dfrac{7}{2}$

When $x = 0$, $-2(0) - 3y = 7$ and $y = -\dfrac{7}{3}$

x-intercept: $\left(-\frac{7}{2}, 0\right)$; y-intercept: $\left(0, -\frac{7}{3}\right)$

87. a. Sample answer. Using the graphical method, the system is consistent when the lines intersect.
b. Sample answer. Using the substitution method, the system is consistent when the slopes of the lines are different.
c. Sample answer. Using the elimination method, the system is consistent when the slopes of the lines are different.

89. a. Sample answer. Using the graphical method, the system is dependent when the graphs of the lines coincide with each other.
b. Sample answer. Using the substitution method, the system is dependent when the slopes of the lines are the same and the y-intercepts are the same.
c. Sample answer. Using the elimination method, the system is dependent when the slopes of the lines are the same and the y-intercepts are the same.

91. Write each equation in standard form:
$5x = 6 - 4y \rightarrow 5x + 4y = 6$
$3y = 4x - 11 \rightarrow -4x + 3y = -11$
To eliminate x, multiply the first equation by 4, multiply the second equation by 5, add the results, and solve:
$5x + 4y = 6 \quad - \quad 20x + 16y = 24$
$-4x + 3y = -11 \quad - \quad -20x + 15y = -55$
$ 31y = -31$

$y = -1$
$5x = 6 - 4(-1) = 10$
$x = 2$
Solution: $(2, -1)$.

93. To eliminate fractions, multiply the first equation by 9, and multiply the second equation by 6.

$$\frac{x}{3} + \frac{y}{9} = \frac{2}{3} \rightarrow 3x + y = 6$$

$$\frac{x}{2} + \frac{y}{6} = 1 \rightarrow 3x + y = 6$$

Both equations are the same so there are infinitely many solutions.

95. To eliminate y, multiply the second equation by 2, add the result to the first equation, and solve:

$$3x + 2y = 8 \;-\; 3x + 2y = 8$$
$$x - y = 1 \;-\; 2x - 2y = 2$$
$$5x = 10$$
$$x = 2$$
$$2 - y = 1 \rightarrow -y = -1 \rightarrow y = 1$$

Solution: $(2, 1)$.

97. Solve the first equation for x and substitute into the second equation:

$$x - 3y = 4 \rightarrow x = 3y + 4$$
$$2x = 8 + 6y$$
$$2(3y + 4) = 8 + 6y$$
$$6y + 8 = 6y + 8$$

This is an identity. There are infinitely many solutions.

99. Solve the second equation for x and substitute into the first equation:

$$2y + x = 8$$
$$x - 3y = 0 \mid x = 3y$$
$$2y + 3y = 8$$
$$5y = 8$$
$$y = \frac{8}{5}$$
$$x = 3\left|\frac{8}{5}\right| = \frac{24}{5}$$

Solution: $\left(\frac{24}{5}, \frac{8}{5}\right)$

101. Graph each equation.
$$2x + y = 4$$
$$2y + 4x = 6$$

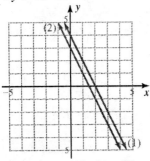

There is no solution. The system is inconsistent.

103. Graph each equation.
$$x + \frac{1}{2}y = -2$$
$$y = -2x - 4$$

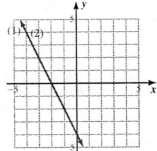

There are infinitely many solution. The system is dependent.

118

105. Graph each equation.

$$x - y = -1$$

$$y = -x - 1$$

The solution is $(-1, 0)$. The system is consistent.

4.2 Systems with Three Variables

Problems 4.2

1.

$$x + y + z = 4$$
$$x - y + z = 2$$
$$2x + 2y - z = -4$$

Add the first equation to the third equation to eliminate z:

$$3x + 3y = 0$$

Add the second equation to the third equation to eliminate z:

$$3x + y = -2$$

Multiply the second result by -1, add to the first result, and solve:

$$3x + 3y = 0$$
$$\underline{-3x \rightarrow y = 2}$$
$$2y = 2 \rightarrow y = 1$$

Substitute and solve for the other variables:

$$3x + 3(1) = 0 \qquad -1 + 1 + z = 4$$
$$3x = -3 \qquad\qquad z = 4$$
$$x = -1$$

The solution is $(-1, 1, 4)$.

2.

$$x + 8y - z = 8$$
$$-x + 2y - z = 4$$
$$2x + y + z = 2$$

Add the first equation to the third equation to eliminate z:

$$3x + 9y = 10$$

Add the second equation to the third equation to eliminate z:

$$x + 3y = 6$$

Multiply the second result by -3, add to the first result, and solve:

$$3x + 9y = 10$$
$$\underline{-3x - 9y = -18}$$
$$0 = -8$$

This result is a contradiction. There is no solution. The system is inconsistent.

3. $x + 2y + z = -10$

$x + y - z = -3$

$5x + 7y - z = -29$

Add the first equation to the second equation to eliminate z:

$$2x + 3y = -13$$

Add the first equation to the third equation to eliminate z:

$$6x + 9y = -39$$

Multiply the first result by –3, add to the second result, and solve:

$$-6x - 9y = 39$$
$$\underline{6x + 9y = -39}$$
$$0 = 0$$

This result is an identity. There are infinitely many solutions. The system is dependent.

4.   Let x = the number of ounces of V_1

Let y = the number of ounces of V_2

Let z = the number of ounces of V_3

$50x + 60y \qquad = 520$

$20x \qquad + 20z = 260$

$10x + 20y + 10z = 170$

Multiply the third equation by –3, and add the first equation to eliminate y:

$$50x + 60y \qquad = \quad 520$$
$$\underline{-30x - 60y - 30z = -510}$$
$$20x \qquad - 30z = \quad 10$$

Multiply this result by –1, add to the original second equation to eliminate x, and solve:

$$20x + 20z = 260$$
$$\underline{-20x + 30z = -10}$$
$$50z = 250 \rightarrow z = 5$$

Substitute and solve for the other variables:

$$20x + 20(5) = 260 \qquad 50(8) + 60y = 520$$
$$20x + 100 = 260 \qquad 400 + 60y = 520$$
$$20x = 160 \qquad 60y = 120$$
$$x = 8 \qquad y = 2$$

The number of ounces of each vegetable is 8 oz of V_1; 2 oz of V_2; and 5 oz of V_3.

Exercises 4.2

1. $x + y + z = 12$

$x - y + z = 6$

$x + 2y - z = 7$

Add the first equation to the third equation to eliminate z:

$$2x + 3y = 19$$

Add the second equation to the third equation to eliminate z:

$$2x + y = 13$$

Multiply the second result by –1, add to the first result, and solve:

$$2x + 3y = \quad 19$$
$$\underline{-2x \rightarrow y = -13}$$
$$2y = \quad 6 \rightarrow y = 3$$

Substitute and solve for the other variables:

$$2x + 3 = 13 \qquad 5 + 3 + z = 12$$
$$2x = 10 \qquad 8 + z = 12$$
$$x = 5 \qquad z = 4$$

The solution is (5, 3, 4). The system is consistent.

3.
$$x + y + z = 4$$
$$2x + 2y - z = -4$$
$$x - y + z = 2$$

Add the first equation to the second equation to eliminate z:
$$3x + 3y = 0$$

Add the second equation to the third equation to eliminate z:
$$3x + y = -2$$

Multiply the second result by -1, add to the first result, and solve:
$$3x + 3y = 0$$
$$\underline{-3x \rightarrow y = 2}$$
$$2y = 2 \rightarrow y = 1$$

Substitute and solve for the other variables:
$$3x + 1 = -2 \qquad -1 + 1 + z = 4$$
$$3x = -3 \qquad\qquad z = 4$$
$$x = -1$$

The solution is $(-1, 1, 4)$. The system is consistent.

5.
$$2x - y + z = 3$$
$$x + 2y + z = 12$$
$$4x - 3y + z = 1$$

Multiply the first equation by -1, add to the second equation to eliminate z:
$$-2x + y - z = -3$$
$$\underline{x + 2y + z = 12}$$
$$-x + 3y = 9$$

Multiply the first equation by -1, add to the third equation to eliminate z:
$$-2x + y - z = -3$$
$$\underline{4x - 3y + z = 1}$$
$$2x - 2y = -2$$

Multiply the first result by 2, add to the second result, and solve:
$$-2x + 6y = 18$$
$$\underline{2x - 2y = -2}$$
$$4y = 16 \rightarrow y = 4$$

Substitute and solve for the other variables:
$$2x - 2(4) = -2 \qquad 2(3) - 4 + z = 3$$
$$2x - 8 = -2 \qquad\qquad 6 - 4 + z = 3$$
$$2x = 6 \qquad\qquad\qquad 2 + z = 3$$
$$x = 3 \qquad\qquad\qquad\quad z = 1$$

The solution is $(3, 4, 1)$. The system is consistent.

7.
$$x - 2y - 3z = 2$$
$$x - 4y - 13z = 14$$
$$-3x + 5y + 4z = 2$$

Multiply the first equation by -1, add to the second equation to eliminate x:
$$-x + 2y + 3z = -2$$
$$\underline{x - 4y - 13z = 14}$$
$$-2y - 10z = 12$$

Multiply the first equation by 3, add to the third equation to eliminate x:
$$3x - 6y - 9z = 6$$
$$\underline{-3x + 5y + 4z = 2}$$
$$-y - 5z = 8$$

Multiply the second result by -2, add to the first result, and solve:

$$-2y - 10z = 12$$
$$\underline{2y + 10z = -16}$$
$$0 = -4$$

This result is a contradiction. There is no solution. The system is inconsistent.

9.
$$2x + 4y + 3z = 3$$
$$10x - 8y - 9z = 0$$
$$4x + 4y - 3z = 2$$

Multiply the first equation by 3, add to the second equation to eliminate z:
$$6x + 12y + 9z = 9$$
$$\underline{10x - 8y - 9z = 0}$$
$$16x + 4y \qquad = 9$$

Add the first equation to the third equation to eliminate z:
$$2x + 4y + 3z = 3$$
$$\underline{4x + 4y - 3z = 2}$$
$$6x + 8y \qquad = 5$$

Multiply the first result by -2, add to the second result, and solve:
$$-32x - 8y = -18$$
$$\underline{6x + 8y = \quad 5}$$
$$-26x \qquad = -13 \rightarrow x = \tfrac{1}{2}$$

Substitute and solve for the other variables:
$$6\left(\tfrac{1}{2}\right) + 8y = 5 \qquad 2\left(\tfrac{1}{2}\right) + 4\left(\tfrac{1}{4}\right) + 3z = 3$$
$$3 + 8y = 5 \qquad 1 + 1 + 3z = 3$$
$$8y = 2 \qquad 3z = 1$$
$$y = \tfrac{1}{4} \qquad z = \tfrac{1}{3}$$

The solution is $\left(\tfrac{1}{2}, \tfrac{1}{4}, \tfrac{1}{3}\right)$ The system is consistent.

13.
$$2x + y + z = 5$$
$$-x + 2y - z = 3$$
$$3x + 4y + z = 10$$

Add the first equation to the second equation to eliminate z:
$$2x + y + z = 5$$
$$\underline{-x + 2y - z = 3}$$
$$x + 3y \qquad = 8$$

Add the second equation to the third equation to eliminate z:
$$-x + 2y - z = 3$$
$$\underline{3x + 4y + z = 10}$$
$$2x + 6y \qquad = 13$$

Multiply the first result by -2, add to the second result, and solve:

11.
$$x - 2y - z = 3$$
$$2x - 5y + z = -1$$
$$x - 2y - z = -3$$

Multiply the first equation by -1, add to the third equation to eliminate z:
$$-x + 2y + z = -3$$
$$\underline{x - 2y - z = -3}$$
$$0 = -6$$

This result is a contradiction. There is no solution. The system is inconsistent.

$$-2x - 6y = -16$$
$$\underline{2x + 6y = \quad 13}$$
$$0 = -3$$

This result is a contradiction. There is no solution. The system is inconsistent.

15.
$$x + y \quad\;\; = 5$$
$$\quad\;\; y + z = 3$$
$$x \quad\;\; + z = 7$$

Multiply the first equation by -1, add to the third equation to eliminate x:
$$-x - y \quad\;\; = -5$$
$$\underline{x \quad\;\; + z = \;\; 7}$$
$$-y + z = \;\; 2$$

Add this result to the second equation to eliminate y:
$$-\cancel{y} + z = 2$$
$$\underline{y + z = 3}$$
$$2z = 5 \rightarrow z = \tfrac{5}{2}$$

Substitute and solve for the other variables:
$$x + \tfrac{5}{2} = 7 \qquad y + \tfrac{5}{2} = 3$$
$$x = \tfrac{9}{2} \qquad\quad y = \tfrac{1}{2}$$

The solution is $\left(\tfrac{9}{2}, \tfrac{1}{2}, \tfrac{5}{2}\right)$. The system is consistent.

17.
$$x - 2y \quad\;\; = 0$$
$$\quad\;\; y - 2z = 5$$
$$x + \; y + \; z = 8$$

Multiply the first equation by -1, add to the third equation to eliminate x:
$$-x + 2y \quad\;\; = 0$$
$$\underline{x + \; y + z = 8}$$
$$3y + z = 8$$

Multiply the second equation by -3, add to this result to eliminate y:
$$-\cancel{3}y + 6z = -\cancel{1}5$$
$$\underline{3y + \; z = \;\; 8}$$
$$7z = -\cancel{7} \rightarrow z = -\cancel{1}$$

Substitute and solve for the other variables:
$$y - 2(-1) = 5 \qquad x - 2(3) = 0$$
$$y + 2 = 5 \qquad\quad x - 6 = 0$$
$$y = 3 \qquad\qquad x = 6$$

The solution is $(6, 3, -1)$. The system is consistent.

19.
$$5x \quad\;\; - 3z = \;\; 2$$
$$\quad -y + 2z = -5$$
$$x + 2y - 4z = \;\; 8$$

Multiply the second equation by 2, add to the third equation to eliminate y:
$$-2y + 4z = -10$$
$$\underline{x + 2y - 4z = \quad 8}$$
$$x \qquad\qquad = -2$$

Substitute and solve for the other variables:
$$5(-2) - 3z = 2 \qquad -y + 2(-4) = -5$$
$$-10 - 3z = 2 \qquad\quad -y - 8 = -5$$
$$-3z = 12 \qquad\qquad -y = 3$$
$$z = -4 \qquad\qquad\;\; y = -3$$

The solution is $(-2, -3, -4)$. The system is consistent.

21.
$$y = ax^2 + bx + c$$
$$0 = a(1)^2 + b(1) + c \quad - \quad 0 = a + b + c$$
$$-4 = a(-1)^2 + b(-1) + c \quad - \quad -4 = a - b + c$$
$$5 = a(2)^2 + b(2) + c \quad - \quad 5 = 4a + 2b + c$$

Multiply the first equation by -1, add to the second equation to eliminate a:
$$0 = -\cancel{a} - \cancel{b} - \cancel{c}$$
$$\underline{-\cancel{4} = \;\; a - \cancel{b} + c}$$
$$-\cancel{4} = \;\; -\cancel{2}b \quad \rightarrow b = 2$$

Multiply the first equation by -1, add to the third equation to eliminate c:
$$0 = -a - \; b - c$$
$$\underline{5 = 4a + 2b + c}$$
$$5 = 3a + \; b$$

Substitute and solve for the other variables:
$$5 = 3a + 2 \qquad 0 = 1 + 2 + c$$
$$3 = 3a \qquad\qquad 0 = 3 + c$$
$$1 = a \qquad\qquad -3 = c$$

The solution is $a = 1$, $b = 2$, $c = -3$. The quadratic equation is $y = x^2 + 2x - 3$.

23.
$$2x \qquad +4z = 6$$
$$3x + y + z = -1$$
$$2y - z = -2$$
$$x - y - 2z = -5$$

Add the second equation to the fourth equation to eliminate y:
$$3x + y + z = -1$$
$$\underline{x - y - 2z = -5}$$
$$4x \qquad -z = -6$$

Multiply the first equation by -2, add to this result to eliminate x:
$$-4x - 8z = -12$$
$$\underline{4x \rightarrow z = -6}$$
$$-9z = -18 \rightarrow z = 2$$

Substitute and solve for the other variables:
$$4x - 2 = -6 \qquad 3(-1) + y + 2 = -1$$
$$4x = -4 \qquad -3 + y + 2 = -1$$
$$x = -1 \qquad y - 1 = -1$$
$$\qquad y = 0$$

Check these results in the third equation:
$$2(0) - 2 = -2 \rightarrow -2 = -2$$
The solution is $(-1, 0, 2)$.

25.
$$2x \qquad +4z = 6$$
$$3x + y + z = -1$$
$$2y - z = -2$$
$$x - y + kz = -5$$

Multiply the second equation by -2, add to the third equation to eliminate y:
$$-6x - 2y - 2z = 2$$
$$\underline{2y - z = -2}$$
$$-6x \qquad -3z = 0$$

Multiply this result by $\frac{1}{3}$, add to the first equation to eliminate x:
$$-2x \rightarrow z = 0$$
$$\underline{2x + 4z = 6}$$
$$3z = 6 \rightarrow z = 2$$

Substitute and solve for the other variables:
$$2x + 4(2) = 6 \qquad 3(-1) + y + 2 = -1$$
$$2x + 8 = 6 \qquad -3 + y + 2 = -1$$
$$2x = -2 \qquad y - 1 = -1$$
$$x = -1 \qquad y = 0$$

The solution is $(-1, 0, 2)$.
$$-1 - 0 + k(2) = -5$$
$$2k = -4$$
$$k = -2$$

27.   Let $x =$ the first number
Let $y =$ the second number
Let $z =$ the third number
$$x + y + z = 49$$
$$y = x + 8$$
$$z = y + 9$$

Substitute the third equation into the first:
$$x + y + y + 9 = 49$$
$$x + 2y = 40$$

Substitute the second equation into this result and solve:
$$x + 2(x + 8) = 40$$
$$x + 2x + 16 = 40$$
$$3x = 24$$
$$x = 8$$
$$y = x + 8 = 8 + 8 = 16$$
$$z = y + 9 = 16 + 9 = 25$$
The numbers are 8, 16, and 25.

29.   Let $x =$ the first angle
Let $y =$ the second angle
Let $z =$ the third angle
$$x + y + z = 180$$
$$x + y = z - 20$$
$$y = x + 20$$

Substitute the third equation into the second:
$$x + x + 20 = z - 20$$
$$2x + 20 = z - 20$$
$$2x + 40 = z$$

Substitute this result and the third equation into the first and solve:
$$x + x + 20 + 2x + 40 = 180$$
$$4x + 60 = 180$$
$$4x = 120$$
$$x = 30$$
$$y = x + 20 = 30 + 20 = 50$$
$$z = 2x + 40 = 2(30) + 40 = 100$$
The angles are 30°, 50°, and 100°.

31. Let x = corn problems
Let y = heel pain
Let z = ingrown toenails

$$x+y+z=150 \quad - \quad x+y+z=150$$
$$y=z+2 \quad - \quad y-z=2$$
$$y+z=x+10 \quad - \quad -x+y+z=10$$

Add the first and third equations:
$$2y+2z=160$$

Multiply the second equation by 2, add to this result, and solve:
$$2y-2z=4$$
$$2y+2z=160$$
$$\overline{}$$
$$4y=164$$
$$y=41$$

Substitute and solve for the other variables:
$$41-z=2 \qquad x+41+39=150$$
$$39=z \qquad x+80=150$$
$$x=70$$

70 people have corns, 41 people have heel pain, and 39 people have ingrown toenails.

35. Let x = Gates fortune
Let y = Kluge fortune
Let z = Walton fortune

$$x+y+z=16.9$$
$$y=z+0.4 \quad - \quad z=y-0.4$$
$$y=x-0.8 \quad - \quad x=y+0.8$$

Substitute the second and third equations into the first and solve:
$$y+0.8+y+y-0.4=16.9$$
$$3y+0.4=16.9$$
$$3y=16.5$$
$$y=5.5$$
$$x=y+0.8=5.5+0.8=6.3$$
$$z=y-0.4=5.5-0.4=5.1$$

Gates: \$6.3 billion; Kluge: \$5.5 billion; Walton: \$5.1 billion

33. Let x = Pizza Hut's share
Let y = Domino's share
Let z = Papa John's share

$$x+y+z=72$$
$$x=y+40$$
$$y=z+4$$

Substitute the third equation into the second:
$$x=z+4+40$$
$$x=z+44$$

Substitute this result and the third equation into the first and solve:
$$z+44+z+4+z=72$$
$$3z+48=72$$
$$3z=24$$
$$z=8$$
$$y=z+4=8+4=12$$
$$x=z+44=8+44=52$$

Pizza Hut's share is 52%, Domino's share is 12%, and Papa John's share is 8%.

37. Let x = amount invested in bonds
Let y = amount invested in certificates
$$x+y=20,000$$
$$0.05x+0.07y=1160$$

Solve the first equation for x and substitute in the second equation:
$$x=20,000-y$$
$$0.05(20,000-y)+0.07y=1160$$
$$1000-0.05y+0.07y=1160$$
$$0.02y=160$$
$$y=8000$$
$$x=12,000$$

Bonds: \$12,000; Certificates: \$8,000

39.

Percent of solution	Amount to be mixed	Amount of Pure solution
30% or 0.30	x	$0.30x$
12% or 0.12	40	0.12(40)
20% or 0.20	$x+40$	0.20(x + 40)

$$0.30x+0.12(40)=0.20(x+40)$$
$$0.30x+4.8=0.20x+8$$
$$0.10x=3.2$$
$$x=32$$

32 gallons of 30% solution must be added.

41.   Let x = plane's speed in still air
$x + 30$ = plane's speed with tail wind
$x - 30$ = plane's speed against the wind
t = the time for the trip

$$(x+30)t = 840 \quad \rightarrow \quad t = \frac{840}{x+30}$$

$$(x-30)t = 660 \quad \rightarrow \quad t = \frac{660}{x-30}$$

$$\frac{840}{x+30} = \frac{660}{x-30}$$

$$840(x-30) = 660(x+30)$$

$$840x - 25200 = 660x + 19800$$

$$180x = 45000$$

$$x = 250$$

The plane's speed in still air is 250 mi/hr.

43.

$$F = \frac{n}{4} + 40 \qquad F = \frac{54d+196}{5}$$

$$\frac{54d+196}{5} = \frac{n}{4} + 40$$

$$4(54d+196) = 5n + 800$$

$$216d + 784 = 5n + 800$$

$$216d = 5n + 16$$

$$d = \frac{5n+16}{216}$$

45.   Sample answer. The interpretation of a three dimensional graph on a two dimensional page is very difficult to read, because each equation is representing a plane.

47.   Sample answer. The graph of $2x + y = 4$ is a line. The graph of $2x + y + z = 4$ is a plane.

49.   Answers will vary.

51.
$$\begin{aligned} x + y + z &= 4 \\ x - 2y - z &= 1 \\ 2x - y - 2z &= -1 \end{aligned}$$

Add the first equation to the second equation to eliminate z:
$$2x - y = 5$$

Multiply the first equation by 2, add to the third equation to eliminate z:
$$\begin{aligned} 2x + 2y + 2z &= 8 \\ 2x - y - 2z &= -1 \\ \hline 4x + y \phantom{{}+2z} &= 7 \end{aligned}$$

Add the first and second results, and solve:
$$\begin{aligned} 2x - y &= 5 \\ 4x + y &= 7 \\ \hline 6x \phantom{{}+y} &= 12 \rightarrow x = 2 \end{aligned}$$

Substitute and solve for the other variables:
$$4(2) + y = 7 \qquad 2 + (-1) + z = 4$$
$$y = -1 \qquad\qquad z = 3$$

The solution is (2, –1, 3). The system is consistent.

53.
$$\begin{aligned} 2x + 2y - 6z &= 5 \\ -x - y + 3z &= 4 \\ 3x - y + z &= 2 \end{aligned}$$

Multiply the second equation by 2, add to the first equation to eliminate z:
$$\begin{aligned} -2x - 2y + 6z &= 8 \\ 2x + 2y - 6z &= 5 \\ \hline 0 &= 13 \end{aligned}$$

This result is a contradiction. There is no solution. The system is inconsistent.

55.
$$x + 2y + z = 4$$
$$-3x + 4y - z = -4$$
$$-2x - 4y - 2z = -8$$

Multiply the first equation by 2, add to the third equation to eliminate z:

$$2x + 4y + 2z = 8$$
$$\underline{-2x - 4y - 2z = -8}$$
$$0 = 0$$

Add the first equation to the second equation to determine the conclusion:

$$x + 2y + z = 4$$
$$\underline{-3x + 4y - z = -4}$$
$$-2x + 6y = 0$$

Since this result is not a contradiction, there are infinitely many solutions. The system is dependent.

4.3   Coin, Distance-Rate-Time, Investment, and Geometry Problems

Problems 4.3

1. Let d = the number of dimes
Let q = the number of quarters
$$10d + 25q = 990$$
$$d = 3q$$

Substitute the second equation into the first equation and solve:
$$10(3q) + 25q = 990$$
$$30q + 25q = 990$$
$$55q = 990$$
$$q = 18 \qquad d = 3q = 3(18) = 54$$

Jill has 54 dimes and 18 quarters.

2. Let r = the height of the redwood
Let m = the height of the silver maple
$$r + m = 436$$
$$r - m = 292$$

Add the two equations and solve:
$$2r = 728$$
$$r = 364$$
$$364 + m = 436$$
$$m = 72$$

The redwood tree is 364 feet tall and the silver maple is 72 feet tall.

3. Let x = the velocity of the plane in still air
Let y = the velocity of the wind
$x + y$ = the velocity with a tail wind
$x - y$ = the velocity against the wind
$$(x + y)3 = 1200 - \quad x + y = 400$$
$$(x - y)4 = 1200 - \quad x - y = 300$$

Add the two equations and solve:

$$2x = 700$$
$$x = 350$$
$$350 + y = 400$$
$$y = 50$$

The wind velocity is 50 mi/hr and the plane's velocity in still air is 350 mi/hr.

4. Let x = the amount invested at 4%
Let y = the amount invested at 6%
Let z = the amount invested at 8%
$$y = 3x$$
$$z = y + 7000$$
$$0.04x + 0.06y + 0.08z = 2860$$
Rewrite the equations:
$$-3x + y \quad = \quad 0$$
$$-y + \ z = \quad 7000$$
$$4x + 6y + 8z = 286{,}000$$
Add the first and second equations:
$$-3x + z = 7000$$
Multiply the second equation by 6 and add to the third equation:
$$-6y + 6z = \ 42{,}000$$
$$\underline{4x + 6y + 8z = 286{,}000}$$
$$4x \quad\ + 14z = 328{,}000$$
Multiply the first result by -14, add to the second result, and solve:
$$42x - 14z = -98{,}000$$
$$\underline{4x + 14z = 328{,}000}$$
$$46x \quad\quad = 230{,}000 \rightarrow x = 5000$$
Substitute and solve for the other variables:
$$y = 3(5000) = 15{,}000$$
$$z = 15{,}000 + 7000 = 22{,}000$$
4% investment is $5,000; 6% investment is $15,000; 8% investment is $22,000
The total inheritance is:
$$\$5{,}000 + \$15{,}000 + \$22{,}000 = \$42{,}000$$

5. Let w = the width of the base
Let l = the length of the base
$$2l + 2w = 183\tfrac{5}{6}$$
$$l = w + 18\tfrac{1}{4}$$
Substitute the second equation into the first equation and solve:
$$2\left(w + 18\tfrac{1}{4}\right) + 2w = 183\tfrac{5}{6}$$
$$2w + 36\tfrac{1}{2} + 2w = 183\tfrac{5}{6}$$
$$4w + 36\tfrac{1}{2} = 183\tfrac{5}{6}$$
$$4w + \frac{73}{2} = \frac{1103}{6}$$
$$24w + 219 = 1103$$
$$24w = 884$$
$$w = \frac{884}{24} = \frac{221}{6} = 36\tfrac{5}{6}$$
$$l = 36\tfrac{5}{6} + 18\tfrac{1}{4} = 55\tfrac{1}{12}$$
The length is $55\tfrac{1}{12}$ ft and the width is $36\tfrac{5}{6}$ ft.

Exercises 4.3

1. Let n = the number of nickels
Let d = the number of dimes
$$5n + 10d = 625$$
$$d = 2n$$
Substitute the second equation into the first equation and solve:
$$5n + 10(2n) = 625$$
$$5n + 20n = 625$$
$$25n = 625$$
$$n = 25$$
$$d = 2n = 2(25) = 50$$
Natasha has 25 nickels and 50 dimes.

3. Let n = the number of nickels
Let q = the number of quarters
$$5n + 25q = 550$$
$$q = 2n$$
Substitute the second equation into the first equation and solve:
$$5n + 25(2n) = 550$$
$$5n + 50n = 550$$
$$55n = 550$$
$$n = 10$$
$$q = 2n = 2(10) = 20$$
Dora has 10 nickels and 20 quarters.

5.   Let n = the number of nickels
Let p = the number of pennies
$$n + p = 10$$
$$5n + 1p = 5p + 1n$$
Solve the first equation for p, substitute into the second equation and solve:
$$p = 10 - n$$
$$5n + 10 - n = 5(10 - n) + n$$
$$4n + 10 = 50 - 5n + n$$
$$4n + 10 = 50 - 4n$$
$$8n = 40$$
$$n = 5$$
$$p = 10 - n = 10 - 5 = 5$$
Desi has 5 nickels and 5 pennies.

9.   Let n = the number of nickels
Let d = the number of dimes
Let q = the number of quarters
$$5n + 10d + 25q = 705$$
$$25q = 10d + 460$$
$$10d = 5n + 25$$
Substitute the second and third equations into the first equation:
$$5n + 5n + 25 + 10d + 460 = 705$$
$$10n + 10d + 485 = 705$$
$$10n + 10d = 220$$
Substitute the third equation into this result and solve:
$$10n + 5n + 25 = 220$$
$$15n + 25 = 220$$
$$15n = 195$$
$$n = 13$$
$$10d = 5(13) + 25 = 65 + 25 = 90$$
$$d = 9$$
$$25q = 10(9) + 460 = 90 + 460 = 550$$
$$q = 22$$
There are 13 nickels, 9 dimes, 22 quarters.

7.   Let x = the number of 10-dollar bills
Let y = the number of 20-dollar bills
$$x + y = 25$$
$$10x + 20y = 300$$
Multiply the first equation by -10, add to the second equation and solve:
$$-10x - 10y = -250$$
$$\underline{10x + 20y = 300}$$
$$10y = 50 \rightarrow y = 5$$
$$x + 5 = 25 \rightarrow x = 20$$
There are 20 ten dollar bills and 5 twenty dollar bills.

11.   Let x = the first number
Let y = the second number
$$x + y = 102$$
$$x - y = 16$$
Add the two equations and solve:
$$2x = 118$$
$$x = 59$$
$$59 + y = 102$$
$$y = 43$$
The numbers are 43 and 59.

13. Let x = the first integer
Let y = the second integer
$$x + y = 126$$
$$y = 5x$$
Substitute the second equation into the first and solve:
$$x + 5x = 126$$
$$6x = 126 \qquad y = 5(21)$$
$$x = 21 \qquad y = 105$$
The integers are 21 and 105.

15. Let x = the first number
Let y = the second number
$$x - y = 16$$
$$x = 5y$$
Substitute the second equation into the first and solve:
$$5y - y = 16$$
$$4y = 16 \qquad x = 5(4)$$
$$y = 4 \qquad x = 20$$
The numbers are 20 and 4.

17. Let x = elevation of Longs Peak
Let y = elevation of Pikes Peak
$$x = y + 145$$
$$x + y + 637 = 29002$$
Substitute the first equation into the second and solve:
$$y + 145 + y + 637 = 29002$$
$$2y + 782 = 29002$$
$$2y = 28220$$
$$y = 14110$$
$$x = 14110 + 145 = 14255$$
Longs Peak is 14,255 ft high and Pikes Peak is 14,110 ft high.

19. Let x = the amount of chocolate
Let y = the amount of butterscotch
Let z = the amount of caramel
$$x + y + z = 6700$$
$$y = z$$
$$x = y + 600$$
Substitute the second and third equations into the first equation and solve:
$$y + 600 + y + y = 6700$$
$$3y + 600 = 6700$$
$$3y = 6100$$
$$y = 2033\tfrac{1}{3}$$
$$z = 2033\tfrac{1}{3}$$
$$x = 2033\tfrac{1}{3} + 600 = 2633\tfrac{1}{3}$$
There were $2633\tfrac{1}{3}$ lb of chocolate, $2033\tfrac{1}{3}$ lb of butterscotch, and $2033\tfrac{1}{3}$ lb of caramel.

21. Let x = the velocity of the plane in still air
Let y = the velocity of the wind
$x + y$ = the velocity with a tail wind
$x - y$ = the velocity against the wind
$$(x + y)\frac{9}{4} = 540 \;-\; x + y = 240$$
$$(x - y)3 = 540 \;-\; x - y = 180$$
Add the two equations and solve:
$$2x = 420$$
$$x = 210$$
$$210 + y = 240$$
$$y = 30$$
The wind velocity is 30 mi/hr and the plane's velocity in still air is 210 mi/hr.

23. Let x = the speed of the boat in still water
Let y = the speed of the current
$x + y$ = the speed downstream
$x - y$ = the speed upstream
$$x + y = 15$$
$$x - y = 9$$
Add the two equations and solve:
$$2x = 24$$
$$x = 12$$
$$12 + y = 15$$
$$y = 3$$
The speed of the boat in still water is 12 mi/hr and the speed of the current is 3 mi/hr.

25. Let x = the velocity of the plane in still air
Let y = the velocity of the wind
$x + y$ = the velocity with a tail wind

$x - y$ = the velocity against the wind

$$(x+y)2 = 1000 - \quad x + y = 500$$

$$(x-y)\frac{5}{2} = 1000 - \quad x - y = 400$$

Add the two equations and solve:

$$2x = 900 \qquad\qquad 450 + y = 500$$

$$x = 450 \qquad\qquad\quad y = 50$$

The wind velocity is 50 mi/hr and the plane's velocity in still air is 450 mi/hr.

27. Let x = the amount invested at 6%
Let y = the amount invested at 8%

$$x + y = 10,000$$

$$0.06x + 0.08y = 720$$

Solve the first equation for x, substitute into the second equation and solve:

$$x = 10,000 - y$$

$$0.06(10,000 - y) + 0.08y = 720$$

$$600 - 0.06y + 0.08y = 720$$

$$0.02y = 120$$

$$y = 6,000$$

$$x = 10,000 - 6,000 = 4,000$$

6% investment is $4,000; 8% investment is $6,000.

29. Let x = the amount invested at 6%
Let y = the amount invested at 8%
Let z = the amount invested at 10%

$$x + y + z = 25,000$$

$$0.06x + 0.08y + 0.10z = 2000$$

$$0.06x + 0.08y = 0.10z$$

Substitute the third equation into the second equation and solve:

$$0.10z + 0.10z = 2000$$

$$0.20z = 2000$$

$$z = 10,000$$

Substitute this result into the first and third equations, solve the first equation for x and substitute into the second result:

$$x + y + 10,000 = 25,000 - \quad x + y = 15,000$$

$$x = 15,000 - y$$

$$0.06x + 0.08y = 0.10(10,000)$$

$$0.06x + 0.08y = 1000$$

$$0.06(15,000 - y) + 0.08y = 1000$$

$$900 - 0.06y + 0.08y = 1000$$

$$0.02y = 100$$

$$y = 5,000$$

$$x = 15,000 - 5,000$$

$$x = 10,000$$

6% investment is $10,000; 8% investment is $5,000; 10% investment is $10,000.

31. Let w = the width of the flag
Let l = the length of the flag

$$2l + 2w = 1520$$

$$l = w + 250$$

Substitute the second equation into the first equation and solve:

$$2(w + 250) + 2w = 1520$$

$$2w + 500 + 2w = 1520$$

$$4w = 1020$$

$$w = 255$$

$$l = 255 + 250 = 505$$

The length is 505 ft and the width is 255 ft.

33.   Let w = the width of the quilt
Let l = the length of the quilt
$$2l + 2w = 438$$
$$l = w + 49$$
Substitute the second equation into the first equation and solve:
$$2(w + 49) + 2w = 438$$
$$2w + 98 + 2w = 438$$
$$4w = 340$$
$$w = 85$$
$$l = 85 + 49 = 134$$
The length is 134 ft and the width is 85 ft.

35. $$2x - y + z = 3$$
$$x + y \quad = -1$$
$$3x - y - 2z = 7$$
Multiply the first equation by 2, add to the third equation to eliminate z:
$$4x - 2y + 2z = 6$$
$$\underline{3x - y - 2z = 7}$$
$$7x - 3y \quad = 13$$
Multiply the second equation by 3 and add to the above result to eliminate y:
$$3x + 3y = -3$$
$$\underline{7x - 3y = 13}$$
$$10x \quad = 10 \rightarrow x = 1$$
Substitute and solve for the other variables:
$$1 + y = -1 \qquad 2(1) - (-2) + z = 3$$
$$y = -2 \qquad 2 + 2 + z = 3$$
$$z = -1$$
The solution is $(1, -2, -1)$.

37.   The amount of calcium needed is $15a + 5b + 20c = 170$.

39.   The amount of vitamin B needed is $10a + 15b + 10c = 110$.

41.   Answers may vary.

43.   Let w = the width of the rectangle
Let l = the length of the rectangle
$$2l + 2w = 170$$
$$l = w + 15$$
Substitute the second equation into the first equation and solve:
$$2(w + 15) + 2w = 170$$
$$2w + 30 + 2w = 170$$
$$4w = 140$$
$$w = 35$$
$$l = 35 + 15 = 50$$
The length is 50 cm and the width is 35 cm.

45.   Let x = the velocity of the plane in still air
Let y = the velocity of the wind
$x + y$ = the velocity with a tail wind
$x - y$ = the velocity against the wind
$$(x + y)3 = 1200 - \quad x + y = 400$$
$$(x - y)4 = 1200 - \quad x - y = 300$$
Add the two equations and solve:
$$2x = 700 \qquad 350 + y = 400$$
$$x = 350 \qquad y = 50$$
The wind velocity is 50 mi/hr and the plane's velocity in still air is 350 mi/hr.

47.   Let x = the weight of the first twin
Let y = the weight of the second twin
$$x + y = 1466$$
$$x - y = 20$$
Add the two equations and solve:
$$2x = 1486$$
$$x = 743$$
$$743 + y = 1466$$
$$y = 723$$
The twins weigh 743 lb and 723 lb.

4.4 Matrices

Problems 4.4

1. $2x+3y+z=14$

$x-y+2z=8$

$-x+4y-z=2$

$$\begin{bmatrix} 2 & 3 & 1 \\ 1 & -1 & 2 \\ -1 & 4 & -1 \end{bmatrix}\begin{matrix} 14 \\ 8 \\ 2 \end{matrix} \rightarrow \sim \begin{bmatrix} 1 & -1 & 2 \\ 2 & 3 & 1 \\ -1 & 4 & -1 \end{bmatrix}\begin{matrix} 8 \\ 14 \\ 2 \end{matrix} \rightarrow \sim \begin{bmatrix} 1 & -1 & 2 \\ 0 & 5 & -3 \\ 0 & 3 & 1 \end{bmatrix}\begin{matrix} 8 \\ -2 \\ 10 \end{matrix} \rightarrow \sim \begin{bmatrix} 1 & -1 & 2 \\ 0 & 5 & -3 \\ 0 & 0 & 14 \end{bmatrix}\begin{matrix} 8 \\ -2 \\ 56 \end{matrix}$$

$R_1 \to R_2 \qquad -2R_1+R_2 \to R_2 \qquad -3R_2+5R_3 \to R_3$

$R_1+R_3 \to R_3$

$x-y+2z=8 \qquad\qquad\qquad 5y-3(4)=-2 \qquad x-2+2(4)=8$

$5y-3z=-2 \qquad 14z=56 \qquad 5y=10 \qquad x-2+8=8$

$14z=56 \qquad z=4 \qquad y=2 \qquad x=2$

The solution is $(2, 2, 4)$.

2. $x+2y-z=4$

$5x-3y+2z=1$

$6x-y+z=3$

$$\begin{bmatrix} 1 & 2 & -1 \\ 5 & -3 & 2 \\ 6 & -1 & 1 \end{bmatrix}\begin{matrix} 4 \\ 1 \\ 3 \end{matrix} \rightarrow \sim \begin{bmatrix} 1 & 2 & -1 \\ 0 & -13 & 7 \\ 0 & -13 & 7 \end{bmatrix}\begin{matrix} 4 \\ -19 \\ -21 \end{matrix} \rightarrow \sim \begin{bmatrix} 1 & 2 & -1 \\ 0 & -13 & 7 \\ 0 & 0 & 0 \end{bmatrix}\begin{matrix} 4 \\ -19 \\ -2 \end{matrix}$$

$-5R_1+R_2 \to R_2 \qquad -1R_2+R_3 \to R_3$

$-6R_1+R_3 \to R_3$

The last line corresponds to the equation $0x+0y+0z=-2$ which is false. The system has no solution.

3. $-4x+y+z=4$

$2x-y+3z=5$

$6x-2y+2z=1$

$$\begin{bmatrix} -4 & 1 & 1 \\ 2 & -1 & 3 \\ 6 & -2 & 2 \end{bmatrix}\begin{matrix} 4 \\ 5 \\ 1 \end{matrix} \rightarrow \sim \begin{bmatrix} -4 & 1 & 1 \\ 0 & -1 & 7 \\ 0 & 1 & -7 \end{bmatrix}\begin{matrix} 4 \\ 14 \\ -14 \end{matrix} \rightarrow \sim \begin{bmatrix} -4 & 1 & 1 \\ 0 & -1 & 7 \\ 0 & 0 & 0 \end{bmatrix}\begin{matrix} 4 \\ 14 \\ 0 \end{matrix}$$

$-3R_2+R_3 \to R_3 \qquad R_2+R_3 \to R_3$

$R_1+2R_2 \to R_2$

$-4x+y+z=4 \qquad\qquad\qquad -4x+7z-14+z=4$

$-y+7z=14 \qquad -y=-7z+14 \qquad -4x=-8z+18$

$0=0 \qquad y=7z-14 \qquad x=2z-\frac{9}{2}$

There are infinitely many solutions such that for any real number, $z=k$, solutions are of the form $\left(2k-\frac{9}{2}, 7k-14, k\right)$.

4. Let $x =$ the cost of a bush
Let $y =$ the cost of a flowering plant
Let $z =$ the cost of a tree

$24x + 16y + 10z = 374$

$36x + 48y + 20z = 828$

$50x + 40y + 30z = 970$

$$\begin{array}{ccc|c}
24 & 16 & 10 & 374 \\
36 & 48 & 20 & 828 \\
50 & 40 & 30 & 970
\end{array}
\rightarrow
\begin{array}{ccc|c}
24 & 16 & 10 & 374 \\
0 & 48 & 10 & 534 \\
0 & 80 & 110 & 2290
\end{array}
\rightarrow
\begin{array}{ccc|c}
24 & 16 & 10 & 374 \\
0 & 48 & 10 & 534 \\
0 & 0 & 280 & 4200
\end{array}$$

$-3 \to R_1 + 2 \to R_2 \to R_2$ $\qquad$ $-5 \to R_2 + 3 \to R_3 \to R_3$

$-25 \to R_1 + 12 \to R_3 \to R_3$

$24x + 16y + 10z = 374$ $\qquad$ $280z = 4200$ $\qquad$ $48y + 10(15) = 534$ $\qquad$ $24x + 16(8) + 10(15) = 374$

$48y + 10z = 534$ $\qquad\qquad$ $z = 15$ $\qquad\qquad$ $48y = 384$ $\qquad\qquad$ $24x + 128 + 150 = 374$

$280z = 4200$ $\qquad\qquad\qquad\qquad$ $y = 8$ $\qquad\qquad\qquad$ $24x = 96$

$x = 4$

The price of one bush is \$4, one flowering plant is \$8, and one tree is \$15.

5. Let $x =$ the cost of one pound of small nails
Let $y =$ the cost of one pound of medium nails
Let $z =$ the cost of one pound of large nails

$x + y + z = 215$

$4x + y \quad = 285$

$y + 2z = 255$

$$\begin{array}{ccc|c}
1 & 1 & 1 & 215 \\
4 & 1 & 0 & 285 \\
0 & 1 & 2 & 255
\end{array}
\rightarrow
\begin{array}{ccc|c}
1 & 1 & 1 & 215 \\
0 & -3 & -4 & -575 \\
0 & 1 & 2 & 255
\end{array}
\rightarrow
\begin{array}{ccc|c}
1 & 1 & 1 & 215 \\
0 & -3 & -4 & -575 \\
0 & 0 & 2 & 190
\end{array}$$

$-4 \to R_1 + R_2 \to R_2$ $\qquad$ $R_2 + 3 \to R_3 \to R_3$

$x + y + z = 215$ $\qquad$ $2z = 190$ $\qquad$ $-3y - 4(95) = -575$ $\qquad$ $x + 65 + 95 = 215$

$-3y - 4z = -575$ $\qquad\qquad$ $z = 95$ $\qquad\qquad$ $-3y - 380 = -575$ $\qquad\qquad$ $x + 160 = 215$

$2z = 190$ $\qquad\qquad\qquad\qquad\qquad$ $-3y = -195$ $\qquad\qquad\qquad$ $x = 55$

$y = 65$

The price of one pound of small nails is \$0.55, medium nails is \$0.65, and large nails is \$0.95.

Exercises 4.4

1.
$$x + y - z = 3$$
$$x - 2y + z = -3$$
$$2x + y + z = 4$$

$$\left[\begin{array}{ccc|c} 1 & 1 & -1 & 3 \\ 1 & -2 & 1 & -3 \\ 2 & 1 & 1 & 4 \end{array}\right] \rightarrow \left[\begin{array}{ccc|c} 1 & 1 & -1 & 3 \\ 0 & -3 & 2 & -6 \\ 0 & -1 & 3 & -2 \end{array}\right] \sim \rightarrow \left[\begin{array}{ccc|c} 1 & 1 & -1 & 3 \\ 0 & -3 & 2 & -6 \\ 0 & 0 & -7 & 0 \end{array}\right]$$

$$-1 \cdot R_1 + R_2 \rightarrow R_2 \qquad R_2 - 3 \cdot R_3 \rightarrow R_3$$
$$-2 \cdot R_1 + R_3 \rightarrow R_3$$

$$\begin{array}{ll} x + y - z = 3 & \\ -3y + 2z = -6 & -7z = 0 \\ -7z = 0 & z = 0 \end{array} \qquad \begin{array}{l} -3y + 2(0) = -6 \\ -3y = -6 \\ y = 2 \end{array} \qquad \begin{array}{l} x + 2 - 0 = 3 \\ x = 1 \end{array}$$

The solution is $(1, 2, 0)$.

3.
$$2x - y + 2z = 5$$
$$2x + y - z = -6$$
$$3x + 2z = 3$$

$$\left[\begin{array}{ccc|c} 2 & -1 & 2 & 5 \\ 2 & 1 & -1 & -6 \\ 3 & 0 & 2 & 3 \end{array}\right] \rightarrow \left[\begin{array}{ccc|c} 2 & -1 & 2 & 5 \\ 0 & 2 & -3 & -11 \\ 0 & 3 & -2 & -9 \end{array}\right] \sim \rightarrow \left[\begin{array}{ccc|c} 2 & -1 & 2 & 5 \\ 0 & 2 & -3 & -11 \\ 0 & 0 & 5 & 15 \end{array}\right]$$

$$-1 \cdot R_1 + R_2 \rightarrow R_2 \qquad -3 \cdot R_2 + 2 \cdot R_3 \rightarrow R_3$$
$$-3 \cdot R_1 + 2 \cdot R_3 \rightarrow R_3$$

$$\begin{array}{ll} 2x - y + 2z = 5 & \\ 2y - 3z = -11 & 5z = 15 \\ 5z = 15 & z = 3 \end{array} \qquad \begin{array}{l} 2y - 3(3) = -11 \\ 2y = -2 \\ y = -1 \end{array} \qquad \begin{array}{l} 2x - (-1) + 2(3) = 5 \\ 2x = -2 \\ x = -1 \end{array}$$

The solution is $(-1, -1, 3)$.

5.
$$3x + 2y + z = -5$$
$$2x - y - z = -6$$
$$2x + y + 3z = 4$$

$$\left[\begin{array}{ccc|c} 3 & 2 & 1 & -5 \\ 2 & -1 & -1 & -6 \\ 2 & 1 & 3 & 4 \end{array}\right] \rightarrow \left[\begin{array}{ccc|c} 3 & 2 & 1 & -5 \\ 0 & 7 & 5 & 8 \\ 0 & 1 & -7 & -22 \end{array}\right] \sim \rightarrow \left[\begin{array}{ccc|c} 3 & 2 & 1 & -5 \\ 0 & 7 & 5 & 8 \\ 0 & 0 & 54 & 162 \end{array}\right]$$

$$2 \cdot R_1 - 3 \cdot R_2 \rightarrow R_2 \qquad R_2 - 7 \cdot R_3 \rightarrow R_3$$
$$2 \cdot R_1 - 3 \cdot R_3 \rightarrow R_3$$

$$\begin{array}{ll} 3x + 2y + z = -5 & \\ 7y + 5z = 8 & 54z = 162 \\ 54z = 162 & z = 3 \end{array} \qquad \begin{array}{l} 7y + 5(3) = 8 \\ 7y = -7 \\ y = -1 \end{array} \qquad \begin{array}{l} 3x + 2(-1) + 3 = -5 \\ 3x = -6 \\ x = -2 \end{array}$$

The solution is $(-2, -1, 3)$.

7.
$$x + y + z = 3$$
$$x - 2y + z = -3$$
$$3x \quad + 3z = 5$$

$$\begin{bmatrix} 1 & 1 & 1 & | & 3 \\ 1 & -2 & 1 & | & -3 \\ 3 & 0 & 3 & | & 5 \end{bmatrix} \rightarrow \begin{bmatrix} 1 & 1 & 1 & | & 3 \\ 0 & -3 & 0 & | & -6 \\ 0 & -3 & 0 & | & -4 \end{bmatrix} \rightarrow \begin{bmatrix} 1 & 1 & 1 & | & 3 \\ 0 & -3 & 0 & | & -6 \\ 0 & 0 & 0 & | & 2 \end{bmatrix}$$

$$-1 \rightarrow R_1 + R_2 \rightarrow R_2 \qquad -1 \rightarrow R_2 + R_3 \rightarrow R_3$$
$$-3 \rightarrow R_1 + R_3 \rightarrow R_3$$

The last line corresponds to the equation $0x + 0y + 0z = 2$ which is false. The system has no solution.

9.
$$x + y + z = 3$$
$$x - 2y + z = -3$$
$$x \quad + z = 1$$

$$\begin{bmatrix} 1 & 1 & 1 & | & 3 \\ 1 & -2 & 1 & | & -3 \\ 1 & 0 & 1 & | & 1 \end{bmatrix} \rightarrow \begin{bmatrix} 1 & 1 & 1 & | & 3 \\ 0 & -3 & 0 & | & -6 \\ 0 & -1 & 0 & | & -2 \end{bmatrix} \rightarrow \begin{bmatrix} 1 & 1 & 1 & | & 3 \\ 0 & -3 & 0 & | & -6 \\ 0 & 0 & 0 & | & 0 \end{bmatrix}$$

$$-1 \rightarrow R_1 + R_2 \rightarrow R_2 \qquad R_2 -3 \rightarrow R_3 \rightarrow R_3$$
$$-1 \rightarrow R_1 + R_3 \rightarrow R_3$$

$$\begin{aligned} x + y + z &= 3 \\ -3y &= -6 \\ 0 &= 0 \end{aligned} \qquad \begin{aligned} -3y &= -6 \\ y &= 2 \end{aligned} \qquad \begin{aligned} x + 2 + z &= 3 \\ x + z &= 1 \\ x &= 1 - z \end{aligned}$$

There are infinitely many solutions such that for any real number, $z = k$, solutions are of the form $(1 - k, 2, k)$.

11. Let $x =$ the number of dimes
Let $y =$ the number of quarters
Let $z =$ the number of one-dollar coins

$$\begin{aligned} 10x + 25y + 100z &= 5800 \\ 2y &= z + 80 \\ y + 2x &= 3z \end{aligned} \quad - \quad \begin{aligned} 10x + 25y + 100z &= 5800 \\ 2y - z &= 80 \\ -2x + y - 3z &= 0 \end{aligned}$$

$$\begin{bmatrix} 10 & 25 & 100 & | & 5800 \\ 0 & 2 & -1 & | & 80 \\ 2 & 1 & -3 & | & 0 \end{bmatrix} \rightarrow \begin{bmatrix} 10 & 25 & 100 & | & 5800 \\ 0 & 2 & -1 & | & 80 \\ 0 & 20 & 115 & | & 5800 \end{bmatrix} \rightarrow \begin{bmatrix} 10 & 25 & 100 & | & 5800 \\ 0 & 2 & -1 & | & 80 \\ 0 & 0 & 125 & | & 5000 \end{bmatrix}$$

$$R_1 - 5 \rightarrow R_3 \rightarrow R_3 \qquad -10 \rightarrow R_2 + R_3 \rightarrow R_3$$

$$\begin{aligned} 10x + 25y + 100z &= 5800 \\ 2y - z &= 80 \\ 125z &= 5000 \end{aligned} \qquad \begin{aligned} 125z &= 5000 \\ z &= 40 \end{aligned} \qquad \begin{aligned} 2y - 40 &= 80 \\ 2y &= 120 \\ y &= 60 \end{aligned} \qquad \begin{aligned} 10x + 25(60) + 100(40) &= 5800 \\ 10x &= 300 \\ x &= 30 \end{aligned}$$

Nancy has 30 dimes, 60 quarters, and 40 one-dollar coins.

13. Let x = the number of Type I machine
Let y = the number of Type II machine
Let z = the number of Type III machine
$$20x + 24y + 30z = 760$$
$$10x + 18y + 10z = 380$$
$$30y + 30z = 660$$

$$\begin{array}{ccc|c} -20 & 24 & 30 & 760 \\ 10 & 18 & 10 & 380 \\ 0 & 30 & 30 & 660 \end{array} \rightarrow \begin{array}{ccc|c} -20 & 24 & 30 & 760 \\ 0 & -12 & 10 & 0 \\ 0 & 30 & 30 & 660 \end{array} \rightarrow \begin{array}{ccc|c} -20 & 24 & 30 & 760 \\ 0 & -12 & 10 & 0 \\ 0 & 0 & 110 & 1320 \end{array}$$

$$R_1 - 2 \to R_2 \to R_2 \qquad 5 \to R_2 + 2 \to R_3 \to R_3$$

$$20x + 24y + 30z = 760 \qquad\qquad -12y + 10(12) = 0 \qquad\qquad 20x + 24(10) + 30(12) = 760$$
$$-12y + 10z = 0 \qquad 110z = 1320 \qquad -12y = -120 \qquad\qquad 20x = 160$$
$$110z = 1320 \qquad z = 12 \qquad\qquad y = 10 \qquad\qquad x = 8$$

Mechano has 8 Type I machines, 10 Type II machines, and 12 Type III machines.

15. Let x = the percent of Type I fertilizer
Let y = the percent of Type II fertilizer
Let z = the percent of Type III fertilizer
$$6x + 8y + 12z = 8$$
$$6x + 12y + 8z = 8$$
$$8x + 4y + 12z = 8$$

$$\begin{array}{ccc|c} 6 & 8 & 12 & 8 \\ 6 & 12 & 8 & 8 \\ 8 & 4 & 12 & 8 \end{array} \sim \begin{array}{ccc|c} 6 & 8 & 12 & 8 \\ 0 & 4 & 4 & 0 \\ 0 & 20 & 12 & 8 \end{array} \sim \begin{array}{ccc|c} 6 & 8 & 12 & 8 \\ 0 & 4 & 4 & 0 \\ 0 & 0 & 32 & 8 \end{array}$$

$$R_1 \times R_2 \times R_2 \qquad 5 \times R_2 + R_3 \times R_3$$
$$4 \times R_1 \times 3 \times R_3 \times R_3$$

$$6x + 8y + 12z = 8 \qquad\qquad -4y + 4(0.25) = 0 \qquad 6x + 8(0.25) + 12(0.25) = 8$$
$$-4y + 4z = 0 \qquad 32z = 8 \qquad -4y = -1 \qquad\qquad 6x = 3$$
$$32z = 8 \qquad z = 0.25 \qquad\qquad y = 0.25 \qquad\qquad x = 0.50$$

They should use 50% of Type I, 25% of Type II, and 25% of Type III.

17. a. $\begin{array}{cc} 1 & 2 \\ 2 & 4 \end{array}$ $ad - bc = 1(4) - 2(2) = 4 - 4 = 0$ singular

b. $\begin{vmatrix} 2 & 3 \\ 3 & 5 \end{vmatrix}$ $ad - bc = 2(5) - (-3)(3) = 10 + 9 = 19$ nonsingular

c. $\begin{array}{cc} 0 & 2 \\ 2 & 4 \end{array}$ $ad - bc = 0(4) - 2(2) = 0 - 4 = -4$ nonsingular

19. $(8)(-5) - (22)(5) = -40 - 110 = -150$

21. $(2)(-4)-(5)(93)=-8-465=-473$

23. $\dfrac{(1)(3)-(3)(5)}{(1)(-3)-(1)(93)}=\dfrac{3-15}{-3-93}=\dfrac{-12}{-96}=\dfrac{1}{8}$

25. Sample answer. A system of equations is consistent when the final row of the matrix has a single non-zero number to the left of the bar which gives an equation with one variable.

27. Sample answer. A system of equations is dependent when the final row of the matrix contains only zeros (both sides of the bar).

29. $2x+3y-\ z=-1$
  $3x+4y+2z=14$
  $\ x-6y-5z=\ 4$

$$\left[\begin{array}{ccc|c} 2 & 3 & -1 & -1 \\ 3 & 4 & 2 & 14 \\ 1 & -6 & -5 & 4 \end{array}\right] \rightarrow \left[\begin{array}{ccc|c} 2 & 3 & -1 & -1 \\ 0 & 1 & -7 & -31 \\ 0 & -9 & 15 & 9 \end{array}\right] \rightarrow \left[\begin{array}{ccc|c} 2 & 3 & -1 & -1 \\ 0 & 1 & -7 & -31 \\ 0 & 0 & 114 & 456 \end{array}\right]$$

$3\cdot R_1-2\cdot R_2\rightarrow R_2 \qquad -15\cdot R_2+R_3\rightarrow R_3$

$R_1-2\cdot R_3\rightarrow R_3$

$2x+3y-\ z=\ -1$ 　　 $114z=456$ 　　 $y-7(4)=-31$ 　　 $2x+3(-3)-4=-1$

　$y-7z=-31$ 　　　 $z=4$ 　　　 $y-28=-31$ 　　　 $2x=12$

　　$114z=456$ 　　　　　　　　 $y=-3$ 　　　　　　 $x=6$

The solution is $(6, -3, 4)$.

31. $5x+6y-30z=13$
  $2x+4y-12z=\ 6$
  $\ x+2y-\ 6z=\ 8$

$$\left[\begin{array}{ccc|c} 5 & 6 & -30 & 13 \\ 2 & 4 & -12 & 6 \\ 1 & 2 & -6 & 8 \end{array}\right] \rightarrow \left[\begin{array}{ccc|c} 5 & 6 & -30 & 13 \\ 0 & 8 & 0 & 4 \\ 0 & -4 & 0 & -27 \end{array}\right] \rightarrow \left[\begin{array}{ccc|c} 5 & 6 & -30 & 13 \\ 0 & 8 & 0 & 4 \\ 0 & 0 & 0 & -50 \end{array}\right]$$

$-2\cdot R_1+5\cdot R_2\rightarrow R_2 \qquad R_2+2\cdot R_3\rightarrow R_3$

$R_1-5\cdot R_3\rightarrow R_3$

The last line corresponds to the equation $0x+0y+0z=-50$ which is false. The system has no solution.

33. $3x - y + z = 3$
$x - 2y + 2z = 1$
$2x + y - z = 2$

$$\begin{bmatrix} 3 & -1 & 1 \\ 1 & -2 & 2 \\ 2 & 1 & -1 \end{bmatrix}\begin{matrix} 3 \\ 1 \\ 2 \end{matrix} \rightarrow \begin{bmatrix} 3 & -1 & 1 \\ 0 & 5 & -5 \\ 0 & 5 & -5 \end{bmatrix}\begin{matrix} 3 \\ 0 \\ 0 \end{matrix} \rightarrow \begin{bmatrix} 3 & -1 & 1 \\ 0 & 5 & -5 \\ 0 & 0 & 0 \end{bmatrix}\begin{matrix} 3 \\ 0 \\ 0 \end{matrix}$$

$R_1 - 3 \to R_2 \to R_2 \qquad R_2 - R_3 \to R_3$

$-2 \to R_1 + 3 \to R_3 \to R_3$

$3x - y + z = 3$	$5y - 5z = 0$	$3x - z + z = 3$
$5y - 5z = 0$	$5y = 5z$	$3x = 3$
$0 = 0$	$y = z$	$x = 1$

There are infinitely many solutions such that for any real number, $z = k$, solutions are of the form $(1, k, k)$.

35.   Let $x = $ the number of nickels
Let $y = $ the number of dimes
Let $z = $ the number of quarters

$5x + 10y + 25z = 1410$	$- \quad 5x + 10y + 25z = 1410$
$z = y + 3x$	$- \quad -3x - y + z = 0$
$25z = 5x + 10y + 890$	$- \quad -5x - 10y + 25z = 890$

$$\begin{bmatrix} 5 & 10 & 25 \\ -3 & -1 & 1 \\ -5 & -10 & 25 \end{bmatrix}\begin{matrix} 1410 \\ 0 \\ 890 \end{matrix} \rightarrow \begin{bmatrix} 5 & 10 & 25 \\ 0 & 25 & 80 \\ 0 & 0 & 50 \end{bmatrix}\begin{matrix} 1410 \\ 4230 \\ 2300 \end{matrix}$$

$3 \to R_1 + 5 \to R_2 \to R_2$

$R_1 + R_3 \to R_3$

$5x + 10y + 25z = 1410$		$25y + 80(46) = 4230$	$5x + 10(22) + 25(46) = 1410$
$25y + 80z = 4230$	$50z = 2300$	$25y = 550$	$5x = 40$
$50z = 2300$	$z = 46$	$y = 22$	$x = 8$

There are 8 nickels, 22 dimes, and 46 quarters.

4.5   Determinants and Cramer's Rule

Problems 4.5

1.   a. $\begin{vmatrix} -2 & 6 \\ -3 & -5 \end{vmatrix} = (-2)(-5) - (-3)(6)$

$\qquad = 10 + 18 = 28$

b. $\begin{vmatrix} 2 & -8 \\ -3 & -12 \end{vmatrix} = (2)(-12) - (-3)(-8)$

$\qquad = -24 - 24 = -48$

2.

$$D = \begin{vmatrix} 2 & 3 \\ 5 & -4 \end{vmatrix} = -8 - 15 = -23$$

$$D_x = \begin{vmatrix} 16 & 3 \\ 17 & -4 \end{vmatrix} = -64 - 51 = -115$$

$$D_y = \begin{vmatrix} 2 & 16 \\ 5 & 17 \end{vmatrix} = 34 - 80 = -46$$

$$x = \frac{D_x}{D} = \frac{-115}{-23} = 5 \quad y = \frac{D_y}{D} = \frac{-46}{-23} = 2$$

The solution is (5, 2).

3.

$$\begin{vmatrix} 2 & -3 & 4 \\ -1 & 2 & 1 \\ 1 & 1 & -2 \end{vmatrix}$$

$$= 2\begin{vmatrix} 2 & 1 \\ 1 & -2 \end{vmatrix} - (-3)\begin{vmatrix} -1 & 1 \\ 1 & -2 \end{vmatrix} + 4\begin{vmatrix} -1 & 2 \\ 1 & 1 \end{vmatrix}$$

$$= 2(-4-1) + 3(2-1) + 4(-1-2)$$

$$= -10 + 3 - 12 = -19$$

4.   a.

$$\begin{vmatrix} -2 & -1 & 4 \\ 0 & 1 & -1 \\ -1 & 2 & 1 \end{vmatrix} = 4\begin{vmatrix} 0 & 1 \\ -1 & 2 \end{vmatrix} - (-1)\begin{vmatrix} -2 & -1 \\ -1 & 2 \end{vmatrix} + 1\begin{vmatrix} -2 & -1 \\ 0 & 1 \end{vmatrix} = 4(0+1) + 1(-4-1) + 1(-2-0)$$

$$= 4 - 5 - 2 = -3$$

b.

$$\begin{vmatrix} -1 & 3 & -2 \\ -4 & 2 & 0 \\ 0 & -1 & 1 \end{vmatrix} = -2\begin{vmatrix} -4 & 2 \\ 0 & -1 \end{vmatrix} - (0)\begin{vmatrix} -1 & 3 \\ 0 & -1 \end{vmatrix} + 1\begin{vmatrix} -1 & 3 \\ -4 & 2 \end{vmatrix} = -2(4-0) + 0(1-0) + 1(-2+12)$$

$$= -8 + 0 + 10 = 2$$

5.

$$D = \begin{vmatrix} 1 & 1 & 1 \\ 1 & -1 & 1 \\ 1 & 1 & -2 \end{vmatrix} = 1\begin{vmatrix} -1 & 1 \\ 1 & -2 \end{vmatrix} - 1\begin{vmatrix} 1 & 1 \\ 1 & -2 \end{vmatrix} + 1\begin{vmatrix} 1 & -1 \\ 1 & 1 \end{vmatrix} = 1(2-1) - 1(-2-1) + 1(1+1) = 1 + 3 + 2 = 6$$

$$D_x = \begin{vmatrix} 6 & 1 & 1 \\ 2 & -1 & 1 \\ -3 & 1 & -2 \end{vmatrix} = 6\begin{vmatrix} -1 & 1 \\ 1 & -2 \end{vmatrix} - 1\begin{vmatrix} 2 & 1 \\ -3 & -2 \end{vmatrix} + 1\begin{vmatrix} 2 & -1 \\ -3 & 1 \end{vmatrix} = 6(2-1) - 1(-4+3) + 1(2-3)$$

$$= 6 + 1 - 1 = 6$$

$$D_y = \begin{vmatrix} 1 & 6 & 1 \\ 1 & 2 & 1 \\ 1 & -3 & -2 \end{vmatrix} = 1\begin{vmatrix} 2 & 1 \\ -3 & -2 \end{vmatrix} - 6\begin{vmatrix} 1 & 1 \\ 1 & -2 \end{vmatrix} + 1\begin{vmatrix} 1 & 2 \\ 1 & -3 \end{vmatrix} = 1(-4+3) - 6(-2-1) + 1(-3-2)$$

$$= -1 + 18 - 5 = 12$$

$$D_z = \begin{vmatrix} 1 & 1 & 6 \\ 1 & -1 & 2 \\ 1 & 1 & -3 \end{vmatrix} = 1\begin{vmatrix} -1 & 2 \\ 1 & -3 \end{vmatrix} - 1\begin{vmatrix} 1 & 2 \\ 1 & -3 \end{vmatrix} + 6\begin{vmatrix} 1 & -1 \\ 1 & 1 \end{vmatrix} = 1(3-2) - 1(-3-2) + 6(1+1) = 1 + 5 + 12 = 18$$

$$x = \frac{D_x}{D} = \frac{6}{6} = 1 \qquad y = \frac{D_y}{D} = \frac{12}{6} = 2 \qquad z = \frac{D_z}{D} = \frac{18}{6} = 3$$

The solution is (1, 2, 3).

6.

$$D = \begin{vmatrix} 1 & \neq 1 & \neq 1 \\ 1 & 2 & 1 \\ \neq 1 & 1 & 1 \end{vmatrix} = 1\begin{vmatrix} 2 & 1 \\ 1 & 1 \end{vmatrix} \neq (\neq 1)\begin{vmatrix} 1 & 1 \\ \neq 1 & 1 \end{vmatrix} + (\neq 1)\begin{vmatrix} 1 & 2 \\ \neq 1 & 1 \end{vmatrix} = 1(2 \neq 1) + 1(1+1) \neq 1(1+2)$$

$$= 1 + 2 \neq 3 = 0$$

$$D_x = \begin{vmatrix} 2 & \neq 1 & \neq 1 \\ 6 & 2 & 1 \\ 4 & 1 & 1 \end{vmatrix} = 2\begin{vmatrix} 2 & 1 \\ 1 & 1 \end{vmatrix} \neq (\neq 1)\begin{vmatrix} 6 & 1 \\ 4 & 1 \end{vmatrix} + (\neq 1)\begin{vmatrix} 6 & 2 \\ 4 & 1 \end{vmatrix} = 2(2 \neq 1) + 1(6 \neq 4) \neq 1(6 \neq 8)$$

$$= 2 + 2 + 2 = 6 \neq 0$$

Since $D = 0$ and $D_x \neq 0$, the system is inconsistent. There is no solution.

Exercises 4.5

1. $\begin{vmatrix} 1 & 1 \\ 0 & 2 \end{vmatrix} = (1)(2) - (0)(1) = 2 - 0 = 2$

3. $\begin{vmatrix} -3 & -2 \\ 5 & 1 \end{vmatrix} = (-3)(1) - (5)(-2) = -3 + 10 = 7$

5. $\begin{vmatrix} -2 & 0 \\ 5 & -3 \end{vmatrix} = (-2)(-3) - (5)(0) = 6 - 0 = 6$

7. $\begin{vmatrix} \dfrac{1}{2} & |\dfrac{1}{4} \\ \dfrac{1}{2} & \dfrac{3}{4} \end{vmatrix} = \begin{vmatrix} \dfrac{1}{2} \end{vmatrix}\begin{vmatrix} \dfrac{3}{4} \end{vmatrix} | \begin{vmatrix} \dfrac{1}{2} \end{vmatrix}|\begin{vmatrix} \dfrac{1}{4} \end{vmatrix}$

$$= \dfrac{3}{8} + \dfrac{1}{8} = \dfrac{4}{8} = \dfrac{1}{2}$$

9. $\begin{vmatrix} \dfrac{3}{5} & \dfrac{1}{2} \\ |\dfrac{1}{4} & |\dfrac{1}{2} \end{vmatrix} = \begin{vmatrix} \dfrac{3}{5} \end{vmatrix}|\begin{vmatrix} \dfrac{1}{2} \end{vmatrix} | \begin{vmatrix} \dfrac{1}{4} \end{vmatrix}\begin{vmatrix} \dfrac{1}{2} \end{vmatrix}$

$$= |\dfrac{3}{10} + \dfrac{1}{8} = |\dfrac{12}{40} + \dfrac{5}{40} = |\dfrac{7}{40}$$

11.

$$D = \begin{vmatrix} 1 & 1 \\ 3 & -1 \end{vmatrix} = -1 - 3 = -4$$

$$D_x = \begin{vmatrix} 5 & 1 \\ 3 & -1 \end{vmatrix} = -5 - 3 = -8$$

$$D_y = \begin{vmatrix} 1 & 5 \\ 3 & 3 \end{vmatrix} = 3 - 15 = -12$$

$$x = \dfrac{D_x}{D} = \dfrac{-8}{-4} = 2 \quad y = \dfrac{D_y}{D} = \dfrac{-12}{-4} = 3$$

The solution is (2, 3).

13.

$$D = \begin{vmatrix} 1 & 1 \\ 1 & -1 \end{vmatrix} = -1 - 1 = -2$$

$$D_x = \begin{vmatrix} 9 & 1 \\ -1 & -1 \end{vmatrix} = -9 + 1 = -8$$

$$D_y = \begin{vmatrix} 1 & 9 \\ 1 & -1 \end{vmatrix} = -1 - 9 = -10$$

$$x = \dfrac{D_x}{D} = \dfrac{-8}{-2} = 4 \quad y = \dfrac{D_y}{D} = \dfrac{-10}{-2} = 5$$

The solution is (4, 5).

15.
$$D = \begin{vmatrix} 4 & 9 \\ 3 & 7 \end{vmatrix} = 28 - 27 = 1$$

$$D_x = \begin{vmatrix} 3 & 9 \\ 2 & 7 \end{vmatrix} = 21 - 18 = 3$$

$$D_y = \begin{vmatrix} 4 & 3 \\ 3 & 2 \end{vmatrix} = 8 - 9 = -1$$

$$x = \frac{D_x}{D} = \frac{3}{1} = 3 \quad y = \frac{D_y}{D} = \frac{-1}{1} = -1$$

The solution is $(3, -1)$.

17.
$$D = \begin{vmatrix} 1 & -1 \\ 1 & -2 \end{vmatrix} = -2 + 1 = -1$$

$$D_x = \begin{vmatrix} -1 & -1 \\ -6 & -2 \end{vmatrix} = 2 - 6 = -4$$

$$D_y = \begin{vmatrix} 1 & -1 \\ 1 & -6 \end{vmatrix} = -6 + 1 = -5$$

$$x = \frac{D_x}{D} = \frac{-4}{-1} = 4 \quad y = \frac{D_y}{D} = \frac{-5}{-1} = 5$$

The solution is $(4, 5)$.

19.
$$D = \begin{vmatrix} 2 & 3 \\ 6 & 9 \end{vmatrix} = 18 - 18 = 0$$

$$D_x = \begin{vmatrix} -13 & 3 \\ -39 & 9 \end{vmatrix} = -117 + 117 = 0$$

$$2x + 3y = -13$$
$$3y = -2x - 13$$
$$y = \frac{-2x - 13}{3}$$

There are infinitely many solutions of the form $\left(x, \frac{-2x-13}{3}\right)$. The system is dependent.

21.
$$D = \begin{vmatrix} 1 & -1 \\ 1 & -2 \end{vmatrix} = -2 + 1 = -1$$

$$D_x = \begin{vmatrix} 1 & -1 \\ 4 & -2 \end{vmatrix} = -2 + 4 = 2$$

$$D_y = \begin{vmatrix} 1 & 1 \\ 1 & 4 \end{vmatrix} = 4 - 1 = 3$$

$$x = \frac{D_x}{D} = \frac{2}{-1} = -2 \quad y = \frac{D_y}{D} = \frac{3}{-1} = -3$$

The solution is $(-2, -3)$.

23.
$$D = \begin{vmatrix} 1 & 3 \\ 2 & 6 \end{vmatrix} = 6 - 6 = 0$$

$$D_x = \begin{vmatrix} 6 & 3 \\ 5 & 6 \end{vmatrix} = 36 - 15 = 21$$

Since $D = 0$ and $D_x \neq 0$, the system is inconsistent. There is no solution.

25.
$$x = 7y + 3 \quad - \quad x - 7y = 3$$
$$2x + 3y = 23 \quad - \quad 2x + 3y = 23$$

$$D = \begin{vmatrix} 1 & -7 \\ 2 & 3 \end{vmatrix} = 3 + 14 = 17$$

$$D_x = \begin{vmatrix} 3 & -7 \\ 23 & 3 \end{vmatrix} = 9 + 161 = 170$$

$$D_y = \begin{vmatrix} 1 & 3 \\ 2 & 23 \end{vmatrix} = 23 - 6 = 17$$

$$x = \frac{D_x}{D} = \frac{170}{17} = 10 \quad y = \frac{D_y}{D} = \frac{17}{17} = 1$$

The solution is $(10, 1)$.

27.
$$y = -3x + 17 \rightarrow 3x + y = 17$$
$$2x - y = 8 \qquad \rightarrow 2x - y = 8$$

$$D = \begin{vmatrix} 3 & 1 \\ 2 & -1 \end{vmatrix} = -3 - 2 = -5$$

$$D_x = \begin{vmatrix} 17 & 1 \\ 8 & -1 \end{vmatrix} = -17 - 8 = -25$$

$$D_y = \begin{vmatrix} 3 & 17 \\ 2 & 8 \end{vmatrix} = 24 - 34 = -10$$

$$x = \frac{D_x}{D} = \frac{-25}{-5} = 5 \quad y = \frac{D_y}{D} = \frac{-10}{-5} = 2$$

The solution is (5, 2).

29.
$$D = \begin{vmatrix} \dfrac{1}{2} & -\dfrac{1}{3} \\ \dfrac{1}{3} & \dfrac{1}{4} \end{vmatrix} = \frac{1}{8} + \frac{1}{9} = \frac{9}{72} + \frac{8}{72} = \frac{17}{72}$$

$$D_x = \begin{vmatrix} -\dfrac{1}{6} & -\dfrac{1}{3} \\ -\dfrac{7}{12} & \dfrac{1}{4} \end{vmatrix} = -\frac{1}{24} - \frac{7}{36}$$

$$= -\frac{3}{72} - \frac{14}{72} = -\frac{17}{72}$$

$$D_y = \begin{vmatrix} \dfrac{1}{2} & -\dfrac{1}{6} \\ \dfrac{1}{3} & -\dfrac{7}{12} \end{vmatrix} = -\frac{7}{24} + \frac{1}{18}$$

$$= -\frac{21}{72} + \frac{4}{72} = -\frac{17}{72}$$

$$x = \frac{D_x}{D} = \frac{-\dfrac{17}{72}}{\dfrac{17}{72}} = -1 \quad y = \frac{D_y}{D} = \frac{-\dfrac{17}{72}}{\dfrac{17}{72}} = -1$$

The solution is (−1, −1).

31.
$$\begin{vmatrix} 1 & 3 & 2 \\ 2 & 4 & 1 \\ 3 & 6 & 5 \end{vmatrix} = 1 \begin{vmatrix} 4 & 1 \\ 6 & 5 \end{vmatrix} - 3 \begin{vmatrix} 2 & 1 \\ 3 & 5 \end{vmatrix} + 2 \begin{vmatrix} 2 & 4 \\ 3 & 6 \end{vmatrix} = 1(20 - 6) - 3(10 - 3) + 2(12 - 12) = 14 - 21 + 0 = -7$$

33.
$$\begin{vmatrix} 1 & 2 & 3 \\ 4 & 5 & 6 \\ 7 & 8 & 9 \end{vmatrix} = 1 \begin{vmatrix} 5 & 6 \\ 8 & 9 \end{vmatrix} - 2 \begin{vmatrix} 4 & 6 \\ 7 & 9 \end{vmatrix} + 3 \begin{vmatrix} 4 & 5 \\ 7 & 8 \end{vmatrix} = 1(45 - 48) - 2(36 - 42) + 3(32 - 35) = -3 + 12 - 9 = 0$$

35.
$$\begin{vmatrix} 2 & 1 & 3 \\ 1 & 2 & -1 \\ 3 & 1 & 5 \end{vmatrix} = 2 \begin{vmatrix} 2 & -1 \\ 1 & 5 \end{vmatrix} - 1 \begin{vmatrix} 1 & -1 \\ 3 & 5 \end{vmatrix} + 3 \begin{vmatrix} 1 & 2 \\ 3 & 1 \end{vmatrix} = 2(10 + 1) - 1(5 + 3) + 3(1 - 6) = 22 - 8 - 15 = -1$$

37.
$$\begin{vmatrix} 1 & 1 & 6 \\ 1 & 1 & 4 \\ 1 & -1 & 2 \end{vmatrix} = 1 \begin{vmatrix} 1 & 4 \\ -1 & 2 \end{vmatrix} - 1 \begin{vmatrix} 1 & 4 \\ 1 & 2 \end{vmatrix} + 6 \begin{vmatrix} 1 & 1 \\ 1 & -1 \end{vmatrix} = 1(2 + 4) - 1(2 - 4) + 6(-1 - 1) = 6 + 2 - 12 = -4$$

39.
$$\begin{vmatrix} 0 & -1 & 2 \\ 2 & 1 & -3 \\ 1 & -3 & 1 \end{vmatrix} = 0 \begin{vmatrix} 1 & -3 \\ -3 & 1 \end{vmatrix} - (-1) \begin{vmatrix} 2 & -3 \\ 1 & 1 \end{vmatrix} + 2 \begin{vmatrix} 2 & 1 \\ 1 & -3 \end{vmatrix} = 0(1 - 9) + 1(2 + 3) + 2(-6 - 1)$$

$$= 0 + 5 - 14 = -9$$

41.

$$D = \begin{vmatrix} 1 & 1 & 1 \\ 2 & -3 & 3 \\ 3 & -2 & -1 \end{vmatrix} = 1\begin{vmatrix} -3 & 3 \\ -2 & -1 \end{vmatrix} - 1\begin{vmatrix} 2 & 3 \\ 3 & -1 \end{vmatrix} + 1\begin{vmatrix} 2 & -3 \\ 3 & -2 \end{vmatrix} = 1(3+6) - 1(-2-9) + 1(-4+9)$$

$$= 9 + 11 + 5 = 25$$

$$D_x = \begin{vmatrix} 6 & 1 & 1 \\ 5 & -3 & 3 \\ -4 & -2 & -1 \end{vmatrix} = 6\begin{vmatrix} -3 & 3 \\ -2 & -1 \end{vmatrix} - 1\begin{vmatrix} 5 & 3 \\ -4 & -1 \end{vmatrix} + 1\begin{vmatrix} 5 & -3 \\ -4 & -2 \end{vmatrix} = 6(3+6) - 1(-5+12) + 1(-10-12)$$

$$= 54 - 7 - 22 = 25$$

$$D_y = \begin{vmatrix} 1 & 6 & 1 \\ 2 & 5 & 3 \\ 3 & -4 & -1 \end{vmatrix} = 1\begin{vmatrix} 5 & 3 \\ -4 & -1 \end{vmatrix} - 6\begin{vmatrix} 2 & 3 \\ 3 & -1 \end{vmatrix} + 1\begin{vmatrix} 2 & 5 \\ 3 & -4 \end{vmatrix} = 1(-5+12) - 6(-2-9) + 1(-8-15)$$

$$= 7 + 66 - 23 = 50$$

$$D_z = \begin{vmatrix} 1 & 1 & 6 \\ 2 & -3 & 5 \\ 3 & -2 & -4 \end{vmatrix} = 1\begin{vmatrix} -3 & 5 \\ -2 & -4 \end{vmatrix} - 1\begin{vmatrix} 2 & 5 \\ 3 & -4 \end{vmatrix} + 6\begin{vmatrix} 2 & -3 \\ 3 & -2 \end{vmatrix} = 1(12+10) - 1(-8-15) + 6(-4+9)$$

$$= 22 + 23 + 30 = 75$$

$$x = \frac{D_x}{D} = \frac{25}{25} = 1 \qquad y = \frac{D_y}{D} = \frac{50}{25} = 2 \qquad z = \frac{D_z}{D} = \frac{75}{25} = 3$$

The solution is (1, 2, 3).

43.

$$D = \begin{vmatrix} 6 & 5 & 4 \\ 5 & 4 & 3 \\ 4 & 3 & 1 \end{vmatrix} = 6\begin{vmatrix} 4 & 3 \\ 3 & 1 \end{vmatrix} - 5\begin{vmatrix} 5 & 3 \\ 4 & 1 \end{vmatrix} + 4\begin{vmatrix} 5 & 4 \\ 4 & 3 \end{vmatrix} = 6(4-9) - 5(5-12) + 4(15-16) = -30 + 35 - 4 = 1$$

$$D_x = \begin{vmatrix} 5 & 5 & 4 \\ 5 & 4 & 3 \\ 7 & 3 & 1 \end{vmatrix} = 5\begin{vmatrix} 4 & 3 \\ 3 & 1 \end{vmatrix} - 5\begin{vmatrix} 5 & 3 \\ 7 & 1 \end{vmatrix} + 4\begin{vmatrix} 5 & 4 \\ 7 & 3 \end{vmatrix} = 5(4-9) - 5(5-21) + 4(15-28)$$

$$= -25 + 80 - 52 = 3$$

$$D_y = \begin{vmatrix} 6 & 5 & 4 \\ 5 & 5 & 3 \\ 4 & 7 & 1 \end{vmatrix} = 6\begin{vmatrix} 5 & 3 \\ 7 & 1 \end{vmatrix} - 5\begin{vmatrix} 5 & 3 \\ 4 & 1 \end{vmatrix} + 4\begin{vmatrix} 5 & 5 \\ 4 & 7 \end{vmatrix} = 6(5-21) - 5(5-12) + 4(35-20)$$

$$= -96 + 35 + 60 = -1$$

$$D_z = \begin{vmatrix} 6 & 5 & 5 \\ 5 & 4 & 5 \\ 4 & 3 & 7 \end{vmatrix} = 6\begin{vmatrix} 4 & 5 \\ 3 & 7 \end{vmatrix} - 5\begin{vmatrix} 5 & 5 \\ 4 & 7 \end{vmatrix} + 5\begin{vmatrix} 5 & 4 \\ 4 & 3 \end{vmatrix} = 6(28-15) - 5(35-20) + 5(15-16)$$

$$= 78 - 75 - 5 = -2$$

$$x = \frac{D_x}{D} = \frac{3}{1} = 3 \qquad y = \frac{D_y}{D} = \frac{-1}{1} = -1 \qquad z = \frac{D_z}{D} = \frac{-2}{1} = -2$$

The solution is (3, -1, -2).

45.

$$D = \begin{vmatrix} 1 & -2 & 3 \\ 5 & 7 & -11 \\ -13 & 17 & 19 \end{vmatrix} = 1\begin{vmatrix} 7 & -11 \\ 17 & 19 \end{vmatrix} - (-2)\begin{vmatrix} 5 & -11 \\ -13 & 19 \end{vmatrix} + 3\begin{vmatrix} 5 & 7 \\ -13 & 17 \end{vmatrix}$$

$$= 1(133+187) + 2(95-143) + 3(85+91) = 320 - 96 + 528 = 752$$

$$D_x = \begin{vmatrix} 15 & -2 & 3 \\ -29 & 7 & -11 \\ 37 & 17 & 19 \end{vmatrix} = 15\begin{vmatrix} 7 & -11 \\ 17 & 19 \end{vmatrix} - (-2)\begin{vmatrix} -29 & -11 \\ 37 & 19 \end{vmatrix} + 3\begin{vmatrix} -29 & 7 \\ 37 & 17 \end{vmatrix}$$

$$= 15(133+187) + 2(-551+407) + 3(-493-259) = 4800 - 288 - 2256 = 2256$$

$$D_y = \begin{vmatrix} 1 & 15 & 3 \\ 5 & -29 & -11 \\ -13 & 37 & 19 \end{vmatrix} = 1\begin{vmatrix} -29 & -11 \\ 37 & 19 \end{vmatrix} - 15\begin{vmatrix} 5 & -11 \\ -13 & 19 \end{vmatrix} + 3\begin{vmatrix} 5 & -29 \\ -13 & 37 \end{vmatrix}$$

$$= 1(-551+407) - 15(95-143) + 3(185-377) = -144 + 720 - 576 = 0$$

$$D_z = \begin{vmatrix} 1 & -2 & 15 \\ 5 & 7 & -29 \\ -13 & 17 & 37 \end{vmatrix} = 1\begin{vmatrix} 7 & -29 \\ 17 & 37 \end{vmatrix} - (-2)\begin{vmatrix} 5 & -29 \\ -13 & 37 \end{vmatrix} + 15\begin{vmatrix} 5 & 7 \\ -13 & 17 \end{vmatrix}$$

$$= 1(259+493) + 2(185-377) + 15(85+91) = 752 - 384 + 2640 = 3008$$

$$x = \frac{D_x}{D} = \frac{2256}{752} = 3 \qquad y = \frac{D_y}{D} = \frac{0}{752} = 0 \qquad z = \frac{D_z}{D} = \frac{3008}{752} = 4$$

The solution is (3, 0, 4).

47.

$$D = \begin{vmatrix} 5 & 3 & 5 \\ 3 & 5 & 1 \\ 2 & 2 & 3 \end{vmatrix} = 5\begin{vmatrix} 5 & 1 \\ 2 & 3 \end{vmatrix} - 3\begin{vmatrix} 3 & 1 \\ 2 & 3 \end{vmatrix} + 5\begin{vmatrix} 3 & 5 \\ 2 & 2 \end{vmatrix} = 5(15-2) - 3(9-2) + 5(6-10) = 65 - 21 - 20 = 24$$

$$D_x = \begin{vmatrix} 3 & 3 & 5 \\ -5 & 5 & 1 \\ 7 & 2 & 3 \end{vmatrix} = 3\begin{vmatrix} 5 & 1 \\ 2 & 3 \end{vmatrix} - 3\begin{vmatrix} -5 & 1 \\ 7 & 3 \end{vmatrix} + 5\begin{vmatrix} -5 & 5 \\ 7 & 2 \end{vmatrix} = 3(15-2) - 3(-15-7) + 5(-10-35)$$

$$= 39 + 66 - 225 = -120$$

$$D_y = \begin{vmatrix} 5 & 3 & 5 \\ 3 & -5 & 1 \\ 2 & 7 & 3 \end{vmatrix} = 5\begin{vmatrix} -5 & 1 \\ 7 & 3 \end{vmatrix} - 3\begin{vmatrix} 3 & 1 \\ 2 & 3 \end{vmatrix} + 5\begin{vmatrix} 3 & -5 \\ 2 & 7 \end{vmatrix} = 5(-15-7) - 3(9-2) + 5(21+10)$$

$$= -110 - 21 + 155 = 24$$

$$D_z = \begin{vmatrix} 5 & 3 & 3 \\ 3 & 5 & -5 \\ 2 & 2 & 7 \end{vmatrix} = 5\begin{vmatrix} 5 & -5 \\ 2 & 7 \end{vmatrix} - 3\begin{vmatrix} 3 & -5 \\ 2 & 7 \end{vmatrix} + 3\begin{vmatrix} 3 & 5 \\ 2 & 2 \end{vmatrix} = 5(35+10) - 3(21+10) + 3(6-10)$$

$$= 225 - 93 - 12 = 120$$

$$x = \frac{D_x}{D} = \frac{-120}{24} = -5 \qquad y = \frac{D_y}{D} = \frac{24}{24} = 1 \qquad z = \frac{D_z}{D} = \frac{120}{24} = 5$$

The solution is (-5, 1, 5).

49.

$$D = \begin{vmatrix} 0 & 2 & 1 \\ 0 & -2 & 1 \\ 1 & 1 & 1 \end{vmatrix} = 0 \begin{vmatrix} -2 & 1 \\ 1 & 1 \end{vmatrix} - 0 \begin{vmatrix} 2 & 1 \\ 1 & 1 \end{vmatrix} + 1 \begin{vmatrix} 2 & 1 \\ -2 & 1 \end{vmatrix} = 0(-2-1) - 0(2-1) + 1(2+2) = 0 - 0 + 4 = 4$$

$$D_x = \begin{vmatrix} 9 & 2 & 1 \\ 1 & -2 & 1 \\ 1 & 1 & 1 \end{vmatrix} = 9 \begin{vmatrix} -2 & 1 \\ 1 & 1 \end{vmatrix} - 1 \begin{vmatrix} 2 & 1 \\ 1 & 1 \end{vmatrix} + 1 \begin{vmatrix} 2 & 1 \\ -2 & 1 \end{vmatrix} = 9(-2-1) - 1(2-1) + 1(2+2)$$

$$= -27 - 1 + 4 = -24$$

$$D_y = \begin{vmatrix} 0 & 9 & 1 \\ 0 & 1 & 1 \\ 1 & 1 & 1 \end{vmatrix} = 0 \begin{vmatrix} 1 & 1 \\ 1 & 1 \end{vmatrix} - 0 \begin{vmatrix} 9 & 1 \\ 1 & 1 \end{vmatrix} + 1 \begin{vmatrix} 9 & 1 \\ 1 & 1 \end{vmatrix} = 0(1-1) - 0(9-1) + 1(9-1) = 0 - 0 + 8 = 8$$

$$D_z = \begin{vmatrix} 0 & 2 & 9 \\ 0 & -2 & 1 \\ 1 & 1 & 1 \end{vmatrix} = 0 \begin{vmatrix} -2 & 1 \\ 1 & 1 \end{vmatrix} - 0 \begin{vmatrix} 2 & 9 \\ 1 & 1 \end{vmatrix} + 1 \begin{vmatrix} 2 & 9 \\ -2 & 1 \end{vmatrix} = 0(-2-1) - 0(2-9) + 1(2+18)$$

$$= 0 - 0 + 20 = 20$$

$$x = \frac{D_x}{D} = \frac{-24}{4} = -6 \qquad y = \frac{D_y}{D} = \frac{8}{4} = 2 \qquad z = \frac{D_z}{D} = \frac{20}{4} = 5$$

The solution is (–6, 2, 5).

51.

$$\begin{vmatrix} a & b & 0 \\ c & d & 0 \\ e & f & 0 \end{vmatrix} = 0 \begin{vmatrix} c & d \\ e & f \end{vmatrix} - 0 \begin{vmatrix} a & b \\ e & f \end{vmatrix} + 0 \begin{vmatrix} a & b \\ c & d \end{vmatrix} = 0 - 0 + 0 = 0$$

53.

$$\begin{vmatrix} a & b & c \\ 1 & 2 & 3 \\ a & b & c \end{vmatrix} = -1 \begin{vmatrix} b & c \\ b & c \end{vmatrix} + 2 \begin{vmatrix} a & c \\ a & c \end{vmatrix} - 3 \begin{vmatrix} a & b \\ a & b \end{vmatrix} = -1(bc - bc) + 2(ac - ac) - 3(ab - ab)$$

$$= -1(0) + 2(0) - 3(0) = 0 + 0 + 0 = 0$$

55.

$$\begin{vmatrix} 1 & 2 & 3 \\ 3 & 1 & 2 \\ 3k & 2k & k \end{vmatrix} = 3k \begin{vmatrix} 2 & 3 \\ 1 & 2 \end{vmatrix} \therefore 2k \begin{vmatrix} 1 & 3 \\ 3 & 2 \end{vmatrix} + k \begin{vmatrix} 1 & 2 \\ 3 & 1 \end{vmatrix} = 3k(4 \therefore 3) \therefore 2k(2 \therefore 9) + k(1 \therefore 6)$$

$$= 3k + 14k \therefore 5k = 12k$$

$$k \begin{vmatrix} 1 & 2 & 3 \\ 3 & 1 & 2 \\ 3 & 2 & 1 \end{vmatrix} = k \cdot 1 \begin{vmatrix} 1 & 2 \\ 2 & 1 \end{vmatrix} \therefore 2 \begin{vmatrix} 3 & 2 \\ 3 & 1 \end{vmatrix} + 3 \begin{vmatrix} 3 & 1 \\ 3 & 2 \end{vmatrix} = k \cdot 1(1 \therefore 4) \therefore 2(3 \therefore 6) + 3(6 \therefore 3)$$

$$= k(\therefore 3 + 6 + 9) = 12k$$

$$\therefore \begin{vmatrix} 1 & 2 & 3 \\ 3 & 1 & 2 \\ 3k & 2k & k \end{vmatrix} = k \begin{vmatrix} 1 & 2 & 3 \\ 3 & 1 & 2 \\ 3 & 2 & 1 \end{vmatrix}$$

57.

$$\begin{vmatrix} kb_1 & b_1 & 1 \\ kb_2 & b_2 & 2 \\ kb_3 & b_3 & 3 \end{vmatrix} = 1\begin{vmatrix} kb_2 & b_2 \\ kb_3 & b_3 \end{vmatrix} - 2\begin{vmatrix} kb_1 & b_1 \\ kb_3 & b_3 \end{vmatrix} + 3\begin{vmatrix} kb_1 & b_1 \\ kb_2 & b_2 \end{vmatrix}$$

$$= 1(kb_2b_3 - kb_2b_3) - 2(kb_1b_3 - kb_1b_3) + 3(kb_1b_2 - kb_1b_2)$$

$$= 1(0) - 2(0) + 3(0) = 0 + 0 + 0 = 0$$

59.

$$\begin{vmatrix} 1 & 1 & 1 \\ 2 & a & a \\ 3 & b & b \end{vmatrix} = 1\begin{vmatrix} a & a \\ b & b \end{vmatrix} - 2\begin{vmatrix} 1 & 1 \\ b & b \end{vmatrix} + 3\begin{vmatrix} 1 & 1 \\ a & a \end{vmatrix} = 1(ab - ab) - 2(b - b) + 3(a - a) = 1(0) - 2(0) + 3(0) = 0$$

61. $3x + 2y > 6$

$3(-3) + 2(1) = -9 + 2 = -7 \not> 6$

(−3, 1) does not satisfy the inequality.

63. $-2y < 6$

$-2(1) = -2 < 6$

(−3, 1) does satisfy the inequality.

65. $y < x$

$1 \not< -3$

(−3, 1) does not satisfy the inequality.

67.

$$\begin{vmatrix} x & y & 1 \\ 2 & 7 & 1 \\ 0 & 3 & 1 \end{vmatrix} = x\begin{vmatrix} 7 & 1 \\ 3 & 1 \end{vmatrix} - y\begin{vmatrix} 2 & 1 \\ 0 & 1 \end{vmatrix} + 1\begin{vmatrix} 2 & 7 \\ 0 & 3 \end{vmatrix} = x(7 - 3) - y(2 - 0) + 1(6 - 0) = 4x - 2y + 6$$

The equivalent equation is: $2x - y + 3 = 0$

69.

$$\begin{vmatrix} x & y & 1 \\ -1 & 4 & 1 \\ 8 & 2 & 1 \end{vmatrix} = x\begin{vmatrix} 4 & 1 \\ 2 & 1 \end{vmatrix} - y\begin{vmatrix} -1 & 1 \\ 8 & 1 \end{vmatrix} + 1\begin{vmatrix} -1 & 4 \\ 8 & 2 \end{vmatrix} = x(4 - 2) - y(-1 - 8) + 1(-2 - 32) = 2x + 9y - 34$$

The equation is: $2x + 9y - 34 = 0$

71.

$$\begin{vmatrix} x & y & 1 \\ a & 0 & 1 \\ 0 & b & 1 \end{vmatrix} = x\begin{vmatrix} 0 & 1 \\ b & 1 \end{vmatrix} - y\begin{vmatrix} a & 1 \\ 0 & 1 \end{vmatrix} + 1\begin{vmatrix} a & 0 \\ 0 & b \end{vmatrix} = x(0 - b) - y(a - 0) + 1(ab - 0) = -bx - ay + ab$$

The equivalent equation is: $bx + ay - ab = 0$

73. Sample answer. A system of equations is consistent when the value of the determinant D does not equal zero.

75. Sample answer. A system of equations is dependent when the value of the determinant D equals zero and the values of the variable determinants are also zero.

77. Sample answer. When a determinant has a row or column of zeros, the expansion of the minors along that row or column will have a value of zero because each minor is multiplied by zero.

79.
$$\begin{vmatrix} 1 & 1 & 0 \\ 0 & -1 & 1 \\ 2 & -1 & -3 \end{vmatrix} = 2\begin{vmatrix} 1 & 0 \\ -1 & 1 \end{vmatrix} - (-1)\begin{vmatrix} 1 & 0 \\ 0 & 1 \end{vmatrix} + (-3)\begin{vmatrix} 1 & 1 \\ 0 & -1 \end{vmatrix} = 2(1-0)+1(1-0)-3(-1-0) = 2+1+3 = 6$$

81.
$$D = \begin{vmatrix} 1 & \neq 1 & \neq 1 \\ 1 & 2 & 1 \\ \neq 1 & 1 & 1 \end{vmatrix} = 1\begin{vmatrix} 2 & 1 \\ 1 & 1 \end{vmatrix} \neq (\neq 1)\begin{vmatrix} 1 & 1 \\ \neq 1 & 1 \end{vmatrix} + (\neq 1)\begin{vmatrix} 1 & 2 \\ \neq 1 & 1 \end{vmatrix} = 1(2 \neq 1)+1(1+1) \neq 1(1+2)$$
$$= 1+2 \neq 3 = 0$$

$$D_x = \begin{vmatrix} 2 & \neq 1 & \neq 1 \\ 6 & 2 & 1 \\ 4 & 1 & 1 \end{vmatrix} = 2\begin{vmatrix} 2 & 1 \\ 1 & 1 \end{vmatrix} \neq (\neq 1)\begin{vmatrix} 6 & 1 \\ 4 & 1 \end{vmatrix} + (\neq 1)\begin{vmatrix} 6 & 2 \\ 4 & 1 \end{vmatrix} = 2(2 \neq 1)+1(6 \neq 4) \neq 1(6 \neq 8)$$
$$= 2+2+2 = 6 \neq 0$$

Since $D = 0$ and $D_x \neq 0$, the system is inconsistent. There is no solution.

83.
$$\begin{vmatrix} 2 & -3 & 4 \\ -1 & 2 & 1 \\ 1 & 1 & -2 \end{vmatrix} = 2\begin{vmatrix} 2 & 1 \\ 1 & -2 \end{vmatrix} - (-3)\begin{vmatrix} -1 & 1 \\ 1 & -2 \end{vmatrix} + 4\begin{vmatrix} -1 & 2 \\ 1 & 1 \end{vmatrix} = 2(-4-1)+3(2-1)+4(-1-2)$$
$$= -10+3-12 = -19$$

85.
$$\begin{vmatrix} -2 & 6 \\ -3 & 5 \end{vmatrix} = -10+18 = 8$$

4.6   Systems of Linear Inequalities

Problems 4.6

1. $y \geq 0$

$x < 3$

Graph the line $y = 0$. The boundary is a solid line.

Shade above the line.

Graph the border line $x = 3$. The boundary is a dashed line.

Shade to the left of the line.

The solution is the intersection of the graphs.

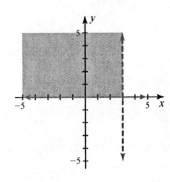

2. $y > x + 2$

$x + 2y < 4$

Graph the line $y = x + 2$. The boundary is a dashed line.

Test point $(0,0)$: $0 > 0 + 2$, $0 > 2$.

False, so shade the opposite side of the line.

Graph the line $x + 2y = 4$. The boundary is a dashed line.

Test point $(0,0)$: $0 + 2(0) < 4$, $0 < 4$.

True, so shade this side of the line.

The solution is the intersection of the graphs.

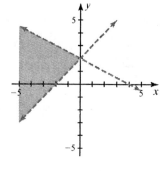

3. $x - 0$

$y - 0$

$y > -x - 2$

Graph the line $x = 0$. The boundary is a solid line.

Shade to the left of the line.

Graph the line $y = 0$. The boundary is a solid line.

Shade below the line.

Graph the line $y = -x - 2$. The boundary is a dashed line.

Test point $(0,0)$: $0 > -0 - 2$, $0 > -2$.

True, so shade this side of the line.

The solution is the intersection of the graphs.

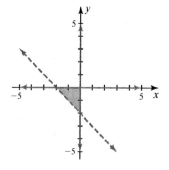

Exercises 4.6

1. $x \leq y \leq 2$ and $x + y \leq 6$

Graph the line $x \leq y = 2$. The boundary is a solid line.

Test point $(0,0)$: $0 \leq 0 \leq 2$, $0 \leq 2$.

False, so shade the opposite side of the line.

Graph the line $x + y = 6$. The boundary is a solid line.

Test point $(0,0)$: $0 + 0 \leq 6$, $0 \leq 6$.

True, so shade this side of the line.

The solution is the intersection of the graphs.

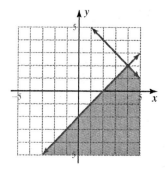

3. $2x \geq 3y \geq 6$ and $4x \geq 3y \geq 12$

Graph the line $2x \geq 3y = 6$. The boundary is a solid line.

Test point $(0,0)$: $2(0) \geq 3(0) \geq 6$, $0 \geq 6$.

True, so shade this side of the line.

Graph the line $4x \geq 3y = 12$. The boundary is a solid line.

Test point $(0,0)$: $4(0) \geq 3(0) \geq 12$, $0 \geq 12$.

False, so shade the opposite side of the line.

The solution is the intersection of the graphs.

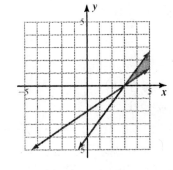

5. $2x \geq 3y \geq 5$ and $x \geq y$

Graph the line $2x \geq 3y = 5$. The boundary is a solid line.

Test point $(0,0)$: $2(0) \geq 3(0) \geq 5$, $0 \geq 5$.

True, so shade this side of the line.

Graph the line $x = y$. The boundary is a solid line.

Test point $(0,1)$: $0 \geq 1$.

False, so shade the opposite side of the line.

The solution is the intersection of the graphs.

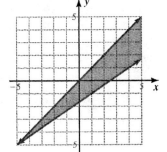

7. $x + 3y \geq 6$ and $x \geq y$

Graph the line $x + 3y = 6$. The boundary is a solid line.

Test point $(0,0)$: $(0) + 3(0) \geq 6$, $0 \geq 6$.

True, so shade this side of the line.

Graph the line $x = y$. The boundary is a solid line.

Test point $(0,1)$: $0 \geq 1$.

False, so shade the opposite side of the line.

The solution is the intersection of the graphs.

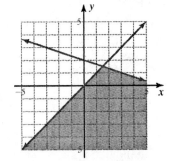

9. $x \leq y \leq 1$ and $3x \leq y < 3$

Graph the line $x \leq y = 1$. The boundary is a solid line.

Test point $(0,0)$: $(0) \leq (0) \leq 1$, $0 \leq 1$.

True, so shade this side of the line.

Graph the line $3x \leq y = 3$. The boundary is a dashed line.

Test point $(0, 0)$: $3(0) \leq 0 < 3$, $0 < 3$.

True, so shade this side of the line.

The solution is the intersection of the graphs.

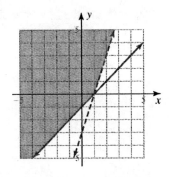

11. $x - 1$ and $x - 4$ and $y - 4$ and $x - 3y - -2$

Graph the line $x = 1$. The boundary is a solid line. Shade to the right of the line. Graph the line $x = 4$. The boundary is a solid line. Shade to the left of the line. Graph the line $y = 4$. The boundary is a solid line. Shade below the line. Graph the line $x - 3y = -2$. The boundary is a solid line.

Test point $(0,0)$: $0 - 3(0) - -2$, $0 - -2$.

False, so shade the opposite side of the line.

The solution is the intersection of these graphs.

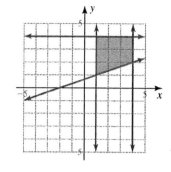

13. $x + y \le 1$ and $2y \le x \le 1$ and $x \le 1$

Graph the line $x + y = 1$. The boundary is a solid line.

Test point $(0, 0)$: $0 + 0 \le 1$, $0 \le 1$.

False, so shade the opposite side of the line.

Graph the line $2y \le x = 1$. The boundary is a solid line.

Test point $(0, 0)$: $2(0) \le 0 \le 1$, $0 \le 1$.

True, so shade this side of the line.

Graph the line $x = 1$. The boundary is a solid line. Shade to the left of the line.

The solution is the intersection of these graphs.

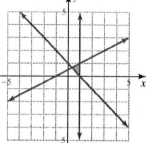

15. $x \le 1$ and $y \le 2$ and $4 \le 2x + y$ and $2x + y \le 6$

Graph the line $x = 1$. The boundary is a solid line. Shade to the right of the line. Graph the line $y = 2$. The boundary is a solid line. Shade above the line.

Graph the line $4 = 2x + y$. The boundary is a solid line.

Test point $(0, 0)$: $4 \le 2(0) + 0$, $4 \le 0$.

False, so shade the opposite side of the line.

Graph the line $2x + y = 6$. The boundary is a solid line.

Test point $(0, 0)$: $2(0) + 0 \le 6$, $0 \le 6$.

True, so shade this side of the line.

The solution is the intersection of these graphs.

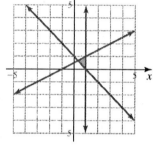

17. $2x - 4y = 12$
$$-4y = -2x + 12$$
$$y = \frac{1}{2}x - 3$$

19. $x - y = 5$
$$-y = -x + 5$$
$$y = x - 5$$

21.   Let $x =$ the number of cars

Let $y =$ the number of trucks

Objective function: $R = 20x + 35y$

Constraints: $100x + 200y \geq 12,000$ $x \geq 0$

$\qquad\qquad\quad x + y \geq 100$ $y \geq 0$

Graph the line $100x + 200y = 12,000$ or $x + 2y = 120$. Test point

$(0, 0)$: $0 + 2(0) \geq 120,\ 0 \geq 120$. True, so shade this side of the line.

Graph the line $x + y = 100$. Test point $(0, 0)$: $0 + 0 \geq 100,\ 0 \geq 100$.

True, so shade this side of the line. Graph the lines $x = 0$. Shade to

the right of the line. Graph the line $y = 0$. Shade above the line.

The solution is the intersection of these graphs.

The vertices are $(0, 0), (0, 60), (80, 20),$ and $(100, 0)$.

Evaluating the objective function:

At $(0, 0)$, $R = 20(0) + 35(0) = 0$

At $(0, 60)$, $R = 20(0) + 35(60) = 2100$

At $(80, 20)$, $R = 20(80) + 35(20) = 2300$

At $(100, 0)$, $R = 20(100) + 35(0) = 2000$

The maximum revenue is \$2300 when 80 cars and 20 trucks

are stored.

23.   Sample answer. It is possible to have no solution if none of the shaded regions intersect. Answers
will vary.

25. $y > 2x + 1$ and $x - -1$

Graph the line $y = 2x + 1$. The boundary is a dashed line.

Test point $(0,0)$: $0 > 2(0) + 1,\ 0 > 1$.

False, so shade the opposite side of the line.

Graph the line $x = -1$. The boundary is a solid line.

Shade the left side of the line.

The solution is the intersection of the graphs.

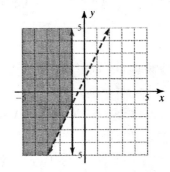

27. $x + y \geq 5$ and $2x \geq y \geq 4$ and $x \geq 0$ and $y \geq 0$

Graph the line $x + y = 5$. The boundary is a solid line.

Test point $(0, 0)$: $0 + 0 \geq 5$, $0 \geq 5$.

True, so shade this side of the line.

Graph the line $2x \geq y = 4$. The boundary is a solid line.

Test point $(0, 0)$: $2(0) \geq 0 \geq 4$, $0 \geq 4$.

True, so shade this side of the line.

Graph the line $x = 0$. The boundary is a solid line.

Shade to the right of the line. Graph the line $y = 0$.

The boundary is a solid line. Shade above the line.

The solution is the intersection of the graphs.

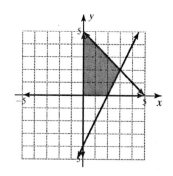

Review Exercises

1. **a.** Graph each equation.
$2x - y = 2$

$y = 3x - 4$

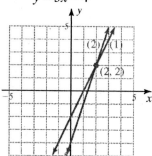

The solution is $(2, 2)$.

b. Graph each equation.
$x - 2y = 0$

$y = x - 2$

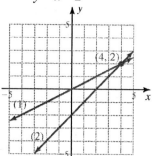

The solution is $(4, 2)$.

2. **a.** Graph each equation.
$2y - x = 3$

$4y = 2x + 8$

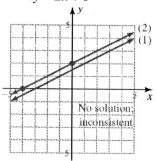

There is no solution. The system is inconsistent.

b. Graph each equation.
$2y + x = 4$

$2x = 10 - 4y$

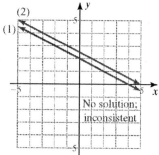

There is no solution. The system is inconsistent.

3. **a.** Graph each equation.

$3x + 2y = 6$

$y = 3 - \dfrac{3}{2}x$

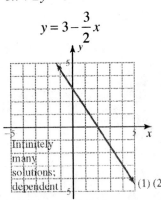

There are infinitely many solutions. The system is dependent.

b. Graph each equation.

$x + 2y = 4$

$2x = 8 - 4y$

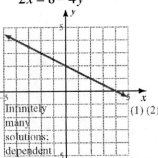

There are infinitely many solutions. The system is dependent.

4. **a.** Solve the second equation for y and substitute into the first equation:

$$2x - y = 4$$
$$x + y = 5 \rightarrow y = 5 - x$$
$$2x - (5 - x) = 4$$
$$2x - 5 + x = 4$$
$$3x = 9$$
$$x = 3$$
$$y = 5 - 3 = 2$$

The solution is $(3, 2)$.

b. Solve the second equation for x and substitute into the first equation:

$$2x + 3y = 10$$
$$x - y = -1 \rightarrow x = y - 1$$
$$2(y - 1) + 3y = 10$$
$$2y - 2 + 3y = 10$$
$$5y = 12$$
$$y = \dfrac{12}{5}$$
$$x = \dfrac{12}{5} - 1 = \dfrac{7}{5}$$

The solution is $\left(\frac{7}{5}, \frac{12}{5}\right)$.

5. **a.** $2x + 4y = 7$

$x = -2y - 1$

Substitute the second equation into the first equation and solve:

$$2(-2y - 1) + 4y = 7$$
$$-4y - 2 + 4y = 7$$
$$-2 = 7$$

This is a contradiction. There is no solution.

b. Solve the first equation for x, substitute into the second equation, and solve:

$$2y + x = 5 - x = 5 - 2y$$
$$3x = 10 - 6y$$
$$3(5 - 2y) = 10 - 6y$$
$$15 - 6y = 10 - 6y$$
$$15 = 10$$

This is a contradiction. There is no solution.

6. **a.** Solve the second equation for x, substitute into the first equation, and solve:

$$2y - x = 5$$
$$2x = 4y - 10 \rightarrow x = 2y - 5$$
$$2y - (2y - 5) = 5$$
$$2y - 2y + 5 = 5$$
$$5 = 5$$

This is an identity. There are infinitely many solutions of the form $\left(x, \frac{1}{2}x + \frac{5}{2}\right)$.

b. Substitute the second equation into the first equation, and solve:

$$x + 5y = 5$$
$$y = 1 - \frac{x}{5}$$
$$x + 5\left[1 - \frac{x}{5}\right] = 5$$
$$x + 5 - x = 5$$
$$5 = 5$$

This is an identity. There are infinitely many solutions of the form $\left(x, 1 - \frac{1}{5}x\right)$.

7. **a.** To eliminate x, multiply the first equation by -2, add the result to the second equation, and solve:

$$x - 3y = 7 \rightarrow -2x + 6y = -14$$
$$2x - y = 9 \rightarrow \underline{\;\;2x - y = \;\;\;9\;\;}$$
$$5y = -5$$
$$y = -1$$
$$x - 3(-1) = 7 \rightarrow x = 4$$

Solution: $(4, -1)$.

b. To eliminate x, multiply the second equation by -2, add the result to the first equation, and solve:

$$2x + 3y = 4 - \quad 2x + 3y = \;\;4$$
$$x + y = 1 - \quad \underline{-2x - 2y = -2}$$
$$y = 2$$
$$x + 2 = 1 - \quad x = -1$$

Solution: $(-1, 2)$.

8. **a.** To eliminate x, multiply the first equation by -3, add the result to the second equation, and solve:

$$2x + 3y = 7 - \quad -6x - 9y = -21$$
$$6x + 9y = 14 - \quad \underline{\;\;6x + 9y = \;\;14\;\;}$$
$$0 = -7$$

This is a contradiction. There is no solution.

b. To eliminate x, multiply the first equation by -2, add the result to the second equation, and solve:

$$3x - 4y = 5 \rightarrow -6x + 8y = -10$$
$$6x - 8y = 15 \rightarrow \underline{\;\;6x - 8y = \;\;15\;\;}$$
$$0 = 5$$

This is a contradiction. There is no solution.

9. **a.** To eliminate x, multiply the first equation by -10, add the result to the second equation, and solve:

$$\frac{x}{5} + \frac{y}{2} = \frac{1}{5} - \quad -2x - 5y = -2$$
$$2x + 5y = 2 - \quad \underline{\;\;2x + 5y = \;\;2\;\;}$$
$$0 = 0$$

This is an identity. There are infinitely many solutions of the form $\left(x, \frac{2-2x}{5}\right)$.

b. To eliminate x, multiply the second equation by -2, add the result to the first equation, and solve:

$$\frac{x}{3} - \frac{y}{4} = 2 \rightarrow \frac{x}{3} - \frac{y}{4} = \;\;2$$
$$\frac{x}{6} - \frac{y}{8} = 1 \rightarrow \underline{-\frac{x}{3} + \frac{y}{4} = -2}$$
$$0 = 0$$

This is an identity. There are infinitely many solutions of the form $\left(x, \frac{4}{3}x - 8\right)$.

10. a. Move all the variables to one side of each equation:

$$4y = -5x - 2 \rightarrow 5x + 4y = -2$$
$$4x = -3y - 1 \rightarrow 4x + 3y = -1$$

To eliminate y, multiply the first equation by -3 and the second equation by 4, add the results and solve:

$$5x + 4y = -2 \rightarrow -15x - 12y = 6$$
$$4x + 3y = -1 \rightarrow \underline{16x + 12y = -4}$$
$$x = 2$$
$$4(2) + 3y = -1$$
$$3y = -9$$
$$y = -3$$

The solution is $(2, -3)$.

b. Move all the variables to one side of each equation:

$$2x = 3y - 1 \rightarrow 2x - 3y = -1$$
$$2y = 3x - 1 \rightarrow -3x + 2y = -1$$

To eliminate x, multiply the first equation by 3 and the second equation by 2, add the results and solve:

$$2x - 3y = -1 \rightarrow 6x - 9y = -3$$
$$-3x + 2y = -1 \rightarrow \underline{-6x + 4y = -2}$$
$$-5y = -5$$
$$y = 1$$
$$2x = 3(1) - 1 = 2$$
$$x = 1$$

The solution is $(1, 1)$.

11. a.

$$x - y + z = 1$$
$$x + y - z = 7$$
$$2x + y + z = 9$$

Add the first equation to the second equation to eliminate z and solve:

$$2x = 8 \rightarrow x = 4$$

Add the second equation to the third equation to eliminate z:

$$3x + 2y = 16$$

Substitute and solve for the other variables:

$$3(4) + 2y = 16 \qquad 4 - 2 + z = 1$$
$$2y = 4 \qquad\qquad z = -1$$
$$y = 2$$

The solution is $(4, 2, -1)$.

b.

$$2x - 3y + z = -4$$
$$2x + y + z = 4$$
$$4x + 9y + 3z = 10$$

Multiply the first equation by -1, add to the second equation to eliminate x:

$$-2x + 3y - z = 4$$
$$\underline{2x + y + z = 4}$$
$$4y = 8 \rightarrow y = 2$$

Multiply the first equation by -2, add to the third equation to eliminate x:

$$-4x + 6y - 2z = 8$$
$$\underline{4x + 9y + 3z = 10}$$
$$15y + z = 18$$

Substitute and solve for the other variables:

$$15(2) + z = 18 \qquad 2x + 2 - 12 = 4$$
$$z = -12 \qquad\qquad 2x = 14$$
$$x = 7$$

The solution is $(7, 2, -12)$.

12. a.
$$2x + y - 2z = 4$$
$$-x + y + 2z = 2$$
$$3y + 2z = 12$$

Add the first equation to the second equation to eliminate z:
$$x + 2y = 6$$

Add the first equation to the third equation to eliminate z:
$$2x + 4y = 16$$

Multiply the first result by -2 and add to the second result:
$$-2x - 4y = -12$$
$$\underline{2x + 4y = 16}$$
$$0 = 4$$

This is a contradiction. There is no solution.

b.
$$2x + 2y + z = 4$$
$$-2x - y + z = 2$$
$$-2x + 3z = 9$$

Add the first equation to the second equation to eliminate x:
$$y + 2z = 6$$

Add the first equation to the third equation to eliminate x:
$$2y + 4z = 13$$

Multiply the first result by -2 and add to the second result:
$$-2y - 4z = -12$$
$$\underline{2y + 4z = 13}$$
$$0 = 1$$

This is a contradiction. There is no solution.

13. a.
$$x + 2y + 3z = 6$$
$$x - 2y - z = -2$$
$$x + z = 2$$

Add the first equation to the second equation to eliminate y:
$$2x + 2z = 4$$

Multiply the third equation by -2 and add to this result to eliminate z:
$$-2x - 2z = -4$$
$$\underline{2x + 2z = 4}$$
$$0 = 0$$

This is an identity. There are infinitely many solutions of the form $(2 - k, 2 - k, k)$.

b.
$$x + 2y = 4$$
$$y + 2z = 6$$
$$2x + 2y - 4z = -4$$

Multiply the second equation by 2 and add to the third equation to eliminate z:
$$2y + 4z = 12$$
$$\underline{2x + 2y - 4z = -4}$$
$$2x + 4y = 8$$

Multiply the first equation by -2 and add to this result to eliminate x:
$$-2x - 4y = -8$$
$$\underline{2x + 4y = 8}$$
$$0 = 0$$

This is an identity. There are infinitely many solutions of the form $(4k - 8, 6 - 2k, k)$.

14. a. Let n = the number of nickels
Let d = the number of dimes
$$5n + 10d = 400$$
$$n = d + 5$$
Substitute the second equation into the first equation and solve:
$$5(d + 5) + 10d = 400$$
$$5d + 25 + 10d = 400$$
$$15d = 375$$
$$d = 25$$
$$n = 25 + 5 = 30$$
Joey has 30 nickels and 25 dimes.

b. Let n = the number of nickels
Let d = the number of dimes
$$5n + 10d = 200$$
$$n = d - 5$$
Substitute the second equation into the first equation and solve:
$$5(d - 5) + 10d = 200$$
$$5d - 25 + 10d = 200$$
$$15d = 225$$
$$d = 15$$
$$n = 15 - 5 = 10$$
Alice has 10 nickels and 15 dimes.

15. a. Let x = the height of the building
Let y = the height of the flagpole
$$x + y = 200$$
$$x = 9y$$
Substitute the second equation into the first equation and solve:
$$9y + y = 200$$
$$10y = 200$$
$$y = 20$$
$$x = 9(20) = 180$$
The height of the building is 180 ft.

b. Let x = the height of the building
Let y = the height of the flagpole
$$x + y = 180$$
$$x = 8y$$
Substitute the second equation into the first equation and solve:
$$8y + y = 180$$
$$9y = 180$$
$$y = 20$$
$$x = 8(20) = 160$$
The height of the building is 160 ft.

16. a. Let x = the speed of the boat in still water
Let y = the speed of the current
$x + y$ = the speed downstream
$x - y$ = the speed upstream
$$(x+y)\frac{1}{3} = 12 \ - \ x+y = 36$$
$$(x-y)\frac{1}{2} = 12 \ - \ x-y = 24$$
Add the two equations and solve:
$$2x = 60$$
$$x = 30$$
$$30 + y = 36$$
$$y = 6$$
The speed of the current is 6 mi/hr.

b. Let x = the speed of the boat in still water
Let y = the speed of the current
$x + y$ = the speed downstream
$x - y$ = the speed upstream
$$(x+y)\frac{1}{4} = 6 \ - \ x+y = 24$$
$$(x-y)\frac{1}{3} = 6 \ - \ x-y = 18$$
Add the two equations and solve:
$$2x = 42$$
$$x = 21$$
$$21 + y = 24$$
$$y = 3$$
The speed of the current is 3 mi/hr.

17. a. Let x = the amount invested at 4%
Let y = the amount invested at 6%
Let z = the amount invested at 8%
$$x + y + z = 40,000$$
$$0.04x + 0.06y + 0.08z = 2600$$
$$0.08z - 600 = 0.04x + 0.06y$$
Substitute the third equation into the second equation and solve:
$$0.08z - 600 + 0.08z = 2600$$
$$0.16z = 3200$$
$$z = 20,000$$
Substitute this result into the first and third equations,
$$x + y + 20,000 = 40,000 \rightarrow x + y = 20,000$$

$$0.08(20,000) - 600 = 0.04x + 0.06y$$
$$1600 - 600 = 0.04x + 0.06y$$
$$1000 = 0.04x + 0.06y$$
Solve the first equation for x and substitute into the last result:
$$x = 20,000 - y$$
$$0.04(20,000 - y) + 0.06y = 1000$$
$$800 - 0.04y + 0.06y = 1000$$
$$0.02y = 200$$
$$y = 10,000$$
$$x = 20,000 - 10,000$$
$$x = 10,000$$
4% investment is $10,000; 6% investment is $10,000; 8% investment is $20,000.

158

b. Let x = the amount invested at 4%
Let y = the amount invested at 6%
Let z = the amount invested at 8%
$$x + y + z = 45,000$$
$$0.04x + 0.06y + 0.08z = 2900$$
$$0.08z - 300 = 0.04x + 0.06y$$
Substitute the third equation into the second equation and solve:
$$0.08z - 300 + 0.08z = 2900$$
$$0.16z = 3200$$
$$z = 20,000$$
Substitute this result into the first and third equations:
$$x + y + 20,000 = 45,000 \rightarrow x + y = 25,000$$

$$0.08(20,000) - 300 = 0.04x + 0.06y$$
$$1600 - 300 = 0.04x + 0.06y$$
$$1300 = 0.04x + 0.06y$$
Solve the first equation for x and substitute into the last result:
$$x = 25,000 - y$$
$$0.04(25,000 - y) + 0.06y = 1300$$
$$1000 - 0.04y + 0.06y = 1300$$
$$0.02y = 300$$
$$y = 15,000$$
$$x = 25,000 - 15,000$$
$$x = 10,000$$
4% investment is $10,000; 6% investment is $15,000; 8% investment is $20,000.

18. **a.** Let w = the width of the rectangle
Let l = the length of the rectangle
$$2l + 2w = 100$$
$$l = w + 30$$
Substitute the second equation into the first equation and solve:
$$2(w + 30) + 2w = 100$$
$$2w + 60 + 2w = 100$$
$$4w = 40$$
$$w = 10$$
$$l = 10 + 30 = 40$$
The length is 40 in and the width is 10 in.

b. Let w = the width of the rectangle
Let l = the length of the rectangle
$$2l + 2w = 80$$
$$l = 3w$$
Substitute the second equation into the first equation and solve:
$$2(3w) + 2w = 80$$
$$6w + 2w = 80$$
$$8w = 80$$
$$w = 10$$
$$l = 3(10) = 30$$
The length is 30 in and the width is 10 in.

19. **a.**
$$2x - y - z = 3$$
$$x + y + z = 6$$
$$3x + 2y + z = 10$$

$$\begin{vmatrix} 2 & -1 & -1 \\ 1 & 1 & 1 \\ 3 & 2 & 1 \end{vmatrix} \begin{matrix} 3 \\ 6 \\ 10 \end{matrix} \rightarrow \sim \begin{vmatrix} 2 & -1 & -1 \\ 0 & -3 & -3 \\ 0 & 7 & 5 \end{vmatrix} \begin{matrix} 3 \\ -9 \\ 11 \end{matrix} \rightarrow \sim \begin{vmatrix} 2 & -1 & -1 \\ 0 & -3 & -3 \\ 0 & 0 & -6 \end{vmatrix} \begin{matrix} 3 \\ -9 \\ -30 \end{matrix}$$

$$R_1 - 2 \rightarrow R_2 \rightarrow R_2 \qquad 7 \rightarrow R_2 + 3 \rightarrow R_3 \rightarrow R_3$$
$$-3 \rightarrow R_1 + 2 \rightarrow R_3 \rightarrow R_3$$

$$2x - y - z = 3$$
$$-3y - 3z = -9$$
$$-6z = -30$$

$$-6z = -30$$
$$z = 5$$

$$-3y - 3(5) = -9$$
$$-3y = 6$$
$$y = -2$$

$$2x - (-2) - 5 = 3$$
$$2x = 6$$
$$x = 3$$

The solution is $(3, -2, 5)$.

b. $2x - 6y + 2z = -6$

$2x + y - 4z = -12$

$-x + 3y + z = 5$

$$\begin{bmatrix} 2 & -6 & 2 \\ 2 & 1 & -4 \\ -1 & 3 & 1 \end{bmatrix} \begin{matrix} -6 \\ -12 \\ 5 \end{matrix} \sim \begin{bmatrix} 2 & -6 & 2 \\ 0 & 7 & 6 \\ 0 & 0 & 4 \end{bmatrix} \begin{matrix} -6 \\ 6 \\ 4 \end{matrix}$$

$$R_1 \times R_2 \times R_2$$

$$R_1 + 2 \times R_3 \times R_3$$

$$\begin{matrix} 2x - 6y + 2z = -6 \\ -7y + 6z = 6 \\ 4z = 4 \end{matrix} \qquad \begin{matrix} 4z = 4 \\ z = 1 \end{matrix} \qquad \begin{matrix} -7y + 6(1) = 6 \\ -7y = 0 \\ y = 0 \end{matrix} \qquad \begin{matrix} 2x - 6(0) + 2(1) = -6 \\ 2x = -8 \\ x = -4 \end{matrix}$$

The solution is $(-4, 0, 1)$.

20. a. $3x + y - 2z = 1$

$9x + 3y - 6z = 6$

$-2x - y + 3z = -1$

$$\begin{bmatrix} 3 & 1 & -2 \\ 9 & 3 & -6 \\ -2 & -1 & 3 \end{bmatrix} \begin{matrix} 1 \\ 6 \\ -1 \end{matrix} \sim \begin{bmatrix} 3 & 1 & -2 \\ 0 & 0 & 0 \\ -2 & -1 & 3 \end{bmatrix} \begin{matrix} 1 \\ 3 \\ -1 \end{matrix}$$

$$-3 - R_1 + R_2 \to R_2$$

The second line corresponds to the equation $0x + 0y + 0z = 3$ which is false. The system has no solution.

b. $x + y + z = 2$

$2x - y + z = -1$

$x - y - z = 0$

$$\begin{bmatrix} 1 & 1 & 1 \\ 2 & -1 & 1 \\ 1 & -1 & -1 \end{bmatrix} \begin{matrix} 2 \\ -1 \\ 0 \end{matrix} \sim \begin{bmatrix} 1 & 1 & 1 \\ 0 & -3 & -1 \\ 0 & 0 & 2 \end{bmatrix} \begin{matrix} 2 \\ -5 \\ 2 \end{matrix} \sim \begin{bmatrix} 1 & 1 & 1 \\ 0 & -3 & -1 \\ 0 & 0 & 4 \end{bmatrix} \begin{matrix} 2 \\ -5 \\ -4 \end{matrix}$$

$$-2 - R_1 + R_2 \to R_2 \qquad 2 - R_2 + 3 - R_3 \to R_3$$

$$R_1 - R_3 \to R_3$$

$$\begin{matrix} x + y + z = 2 \\ -3y - z = -5 \\ 4z = -4 \end{matrix} \qquad \begin{matrix} 4z = -4 \\ z = -1 \end{matrix} \qquad \begin{matrix} -3y - (-1) = -5 \\ -3y = -6 \\ y = 2 \end{matrix} \qquad \begin{matrix} x + 2 + (-1) = 2 \\ x = 1 \end{matrix}$$

The solution is $(1, 2, -1)$.

21. a. $\begin{vmatrix} 3 & 5 \\ 2 & -4 \end{vmatrix} = -12 - 10 = -22$

b. $\begin{vmatrix} -4 & 5 \\ -6 & 4 \end{vmatrix} = -16 + 30 = 14$

22. a.

$$D = \begin{vmatrix} 2 & 5 \\ 3 & -4 \end{vmatrix} = -8 - 15 = -23$$

$$D_x = \begin{vmatrix} -8 & 5 \\ 11 & -4 \end{vmatrix} = 32 - 55 = -23$$

$$D_y = \begin{vmatrix} 2 & -8 \\ 3 & 11 \end{vmatrix} = 22 + 24 = 46$$

$$x = \frac{D_x}{D} = \frac{-23}{-23} = 1 \quad y = \frac{D_y}{D} = \frac{46}{-23} = -2$$

The solution is $(1, -2)$.

b.

$$D = \begin{vmatrix} 4 & 2 \\ 2 & -6 \end{vmatrix} = -24 - 4 = -28$$

$$D_x = \begin{vmatrix} 1 & 2 \\ 4 & -6 \end{vmatrix} = -6 - 8 = -14$$

$$D_y = \begin{vmatrix} 4 & 1 \\ 2 & 4 \end{vmatrix} = 16 - 2 = 14$$

$$x = \frac{D_x}{D} = \frac{-14}{-28} = \frac{1}{2} \quad y = \frac{D_y}{D} = \frac{14}{-28} = -\frac{1}{2}$$

The solution is $\left(\frac{1}{2}, -\frac{1}{2}\right)$.

23. a.

$$\begin{vmatrix} 1 & -2 & -2 \\ 3 & 0 & -1 \\ 4 & 1 & 2 \end{vmatrix} = 1\begin{vmatrix} 0 & -1 \\ 1 & 2 \end{vmatrix} - (-2)\begin{vmatrix} 3 & -1 \\ 4 & 2 \end{vmatrix} + (-2)\begin{vmatrix} 3 & 0 \\ 4 & 1 \end{vmatrix} = 1(0+1) + 2(6+4) - 2(3-0)$$

$$= 1 + 20 - 6 = 15$$

b.

$$\begin{vmatrix} 0 & 2 & 4 \\ 1 & 2 & 0 \\ 2 & 1 & 3 \end{vmatrix} = 0\begin{vmatrix} 2 & 0 \\ 1 & 3 \end{vmatrix} - 2\begin{vmatrix} 1 & 0 \\ 2 & 3 \end{vmatrix} + 4\begin{vmatrix} 1 & 2 \\ 2 & 1 \end{vmatrix} = 0(6-0) - 2(3-0) + 4(1-4) = 0 - 6 - 12 = -18$$

24. a.

$$\begin{vmatrix} 1 & -1 & 1 \\ 2 & 3 & 1 \\ 1 & 3 & 2 \end{vmatrix} = 1\begin{vmatrix} 3 & 1 \\ 3 & 2 \end{vmatrix} - (-1)\begin{vmatrix} 2 & 1 \\ 1 & 2 \end{vmatrix} + 1\begin{vmatrix} 2 & 3 \\ 1 & 3 \end{vmatrix} = 1(6-3) + 1(4-1) + 1(6-3) = 3 + 3 + 3 = 9$$

b.

$$\begin{vmatrix} 4 & -2 & -1 \\ 2 & 5 & -2 \\ 1 & -2 & 2 \end{vmatrix} = 4\begin{vmatrix} 5 & -2 \\ -2 & 2 \end{vmatrix} - (-2)\begin{vmatrix} 2 & -2 \\ 1 & 2 \end{vmatrix} + (-1)\begin{vmatrix} 2 & 5 \\ 1 & -2 \end{vmatrix} = 4(10-4) + 2(4+2) - 1(-4-5)$$

$$= 24 + 12 + 9 = 45$$

25. a.

$$\begin{vmatrix} 1 & 0 & 5 \\ 3 & 2 & 1 \\ 5 & 3 & -1 \end{vmatrix} = 0\begin{vmatrix} 3 & 1 \\ 5 & -1 \end{vmatrix} + 2\begin{vmatrix} 1 & 5 \\ 5 & -1 \end{vmatrix} - 3\begin{vmatrix} 1 & 5 \\ 3 & 1 \end{vmatrix} = 0(-3-5) + 2(-1-25) - 3(1-15)$$

$$= 0 - 52 + 42 = -10$$

b.

$$\begin{vmatrix} 1 & 2 & 1 \\ 0 & 4 & -2 \\ 3 & 6 & -2 \end{vmatrix} = -2\begin{vmatrix} 0 & -2 \\ 3 & -2 \end{vmatrix} + 4\begin{vmatrix} 1 & 1 \\ 3 & -2 \end{vmatrix} - 6\begin{vmatrix} 1 & 1 \\ 0 & -2 \end{vmatrix} = -2(0+6) + 4(-2-3) - 6(-2-0)$$

$$= -12 - 20 + 12 = -20$$

26. a.

$$\begin{vmatrix} 1 & 3 & 0 \\ 0 & 1 & -2 \\ 2 & 4 & 3 \end{vmatrix} = 0\begin{vmatrix} 0 & 1 \\ 2 & 4 \end{vmatrix} - (-2)\begin{vmatrix} 1 & 3 \\ 2 & 4 \end{vmatrix} + 3\begin{vmatrix} 1 & 3 \\ 0 & 1 \end{vmatrix} = 0(0-2)+2(4-6)+3(1-0) = 0-4+3 = -1$$

b.

$$\begin{vmatrix} 3 & 1 & 5 \\ 1 & 0 & -2 \\ 6 & 1 & 3 \end{vmatrix} = 5\begin{vmatrix} 1 & 0 \\ 6 & 1 \end{vmatrix} - (-2)\begin{vmatrix} 3 & 1 \\ 6 & 1 \end{vmatrix} + 3\begin{vmatrix} 3 & 1 \\ 1 & 0 \end{vmatrix} = 5(1-0)+2(3-6)+3(0-1) = 5-6-3 = -4$$

27. a.

$$D = \begin{vmatrix} 1 & 2 & 1 \\ 1 & 1 & -1 \\ 2 & -1 & 2 \end{vmatrix} = 1\begin{vmatrix} 1 & -1 \\ -1 & 2 \end{vmatrix} - 2\begin{vmatrix} 1 & -1 \\ 2 & 2 \end{vmatrix} + 1\begin{vmatrix} 1 & 1 \\ 2 & -1 \end{vmatrix} = 1(2-1)-2(2+2)+1(-1-2)$$

$$= 1-8-3 = -10$$

$$D_x = \begin{vmatrix} 6 & 2 & 1 \\ 7 & 1 & -1 \\ -3 & -1 & 2 \end{vmatrix} = 6\begin{vmatrix} 1 & -1 \\ -1 & 2 \end{vmatrix} - 2\begin{vmatrix} 7 & -1 \\ -3 & 2 \end{vmatrix} + 1\begin{vmatrix} 7 & 1 \\ -3 & -1 \end{vmatrix} = 6(2-1)-2(14-3)+1(-7+3)$$

$$= 6-22-4 = -20$$

$$D_y = \begin{vmatrix} 1 & 6 & 1 \\ 1 & 7 & -1 \\ 2 & -3 & 2 \end{vmatrix} = 1\begin{vmatrix} 7 & -1 \\ -3 & 2 \end{vmatrix} - 6\begin{vmatrix} 1 & -1 \\ 2 & 2 \end{vmatrix} + 1\begin{vmatrix} 1 & 7 \\ 2 & -3 \end{vmatrix} = 1(14-3)-6(2+2)+1(-3-14)$$

$$= 11-24-17 = -30$$

$$D_z = \begin{vmatrix} 1 & 2 & 6 \\ 1 & 1 & 7 \\ 2 & -1 & -3 \end{vmatrix} = 1\begin{vmatrix} 1 & 7 \\ -1 & -3 \end{vmatrix} - 2\begin{vmatrix} 1 & 7 \\ 2 & -3 \end{vmatrix} + 6\begin{vmatrix} 1 & 1 \\ 2 & -1 \end{vmatrix} = 1(-3+7)-2(-3-14)+6(-1-2)$$

$$= 4+34-18 = 20$$

$$x = \frac{D_x}{D} = \frac{-20}{-10} = 2 \qquad y = \frac{D_y}{D} = \frac{-30}{-10} = 3 \qquad z = \frac{D_z}{D} = \frac{20}{-10} = -2$$

The solution is (2, 3, –2).

b.

$$D = \begin{vmatrix} 1 & 1 & 2 \\ 1 & -1 & 2 \\ 1 & 2 & -1 \end{vmatrix} = 1\begin{vmatrix} -1 & 2 \\ 2 & -1 \end{vmatrix} - 1\begin{vmatrix} 1 & 2 \\ 1 & -1 \end{vmatrix} + 2\begin{vmatrix} 1 & -1 \\ 1 & 2 \end{vmatrix} = 1(1-4) - 1(-1-2) + 2(2+1)$$

$$= -3 + 3 + 6 = 6$$

$$D_x = \begin{vmatrix} -3 & 1 & 2 \\ 1 & -1 & 2 \\ -2 & 2 & -1 \end{vmatrix} = -3\begin{vmatrix} -1 & 2 \\ 2 & -1 \end{vmatrix} - 1\begin{vmatrix} 1 & 2 \\ -2 & -1 \end{vmatrix} + 2\begin{vmatrix} 1 & -1 \\ -2 & 2 \end{vmatrix} = -3(1-4) - 1(-1+4) + 2(2-2)$$

$$= 9 - 3 + 0 = 6$$

$$D_y = \begin{vmatrix} 1 & -3 & 2 \\ 1 & 1 & 2 \\ 1 & -2 & -1 \end{vmatrix} = 1\begin{vmatrix} 1 & 2 \\ -2 & -1 \end{vmatrix} - (-3)\begin{vmatrix} 1 & 2 \\ 1 & -1 \end{vmatrix} + 2\begin{vmatrix} 1 & 1 \\ 1 & -2 \end{vmatrix} = 1(-1+4) + 3(-1-2) + 2(-2-1)$$

$$= 3 - 9 - 6 = -12$$

$$D_z = \begin{vmatrix} 1 & 1 & -3 \\ 1 & -1 & 1 \\ 1 & 2 & -2 \end{vmatrix} = 1\begin{vmatrix} -1 & 1 \\ 2 & -2 \end{vmatrix} - 1\begin{vmatrix} 1 & 1 \\ 1 & -2 \end{vmatrix} + (-3)\begin{vmatrix} 1 & -1 \\ 1 & 2 \end{vmatrix} = 1(2-2) - 1(-2-1) - 3(2+1)$$

$$= 0 + 3 - 9 = -6$$

$$x = \frac{D_x}{D} = \frac{6}{6} = 1 \qquad y = \frac{D_y}{D} = \frac{-12}{6} = -2 \qquad z = \frac{D_z}{D} = \frac{-6}{6} = -1$$

The solution is $(1, -2, -1)$.

28. a. $y - x$ and $x > -2$

Graph the line $y = x$. The boundary is a solid line.

Test point $(0, 1)$: $1 - 0$.

True, so shade this side of the line.

Graph the line $x = -2$. The boundary is a dashed line.

Shade the right side of the line.

The solution is the intersection of the graphs.

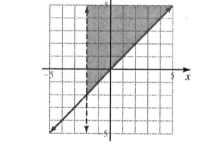

b. $3x \geq 4y \geq \geq 12$ and $x < 1$ and $y \geq 0$

Graph the line $3x \geq 4y = \geq 12$. The boundary is a solid line.

Test point $(0, 0)$: $3(0) \geq 4(0) \geq \geq 12$, $0 \geq \geq 12$.

True, so shade this side of the line.

Graph the line $x = 1$. The boundary is a dashed line.

Shade the left side of the line. Graph the line $y = 0$.

The boundary is a solid line. Shade above the line.

The solution is the intersection of the graphs.

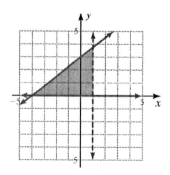

Cumulative Review Chapters 1–4

1. $\sqrt{13}$ is an irrational number and a real number.

2.
$$\lfloor(2x^2 \rfloor 6)+(6x+6)\rfloor\rfloor \lfloor(x\rfloor 2)+(4x^2 \rfloor 5)\rfloor$$
$$=\lfloor 2x^2 \rfloor 6+6x+6\rfloor\rfloor \lfloor x\rfloor 2+4x^2 \rfloor 5\rfloor$$
$$=\lfloor 2x^2 +6x\rfloor\rfloor \lfloor 4x^2 +x\rfloor 7\rfloor$$
$$=2x^2 +6x\rfloor 4x^2 \rfloor x+7=\rfloor 2x^2 +5x+7$$

3.
$$\frac{36x^6}{9x^{\,7}}=\frac{36}{9}\cdot\frac{x^6}{x^{\,7}}=4x^{6\cdot(-7)}=4x^{13\cdot}$$

4.
$$\left(3x^3y^{-4}\right)^2=9x^6y^{-8}=\frac{9x^6}{y^8}$$

5. $\therefore 7(2+6)\therefore +6=\left[\cdot\,7\cdot8\right]+6=\cdot\,56+6=\cdot\,50$

6.
$$x+6=3(5x-2)$$
$$x+6=15x-6$$
$$-14x=-12$$
$$x=\frac{6}{7}$$

7.
$$0.03P+0.08(1300-P)=75$$
$$0.03P+104-0.08P=75$$
$$-0.05P=-29$$
$$P=580$$

8.
$$|x-6|=|x-8|$$
$$x-6=x-8 \text{ or } x-6=-(x-8)$$
$$-6=-8 \quad\text{ or } x-6=-x+8$$
$$\text{Contradiction} \qquad 2x=14$$
$$x=7$$

9.
$$2(x\geq3)\geq3x\geq5$$
$$2x\geq6\geq3x\geq5$$
$$\geq x\geq1$$
$$x\geq\geq1$$

10.
$$\geq5\geq\geq5x\geq15<5$$
$$10\geq\geq5x<20$$
$$\geq2\geq x>\geq4$$
$$\geq4<x\geq\geq2$$

11. $|2x+3|>4$
$$2x+3<-4 \text{ or } 2x+3>4$$
$$2x<-7 \text{ or } \quad 2x>1$$
$$x<-\frac{7}{2} \text{ or } \quad x>\frac{1}{2}$$

12. $H=2.85h+73.82; \quad H=139.37$
$$139.37=2.85h+73.82$$
$$65.55=2.85h$$
$$23=h$$

13. Let x = original salary
$$x + 0.10x = 29,700$$
$$1.10x = 29,700$$
$$x = 27,000$$
The salary before the increase was \$27,000.

14.

Percent of solution	Amount to be mixed	Amount of Pure solution
40% or 0.40	x	$0.40x$
16% or 0.16	30	$0.16(30)$
30% or 0.30	$x + 30$	$0.30(x + 30)$

$$0.40x + 0.16(30) = 0.30(x + 30)$$
$$0.40x + 4.8 = 0.30x + 9$$
$$0.10x = 4.2$$
$$x = 42$$
42 gallons of 40% solution must be added.

15. $x - y = 3$

When $x = 0$, $-y = 3$ and $y = -3$

When $y = 0$, $x = 3$

Graph $(0, -3)$ and $(3, 0)$

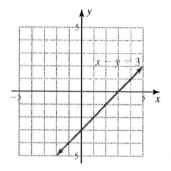

16. $$m = \frac{8 - (-7)}{0 - (-7)} = \frac{15}{7}$$

17.

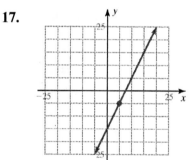

18. $9x + 3y = -54$
$$3y = -9x - 54$$
$$y = -3x - 18$$

19. $8x + 6y = -2$
$$6y = -8x - 2$$
$$y = -\frac{4}{3}x - \frac{1}{3}$$
The slope of the parallel line is $-\frac{4}{3}$.
$$y - 6 = -\frac{4}{3}(x - 3)$$
$$y - 6 = -\frac{4}{3}x + 4$$
$$y = -\frac{4}{3}x + 10$$
$$3y = -4x + 30$$
$$4x + 3y = 30$$

20. $y - x + 1$

When $x = 0$, $y = 1$

When $y = 0$, $0 = x + 1$, and $x = -1$

Graph the line $y = x + 1$.

The boundary is a solid line.

Test point $(0,0)$: $0 - 0 + 1$, $0 - 1$

True, so shade this side of the line.

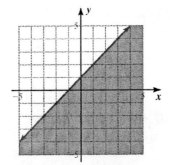

21. $|y| - 4$

$-4 - y - 4$

Shade between the horizontal lines

$y = -4$ and $y = 4$.

The boundary lines are solid lines.

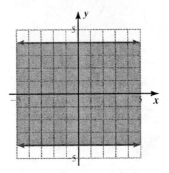

22. $D = \{x \mid x \text{ is a real number}\}$

$R = \{y \mid y \text{ is a real number}\}$

23. $f(2) = 3(2) - 2 = 6 - 2 = 4$

24. Graph each equation.

$x - 3y = -3$

$x = -3 - 3y$

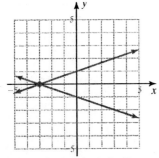

The solution is $(-3, 0)$.

25. $x - 2y = 3$

$\quad\quad 6y = 3x - 9$

Solve the second equation for y, substitute into the first equation, and solve:

$$y = \frac{1}{2}x \mid \frac{3}{2}$$

$$x \mid 2 \left| \frac{1}{2}x \mid \frac{3}{2} \right| = 3$$

$$x \mid x + 3 = 3$$

$$3 = 3$$

This is an identity. There are infinitely many solutions of the form $\left(k, \frac{1}{2}k - \frac{3}{2}\right)$.

26. To eliminate x, multiply the second equation by -3, add the result to the first equation, and solve:

$$\frac{x}{2}+\frac{y}{5}=6 \; - \quad \frac{x}{2}+\frac{y}{5}=6$$

$$\frac{x}{6}+\frac{y}{3}=6 \; - \quad -\frac{x}{2}-y=-18$$

$$\frac{4}{5}y=-12 \to y=15$$

$$\frac{x}{2}+\frac{15}{5}=6$$

$$\frac{x}{2}=3 \to x=6$$

The solution is (6, 15).

27.
$$2x+y-3z=5$$
$$-4x+y-2z=4$$
$$3y-8z=-6$$

Multiply the first equation by 2 and add to the second equation to eliminate x:

$$4x+2y-6z=10$$
$$\underline{-4x+y-2z=4}$$
$$3y-8z=14$$

Multiply the first result by -1 and add to the third equation:

$$-3y+8z=-14$$
$$\underline{3y-8z=-6}$$
$$0=-20$$

This is a contradiction. There is no solution.

28.
$$\begin{vmatrix} 2 & -5 \\ 1 & 1 \end{vmatrix} = 2+5=7$$

29.
$$\begin{vmatrix} 2 & 3 & -1 \\ 1 & 2 & 3 \\ 1 & 1 & -1 \end{vmatrix} = 2\begin{vmatrix} 2 & 3 \\ 1 & -1 \end{vmatrix} - 3\begin{vmatrix} 1 & 3 \\ 1 & -1 \end{vmatrix} + (-1)\begin{vmatrix} 1 & 2 \\ 1 & 1 \end{vmatrix} = 2(-2-3)-3(-1-3)-1(1-2) = -10+12+1 = 3$$

30. Let x = the speed of the boat in still water
Let y = the speed of the current
$x+y$ = the speed downstream
$x-y$ = the speed upstream

$$(x+y)\frac{1}{3}=12 \; - \quad x+y=36$$

$$(x-y)\frac{1}{2}=12 \; - \quad x-y=24$$

Add the two equations and solve:
$$2x=60$$
$$x=30$$
$$30+y=36$$
$$y=6$$
The speed of the boat is 30 mi/hr.

Chapter 5   Polynomials

5.1   Polynomials: Addition and Subtraction

Problems 5.1

1. **a.**   monomial   **b.**   trinomial   **c.**   polynomial   **d.**   binomial

2. **a.**   degree 1   **b.**   degree 4   **c.**   degree 7   **d.**   degree 0
    e. $9x^4 - 7x + 4$

3. $H = -16t^2 + 118;\quad t = 2.5$
    $H = -16(2.5)^2 + 118 = -16(6.25) + 118$
    $\quad = -100 + 118 = 18$
    The altitude after 2.5 seconds is 18 ft.

4. **a.** $P(x) = x^2 - 2x + 3$
    $\qquad P(1) = 1^2 - 2(1) + 3 = 1 - 2 + 3 = 2$

    b. $Q(x) = x^2 + 3x - 5$
    $\qquad Q(-2) = (-2)^2 + 3(-2) - 5$
    $\qquad\qquad = 4 - 6 - 5 = -7$

    c. $P(1) - Q(-2) = 2 - (-7) = 9$

5. $\left(9z^4 - 6z^3 - z^2 + 12z - 8\right) + \left(5 + z - 7z^2 + 4z^3 - 8z^4\right)$
    $= 9z^4 - 6z^3 - z^2 + 12z - 8 + 5 + z - 7z^2 + 4z^3 - 8z^4 = z^4 - 2z^3 - 8z^2 + 13z - 3$

6. $P(t) + Q(t) = -1.2t^3 - 4.5t^2 + 3.6t - 15 + 9.18t^2 + 1.4t - 7.5 = -1.2t^3 + 4.68t^2 + 5t - 22.5$

7. $\left(4y^3 + 4y^2 - 3\right) - \left(8y^3 - 6y^2 + 5y - 3\right) = 4y^3 + 4y^2 - 3 - 8y^3 + 6y^2 - 5y + 3 = -4y^3 + 10y^2 - 5y$

8. $P(x) = R(x) - C(x) = \left(50x - 0.2x^2\right) - \left(5x + 400\right) = 50x - 0.2x^2 - 5x - 400 = -0.2x^2 + 45x - 400$
    $P(200) = -0.2(200)^2 + 45(200) - 400 = -0.2(40000) + 9000 - 400 = -8000 + 9000 - 400 = 600$
    The profit is \$600.

Exercises 5.1

1. Monomial; degree 4

3. Binomial; degree 3

5. Trinomial; degree 3

7. Trinomial; degree 5

9. Zero polynomial; no degree

11. $-x^4 + 3x^2 + 5x + 7$

13. $-8x^2 + 9x + 20$

15. $-5x^2 + x + 2$

17. z^3; $z = -2$

$(-2)^3 = -8$

19. $(x - 2y + z)^2$; $x = 2, y = -1, z = -2$

$(2 - 2(-1) + (-2))^2 = (2 + 2 - 2)^2 = 2^2 = 4$

21. $P(x) = 4x^2 + 4x - 1$

$P(-1) = 4(-1)^2 + 4(-1) - 1 = 4 - 4 - 1 = -1$

23. $Q(y) = y^2 - 7y - 2$

$Q(-2) = (-2)^2 - 7(-2) - 2 = 4 + 14 - 2 = 16$

25. $S(u) = -16u^2 + 120$

$S(4) = -16(4)^2 + 120 = -16(16) + 120$

$= -256 + 120 = -136$

27. $P(x) = 2x^2 + 3x$ $Q(y) = -3y^2 - 7y + 1$

   a. $P(0) = 2(0)^2 + 3(0) = 0 + 0 = 0$

   b. $Q(1) = -3(1)^2 - 7(1) + 1$

        $= -3 - 7 + 1 = -9$

   c. $P(0) + Q(1) = 0 + (-9) = -9$

29. $P(x) = x^2 - 2x + 5$

       $Q(y) = -2y^2 + 5y - 1$

   a. $P(-1) = (-1)^2 - 2(-1) + 5 = 1 + 2 + 5 = 8$

   b. $Q(1) = -2(1)^2 + 5(1) - 1 = -2 + 5 - 1 = 2$

   c. $P(-1) - Q(1) = 8 - 2 = 6$

31. $(x^2 + 4x - 8) + (5x^2 - 4x + 3) = x^2 + 4x - 8 + 5x^2 - 4x + 3 = 6x^2 - 5$

33.

$$
\begin{array}{r}
5x^2 + 3x + 4 \\
(+)\ \underline{-4x^2 - 5x - 8} \\
x^2 - 2x - 4
\end{array}
$$

35. $(4x^2 + 7x - 5) - (3x + x^2 + 4) = 4x^2 + 7x - 5 - 3x - x^2 - 4 = 3x^2 + 4x - 9$

37.

$$
\begin{array}{r}
-3y^2 + 6y - 5 \\
(-)\ \underline{8y^2 + 7y - 2} \\
-11y^2 - y - 3
\end{array}
$$

39. $(x^3 - 6x^2 + 4x - 2) + (3x^3 - 6x^2 + 5x - 4) = x^3 - 6x^2 + 4x - 2 + 3x^3 - 6x^2 + 5x - 4$

$= 4x^3 - 12x^2 + 9x - 6$

41. $(-8y^3 + 5y + 7y^2 - 5) + (8y^3 + 7y - 6) = -8y^3 + 5y + 7y^2 - 5 + 8y^3 + 7y - 6 = 7y^2 + 12y - 11$

43. $(6v^3 - 3v^2 + 2v - 5) - (3v^3 + v - v^2 + 2) = 6v^3 - 3v^2 + 2v - 5 - 3v^3 - v + v^2 - 2 = 3v^3 - 2v^2 + v - 7$

45. $(4u^3 - 5u^2 - u + 3) - (2u + 9u^3 - 7) = 4u^3 - 5u^2 - u + 3 - 2u - 9u^3 + 7 = -5u^3 - 5u^2 - 3u + 10$

47. $\left(x^3 + y^3 - 6xy + 7\right) + \left(3x^3 - y^3 + 8xy - 8\right) = x^3 + y^3 - 6xy + 7 + 3x^3 - y^3 + 8xy - 8 = 4x^3 + 2xy - 1$

49. $\left(x^3 - y^3 + 5xy - 2\right) - \left(x^2 - y^3 + 5xy + 2\right) = x^3 - y^3 + 5xy - 2 - x^2 + y^3 - 5xy - 2 = x^3 - x^2 - 4$

51. $\left(a + a^2\right) + \left(9a - 4a^2\right) + \left(a^2 - 5a\right) = a + a^2 + 9a - 4a^2 + a^2 - 5a = -2a^2 + 5a$

53. $2y + (x + 3y) - (x + y) = 2y + x + 3y - x - y = 4y$

55. $\left(3x^2 + y\right) - \left(x^2 - 3y\right) + \left(3y + x^2\right) = 3x^2 + y - x^2 + 3y + 3y + x^2 = 3x^2 + 7y$

57. $P(0) + Q(0) = \left(0^2 - 2(0) + 3\right) + \left(2(0)^2 + 3(0) - 1\right) = (0 - 0 + 3) + (0 + 0 - 1) = 3 - 1 = 2$

59. $P(x) - P(x) = \left(x^2 - 2x + 3\right) - \left(x^2 - 2x + 3\right) = 0$

61. Commutative property of addition

63. Distributive property

65. Associative property of addition

67. Commutative property of addition

69. Distributive property

71. $H(t) = -16t^2 + 64t$

a. $H(1) = -16(1)^2 + 64(1)$
$= -16 + 64 = 48$ ft

b. $H(2) = -16(2)^2 + 64(2)$
$= -64 + 128 = 64$ ft

73. $R(x) = 100x - 0.03x^2$

a. $R(500) = 100(500) - 0.03(500)^2 = 50,000 - 7500 = \$42,500$

b. $R(1000) = 100(1000) - 0.03(1000)^2 = 100,000 - 30,000 = \$70,000$

75. $V(t) = 100,000(1 - 0.10t)$

a. $V(5) = 100,000(1 - 0.10(5)) = 100,000(1 - 0.50) = 100,000(0.50) = \$50,000$

b. $V(10) = 100,000(1 - 0.10(10)) = 100,000(1 - 1.00) = 100,000(0) = \$0$ (no value)

77. $P(x) = R(x) - C(x) = 1.50x - (100 + 0.3x) = 1.50x - 100 - 0.3x = 1.2x - 100$
$P(100) = 1.2(100) - 100 = 120 - 100 = \$20$

79.
$P(x) = R(x) - C(x) = \left[50x - \dfrac{x^2}{20}\right] - (1500 + 20x) = 50x - \dfrac{x^2}{20} - 1500 - 20x = -\dfrac{x^2}{20} + 30x - 1500$

$P(100) = -\dfrac{(100)^2}{20} + 30(100) - 1500 = -500 + 3000 - 1500 = \$1000$

81.

$$P(x) = R(x) \mid C(x) = \left|300x \mid \frac{x^2}{50}\right| \mid (100,000 + 30x) = 300x \mid \frac{x^2}{50} \mid 100,000 \mid 30x$$

$$= \mid \frac{x^2}{50} + 270x \mid 100,000$$

$$P(500) = \mid \frac{(500)^2}{50} + 270(500) \mid 100,000 = \mid 5000 + 135,000 \mid 100,000 = \$30,000$$

83. $6x^2y \cdot 3xy = 18x^3y^2$

85. $\cdot 2x^3y \cdot (\cdot 7xy) = 14x^4y^2$

87. $x^2 + 2x + 4x + x^2 = 2x^2 + 6x$

89. $x^2 + x^2 + 4x + x^2 = 3x^2 + 4x$

91. $(3x)^2 + 5(2x) + (3x)^2 + (3x)^2 = 9x^2 + 10x + 9x^2 + 9x^2 = 27x^2 + 10x$

93. Sample answer. The degree of a polynomial is found by adding the exponents on each variable in each term and choosing the largest of these sums.

95. **a.** $P = R - C = (50x - 0.2x^2) - (3x + 100) = -0.2x^2 + 47x - 100$

  b. $P(100) = -0.2(100)^2 + 47(100) - 100 = -2000 + 4700 - 100 = \$2600$

97. Degree 4

99. a. binomial   **b.** polynomial   **c.** monomial   **d.** trinomial

101. $P(2) - Q(2) = 0 - 3 = -3$

103. $P(x) - Q(x) = (x^2 - 3x + 2) - (x^2 + 2x - 5) = x^2 - 3x + 2 - x^2 - 2x + 5 = -5x + 7$

5.2   Multiplication of Polynomials

Problems 5.2

1. **a.** $4x^3(5x^4 - 7x^3 + x^2 + 2x - 1) = 20x^7 - 28x^6 + 4x^5 + 8x^4 - 4x^3$

  b. $-3xy^2(9x^2 - 6xy + 10y^2) = -27x^3y^2 + 18x^2y^3 - 30xy^4$

2. a.

$$3x^2 + x - 5$$
$$\underline{x + 4}$$
$$12x^2 + 4x - 20$$
$$\underline{3x^3 + \quad x^2 - 5x}$$
$$3x^3 + 13x^2 - \quad x - 20$$

b.

$$9x^2 - 6x + 2$$
$$\underline{3x - 1}$$
$$-9x^2 + 6x - 2$$
$$\underline{27x^3 - 18x^2 + 6x}$$
$$27x^3 - 27x^2 + 12x - 2$$

3. **a.** $(x+4)(7x-5)$

$\qquad = x(7x)+x(-5)+4(7x)+4(-5)$

$\qquad = 7x^2 - 5x + 28x - 20$

$\qquad = 7x^2 + 23x - 20$

b. $(3x-y)(9x+2y)$

$\qquad = 3x(9x)+3x(2y)+(-y)(9x)+(-y)(2y)$

$\qquad = 27x^2 + 6xy - 9xy - 2y^2$

$\qquad = 27x^2 - 3xy - 2y^2$

c. $(B-3A)(B-8A) = B(B)+B(-8A)+(-3A)(B)+(-3A)(-8A)$

$\qquad = B^2 - 8AB - 3AB + 24A^2$

$\qquad = B^2 - 11AB + 24A^2$

4. **a.** $(x+4y)^2 = (x)^2 + 2\cdot 4y\cdot x + (4y)^2$

$\qquad = x^2 + 8xy + 16y^2$

b. $(7x\cdot\ 5y)^2 = (7x)^2\cdot\ 2\cdot 5y\cdot 7x + (\cdot\ 5y)^2$

$\qquad = 49x^2\cdot\ 70xy + 25y^2$

c. $\rfloor 6x(2x+3)^2$

$\qquad = \rfloor 6x\rfloor (2x)^2 + 2\rfloor 3\rfloor 2x + (3)^2\rfloor$

$\qquad = \rfloor 6x\rfloor 4x^2 + 12x + 9\rfloor$

$\qquad = \rfloor 24x^3\rfloor 72x^2\rfloor 54x$

d. $:(x\cdot\ 8)+y:^2 = (x\cdot\ 8)^2 + 2\cdot y\cdot (x\cdot\ 8)+(y)^2$

$\qquad = x^2\cdot\ 16x + 64 + 2xy\cdot\ 16y + y^2$

$\qquad = x^2 + 2xy + y^2\cdot\ 16x\cdot\ 16y + 64$

5. **a.** $(z-9)(z+9) = (z)^2 - (9)^2$

$\qquad = z^2 - 81$

b. $(y+3x)(y-3x) = (y)^2 - (3x)^2$

$\qquad = y^2 - 9x^2$

c. $\rfloor 2(4x+3y)(4x\rfloor 3y)$

$\qquad = \rfloor 2\rfloor (4x)^2\rfloor (3y)^2\rfloor$

$\qquad = \rfloor 2\rfloor 16x^2\rfloor 9y^2\rfloor$

$\qquad = \rfloor 32x^2 + 18y^2$

$\qquad = 18y^2\rfloor 32x^2$

d. $\rfloor 6x+(2y\rfloor 5)\rfloor\rfloor 6x\rfloor (2y\rfloor 5)\rfloor$

$\qquad = (6x)^2\rfloor (2y\rfloor 5)^2$

$\qquad = 36x^2\rfloor (4y^2\rfloor 20y + 25)$

$\qquad = 36x^2\rfloor 4y^2 + 20y\rfloor 25$

6. $R = 1000p - 10p^2;\qquad p = 75$

$R = 1000(75) - 10(75)^2 = 75{,}000 - 56{,}250 = \$18{,}750$

Exercises 5.2

1. $3x(4x-2) = 12x^2 - 6x$

3. $-3x^2(x-3) = -3x^3 + 9x^2$

5. $-8x(3x^2 - 2x + 1) = -24x^3 + 16x^2 - 8x$

7. $-3xy^2(6x^2 + 3y^2 - 7)$

$\qquad = -18x^3y^2 - 9xy^4 + 21xy^2$

9. $2xy^3(3x^2y^3 - 5xy^2 + xy) = 6x^3y^6 - 10x^2y^5 + 2x^2y^4$

11. $(x+3)(x^2+x+5)=x(x^2+x+5)+3(x^2+x+5)=x^3+x^2+5x+3x^2+3x+15=x^3+4x^2+8x+15$

13. $(x+4)(x^2-x+3)=x(x^2-x+3)+4(x^2-x+3)=x^3-x^2+3x+4x^2-4x+12=x^3+3x^2-x+12$

15.
$$\begin{array}{r} x^2-x-2 \\ \underline{x+3} \\ 3x^2-3x-6 \\ \underline{x^3-\ x^2-2x} \\ x^3+2x^2-5x-6 \end{array}$$

17. $(x-2)(x^2+2x+4)$
$$=x(x^2+2x+4)-2(x^2+2x+4)$$
$$=x^3+2x^2+4x-2x^2-4x-8$$
$$=x^3-8$$

19.
$$\begin{array}{r} x^2-x+2 \\ \underline{x^2-1} \\ -x^2+x-2 \\ \underline{x^4-x^3+2x^2} \\ x^4-x^3+\ x^2+x-2 \end{array}$$

21. $(3x+2)(3x+1)$
$$=3x(3x)+3x(1)+2(3x)+2(1)$$
$$=9x^2+3x+6x+2$$
$$=9x^2+9x+2$$

23. $(5x-4)(x+3)$
$$=5x(x)+5x(3)-4(x)-4(3)$$
$$=5x^2+15x-4x-12$$
$$=5x^2+11x-12$$

25. $(3a-1)(a+5)$
$$=3a(a)+3a(5)-1(a)-1(5)$$
$$=3a^2+15a-a-5$$
$$=3a^2+14a-5$$

27. $(y+5)(2y-3)$
$$=y(2y)+y(-3)+5(2y)+5(-3)$$
$$=2y^2-3y+10y-15$$
$$=2y^2+7y-15$$

29. $(x-3)(x-5)$
$$=x(x)+x(-5)-3(x)-3(-5)$$
$$=x^2-5x-3x+15$$
$$=x^2-8x+15$$

31. $(2x-1)(3x-2)$
$$=2x(3x)+2x(-2)-1(3x)-1(-2)$$
$$=6x^2-4x-3x+2$$
$$=6x^2-7x+2$$

33. $(2x-3a)(2x+5a)$
$$=2x(2x)+2x(5a)-3a(2x)-3a(5a)$$
$$=4x^2+10ax-6ax-15a^2$$
$$=4x^2+4ax-15a^2$$

35. $(x+7)(x+8)$
$$=x(x)+x(8)+7(x)+7(8)$$
$$=x^2+8x+7x+56$$
$$=x^2+15x+56$$

37. $(2a+b)(2a+4b)$
$$=2a(2a)+2a(4b)+b(2a)+b(4b)$$
$$=4a^2+8ab+2ab+4b^2$$
$$=4a^2+10ab+4b^2$$

39. $(4u+v)^2=(4u)^2+2\cdot v\cdot 4u+(v)^2$
$$=16u^2+8uv+v^2$$

41. $(2y+z)^2=(2y)^2+2\cdot z\cdot 2y+(z)^2$
$$=4y^2+4yz+z^2$$

43. $(3a \cdot b)^2 = (3a)^2 \cdot 2 \cdot b \cdot 3a + (b)^2$
$$= 9a^2 \cdot 6ab + b^2$$

45. $(a+b)(a-b) = (a)^2 - (b)^2$
$$= a^2 - b^2$$

47. $(5x-2y)(5x+2y) = (5x)^2 - (2y)^2$
$$= 25x^2 - 4y^2$$

49. $-(3a-b)(3a+b) = -[(3a)^2 - (b)^2]$
$$= -[9a^2 - b^2]$$
$$= -9a^2 + b^2$$
$$= b^2 - 9a^2$$

51. $3x(x+1)(x+2)$
$$= 3x[x(x)+x(2)+1(x)+1(2)]$$
$$= 3x[x^2+2x+x+2]$$
$$= 3x[x^2+3x+2]$$
$$= 3x^3+9x^2+6x$$

53. $-3x(x-1)(x-3)$
$$= -3x[x(x)+x(-3)-1(x)-1(-3)]$$
$$= -3x[x^2-3x-x+3]$$
$$= -3x[x^2-4x+3]$$
$$= -3x^3+12x^2-9x$$

55. $x(x+3)^2 = x[(x)^2+2\cdot3\cdot x+(3)^2]$
$$= x[x^2+6x+9]$$
$$= x^3+6x^2+9x$$

57. $-2x(x-1)^2 = -2x[(x)^2-2\cdot1\cdot x+(-1)^2]$
$$= -2x[x^2-2x+1]$$
$$= -2x^3+4x^2-2x$$

59. $(2x+y)(2x-y)y^2 = [(2x)^2-(y)^2]y^2$
$$= [4x^2-y^2]y^2$$
$$= 4x^2y^2-y^4$$

61. $\left(x+\frac{3}{4}\right)^2 = (x)^2+2\cdot\frac{3}{4}\cdot x+\left(\frac{3}{4}\right)^2$
$$= x^2+\frac{3}{2}x+\frac{9}{16}$$

63. $\left(2y-\frac{1}{5}\right)^2 = (2y)^2 - 2\cdot\frac{1}{5}\cdot2y+\left(\frac{1}{5}\right)^2$
$$= 4y^2 - \frac{4}{5}y+\frac{1}{25}$$

65. $\left(\frac{3}{4}p+\frac{1}{5}q\right)^2 = \left(\frac{3}{4}p\right)^2+2\cdot\frac{1}{5}q\cdot\frac{3}{4}p+\left(\frac{1}{5}q\right)^2$
$$= \frac{9}{16}p^2+\frac{3}{10}pq+\frac{1}{25}q^2$$

67. $[(3x+1)+4y]^2$
$$= (3x+1)^2+2\cdot4y\cdot(3x+1)+(4y)^2$$
$$= 9x^2+6x+1+24xy+8y+16y^2$$
$$= 9x^2+24xy+16y^2+6x+8y+1$$

69. $[(3x-1)-4y]^2$
$$= (3x-1)^2 - 2\cdot4y\cdot(3x-1)+(-4y)^2$$
$$= 9x^2-6x+1-24xy+8y+16y^2$$
$$= 9x^2-24xy+16y^2-6x+8y+1$$

71. $-2y+(3x-1)^2$

$$= (2y)^2 + 2\cdot(3x-1)\cdot 2y + (3x-1)^2$$
$$= 4y^2 + 12xy - 4y + 9x^2 - 6x + 1$$
$$= 9x^2 + 12xy + 4y^2 - 6x - 4y + 1$$

73. $-4p-(3q-1)^2$

$$= (4p)^2 - 2\cdot(3q-1)\cdot 4p + (3q-1)^2$$
$$= 16p^2 - 24pq + 8p + 9q^2 - 6q + 1$$
$$= 16p^2 - 24pq + 9q^2 + 8p - 6q + 1$$

75. a. $R = xp = (1000 - 30p)p = 1000p - 30p^2$

b. $R = 1000(20) - 30(20)^2 = 20{,}000 - 30(400) = 20{,}000 - 12{,}000 = \$8000$

77. $\left(T_1^2 + T_2^2\right)\left(T_1^2 - T_2^2\right) = \left(T_1^2\right)^2 - \left(T_2^2\right)^2$

$$= T_1^4 - T_2^4$$

79. $K\left(t_n - t_a\right)^2 = K\left[\left(t_n\right)^2 - 2\,t_n\,t_a + \left(-t_a\right)^2\right]$

$$= K\left[t_n^2 - 2t_n t_a + t_a^2\right]$$
$$= Kt_n^2 - 2Kt_n t_a + Kt_a^2$$

81. $5x + 5y = 5(x + y)$

83. $3ab + 3ac = 3a(b + c)$

85. a. $(x + y)^2 = (1 + 2)^2 = 3^2 = 9$

b. $x^2 + y^2 = 1^2 + 2^2 = 1 + 4 = 5$

c. $(x + y)^2 = x^2 + y^2 ?\quad 9 \neq 5$

The expressions are not equal.

87. a. x^2
b. xy
c. y^2
d. xy

89. $(x + y)^2 = x^2 + 2xy + y^2$

91. Sample answer. Multiply the integer coefficients and multiply the variables. If the variables are the same add the exponents.

93. Answers will vary.

95. $(4x + 3y)(3x + 2y)$

$$= 4x(3x) + 4x(2y) + 3y(3x) + 3y(2y)$$
$$= 12x^2 + 8xy + 9xy + 6y^2$$
$$= 12x^2 + 17xy + 6y^2$$

97. $(5x - 3y)^2 = (5x)^2 - 2\cdot 3y\cdot 5x + (-3y)^2$

$$= 25x^2 - 30xy + 9y^2$$

99. $(3x + y)(3x - y) = (3x)^2 - (y)^2$

$$= 9x^2 - y^2$$

101. $(x - 2)(x^2 - 4x - 3) = x(x^2 - 4x - 3) - 2(x^2 - 4x - 3) = x^3 - 4x^2 - 3x - 2x^2 + 8x + 6$

$$= x^3 - 6x^2 + 5x + 6$$

103. $4x^3\left(5x^3+3x^2-2x-5\right)$

$\qquad = 20x^6+12x^5-8x^4-20x^3$

105. $:(3x+1)\cdot\ 4y\!:^2$

$\qquad =(3x+1)^2\cdot\ 2\cdot 4y\cdot(3x+1)+(\cdot\ 4y)^2$

$\qquad =9x^2+6x+1\cdot\ 24xy\cdot\ 8y+16y^2$

$\qquad =9x^2\cdot\ 24xy+16y^2+6x\cdot\ 8y+1$

5.3 The Greatest Common Factor and Factoring by Grouping

Problems 5.3

1. **a.** $6x+48=6\cdot x+6\cdot 8=6(x+8)$

b. $\cdot\ 3y^2+21y=\cdot\ 3y\cdot y\cdot\ 3y\cdot(\cdot\ 7)=\cdot\ 3y(y\cdot\ 7)$

c. $4x^2\cdot\ 32x^3=4x^2\cdot 1\cdot\ 4x^2\cdot 8x$

$\qquad =4x^2(1\cdot\ 8x)$

2. **a.** $7x^3+14x^4\cdot\ 49x^2=7x^2\cdot x+7x^2\cdot 2x^2\cdot\ 7x^2\cdot 7$

$\qquad =7x^2\left(x+2x^2\cdot\ 7\right)$

$\qquad =7x^2\left(2x^2+x\cdot\ 7\right)$

b. $3x^6\cdot\ 6x^5+12x^7+27x^2$

$\qquad =3x^2\cdot x^4\cdot\ 3x^2\cdot 2x^3+3x^2\cdot 4x^5+3x^2\cdot 9$

$\qquad =3x^2\left(x^4\cdot\ 2x^3+4x^5+9\right)$

$\qquad =3x^2\left(4x^5+x^4\cdot\ 2x^3+9\right)$

3. $\dfrac{2}{5}x^2-\dfrac{3}{5}x^4+\dfrac{4}{5}x^5=\dfrac{1}{5}x^2\left(2-3x^2+4x^3\right)$

4. **a.** $2x^3-2x^2+3x-3=2x^2(x-1)+3(x-1)$

$\qquad =(x-1)\left(2x^2+3\right)$

b. $6x^3-9x^2-2x+3=3x^2(2x-3)-1(2x-3)$

$\qquad =(2x-3)\left(3x^2-1\right)$

5. **a.** $3x^4\rfloor 6x^3\rfloor x^2+2x$

$\qquad =x\left(3x^3\rfloor 6x^2\rfloor x+2\right)$

$\qquad =x\rfloor 3x^2(x\rfloor 2)\rfloor 1(x\rfloor 2)\rfloor$

$\qquad =x(x\rfloor 2)\left(3x^2\rfloor 1\right)$

b. $6x^6\rfloor 9x^4+2x^3\rfloor 3x$

$\qquad =x\left(6x^5\rfloor 9x^3+2x^2\rfloor 3\right)$

$\qquad =x\rfloor 3x^3\left(2x^2\rfloor 3\right)+1\left(2x^2\rfloor 3\right)\rfloor$

$\qquad =x\left(2x^2\rfloor 3\right)\left(3x^3+1\right)$

Exercises 5.3

1. $8x+16=8\cdot x+8\cdot 2=8(x+2)$

3. $9y\cdot\ 18=9\cdot y\cdot\ 9\cdot 2=9(y\cdot\ 2)$

5. $\cdot\ 5y+25=\cdot\ 5\cdot y\cdot\ 5\cdot(\cdot\ 5)=\cdot\ 5(y\cdot\ 5)$

7. $\cdot\ 8x\cdot\ 24=\cdot\ 8\cdot x\cdot\ 8\cdot 3=\cdot\ 8(x+3)$

9. $4x^2+36x=4x\cdot x+4x\cdot 9=4x(x+9)$

11. $6x\cdot\ 42x^3=6x\cdot 1\cdot\ 6x\cdot 7x^2=6x\left(1\cdot\ 7x^2\right)$

13. $\cdot 5x^2 \cdot 35x^4 = \cdot 5x^2 \cdot 1 \cdot 5x^2 \cdot 7x^2$

$\qquad = \cdot 5x^2\left(1 + 7x^2\right)$

15. $3x^3 + 6x^2 + 39x = 3x \cdot x^2 + 3x \cdot 2x + 3x \cdot 13$

$\qquad = 3x\left(x^2 + 2x + 13\right)$

17. $63y^3 \cdot 18y^2 + 27y = 9y \cdot 7y^2 \cdot 9y \cdot 2y + 9y \cdot 3$

$\qquad = 9y\left(7y^2 \cdot 2y + 3\right)$

19. $36x^6 + 12x^5 \cdot 18x^4 + 30x^2$

$\qquad = 6x^2 \cdot 6x^4 + 6x^2 \cdot 2x^3 \cdot 6x^2 \cdot 3x^2 + 6x^2 \cdot 5$

$\qquad = 6x^2\left(6x^4 + 2x^3 \cdot 3x^2 + 5\right)$

21. $48y^8 + 16y^5 \cdot 24y^4 + 16y^3$

$\qquad = 8y^3 \cdot 6y^5 + 8y^3 \cdot 2y^2 \cdot 8y^3 \cdot 3y + 8y^3 \cdot 2$

$\qquad = 8y^3\left(6y^5 + 2y^2 \cdot 3y + 2\right)$

23. $\dfrac{4}{7}x^3 + \dfrac{3}{7}x^2 \cdot \dfrac{9}{7}x + \dfrac{3}{7}$

$\qquad = \dfrac{1}{7} \cdot 4x^3 + \dfrac{1}{7} \cdot 3x^2 \cdot \dfrac{1}{7} \cdot 9x + \dfrac{1}{7} \cdot 3$

$\qquad = \dfrac{1}{7}\left(4x^3 + 3x^2 \cdot 9x + 3\right)$

25. $\dfrac{7}{8}y^9 + \dfrac{3}{8}y^6 \cdot \dfrac{5}{8}y^4 + \dfrac{5}{8}y^2$

$\qquad = \dfrac{1}{8}y^2 \cdot 7y^7 + \dfrac{1}{8}y^2 \cdot 3y^4 \cdot \dfrac{1}{8}y^2 \cdot 5y^2 + \dfrac{1}{8}y^2 \cdot 5$

$\qquad = \dfrac{1}{8}y^2\left(7y^7 + 3y^4 \cdot 5y^2 + 5\right)$

27. $x^3 + 2x^2 + x + 2 = x^2\left(x + 2\right) + 1\left(x + 2\right)$

$\qquad = \left(x + 2\right)\left(x^2 + 1\right)$

29. $y^3 - 3y^2 + y - 3 = y^2\left(y - 3\right) + 1\left(y - 3\right)$

$\qquad = \left(y - 3\right)\left(y^2 + 1\right)$

31. $4x^3 + 6x^2 + 2x + 3 = 2x^2\left(2x + 3\right) + 1\left(2x + 3\right)$

$\qquad = \left(2x + 3\right)\left(2x^2 + 1\right)$

33. $6x^3 - 2x^2 + 3x - 1 = 2x^2\left(3x - 1\right) + 1\left(3x - 1\right)$

$\qquad = \left(3x - 1\right)\left(2x^2 + 1\right)$

35. $4y^3 + 8y^2 + y + 2 = 4y^2\left(y + 2\right) + 1\left(y + 2\right)$

$\qquad = \left(y + 2\right)\left(4y^2 + 1\right)$

37. $2a^6 + 3a^4 + 2a^2 + 3$

$\qquad = a^4\left(2a^2 + 3\right) + 1\left(2a^2 + 3\right)$

$\qquad = \left(2a^2 + 3\right)\left(a^4 + 1\right)$

39. $3x^5 + 12x^3 + x^2 + 4 = 3x^3\left(x^2 + 4\right) + 1\left(x^2 + 4\right)$

$\qquad = \left(x^2 + 4\right)\left(3x^3 + 1\right)$

41. $6y^5 + 9y^3 + 2y^2 + 3$

$\qquad = 3y^3\left(2y^2 + 3\right) + 1\left(2y^2 + 3\right)$

$\qquad = \left(2y^2 + 3\right)\left(3y^3 + 1\right)$

43. $4y^7 + 12y^5 + y^4 + 3y^2$

$\qquad = y^2 \lfloor 4y^5 + 12y^3 + y^2 + 3 \rfloor$

$\qquad = y^2 \lfloor 4y^3\left(y^2 + 3\right) + 1\left(y^2 + 3\right) \rfloor$

$\qquad = y^2\left(y^2 + 3\right)\left(4y^3 + 1\right)$

45. $3a^7 - 6a^5 - 2a^4 + 4a^2$
$= a^2[3a^5 - 6a^3 - 2a^2 + 4]$
$= a^2[3a^3(a^2 - 2) - 2(a^2 - 2)]$
$= a^2(a^2 - 2)(3a^3 - 2)$

47. $8a^5 - 12a^4 - 10a^3 + 15a^2$
$= a^2[8a^3 - 12a^2 - 10a + 15]$
$= a^2[4a^2(2a - 3) - 5(2a - 3)]$
$= a^2(2a - 3)(4a^2 - 5)$

49. $x^6 - 2x^5 + 2x^4 - 4x^3$
$= x^3[x^3 - 2x^2 + 2x - 4]$
$= x^3[x^2(x - 2) + 2(x - 2)]$
$= x^3(x - 2)(x^2 + 2)$

51. $(x-4)(x+2)+(x-4)(x+3)$
$= (x-4)(x+2+x+3)$
$= (x-4)(2x+5)$

53. $-Lt_2 - aLt_1 = -L(t_2 - t_1)$

55. $R^2 - R - R + 1 = R(R-1) - 1(R-1)$
$= (R-1)(R-1)$ or $(R-1)^2$

57. $(x+3)(x+4) = x^2 + 4x + 3x + 12$
$= x^2 + 7x + 12$

59. $(x+5)(x-2) = x^2 - 2x + 5x - 10$
$= x^2 + 3x - 10$

61. $(5x+2y)^2 = (5x)^2 + 2\cdot 5x \cdot 2y + (2y)^2$
$= 25x^2 + 20xy + 4y^2$

63. $(5x\cdot\ 2y)^2 = (5x)^2 \cdot\ 2\cdot 5x \cdot 2y + (\cdot\ 2y)^2$
$= 25x^2 \cdot\ 20xy + 4y^2$

65. $(u+6)(u-6) = u^2 - 6^2 = u^2 - 36$

67. $-wl + wz = -w(l - z)$

69. $a^2 + 2as = a(a + 2s)$

71. $-16t^2 + 80t + 240 = -16(t^2 - 5t - 15)$

73. Sample answer: The greatest common factor of a list of integers is the largest integer that can be divided into all the numbers in the list without a remainder.

75. Sample answer: The relationship between multiplying and factoring a polynomial is that they are inverse types of procedures – the one undoes the other.

77. $6x^6 - 9x^4 + 2x^3 - 3x$
$= x[6x^5 - 9x^3 + 2x^2 - 3]$
$= x[3x^3(2x^2 - 3) + 1(2x^2 - 3)]$
$= x(2x^2 - 3)(3x^3 + 1)$

79. $6x^3 - 9x^2 - 2x + 3 = 3x^2(2x - 3) - 1(2x - 3)$
$= (2x - 3)(3x^2 - 1)$

81. $3x^6 - 6x^5 + 12x^7 + 27x^2$
$= 3x^2(x^4 - 2x^3 + 4x^5 + 9)$
$= 3x^2(4x^5 + x^4 - 2x^3 + 9)$

83. $6x - 48 = 6(x - 8)$

85. $\dfrac{2}{5}x^2 - \dfrac{4}{5}x^4 - \dfrac{1}{5}x^5 = \dfrac{1}{5}x^2(2 - 4x^2 - x^3)$ or $-\dfrac{1}{5}x^2(x^3 + 4x^2 - 2)$

5.4   Factoring Trinomials

Problems 5.4

1. **a.** Find two integers whose product is 10 and whose sum is 7. They are 2 and 5.
$$x^2 + 7x + 10 = (x+2)(x+5)$$

b. Find two integers whose product is -10 and whose sum is -3. They are 2 and -5.
$$x^2 - 3x - 10 = (x+2)(x-5)$$

c. $x^2 - 6 - 5x = x^2 - 5x - 6$ Find two integers whose product is -6 and whose sum is -5. They are 1 and -6.
$$x^2 - 5x - 6 = (x+1)(x-6)$$

2. **a.** Find two integers whose product is 5 and whose sum is -2. They are no numbers that meet these conditions. The trinomial is Prime.

b. Find two integers whose product is -16 and whose sum is -6. They are 2 and -8.
$$x^2 - 6xy - 16y^2 = (x+2y)(x-8y)$$

3. The factors of 6 are 1 and 6 or 2 and 3. Try the various combinations that yield the correct middle term.
$$6x^2 + 13x + 6 = (2x+3)(3x+2)$$

4. **a.** $5x^2 - 2x + 2$
The key number is 10. Find factors of the key number and rewrite the middle term. There are no integers with a product of 10 and sum of -2. The trinomial is not factorable. It is prime.

b. $3x^2 - 4 - 4x = 3x^2 - 4x - 4$
The key number is -12. Find factors of the key number and rewrite the middle term.
$$3x^2 - 6x + 2x - 4 = 3x(x-2) + 2(x-2)$$
$$= (x-2)(3x+2)$$

5. $2x^2 + xy - 3y^2$
The key number is -6. Find factors of the key number and rewrite the middle term.
$$2x^2 + 3xy - 2xy - 3y^2$$
$$= x(2x+3y) - y(2x+3y)$$
$$= (2x+3y)(x-y)$$

6. $12x^4y + 2x^3y^2 - 4x^2y^3 = 2x^2y(6x^2 + xy - 2y^2)$
The key number is -12. Find factors of the key number and rewrite the middle term.
$$2x^2y(6x^2 + 4xy - 3xy - 2y^2)$$
$$= 2x^2y[2x(3x+2y) - y(3x+2y)]$$
$$= 2x^2y(3x+2y)(2x - y)$$

7. $-3x^2 + 14x + 24 = -1(3x^2 - 14x - 24)$
The key number is -72. Find factors of the key number and rewrite the middle term.
$$-1(3x^2 - 18x + 4x - 24)$$
$$= -1[3x(x-6) + 4(x-6)]$$
$$= -1(x-6)(3x+4)$$

8. $7(z-3)^2 - 2(z-3) - 5$
Let $A = z - 3$; that is $7A^2 - 2A - 5$
The key number is -35. Find factors of the key number and rewrite the middle term.
$$7A^2 - 7A + 5A - 5 = 7A(A-1) + 5(A-1)$$
$$= (A-1)(7A+5)$$
Substitute $z - 3$ for A and simplify:
$$(A-1)(7A+5) = [(z-3)-1][7(z-3)+5]$$
$$= (z-3-1)(7z-21+5)$$
$$= (z-4)(7z-16)$$

Exercises 5.4

1. Find two integers whose product is 6 and whose sum is 5. They are 2 and 3.
$$x^2 + 5x + 6 = (x+2)(x+3)$$

3. Find two integers whose product is 10 and whose sum is 7. They are 2 and 5.
$$a^2 + 7a + 10 = (a+2)(a+5)$$

5. Find two integers whose product is –12 and whose sum is 1. They are 4 and –3.
$$x^2 + x - 12 = (x+4)(x-3)$$

7. $x^2 - 2 + x = x^2 + x - 2$
Find two integers whose product is –2 and whose sum is 1. They are 2 and –1.
$$x^2 + x - 2 = (x+2)(x-1)$$

9. Find two integers whose product is –2 and whose sum is –1. They are 1 and –2.
$$x^2 - x - 2 = (x+1)(x-2)$$

11. Find two integers whose product is –10 and whose sum is –3. They are 2 and –5.
$$x^2 - 3x - 10 = (x+2)(x-5)$$

13. Find two integers whose product is 63 and whose sum is –16. They are –7 and –9.
$$a^2 - 16a + 63 = (a-7)(a-9)$$

15. $y^2 + 22 - 13y = y^2 - 13y + 22$
Find two integers whose product is 22 and whose sum is –13. They are –2 and –11.
$$y^2 - 13y + 22 = (y-2)(y-11)$$

17. $9x^2 + 37x + 4$
The key number is 36. Find factors of the key number and rewrite the middle term.
$$9x^2 + 36x + x + 4 = 9x(x+4) + 1(x+4)$$
$$= (x+4)(9x+1)$$

19. $3a^2 - 5a - 2$
The key number is –6. Find factors of the key number and rewrite the middle term.
$$3a^2 - 6a + a - 2 = 3a(a-2) + 1(a-2)$$
$$= (a-2)(3a+1)$$

21. $2y^2 - 3y - 20$
The key number is –40. Find factors of the key number and rewrite the middle term.
$$2y^2 - 8y + 5y - 20 = 2y(y-4) + 5(y-4)$$
$$= (y-4)(2y+5)$$

23. $4x^2 - 11x + 6$
The key number is 24. Find factors of the key number and rewrite the middle term.
$$4x^2 - 8x - 3x + 6 = 4x(x-2) - 3(x-2)$$
$$= (x-2)(4x-3)$$

25. $6x^2 + x - 12$
The key number is –72. Find factors of the key number and rewrite the middle term.
$$6x^2 + 9x - 8x - 12 = 3x(2x+3) - 4(2x+3)$$
$$= (2x+3)(3x-4)$$

27. $21a^2 + 11a - 2$
The key number is –42. Find factors of the key number and rewrite the middle term.
$$21a^2 + 14a - 3a - 2 = 7a(3a+2) - 1(3a+2)$$
$$= (3a+2)(7a-1)$$

29. $6x^2 + 7xy - 3y^2$

The key number is −18. Find factors of the key number and rewrite the middle term.

$6x^2 + 9xy - 2xy - 3y^2$

$$= 3x(2x+3y) - y(2x+3y)$$
$$= (2x+3y)(3x-y)$$

31. $7x^4 - 10x^3y + 3x^2y^2 = x^2\left(7x^2 - 10xy + 3y^2\right)$

The key number is 21. Find factors of the key number and rewrite the middle term.

$x^2\left(7x^2 - 7xy - 3xy + 3y^2\right)$

$$= x^2\left[7x(x - y) - 3y(x - y)\right]$$
$$= x^2(x - y)(7x - 3y)$$

33. $15x^2y^3 - xy^4 - 2y^5 = y^3\left(15x^2 - xy - 2y^2\right)$

The key number is −30. Find factors of the key number and rewrite the middle term.

$y^3\left(15x^2 - 6xy + 5xy - 2y^2\right)$

$$= y^3\left[3x(5x - 2y) + y(5x - 2y)\right]$$
$$= y^3(5x - 2y)(3x + y)$$

35. $15x^3y^2 - 2x^2y^3 - 2xy^4$

$$= xy^2\left(15x^2 - 2xy - 2y^2\right)$$

37. $-2b^2 + 13b - 20 = -1\left(2b^2 - 13b + 20\right)$

The key number is 40. Find factors of the key number and rewrite the middle term.

$-\left(2b^2 - 5b - 8b + 20\right)$

$$= -\left[b(2b - 5) - 4(2b - 5)\right]$$
$$= -(2b - 5)(b - 4)$$

39. $-12y^2 - 7y + 12 = -1\left(12y^2 + 7y - 12\right)$

The key number is −144. Find factors of the key number and rewrite the middle term.

$-\left(12y^2 + 16y - 9y - 12\right)$

$$= -\left[4y(3y + 4) - 3(3y + 4)\right]$$
$$= -(3y + 4)(4y - 3)$$

41. $2(y+2)^2 + (y+2) - 3$

Let $A = y+2$; that is $2A^2 + A - 3$

The key number is −6. Find factors of the key number and rewrite the middle term.

$2A^2 - 2A + 3A - 3 = 2A(A-1) + 3(A-1)$

$$= (A-1)(2A+3)$$

Substitute $y+2$ for A and simplify:

$(A-1)(2A+3) = \left[(y+2)-1\right]\left[2(y+2)+3\right]$

$$= (y+2-1)(2y+4+3)$$
$$= (y+1)(2y+7)$$

43. $2(x+1)^2 - 13(x+1) + 20$

Let $A = x+1$; that is $2A^2 - 13A + 20$

The key number is 40. Find factors of the key number and rewrite the middle term.

$2A^2 - 8A - 5A + 20 = 2A(A-4) - 5(A-4)$

$$= (A-4)(2A-5)$$

Substitute $x+1$ for A and simplify:

$(A-4)(2A-5) = \left[(x+1)-4\right]\left[2(x+1)-5\right]$

$$= (x+1-4)(2x+2-5)$$
$$= (x-3)(2x-3)$$

45. $-\left(a^2+2a\right)^2 - 2\left(a^2+2a\right) - 1$

Let $A = a^2+2a$; that is

$$-A^2 - 2A - 1 = -1\left(A^2 + 2A + 1\right)$$

Find two integers whose product is 1 and whose sum is 2. They are 1 and 1.

$-\left(A^2 + 2A + 1\right) = -(A+1)(A+1)$

Substitute a^2+2a for A and simplify:

$-(A+1)(A+1)$

$$= -\left[(a^2+2a)+1\right]\left[(a^2+2a)+1\right]$$
$$= -\left(a^2+2a+1\right)\left(a^2+2a+1\right)$$
$$= -(a+1)(a+1)(a+1)(a+1)$$
$$= -(a+1)^4$$

47. $2g^2 + g - 21$

The key number is –42. Find factors of the key number and rewrite the middle term.

$2g^2 + 7g - 6g - 21 = g(2g + 7) - 3(2g + 7)$
$\qquad\qquad\qquad\quad = (2g + 7)(g - 3)$

49. $2R^2 - 3R + 1$

The key number is 2. Find factors of the key number and rewrite the middle term.

$2R^2 - 2R - R + 1 = 2R(R - 1) - 1(R - 1)$
$\qquad\qquad\qquad\quad = (R - 1)(2R - 1)$

51. $(2a + b)^2 = (2a)^2 + 2 \cdot b \cdot 2a + (b)^2$
$\qquad\qquad\quad = 4a^2 + 4ab + b^2$

53. $(a \cdot 2b)^2 = (a)^2 \cdot 2 \cdot 2b \cdot a + (\cdot 2b)^2$
$\qquad\qquad\quad = a^2 \cdot 4ab + 4b^2$

55. $(a + b)(a - b) = (a)^2 - (b)^2$
$\qquad\qquad\qquad = a^2 - b^2$

57. $(2x - 3y)(2x + 3y) = (2x)^2 - (3y)^2$
$\qquad\qquad\qquad\qquad = 4x^2 - 9y^2$

59. $2L^2 - 9L + 9$

The key number is 18. Find factors of the key number and rewrite the middle term.

$2L^2 - 6L - 3L + 9 = 2L(L - 3) - 3(L - 3)$
$\qquad\qquad\qquad\quad = (L - 3)(2L - 3)$

61. $5t^2 - 12t + 7$

The key number is 35. Find factors of the key number and rewrite the middle term.

$5t^2 - 5t - 7t + 7 = 5t(t - 1) - 7(t - 1)$
$\qquad\qquad\qquad\quad = (t - 1)(5t - 7)$

63. Sample answer: The expression is not completely factored because there is a common factor of 2.

65. $6x^2 + 13x + 6$

The key number is 36. Find factors of the key number and rewrite the middle term.

$6x^2 + 9x + 4x + 6 = 3x(2x + 3) + 2(2x + 3)$
$\qquad\qquad\qquad\quad = (2x + 3)(3x + 2)$

67. Find two integers whose product is –10 and whose sum is –3. They are 2 and –5.

$x^2 - 3x - 10 = (x + 2)(x - 5)$

69. Find two integers whose product is 5 and whose sum is –2. They are no numbers that meet these conditions. The polynomial is prime.

71. $12x^4 y + 2x^3 y^2 - 4x^2 y^3 = 2x^2 y\left(6x^2 + xy - 2y^2\right)$

The key number is –12. Find factors of the key number and rewrite the middle term.

$2x^2 y\left(6x^2 \rfloor 3xy + 4xy \rfloor 2y^2\right)$
$\qquad = 2x^2 y \rfloor 3x(2x \rfloor y) + 2y(2x \rfloor y) \rfloor$
$\qquad = 2x^2 y(2x \rfloor y)(3x + 2y)$

73. $5x^2 - 2x + 2$

The key number is 10. There are no factors of 10 whose sum is –2. The polynomial is prime.

75. $-8x^2 + 2x + 21 = -1\left(8x^2 - 2x - 21\right)$

The key number is –168. Find factors of the key number and rewrite the middle term.

$\rfloor\left(8x^2 \rfloor 14x + 12x \rfloor 21\right)$
$\qquad = \rfloor \rfloor 2x(4x \rfloor 7) + 3(4x \rfloor 7) \rfloor$
$\qquad = \rfloor (4x \rfloor 7)(2x + 3)$

77. $2(y-3)^2 + 5(y-3) + 2$

Let $A = y - 3$; that is $2A^2 + 5A + 2$

The key number is 4. Find factors of the key number and rewrite the middle term.

$2A^2 + 4A + A + 2 = 2A(A+2) + 1(A+2)$

$\qquad\qquad\qquad\quad = (A+2)(2A+1)$

Substitute $y - 3$ for A and simplify:

$(A+2)(2A+1) = [(y-3)+2][2(y-3)+1]$

$\qquad\qquad\qquad = (y-3+2)(2y-6+1)$

$\qquad\qquad\qquad = (y-1)(2y-5)$

5.5   Special Factoring

Problems 5.5

1. **a.** $x^2 - 16x + 64 = x^2 - 2\cdot 8\cdot x + 8^2 = (x-8)^2$

b. $x^2 + 8x + 64$ has perfect square first and last terms. However, the middle term is not $2\cdot 8\cdot x$. There are not two factors of 64 whose sum is 8 either. The polynomial is prime.

c. $x^2 + 18x + 81 = x^2 + 2\cdot 9\cdot x + 81$

$\qquad\qquad\qquad\quad = (x+9)^2$

2. **a.** $6x^2 + 30xy + 25y^2$ does not have a perfect square first terms. The key number is 150. It has no factors whose sum is 30. The polynomial is prime.

b. $4x^2 - 12xy + 9y^2$

$\qquad = (2x)^2 - 2\cdot 2x\cdot 3y + (3y)^2$

$\qquad = (2x - 3y)^2$

b. $x^2 + 64 = x^2 + 8^2$ Sums of squares are not factorable. The polynomial is prime.

3. **a.** $4x^2 - 25 = (2x)^2 - 5^2 = (2x+5)(2x-5)$

c. $16x^4 - 81y^4 = (4x^2)^2 - (9y^2)^2$

$\qquad = (4x^2 + 9y^2)(4x^2 - 9y^2)$

$\qquad = (4x^2 + 9y^2)[(2x)^2 - (3y)^2]$

$\qquad = (4x^2 + 9y^2)(2x+3y)(2x-3y)$

4. **a.** $x^2 + 2xy + y^2 - z^2 = (x+y)^2 - z^2$

$\qquad = [(x+y)+z][(x+y)-z]$

$\qquad = (x+y+z)(x+y-z)$

b. $x^2 + 10xy + 25y^2 - 16 = (x+5y)^2 - 4^2$

$\qquad = [(x+5y)+4][(x+5y)-4]$

$\qquad = (x+5y+4)(x+5y-4)$

5. a.

$$64 + \frac{1}{27}x^3 = 4^3 + \left|\frac{1}{3}x\right|^3$$

$$= \left|4 + \frac{1}{3}x\right|\left|4^2 \mid 4\left|\frac{1}{3}x + \left|\frac{1}{3}x\right|^2\right|\right.$$

$$= \left|4 + \frac{1}{3}x\right|\left|16 \mid \frac{4}{3}x + \frac{1}{9}x^2\right|$$

b.

$$8x^3 + y^3 = (2x)^3 + y^3$$

$$= (2x + y)\left|(2x)^2 \rfloor 2x \rfloor y + y^2 \right|$$

$$= (2x + y)\left(4x^2 \rfloor 2xy + y^2\right)$$

c. $(a \cdot b)^3 + 1$

$$= :(a \cdot b) + 1: :(a \cdot b)^2 \cdot (a \cdot b) \cdot 1 + 1^2:$$

$$= (a \cdot b + 1)(a^2 \cdot 2ab + b^2 \cdot a + b + 1)$$

6.

$$x^6 \cdot 1 = (x^3)^2 \cdot 1^2 = (x^3 + 1)(x^3 \cdot 1)$$

$$= (x + 1)(x^2 \cdot x \cdot 1 + 1^2)(x \cdot 1)(x^2 + x \cdot 1 + 1^2)$$

$$= (x + 1)(x^2 \cdot x + 1)(x \cdot 1)(x^2 + x + 1)$$

Exercises 5.5

1. $x^2 + 2x + 1 = x^2 + 2 \cdot 1 \cdot x + 1^2 = (x + 1)^2$

3. $y^2 + 22y + 121 = y^2 + 2 \cdot 11 \cdot y + 11^2 = (y + 11)^2$

5. $1 + 4x + 4x^2 = 1^2 + 2 \cdot 1 \cdot 2x + (2x)^2 = (1 + 2x)^2$

7. $9x^2 + 30xy + 25y^2 = (3x)^2 + 2 \cdot 3x \cdot 5y + (5y)^2$
$$= (3x + 5y)^2$$

9. $36a^2 + 48a + 16 = 4(9a^2 + 12a + 4)$
$$= 4\left|(3a)^2 + 2 \rfloor 3a \rfloor 2 + 2^2\right|$$
$$= 4(3a + 2)^2$$

11. $y^2 \cdot 2y + 1 = y^2 \cdot 2 \cdot 1 \cdot y + 1^2 = (y \cdot 1)^2$

13. $49 \cdot 14x + x^2 = 7^2 \cdot 2 \cdot 7 \cdot x + x^2 = (7 \cdot x)^2$

15. $49a^2 \cdot 28ax + 4x^2 = (7a)^2 \cdot 2 \cdot 7a \cdot 2x + (2x)^2$
$$= (7a \cdot 2x)^2$$

17. $16x^2 \cdot 24xy + 9y^2 = (4x)^2 \cdot 2 \cdot 4x \cdot 3y + (3y)^2$
$$= (4x \cdot 3y)^2$$

19. $9x^4 + 12x^2 + 4 = (3x^2)^2 + 2 \cdot 2 \cdot 3x^2 + 2^2$
$$= (3x^2 + 2)^2$$

21. $16x^4 \cdot 24x^2 + 9 = (4x^2)^2 \cdot 2 \cdot 3 \cdot 4x^2 + 3^2$
$$= (4x^2 \cdot 3)^2$$

23. $1 + 2x^2 + x^4 = 1^2 + 2 \cdot 1 \cdot x^2 + (x^2)^2 = (1 + x^2)^2$

25. $y^2 - 64 = y^2 - 8^2 = (y + 8)(y - 8)$

27. $a^2 \mid \frac{1}{9} = a^2 \mid \left|\frac{1}{3}\right|^2 = \left|a + \frac{1}{3}\right|\left|a \mid \frac{1}{3}\right|$

29. $64 - b^2 = 8^2 - b^2 = (8+b)(8-b)$

31. $36a^2 - 49b^2 = (6a)^2 - (7b)^2$
$$= (6a+7b)(6a-7b)$$

33. $\dfrac{x^2}{9} - \dfrac{y^2}{16} = \left|\dfrac{x}{3}\right|^2 - \left|\dfrac{y}{4}\right|^2 = \left|\dfrac{x}{3}+\dfrac{y}{4}\right|\left|\dfrac{x}{3}-\dfrac{y}{4}\right|$

35. $a^2 + 4ab + 4b^2 - c^2 = (a+2b)^2 - c^2$
$$= [(a+2b)+c][(a+2b)-c]$$
$$= (a+2b+c)(a+2b-c)$$

37. $4x^2 - 4xy + y^2 - 1 = (2x-y)^2 - 1^2$
$$= [(2x-y)+1][(2x-y)-1]$$
$$= (2x-y+1)(2x-y-1)$$

39. $9y^2 - 12xy + 4x^2 - 25 = (3y-2x)^2 - 5^2$
$$= [(3y-2x)+5][(3y-2x)-5]$$
$$= (3y-2x+5)(3y-2x-5)$$

41. $16a^2 - (x^2 + 6xy + 9y^2) = (4a)^2 - (x+3y)^2$
$$= [4a+(x+3y)][4a-(x+3y)]$$
$$= (4a+x+3y)(4a-x-3y)$$

43. $y^2 - a^2 + 2ab - b^2 = y^2 - (a^2 - 2ab + b^2)$
$$= y^2 - (a-b)^2$$
$$= [y+(a-b)][y-(a-b)]$$
$$= (y+a-b)(y-a+b)$$

45. $x^3 + 125 = x^3 + 5^3 = (x+5)[x^2 - x\cdot5 + 5^2]$
$$= (x+5)(x^2 - 5x + 25)$$

47. $1 + a^3 = 1^3 + a^3 = (1+a)[1^2 - 1\cdot a + a^2]$
$$= (1+a)(1 - a + a^2)$$

49. $8x^3 + y^3 = (2x)^3 + y^3$
$$= (2x+y)[(2x)^2 - 2x\cdot y + y^2]$$
$$= (2x+y)(4x^2 - 2xy + y^2)$$

51. $x^3 - 1 = x^3 - 1^3 = (x-1)[x^2 + x\cdot1 + 1^2]$
$$= (x-1)(x^2 + x + 1)$$

53. $125a^3 - 8b^3 = (5a)^3 - (2b)^3$
$$= (5a-2b)[(5a)^2 + 5a\cdot2b + (2b)^2]$$
$$= (5a-2b)(25a^2 + 10ab + 4b^2)$$

55. $x^6 - 64 = (x^3)^2 - 8^2 = (x^3+8)(x^3 - 8)$
$$= (x+2)(x^2 - x\cdot2 + 2^2)(x - 2)(x^2 + x\cdot2 + 2^2)$$
$$= (x+2)(x^2 - 2x + 4)(x - 2)(x^2 + 2x + 4)$$

57.
$$x^6 \cdot \frac{1}{64} = \left(x^3\right)^2 \cdot \frac{1}{8}^{\,2} = \left(x^3 + \frac{1}{8}\right)\left(x^3 \cdot \frac{1}{8}\right)$$

$$= \left(x + \frac{1}{2}\right)\left(x^2 \cdot x \cdot \frac{1}{2} + \frac{1}{2}^{\,2}\right)\left(x \cdot \frac{1}{2}\right)\left(x^2 + x \cdot \frac{1}{2} + \frac{1}{2}^{\,2}\right)$$

$$= \left(x + \frac{1}{2}\right)\left(x^2 \cdot \frac{1}{2}x + \frac{1}{4}\right)\left(x \cdot \frac{1}{2}\right)\left(x^2 + \frac{1}{2}x + \frac{1}{4}\right)$$

59.
$$\frac{x^6}{64} \cdot 1 = \left(\frac{x^3}{8}\right)^2 \cdot 1^2 = \left(\frac{x^3}{8} + 1\right)\left(\frac{x^3}{8} \cdot 1\right)$$

$$= \left(\frac{x}{2} + 1\right)\left(\frac{x}{2}^{\,2} \cdot \frac{x}{2} \cdot 1 + 1^2\right)\left(\frac{x}{2} \cdot 1\right)\left(\frac{x}{2}^{\,2} + \frac{x}{2} \cdot 1 + 1^2\right)$$

$$= \left(\frac{x}{2} + 1\right)\left(\frac{x^2}{4} \cdot \frac{x}{2} + 1\right)\left(\frac{x}{2} \cdot 1\right)\left(\frac{x^2}{4} + \frac{x}{2} + 1\right)$$

61. $(x \cdot y)^3 + 1$

$$= \left[(x \cdot y) + 1\right]\left[(x \cdot y)^2 \cdot (x \cdot y) \cdot 1 + 1^2\right]$$

$$= \left(x \cdot y + 1\right)\left(x^2 \cdot 2xy + y^2 \cdot x + y + 1\right)$$

63. $1 + (x + 2y)^3$

$$= \left[1 + (x + 2y)\right]\left[1^2 \cdot (x + 2y) \cdot 1 + (x + 2y)^2\right]$$

$$= \left(1 + x + 2y\right)\left(1 \cdot x \cdot 2y + x^2 + 4xy + 4y^2\right)$$

65. $(y \cdot 2x)^3 \cdot 1$

$$= \left[(y \cdot 2x) \cdot 1\right]\left[(y \cdot 2x)^2 + (y \cdot 2x) \cdot 1 + 1^2\right]$$

$$= \left(y \cdot 2x \cdot 1\right)\left(y^2 \cdot 4xy + 4x^2 + y \cdot 2x + 1\right)$$

67. $27 \cdot (x + 2y)^3 = 3^3 \cdot (x + 2y)^3$

$$= \left[3 \cdot (x + 2y)\right]\left[3^2 + 3 \cdot (x + 2y) + (x + 2y)^2\right]$$

$$= \left(3 \cdot x \cdot 2y\right)\left(9 + 3x + 6y + x^2 + 4xy + 4y^2\right)$$

69.
$$64 + \left(x^2 \cdot y^2\right)^3 = 4^3 + \left(x^2 \cdot y^2\right)^3$$

$$= \left[4 + \left(x^2 \cdot y^2\right)\right]\left[4^2 \cdot 4 \cdot \left(x^2 \cdot y^2\right) + \left(x^2 \cdot y^2\right)^2\right]$$

$$= \left(4 + x^2 \cdot y^2\right)\left(16 \cdot 4x^2 + 4y^2 + x^4 \cdot 2x^2y^2 + y^4\right)$$

71. $(x + 3)(x - 5) = x^2 - 5x + 3x - 15$
$$= x^2 - 2x - 15$$

73. $(x - 8)(x + 2) = x^2 + 2x - 8x - 16$
$$= x^2 - 6x - 16$$

75. $(2x \cdot 3y)^2 = (2x)^2 \cdot 2 \cdot 2x \cdot 3y + (3y)^2$
$$= 4x^2 \cdot 12xy + 9y^2$$

77. $100 - x^2 = (10)^2 - x^2 = (10 + x)(10 - x)$

79. $8x^3 + 1 = (2x)^3 + 1^3 = (2x + 1)\left[(2x)^2 \,\rfloor\, 2x \,\rfloor 1 + 1^2\right] = (2x + 1)\left(4x^2 \,\rfloor\, 2x + 1\right)$

81. Sample answer: First check whether two of the terms are perfect squares. Then check whether the other term is twice the product of the square roots of the other two terms.

83. Sample answer: Set twice the product of the square roots of the first and last terms equal to the middle term and solve.

$$2\cdot 2x\cdot\sqrt{k} = 12x$$
$$\sqrt{k} = 3$$
$$k = 9$$

The value of k is 9.

85. Sample answer: The polynomial can be factored as follows:

$$a^4 + 64 = a^4 + 16a^2 + 64 - 16a^2$$
$$= \left(a^2+8\right)^2 - (4a)^2$$
$$= \left(a^2+8+4a\right)\left(a^2+8-4a\right)$$
$$= \left(a^2+4a+8\right)\left(a^2-4a+8\right)$$

87.
$$8x^3 + 27y^3 = (2x)^3 + (3y)^3$$
$$= (2x+3y)\left[(2x)^2 - 2x\cdot 3y + (3y)^2\right]$$
$$= (2x+3y)\left(4x^2 - 6xy + 9y^2\right)$$

89.
$$x^6 + 1 = \left(x^2\right)^3 + 1^3 = \left(x^2+1\right)\left(\left(x^2\right)^2 - x^2\cdot 1 + 1^2\right)$$
$$= \left(x^2+1\right)\left(x^4 - x^2 + 1\right)$$

91.
$$16x^4 - 81y^4 = \left(4x^2\right)^2 - \left(9y^2\right)^2$$
$$= \left(4x^2+9y^2\right)\left(4x^2-9y^2\right)$$
$$= \left(4x^2+9y^2\right)(2x+3y)(2x-3y)$$

93.
$$x^2 + 10xy + 25y^2 - 16 = (x+5y)^2 - 4^2$$
$$= \left[(x+5y)+4\right]\left[(x+5y)-4\right]$$
$$= (x+5y+4)(x+5y-4)$$

95.
$$4x^2 - 12xy + 9y^2 = (2x)^2 - 2\cdot 2x\cdot 3y + (3y)^2$$
$$= (2x-3y)^2$$

97.
$$x^2 + 18x + 81 = x^2 + 2\cdot 9\cdot x + 9^2 = (x+9)^2$$

99.
$$(x+y)^2 - (y-x)^2$$
$$= \left[(x+y)+(y-x)\right]\left[(x+y)-(y-x)\right]$$
$$= (x+y+y-x)(x+y-y+x)$$
$$= (2y)(2x) = 4xy$$

101.
$$125 + (x+y)^3 = 5^3 + (x+y)^3$$
$$= \left[5+(x+y)\right]\left[5^2 - 5\cdot(x+y)+(x+y)^2\right]$$
$$= (5+x+y)\left(25 - 5x - 5y + x^2 + 2xy + y^2\right)$$

5.6   General Methods of Factoring

Problems 5.6

1. **a.** $27x^5 \cdot x^2y^3 = x^2\left(27x^3 \cdot y^3\right)$

$= x^2 \cdot (3x)^3 \cdot y^3 \cdot$

$= x^2(3x \cdot y) \cdot (3x)^2 + 3x \cdot y + y^2 \cdot$

$= x^2(3x \cdot y)\left(9x^2 + 3xy + y^2\right)$

   b. $8x^5 + 72x^3 = 8x^3\left(x^2 + 9\right)$

2. **a.** $36x^5 + 24x^4y + 4x^3y^2$

$= 4x^3\left(9x^2 + 6xy + y^2\right)$

$= 4x^3(3x + y)^2$

   b. $12x^2y^2 - 12xy^3 + 3y^4 = 3y^2\left(4x^2 - 4xy + y^2\right)$

$= 3y^2(2x - y)^2$

3. **a.** $9x^3y \rfloor 15x^2y^2 \rfloor 6xy^3$

$= 3xy\left(3x^2 \rfloor 5xy \rfloor 2y^2\right)$

$= 3xy\left(3x^2 \rfloor 6xy + xy \rfloor 2y^2\right)$

$= 3xy\rfloor 3x(x \rfloor 2y) + y(x \rfloor 2y)\rfloor$

$= 3xy(x \rfloor 2y)(3x + y)$

   b. $24x^4 + x^3y + x^2y^2 = x^2\left(24x^2 + xy + y^2\right)$

4. $2x^4 + x^3y + 2x^2y^2 = x^2\left(2x^2 + xy + 2y^2\right)$

5. $9x^3 + 18x^2 - 25x - 50$

$= 9x^2(x + 2) - 25(x + 2)$

$= (x + 2)\left(9x^2 - 25\right)$

$= (x + 2)(3x + 5)(3x - 5)$

6. $x^2 + 8x + 16 - 81y^2 = (x + 4)^2 - (9y)^2$

$= (x + 4 + 9y)(x + 4 - 9y)$

7. $\rfloor 3x^2 + 17x + 28 = \rfloor\left(3x^2 \rfloor 17x \rfloor 28\right)$

$= \rfloor\left(3x^2 \rfloor 21x + 4x \rfloor 28\right)$

$= \rfloor\rfloor 3x(x \rfloor 7) + 4(x \rfloor 7)\rfloor$

$= \rfloor(x \rfloor 7)(3x + 4)$

Exercises 5.6

1. $3x^4 - 3x^3 - 18x^2 = 3x^2\left(x^2 - x - 6\right)$

$= 3x^2(x - 3)(x + 2)$

3. $5x^4 + 10x^3y - 40x^2y^2 = 5x^2\left(x^2 + 2xy - 8y^2\right)$

$= 5x^2(x + 4y)(x - 2y)$

5. $-3x^6 - 6x^5 - 21x^4 = -3x^4\left(x^2 + 2x + 7\right)$

7. $2x^6y - 4x^5y^2 - 10x^4y^3 = 2x^4y\left(x^2 - 2xy - 5y^2\right)$

9. $-4x^6 - 12x^5y - 18x^4y^2$
$$= -2x^4\left(2x^2 + 6xy + 9y^2\right)$$

11. $6x^3y^2 + 12x^2y^2 + 2xy^2 + 4y^2$
$$= 2y^2\left(3x^3 + 6x^2 + x + 2\right)$$
$$= 2y^2\left\lfloor 3x^2(x+2) + 1(x+2)\right\rfloor$$
$$= 2y^2\left(x+2\right)\left(3x^2 + 1\right)$$

13. $-9x^4y - 9x^3y - 6x^2y - 6xy$
$$= -3xy\left(3x^3 + 3x^2 + 2x + 2\right)$$
$$= -3xy\left\lfloor 3x^2(x+1) + 2(x+1)\right\rfloor$$
$$= -3xy(x+1)\left(3x^2 + 2\right)$$

15. $-4x^4 - 4x^3y + 2x^2y + 2xy^2$
$$= -2x\left(2x^3 + 2x^2y - xy - y^2\right)$$
$$= -2x\left\lfloor 2x^2(x+y) - y(x+y)\right\rfloor$$
$$= -2x(x+y)\left(2x^2 - y\right)$$

17. $3x^2y^2 + 24xy^3 + 48y^4$
$$= 3y^2\left(x^2 + 8xy + 16y^2\right)$$
$$= 3y^2\left(x^2 + 2\cdot x\cdot 4y + (4y)^2\right)$$
$$= 3y^2\left(x+4y\right)^2$$

19. $\cdot\, 18kx^2 \cdot 24kxy \cdot 8ky^2$
$$= \cdot\, 2k\left(9x^2 + 12xy + 4y^2\right)$$
$$= \cdot\, 2k\left((3x)^2 + 2\cdot 3x\cdot 2y + (2y)^2\right)$$
$$= \cdot\, 2k\left(3x+2y\right)^2$$

21. $16x^3y^2 \cdot 48x^2y^3 + 36xy^4$
$$= 4xy^2\left(4x^2 \cdot 12xy + 9y^2\right)$$
$$= 4xy^2\left((2x)^2 \cdot 2\cdot 2x\cdot 3y + (3y)^2\right)$$
$$= 4xy^2\left(2x\cdot 3y\right)^2$$

23. $kx^2 - 12kx + 36$
The polynomial is not factorable. It is prime.

25. $3x^5 + 12x^4y + 12x^3y^2$
$$= 3x^3\left(x^2 + 4xy + 4y^2\right)$$
$$= 3x^3\left(x^2 + 2\cdot x\cdot 2y + (2y)^2\right)$$
$$= 3x^3\left(x+2y\right)^2$$

27. $18x^6 + 12x^5y + 2x^4y^2$
$$= 2x^4\left(9x^2 + 6xy + y^2\right)$$
$$= 2x^4\left((3x)^2 + 2\cdot 3x\cdot y + y^2\right)$$
$$= 2x^4\left(3x+y\right)^2$$

29. $12x^4y^2 \cdot 36x^3y^3 + 27x^2y^4$
$$= 3x^2y^2\left(4x^2 \cdot 12xy + 9y^2\right)$$
$$= 3x^2y^2\left((2x)^2 \cdot 2\cdot 2x\cdot 3y + (3y)^2\right)$$
$$= 3x^2y^2\left(2x\cdot 3y\right)^2$$

31. $6x^3 + 12x^2 - 6x - 12 = 6\left(x^3 + 2x^2 - x - 2\right)$
$$= 6\left(x^2(x+2) - 1(x+2)\right)$$
$$= 6(x+2)\left(x^2 - 1\right)$$
$$= 6(x+2)(x+1)(x-1)$$

33. $7x^4 - 7y^4 = 7\left(x^4 - y^4\right)$
$$= 7\left(x^2 + y^2\right)\left(x^2 - y^2\right)$$
$$= 7\left(x^2 + y^2\right)(x+y)(x-y)$$

35. $2x^6 - 32x^2y^4 = 2x^2\left(x^4 - 16y^4\right)$
$$= 2x^2\left(x^2 + 4y^2\right)\left(x^2 - 4y^2\right)$$
$$= 2x^2\left(x^2 + 4y^2\right)(x+2y)(x-2y)$$

37.
$$\cdot\,2x^2\cdot\,12x\cdot\,18=\cdot\,2\left(x^2+6x+9\right)$$
$$=\cdot\,2\left(x^2+2\cdot x\cdot 3+3^2\right)$$
$$=\cdot\,2\left(x+3\right)^2$$

39.
$$\cdot\,3x^2\cdot\,12x\cdot\,12=\cdot\,3\left(x^2+4x+4\right)$$
$$=\cdot\,3\left(x^2+2\cdot x\cdot 2+2^2\right)$$
$$=\cdot\,3\left(x+2\right)^2$$

41.
$$\cdot\,4x^4\cdot\,4x^3y\cdot\,x^2y^2$$
$$=\cdot\,x^2\left(4x^2+4xy+y^2\right)$$
$$=\cdot\,x^2\left((2x)^2+2\cdot 2x\cdot y+y^2\right)$$
$$=\cdot\,x^2\left(2x+y\right)^2$$

43.
$$\cdot\,9x^2y^2\cdot\,12xy^3\cdot\,4y^4$$
$$=\cdot\,y^2\left(9x^2+12xy+4y^2\right)$$
$$=\cdot\,y^2\left((3x)^2+2\cdot 3x\cdot 2y+(2y)^2\right)$$
$$=\cdot\,y^2\left(3x+2y\right)^2$$

45.
$$\cdot\,8x^2y^2+24xy^3\cdot\,18y^4$$
$$=\cdot\,2y^2\left(4x^2\cdot\,12xy+9y^2\right)$$
$$=\cdot\,2y^2\left((2x)^2\cdot\,2\cdot 2x\cdot 3y+(3y)^2\right)$$
$$=\cdot\,2y^2\left(2x\cdot\,3y\right)^2$$

47.
$$\cdot\,18x^3\cdot\,24x^2y\cdot\,8xy^2$$
$$=\cdot\,2x\left(9x^2+12xy+4y^2\right)$$
$$=\cdot\,2x\left((3x)^2+2\cdot 3x\cdot 2y+(2y)^2\right)$$
$$=\cdot\,2x\left(3x+2y\right)^2$$

49.
$$\cdot\,18x^3\cdot\,60x^2y\cdot\,50xy^2$$
$$=\cdot\,2x\left(9x^2+30xy+25y^2\right)$$
$$=\cdot\,2x\left((3x)^2+2\cdot 3x\cdot 5y+(5y)^2\right)$$
$$=\cdot\,2x\left(3x+5y\right)^2$$

51.
$$-x^3+xy^2=-x\left(x^2-y^2\right)=-x(x+y)(x-y)$$

53.
$$-x^4+4x^2y^2=-x^2\left(x^2-4y^2\right)$$
$$=-x^2(x+2y)(x-2y)$$

55.
$$-4x^4+9x^2y^2=-x^2\left(4x^2-9y^2\right)$$
$$=-x^2(2x+3y)(2x-3y)$$

57.
$$-8x^3+18xy^2=-2x\left(4x^2-9y^2\right)$$
$$=-2x(2x+3y)(2x-3y)$$

59.
$$-18x^4+8x^2y^2=-2x^2\left(9x^2-4y^2\right)$$
$$=-2x^2(3x+2y)(3x-2y)$$

61.
$$27x^2\cdot\,x^5=x^2\left(27\cdot\,x^3\right)=x^2\left(3^3\cdot\,x^3\right)$$
$$=x^2\left(3\cdot\,x\right)\left(3^2+3\cdot x+x^2\right)$$
$$=x^2\left(3\cdot\,x\right)\left(9+3x+x^2\right)$$

63.
$$x^7\cdot\,8x^4=x^4\left(x^3\cdot\,8\right)=x^4\left(x^3\cdot\,2^3\right)$$
$$=x^4\left(x\cdot\,2\right)\left(x^2+2\cdot x+2^2\right)$$
$$=x^4\left(x\cdot\,2\right)\left(x^2+2x+4\right)$$

65.
$$27x^4+8x^7=x^4\left(27+8x^3\right)=x^4\left(3^3+(2x)^3\right)$$
$$=x^4\left(3+2x\right)\left(3^2\cdot\,3\cdot 2x+(2x)^2\right)$$
$$=x^4\left(3+2x\right)\left(9\cdot\,6x+4x^2\right)$$

67.
$$27x^7+64x^4y^3=x^4\left(27x^3+64y^3\right)$$
$$=x^4\left((3x)^3+(4y)^3\right)$$
$$=x^4\left(3x+4y\right)\left((3x)^2\cdot\,3x\cdot 4y+(4y)^2\right)$$
$$=x^4\left(3x+4y\right)\left(9x^2\cdot\,12xy+16y^2\right)$$

69.
$$x^2 + 4x + 4 - y^2 = (x+2)^2 - y^2$$
$$= (x+2+y)(x+2-y)$$
$$= (x+y+2)(x-y+2)$$

71.
$$x^2 + y^2 - 6y + 9 = x^2 + (y-3)^2$$
A sum of two squares is not factorable. This polynomial is prime.

73.
$$x^2 - y^2 - 4y - 4 = x^2 - (y^2 + 4y + 4)$$
$$= x^2 - (y+2)^2$$
$$= [x + (y+2)][x - (y+2)]$$
$$= (x+y+2)(x-y-2)$$

75.
$$-9x^2 + 30xy - 25y^2 = -\left(9x^2 - 30xy + 25y^2\right)$$
$$= -\left((3x)^2 - 2 \cdot 3x \cdot 5y + (5y)^2\right)$$
$$= -(3x - 5y)^2$$

77.
$$18x^3 - 60x^2 y + 50xy^2$$
$$= 2x\left(9x^2 - 30xy + 25y^2\right)$$
$$= 2x\left((3x)^2 - 2 \cdot 3x \cdot 5y + (5y)^2\right)$$
$$= 2x(3x - 5y)^2$$

79.
$$6x^2 - x - 2 = 6x^2 - 4x + 3x - 2$$
$$= 2x(3x-2) + 1(3x-2)$$
$$= (3x-2)(2x+1)$$

81.
$$12x^2 - x - 1 = 12x^2 - 4x + 3x - 1$$
$$= 4x(3x-1) + 1(3x-1)$$
$$= (3x-1)(4x+1)$$

83.
$$\frac{2\pi A}{360} R + \frac{2\pi A}{360} Kt = \frac{2\pi A}{360}(R + Kt)$$

85.
$$\frac{3Sd^2}{2bd^3} - \frac{12Sz^2}{2bd^3} = \frac{3S}{2bd^3}\left(d^2 - 4z^2\right)$$
$$= \frac{3S}{2bd^3}(d + 2z)(d - 2z)$$

87. Sample answer: The expression is not completely factored because there are still two terms added together. The expression also contains a common factor.

89. Sample answer: The statement is factored, but is not completely factored. There is a common factor of x.

91.
$$9x^3 - 18x^2 - x + 2 = 9x^2(x-2) - 1(x-2)$$
$$= (x-2)\left(9x^2 - 1\right)$$
$$= (x-2)(3x+1)(3x-1)$$

93.
$$2x^4 + x^3 y + 2x^2 y^2 = x^2\left(2x^2 + xy + 2y^2\right)$$

95.
$$x^2 - 10x + 25 - y^2 = (x-5)^2 - y^2$$
$$= (x-5+y)(x-5-y)$$
$$= (x+y-5)(x-y-5)$$

97.
$$-8y^2 + 2y + 21 = -\left(8y^2 - 2y - 21\right)$$
$$= -\left(8y^2 - 14y + 12y - 21\right)$$
$$= -\left[2y(4y-7) + 3(4y-7)\right]$$
$$= -(4y-7)(2y+3)$$

99. $36x^5 + 24x^4y + 4x^3y^2$

$$= 4x^3\left(9x^2 + 6xy + y^2\right)$$

$$= 4x^3\left(\left(3x\right)^2 + 2\cdot 3x\cdot y + y^2\right)$$

$$= 4x^3\left(3x + y\right)^2$$

101. $8x^5 + 72x^3 = 8x^3\left(x^2 + 9\right)$

5.7   Solving Equations by Factoring: Applications

Problems 5.7

1.   a.
$$x^2 - 36 = 0$$
$$(x+6)(x-6) = 0$$
$$x+6 = 0 \quad \text{or } x - 6 = 0$$
$$x = -6 \text{ or} \quad x = 6$$

b.
$$2x^2 - 50x = 0$$
$$2x(x-25) = 0$$
$$2x = 0 \text{ or } x - 25 = 0$$
$$x = 0 \text{ or} \quad x = 25$$

2.   a.
$$x^2 = 6x - 5$$
$$x^2 - 6x + 5 = 0$$
$$(x-1)(x-5) = 0$$
$$x-1 = 0 \text{ or } x - 5 = 0$$
$$x = 1 \text{ or} \quad x = 5$$

b.
$$x^2 = 12 - x$$
$$x^2 + x - 12 = 0$$
$$(x+4)(x-3) = 0$$
$$x+4 = 0 \quad \text{or } x - 3 = 0$$
$$x = -4 \text{ or} \quad x = 3$$

3.   a.
$$12x^2 + 13x - 4 = 0$$
$$12x^2 + 16x - 3x - 4 = 0$$
$$4x(3x+4) - 1(3x+4) = 0$$
$$(3x+4)(4x-1) = 0$$
$$3x+4 = 0 \quad \text{or } 4x - 1 = 0$$
$$3x = -4 \text{ or} \quad 4x = 1$$
$$x = -\frac{4}{3} \text{or} \quad x = \frac{1}{4}$$

b.
$$8x^2 - 2x = 1$$
$$8x^2 - 2x - 1 = 0$$
$$8x^2 - 4x + 2x - 1 = 0$$
$$4x(2x-1) + 1(2x-1) = 0$$
$$(2x-1)(4x+1) = 0$$
$$2x-1 = 0 \text{ or } 4x + 1 = 0$$
$$2x = 1 \text{ or} \quad 4x = -1$$
$$x = \frac{1}{2} \text{ or} \quad x = -\frac{1}{4}$$

c.
$$x(x-3)=(4x+7)x-8$$
$$x^2-3x=4x^2+7x-8$$
$$3x^2+10x-8=0$$
$$3x^2+12x-2x-8=0$$
$$3x(x+4)-2(x+4)=0$$
$$(x+4)(3x-2)=0$$
$$x+4=0 \quad \text{or } 3x-2=0$$
$$x=-4 \text{ or} \qquad 3x=2$$
$$x=-4 \text{ or} \qquad x=\frac{2}{3}$$

4.
$$x^3+2x^2-9x-18=0$$
$$x^2(x+2)-9(x+2)=0$$
$$(x+2)(x^2-9)=0$$
$$(x+2)(x+3)(x-3)=0$$
$$x+2=0 \quad \text{or } x+3=0 \quad \text{or } x-3=0$$
$$x=-2 \text{ or} \qquad x=-3 \text{ or } x=3$$

5. Apply the Pythagorean theorem to the triangle at the bottom point of the plate:
$$12^2+12^2=17^2$$
$$144+144=289$$
$$288 \neq 289$$
These measurements are not correct because the Pythagorean theorem is not satisfied.

6. Let x = the width of the pool
Let $x+7$ = the length of the pool
$$x^2+(x+7)^2=17^2$$
$$x^2+x^2+14x+49=289$$
$$2x^2+14x-240=0$$
$$x^2+7x-120=0$$
$$(x-8)(x+15)=0$$
$$x-8=0 \text{ or } x+15=0$$
$$x=8 \text{ or} \qquad x=-15$$
$$x+7=15$$
The length of the pool is 15 m.

7.
$$(6-2x)(8-2x)=15$$
$$48-12x-16x+4x^2=15$$
$$4x^2-28x+33=0$$
$$4x^2-22x-6x+33=0$$
$$2x(2x-11)-3(2x-11)=0$$
$$(2x-11)(2x-3)=0$$
$$2x-11=0 \quad \text{or } 2x-3=0$$
$$2x=11 \text{ or} \qquad 2x=3$$
$$x=\frac{11}{2} \text{ or} \qquad x=\frac{3}{2}$$
The border is 1.5 m wide. It cannot be 5.5 m wide because there is insufficient width for it.

Exercises 5.7

1. $(x+1)(x+2)=0$
$$x+1=0 \quad \text{or} \; x+2=0$$
$$x=-1 \; \text{or} \quad x=-2$$

3. $(x-1)(x+4)(x+3)=0$
$$x-1=0 \; \text{or} \; x+4=0 \quad \text{or} \; x+3=0$$
$$x=1 \; \text{or} \quad x=-4 \; \text{or} \quad x=-3$$

5. $\left|x \left|\frac{1}{2}\right|\right|\left|x \left|\frac{1}{3}\right|\right|=0$
$$x \left|\frac{1}{2}=0 \; \text{or} \; x\right|\frac{1}{3}=0$$
$$x=\frac{1}{2} \; \text{or} \quad x=\frac{1}{3}$$

7. $y(y-3)=0$
$$y=0 \; \text{or} \; y-3=0$$
$$y=0 \; \text{or} \quad y=3$$

9. $y^2-64=0$
$$(y+8)(y-8)=0$$
$$y+8=0 \quad \text{or} \; y-8=0$$
$$y=-8 \; \text{or} \quad y=8$$

11. $y^2-81=0$
$$(y+9)(y-9)=0$$
$$y+9=0 \quad \text{or} \; y-9=0$$
$$y=-9 \; \text{or} \quad y=9$$

13. $x^2+6x=0$
$$x(x+6)=0$$
$$x=0 \; \text{or} \; x+6=0$$
$$x=0 \; \text{or} \quad x=-6$$

15. $x^2-3x=0$
$$x(x-3)=0$$
$$x=0 \; \text{or} \; x-3=0$$
$$x=0 \; \text{or} \quad x=3$$

17. $y^2-12y=-27$
$$y^2-12y+27=0$$
$$(y-9)(y-3)=0$$
$$y-9=0 \; \text{or} \; y-3=0$$
$$y=9 \; \text{or} \quad y=3$$

19. $y^2=-6y-5$
$$y^2+6y+5=0$$
$$(y+5)(y+1)=0$$
$$y+5=0 \quad \text{or} \; y+1=0$$
$$y=-5 \; \text{or} \quad y=-1$$

21. $x^2=2x+15$
$$x^2-2x-15=0$$
$$(x-5)(x+3)=0$$
$$x-5=0 \; \text{or} \; x+3=0$$
$$x=5 \; \text{or} \quad x=-3$$

23. $3y^2+5y+2=0$
$$3y^2+2y+3y+2=0$$
$$y(3y+2)+1(3y+2)=0$$
$$(3y+2)(y+1)=0$$
$$3y+2=0 \quad \text{or} \; y+1=0$$
$$3y=-2 \; \text{or} \quad y=-1$$
$$y=-\frac{2}{3} \; \text{or} \quad y=-1$$

25.
$$2y^2 - 3y + 1 = 0$$
$$2y^2 - 2y - 1y + 1 = 0$$
$$2y(y-1) - 1(y-1) = 0$$
$$(y-1)(2y-1) = 0$$
$$y - 1 = 0 \text{ or } 2y - 1 = 0$$
$$y = 1 \text{ or } 2y = 1$$
$$y = 1 \text{ or } y = \frac{1}{2}$$

27.
$$2y^2 - y - 1 = 0$$
$$2y^2 - 2y + y - 1 = 0$$
$$2y(y-1) + 1(y-1) = 0$$
$$(y-1)(2y+1) = 0$$
$$y - 1 = 0 \text{ or } 2y + 1 = 0$$
$$y = 1 \text{ or } 2y = -1$$
$$y = 1 \text{ or } y = -\frac{1}{2}$$

29.
$$\frac{x^2}{12} + \frac{x}{3} - 1 = 0$$
$$12\left(\frac{x^2}{12} + \frac{x}{3} - 1\right) = 12(0)$$
$$x^2 + 4x - 12 = 0$$
$$(x+6)(x-2) = 0$$
$$x + 6 = 0 \quad \text{or } x - 2 = 0$$
$$x = -6 \text{ or } \quad x = 2$$

31.
$$\frac{x^2}{3} - \frac{x}{2} = \frac{1}{6}$$
$$6\left(\frac{x^2}{3} - \frac{x}{2}\right) = 6\left(\frac{1}{6}\right)$$
$$2x^2 - 3x = 1$$
$$2x^2 - 3x + 1 = 0$$
$$2x^2 - 2x - x + 1 = 0$$
$$2x(x-1) - 1(x-1) = 0$$
$$(x-1)(2x-1) = 0$$
$$x - 1 = 0 \text{ or } 2x - 1 = 0$$
$$x = 1 \text{ or } \quad 2x = 1$$
$$x = 1 \text{ or } \quad x = \frac{1}{2}$$

33.
$$\frac{x^2}{12} + \frac{x}{2} = -\frac{2}{3}$$
$$12\left(\frac{x^2}{12} + \frac{x}{2}\right) = 12\left(-\frac{2}{3}\right)$$
$$x^2 + 6x = -8$$
$$x^2 + 6x + 8 = 0$$
$$(x+4)(x+2) = 0$$
$$x + 4 = 0 \quad \text{or } x + 2 = 0$$
$$x = -4 \text{ or } \quad x = -2$$

35.
$$(2x-1)(x-3) = 3x - 5$$
$$2x^2 - 6x - x + 3 = 3x - 5$$
$$2x^2 - 10x + 8 = 0$$
$$x^2 - 5x + 4 = 0$$
$$(x-4)(x-1) = 0$$
$$x - 4 = 0 \text{ or } x - 1 = 0$$
$$x = 4 \text{ or } \quad x = 1$$

37.
$$(2x+3)(x+4) = 2(x-1)+4$$
$$2x^2 + 8x + 3x + 12 = 2x - 2 + 4$$
$$2x^2 + 11x + 12 = 2x + 2$$
$$2x^2 + 9x + 10 = 0$$
$$2x^2 + 5x + 4x + 10 = 0$$
$$x(2x+5) + 2(2x+5) = 0$$
$$(2x+5)(x+2) = 0$$
$$2x + 5 = 0 \quad \text{or } x + 2 = 0$$
$$2x = -5 \quad \text{or} \quad x = -2$$
$$x = -\frac{5}{2} \quad \text{or} \quad x = -2$$

39.
$$(2x-1)(x-1) = x - 1$$
$$2x^2 - 2x - x + 1 = x - 1$$
$$2x^2 - 4x + 2 = 0$$
$$x^2 - 2x + 1 = 0$$
$$(x-1)(x-1) = 0$$
$$x - 1 = 0 \text{ or } x - 1 = 0$$
$$x = 1 \quad \text{or} \quad x = 1$$

41.
$$x^3 + 4x^2 - 4x - 16 = 0$$
$$x^2(x+4) - 4(x+4) = 0$$
$$(x+4)(x^2-4) = 0$$
$$(x+4)(x+2)(x-2) = 0$$
$$x + 4 = 0 \quad \text{or } x + 2 = 0 \quad \text{or } x - 2 = 0$$
$$x = -4 \quad \text{or} \quad x = -2 \text{ or} \quad x = 2$$

43.
$$x^3 - 5x^2 - 9x + 45 = 0$$
$$x^2(x-5) - 9(x-5) = 0$$
$$(x-5)(x^2-9) = 0$$
$$(x-5)(x+3)(x-3) = 0$$
$$x - 5 = 0 \text{ or } x + 3 = 0 \quad \text{or } x - 3 = 0$$
$$x = 5 \text{ or} \quad x = -3 \text{ or} \quad x = 3$$

45.
$$3x^3 + 3x^2 = 12x + 12$$
$$3x^3 + 3x^2 - 12x - 12 = 0$$
$$x^3 + x^2 - 4x - 4 = 0$$
$$x^2(x+1) - 4(x+1) = 0$$
$$(x+1)(x^2-4) = 0$$
$$(x+1)(x+2)(x-2) = 0$$
$$x + 1 = 0 \quad \text{or } x + 2 = 0 \quad \text{or } x - 2 = 0$$
$$x = -1 \text{ or} \quad x = -2 \text{ or} \quad x = 2$$

47. Let x = the first side
Let $x + 2$ = the second side
Let $x + 4$ = the hypotenuse
$$x^2 + (x+2)^2 = (x+4)^2$$
$$x^2 + x^2 + 4x + 4 = x^2 + 8x + 16$$
$$x^2 - 4x - 12 = 0$$
$$(x-6)(x+2) = 0$$
$$x - 6 = 0 \text{ or } x + 2 = 0$$
$$x = 6 \quad \text{or} \quad x = -2$$
$$x + 2 = 8$$
$$x + 4 = 10$$
The lengths of the sides are 6, 8, and 10.

49. Let x = the first side

Let $x + 14$ = the second side

Let $x + 16$ = the hypotenuse

$$x^2 + (x+14)^2 = (x+16)^2$$

$$x^2 + x^2 + 28x + 196 = x^2 + 32x + 256$$

$$x^2 - 4x - 60 = 0$$

$$(x-10)(x+6) = 0$$

$$x - 10 = 0 \text{ or } x + 6 = 0$$

$$x = 10 \text{ or } \quad x = -6$$

$$x + 14 = 24$$

$$x + 16 = 26$$

The lengths of the sides are 10 in, 24 in, and 26 in.

51. $d = 5t^2 + V_0 t; \quad V_0 = 5, d = 10$

$$10 = 5t^2 + 5t$$

$$5t^2 + 5t - 10 = 0$$

$$t^2 + t - 2 = 0$$

$$(t-1)(t+2) = 0$$

$$t - 1 = 0 \text{ or } t + 2 = 0$$

$$t = 1 \text{ or } \quad t = -2$$

The object hits the ground in 1 sec.

53. $d = 5t^2 + V_0 t; \quad V_0 = 10, d = 15$

$$15 = 5t^2 + 10t$$

$$5t^2 + 10t - 15 = 0$$

$$t^2 + 2t - 3 = 0$$

$$(t-1)(t+3) = 0$$

$$t - 1 = 0 \text{ or } t + 3 = 0$$

$$t = 1 \text{ or } \quad t = -3$$

The object hits the ground in 1 sec.

55. $0.1x^2 + x + 50 = C; \quad C = 250$

$$0.1x^2 + x + 50 = 250$$

$$0.1x^2 + x - 200 = 0$$

$$x^2 + 10x - 2000 = 0$$

$$(x-40)(x+50) = 0$$

$$x - 40 = 0 \text{ or } x + 50 = 0$$

$$x = 40 \text{ or } \quad x = -50$$

40 customers can be served for $250.

57. $x^2 + 25x = p; \quad p = 350$

$$x^2 + 25x = 350$$

$$x^2 + 25x - 350 = 0$$

$$(x-10)(x+35) = 0$$

$$x - 10 = 0 \text{ or } x + 35 = 0$$

$$x = 10 \text{ or } \quad x = -35$$

10 units will be produced when the price is $350 per unit.

59. **a.** $p = 5.50 - x$

b. $R = (100 + 100x)(5.50 - x)$

$$= 550 - 100x + 550x - 100x^2$$

$$= 550 + 450x - 100x^2$$

c. $$750 = 550 + 450x - 100x^2$$

$$100x^2 - 450x + 200 = 0$$

$$2x^2 - 9x + 4 = 0$$

$$(2x-1)(x-4) = 0$$

$$2x - 1 = 0 \text{ or } x - 4 = 0$$

$$x = \frac{1}{2} \text{ or } \quad x = 4$$

d. The price reduction is either $0.50 or $4.

The price reduction is $0.50.

61. $\dfrac{3x^5}{15x^7} = \dfrac{3}{15}\dfrac{x^5}{x^7} = \dfrac{1}{5}x^{5-7} = \dfrac{1}{5}x^{-2} = \dfrac{1}{5x^2}$

63. $\dfrac{20x^7}{10x^{-3}} = \dfrac{20}{10}\cdot\dfrac{x^7}{x^{-3}} = 2x^{7-(-3)} = 2x^{10}$

65. $\dfrac{18x^{10}}{9x^2} = \dfrac{18}{9} \cdot \dfrac{x^{10}}{x^2} = 2x^{10-2} = 2x^{12} = \dfrac{2}{x^{12}}$

67.
$$H(t) = -80t^2 + 239t + 3$$
$$0 = -80t^2 + 239t + 3$$
$$80t^2 - 239t - 3 = 0$$
$$(80t + 1)(t - 3) = 0$$
$$80t + 1 = 0 \quad \text{or } t - 3 = 0$$
$$t = -\frac{1}{80} \text{ or } \quad t = 3$$

The ball hit the ground 3 + 1 or 4 sec after it is hit.

69.
$$D(t) = -5t^2 + 115t - 110; \quad t = 6$$
$$D(6) = -5(6)^2 + 115(6) - 110$$
$$= -5(36) + 690 - 110$$
$$= -180 + 690 - 110$$
$$= 400$$

The ball travels 400 ft.

71. Sample answer: As long as one factor of an equation is zero the value of the equation is zero. Thus it does not matter how many factors there are as long as one of them is zero.

73. Answers will vary.

75.
$$8x^2 - 2x = 1$$
$$8x^2 - 2x - 1 = 0$$
$$(4x + 1)(2x - 1) = 0$$
$$4x + 1 = 0 \quad \text{or } 2x - 1 = 0$$
$$4x = -1 \text{ or } \quad 2x = 1$$
$$x = -\frac{1}{4} \text{ or } \quad x = \frac{1}{2}$$

77.
$$x^2 = 6x - 5$$
$$x^2 - 6x + 5 = 0$$
$$(x - 5)(x - 1) = 0$$
$$x - 5 = 0 \text{ or } x - 1 = 0$$
$$x = 5 \text{ or } \quad x = 1$$

79.
$$x^2 - 16 = 0$$
$$(x + 4)(x - 4) = 0$$
$$x + 4 = 0 \quad \text{or } x - 4 = 0$$
$$x = -4 \text{ or } \quad x = 4$$

81. Let x = the longer leg
Let $x + 1$ = the hypotenuse
Let $x - 7$ = the shorter leg
$$x^2 + (x - 7)^2 = (x + 1)^2$$
$$x^2 + x^2 - 14x + 49 = x^2 + 2x + 1$$
$$x^2 - 16x + 48 = 0$$
$$(x - 12)(x - 4) = 0$$
$$x - 12 = 0 \quad \text{or } x - 4 = 0$$
$$x = 12 \text{ or } \quad x = 4$$
$$x + 1 = 13 \qquad x = 5$$
$$x - 7 = 5 \qquad x = -3$$

The lengths of the sides are 5, 12, and 13 units.

Review Exercises

1. **a.** Binomial; degree 6

   b. Monomial; degree 8

   c. Trinomial; degree 5

2. **a.** $3x^4 - x^2 - 5x + 2$

   b. $4x^3 - x^2 + 3x$

   c. $6x^2 + x - 2$

3. **a.** $v(t) = 50,000(1 - 0.20t)$
$v(3) = 50,000(1 - 0.20(3))$
$= 50,000(1 - 0.60)$
$= 50,000(0.40) = \$20,000$

   b. $v(t) = 50,000(1 - 0.20t)$
$v(3) = 50,000(1 - 0.20(5))$
$= 50,000(1 - 1.00)$
$= 50,000(0) = \$0$

4. **a.** $P(x) = x^2 - 3x + 5$
$P(-1) = (-1)^2 - 3(-1) + 5 = 1 + 3 + 5 = 9$

   b. $P(x) = x^2 - 3x + 5$
$P(2) = (2)^2 - 3(2) + 5 = 4 - 6 + 5 = 3$

   c. $P(x) = x^2 - 3x + 5$
$P(-3) = (-3)^2 - 3(-3) + 5 = 9 + 9 + 5 = 23$

5. **a.** $(2x^3 + 5x^2 - 3x - 1) + (8 - 7x + 3x^2 - x^3) = 2x^3 + 5x^2 - 3x - 1 + 8 - 7x + 3x^2 - x^3$
$$= x^3 + 8x^2 - 10x + 7$$

   b. $(2x^3 + 5x^2 - 3x - 1) + (9 - 8x^2 + 3x^3) = 2x^3 + 5x^2 - 3x - 1 + 9 - 8x^2 + 3x^3 = 5x^3 - 3x^2 - 3x + 8$

   c. $(2x^3 + 5x^2 - 3x - 1) + (7 - 4x + x^3) = 2x^3 + 5x^2 - 3x - 1 + 7 - 4x + x^3 = 3x^3 + 5x^2 - 7x + 6$

6. **a.** $(4x^3 + 2x^2 + 2) - (7x^3 - 5x^2 + 3x - 1) = 4x^3 + 2x^2 + 2 - 7x^3 + 5x^2 - 3x + 1 = -3x^3 + 7x^2 - 3x + 3$

   b. $(7x^3 + 5x^2 + 4x - 7) - (7x^3 - 5x^2 + 3x - 1) = 7x^3 + 5x^2 + 4x - 7 - 7x^3 + 5x^2 - 3x + 1$
$$= 10x^2 + x - 6$$

   c. $(8x^2 - 9x + 3) - (7x^3 - 5x^2 + 3x - 1) = 8x^2 - 9x + 3 - 7x^3 + 5x^2 - 3x + 1 = -7x^3 + 13x^2 - 12x + 4$

7. **a.** $-2x^2y(x^2 + 3xy - 2y^3)$
$= -2x^4y - 6x^3y^2 + 4x^2y^4$

   b. $-3x^2y^2(x^2 + 3xy - 2y^3)$
$= -3x^4y^2 - 9x^3y^3 + 6x^2y^5$

   c. $-4xy^2(x^2 + 3xy - 2y^3)$
$= -4x^3y^2 - 12x^2y^3 + 8xy^5$

8. **a.** $(x - 1)(x^2 - 3x - 2)$
$= x(x^2 - 3x - 2) - 1(x^2 - 3x - 2)$
$= x^3 - 3x^2 - 2x - x^2 + 3x + 2$
$= x^3 - 4x^2 + x + 2$

b. $(x-2)(x^2-3x-2)$

$$= x(x^2-3x-2)-2(x^2-3x-2)$$
$$= x^3-3x^2-2x-2x^2+6x+4$$
$$= x^3-5x^2+4x+4$$

c. $(x+3)(x^2-3x-2)$

$$= x(x^2-3x-2)+3(x^2-3x-2)$$
$$= x^3-3x^2-2x+3x^2-9x-6$$
$$= x^3-11x-6$$

9. a. $(2x+3y)(4x-5y)$

$$= 8x^2-10xy+12xy-15y^2$$
$$= 8x^2+2xy-15y^2$$

b. $(2x+3y)(3x+2y)=6x^2+4xy+9xy+6y^2$

$$= 6x^2+13xy+6y^2$$

c. $(2x+3y)(5x-3y)$

$$= 10x^2-6xy+15xy-9y^2$$
$$= 10x^2+9xy-9y^2$$

10. a. $(2x+5y)^2=(2x)^2+2\cdot 2x\cdot 5y+(5y)^2$

$$= 4x^2+20xy+25y^2$$

b. $(3x+7y)^2=(3x)^2+2\cdot 3x\cdot 7y+(7y)^2$

$$= 9x^2+42xy+49y^2$$

c. $(4x+9y)^2=(4x)^2+2\cdot 4x\cdot 9y+(9y)^2$

$$= 16x^2+72xy+81y^2$$

11. a. $(3x-2y)^2=(3x)^2-2\cdot 3x\cdot 2y+(-2y)^2$

$$= 9x^2-12xy+4y^2$$

b. $(4x-7y)^2=(4x)^2-2\cdot 4x\cdot 7y+(-7y)^2$

$$= 16x^2-56xy+49y^2$$

c. $(5x-6y)^2=(5x)^2-2\cdot 5x\cdot 6y+(-6y)^2$

$$= 25x^2-60xy+36y^2$$

12. a. $(3x+2y)(3x-2y)=(3x)^2-(2y)^2$

$$= 9x^2-4y^2$$

b. $(4x+3y)(4x-3y)=(4x)^2-(3y)^2$

$$= 16x^2-9y^2$$

c. $(5x+3y)(5x-3y)=(5x)^2-(3y)^2$

$$= 25x^2-9y^2$$

13. a. $15x^5-20x^4+10x^3+25x^2$

$$= 5x^2(3x^3-4x^2+2x+5)$$

b. $9x^5-12x^4+6x^3+15x^2$

$$= 3x^2(3x^3-4x^2+2x+5)$$

c. $6x^5-8x^4+4x^3+10x^2$

$$= 2x^2(3x^3-4x^2+2x+5)$$

14. a. $6x^6-2x^4+15x^3-5x$

$$= x(6x^5-2x^3+15x^2-5)$$
$$= x[2x^3(3x^2-1)+5(3x^2-1)]$$
$$= x(3x^2-1)(2x^3+5)$$

b. $6x^6-8x^4+15x^3-20x$

$$= x(6x^5-8x^3+15x^2-20)$$
$$= x[2x^3(3x^2-4)+5(3x^2-4)]$$
$$= x(3x^2-4)(2x^3+5)$$

c. $6x^6-4x^4+9x^3-6x$

$$= x(6x^5-4x^3+9x^2-6)$$
$$= x[2x^3(3x^2-2)+3(3x^2-2)]$$
$$= x(3x^2-2)(2x^3+3)$$

15. **a.** $x^2 - 3xy - 18y^2 = (x-6y)(x+3y)$

b. $x^2 - 4xy - 12y^2 = (x-6y)(x+2y)$

c. $x^2 - 5xy - 6y^2 = (x-6y)(x+y)$

16. **a.** $2x^2 - 7xy - 30y^2$
$$= 2x^2 - 12xy + 5xy - 30y^2$$
$$= 2x(x-6y) + 5y(x-6y)$$
$$= (x-6y)(2x+5y)$$

b. $2x^2 - 3xy - 20y^2$
$$= 2x^2 - 8xy + 5xy - 20y^2$$
$$= 2x(x-4y) + 5y(x-4y)$$
$$= (x-4y)(2x+5y)$$

c. $2x^2 - 5xy - 25y^2$
$$= 2x^2 - 10xy + 5xy - 25y^2$$
$$= 2x(x-5y) + 5y(x-5y)$$
$$= (x-5y)(2x+5y)$$

17. **a.** $-18x^4y - 3x^3y^2 + 6x^2y^3$
$$= -3x^2y(6x^2 + xy - 2y^2)$$
$$= -3x^2y(6x^2 + 4xy - 3xy - 2y^2)$$
$$= -3x^2y[2x(3x+2y) - y(3x+2y)]$$
$$= -3x^2y(3x+2y)(2x - y)$$

b. $-30x^4y - 35x^3y^2 - 10x^2y^3$
$$= -5x^2y(6x^2 + 7xy + 2y^2)$$
$$= -5x^2y(6x^2 + 4xy + 3xy + 2y^2)$$
$$= -5x^2y[2x(3x+2y) + y(3x+2y)]$$
$$= -5x^2y(3x+2y)(2x + y)$$

c. $36x^4y - 30x^3y^2 - 36x^2y^3$
$$= 6x^2y(6x^2 - 5xy - 6y^2)$$
$$= 6x^2y(6x^2 + 4xy - 9xy - 6y^2)$$
$$= 6x^2y[2x(3x+2y) - 3y(3x+2y)]$$
$$= 6x^2y(3x+2y)(2x - 3y)$$

18. **a.** $4x^2 - 28xy + 49y^2$
$$= (2x)^2 - 2\cdot 2x\cdot 7y + (7y)^2$$
$$= (2x - 7y)^2$$

b. $9x^2 - 42xy + 49y^2$
$$= (3x)^2 - 2\cdot 3x\cdot 7y + (7y)^2$$
$$= (3x - 7y)^2$$

c. $16x^2 - 56xy + 49y^2$
$$= (4x)^2 - 2\cdot 4x\cdot 7y + (7y)^2$$
$$= (4x - 7y)^2$$

19. **a.** $9x^2 + 24xy + 16y^2$
$$= (3x)^2 + 2\cdot 3x\cdot 4y + (4y)^2$$
$$= (3x + 4y)^2$$

b. $9x^2 + 30xy + 25y^2$
$$= (3x)^2 + 2\cdot 3x\cdot 5y + (5y)^2$$
$$= (3x + 5y)^2$$

c. $9x^2 + 36xy + 36y^2 = 9(x^2 + 4xy + 4y^2)$
$$= 9[x^2 + 2\cdot x\cdot 2y + (2y)^2]$$
$$= 9(x + 2y)^2$$

20. **a.** $81x^4 - y^4 = (9x^2 + y^2)(9x^2 - y^2)$
$$= (9x^2 + y^2)(3x + y)(3x - y)$$

b. $x^4 - 16y^4 = \left(x^2 + 4y^2\right)\left(x^2 - 4y^2\right)$

$\qquad = \left(x^2 + 4y^2\right)(x + 2y)(x - 2y)$

c. $81x^4 - 16y^4 = \left(9x^2 + 4y^2\right)\left(9x^2 - 4y^2\right)$

$\qquad = \left(9x^2 + 4y^2\right)(3x + 2y)(3x - 2y)$

21. a. $x^2 - 4x + 4 - y^2 = (x - 2)^2 - y^2$

$\qquad = (x - 2 + y)(x - 2 - y)$

$\qquad = (x + y - 2)(x - y - 2)$

b. $x^2 - 6x + 9 - y^2 = (x - 3)^2 - y^2$

$\qquad = (x - 3 + y)(x - 3 - y)$

$\qquad = (x + y - 3)(x - y - 3)$

c. $x^2 + 8x + 16 - y^2 = (x + 4)^2 - y^2$

$\qquad = (x + 4 + y)(x + 4 - y)$

$\qquad = (x + y + 4)(x - y + 4)$

22. a. $27x^3 + 8y^3 = (3x)^3 + (2y)^3$

$\qquad = (3x + 2y)\left((3x)^2 \cdot\ 3x \cdot 2y + (2y)^2\right)$

$\qquad = (3x + 2y)\left(9x^2 \cdot\ 6xy + 4y^2\right)$

b. $27x^3 + 64y^3 = (3x)^3 + (4y)^3$

$\qquad = (3x + 4y)\left((3x)^2 \cdot\ 3x \cdot 4y + (4y)^2\right)$

$\qquad = (3x + 4y)\left(9x^2 \cdot\ 12xy + 16y^2\right)$

c. $64x^3 + 27y^3 = (4x)^3 + (3y)^3$

$\qquad = (4x + 3y)\left((4x)^2 \cdot\ 4x \cdot 3y + (3y)^2\right)$

$\qquad = (4x + 3y)\left(16x^2 \cdot\ 12xy + 9y^2\right)$

23. a. $27x^3 \cdot\ 8y^3 = (3x)^3 \cdot\ (2y)^3$

$\qquad = (3x \cdot\ 2y)\left((3x)^2 + 3x \cdot 2y + (2y)^2\right)$

$\qquad = (3x \cdot\ 2y)\left(9x^2 + 6xy + 4y^2\right)$

b. $27x^3 \cdot\ 64y^3 = (3x)^3 \cdot\ (4y)^3$

$\qquad = (3x \cdot\ 4y)\left((3x)^2 + 3x \cdot 4y + (4y)^2\right)$

$\qquad = (3x \cdot\ 4y)\left(9x^2 + 12xy + 16y^2\right)$

c. $64x^3 \cdot\ 27y^3 = (4x)^3 \cdot\ (3y)^3$

$\qquad = (4x \cdot\ 3y)\left((4x)^2 + 4x \cdot 3y + (3y)^2\right)$

$\qquad = (4x \cdot\ 3y)\left(16x^2 + 12xy + 9y^2\right)$

24. a. $27x^6 \cdot\ 8x^3y^3 = x^3\left(27x^3 \cdot\ 8y^3\right)$

$\qquad = x^3\ \vdots\ (3x)^3 \cdot\ (2y)^3\ \vdots$

$\qquad = x^3 (3x \cdot\ 2y)\left((3x)^2 + 3x \cdot 2y + (2y)^2\right)$

$\qquad = x^3 (3x \cdot\ 2y)\left(9x^2 + 6xy + 4y^2\right)$

b. $27x^7 \cdot\ 64x^4y^3 = x^4\left(27x^3 \cdot\ 64y^3\right)$

$\qquad = x^4\ \vdots\ (3x)^3 \cdot\ (4y)^3\ \vdots$

$\qquad = x^4 (3x \cdot\ 4y)\left((3x)^2 + 3x \cdot 4y + (4y)^2\right)$

$\qquad = x^4 (3x \cdot\ 4y)\left(9x^2 + 12xy + 16y^2\right)$

c. $64x^8 \cdot\ 27x^5y^3 = x^5\left(64x^3 \cdot\ 27y^3\right)$

$\qquad = x^5\ \vdots\ (4x)^3 \cdot\ (3y)^3\ \vdots$

$\qquad = x^5 (4x \cdot\ 3y)\left((4x)^2 + 4x \cdot 3y + (3y)^2\right)$

$\qquad = x^5 (4x \cdot\ 3y)\left(16x^2 + 12xy + 9y^2\right)$

25. a. $27x^6 + 3x^4 = 3x^4\left(9x^2 + 1\right)$

b. $4x^6 + 64x^4 = 4x^4\left(x^2 + 16\right)$

c. $2x^6 - 18x^4 = 2x^4(x^2 - 9)$
$= 2x^4(x+3)(x-3)$

26. a. $27x^4 + 36x^3y + 12x^2y^2$
$= 3x^2(9x^2 + 12xy + 4y^2)$
$= 3x^2\left[(3x)^2 + 2\rfloor3x\rfloor2y + (2y)^2\right]$
$= 3x^2(3x+2y)^2$

b. $36x^4 \rfloor 48x^3y + 16x^2y^2$
$= 4x^2(9x^2 \rfloor 12xy + 4y^2)$
$= 4x^2\left[(3x)^2 \rfloor 2\rfloor3x\rfloor2y + (2y)^2\right]$
$= 4x^2(3x \rfloor 2y)^2$

c. $45x^4 + 30x^3y + 5x^2y^2$
$= 5x^2(9x^2 + 6xy + y^2)$
$= 5x^2\left[(3x)^2 + 2\rfloor3x\rfloor y + y^2\right]$
$= 5x^2(3x+y)^2$

27. a. $27x^4 \rfloor 18x^3y + 3x^2y^2$
$= 3x^2(9x^2 \rfloor 6xy + y^2)$
$= 3x^2\left[(3x)^2 \rfloor 2\rfloor3x\rfloor y + y^2\right]$
$= 3x^2(3x \rfloor y)^2$

b. $36x^4 \rfloor 48x^3y + 16x^2y^2$
$= 4x^2(9x^2 \rfloor 12xy + 4y^2)$
$= 4x^2\left[(3x)^2 \rfloor 2\rfloor3x\rfloor2y + (2y)^2\right]$
$= 4x^2(3x \rfloor 2y)^2$

c. $45x^4 + 60x^3y + 20x^2y^2$
$= 5x^2(9x^2 + 12xy + 4y^2)$
$= 5x^2\left[(3x)^2 + 2\rfloor3x\rfloor2y + (2y)^2\right]$
$= 5x^2(3x+2y)^2$

28. a. $12x^3y \rfloor 44x^2y^2 \rfloor 16xy^3$
$= 4xy(3x^2 \rfloor 11xy \rfloor 4y^2)$
$= 4xy\lfloor3x^2 \rfloor 12xy + xy \rfloor 4y^2\rfloor$
$= 4xy\lfloor3x(x \rfloor 4y) + y(x \rfloor 4y)\rfloor$
$= 4xy(x \rfloor 4y)(3x+y)$

b. $15x^3y + 65x^2y^2 + 20xy^3$
$= 5xy(3x^2 + 13xy + 4y^2)$
$= 5xy\lfloor3x^2 + 12xy + xy + 4y^2\rfloor$
$= 5xy\lfloor3x(x+4y) + y(x+4y)\rfloor$
$= 5xy(x+4y)(3x+y)$

c. $18x^3y \rfloor 60x^2y^2 \rfloor 48xy^3$
$= 6xy(3x^2 \rfloor 10xy \rfloor 8y^2)$
$= 6xy\lfloor3x^2 \rfloor 12xy + 2xy \rfloor 8y^2\rfloor$
$= 6xy\lfloor3x(x \rfloor 4y) + 2y(x \rfloor 4y)\rfloor$
$= 6xy(x \rfloor 4y)(3x+2y)$

29. a. $2x^3 - x^2 - 2x + 1 = x^2(2x-1) - 1(2x-1)$
$= (2x-1)(x^2-1)$
$= (2x-1)(x+1)(x-1)$

b. $18x^3 - 9x^2 - 2x + 1$
$= 9x^2(2x-1) - 1(2x-1)$
$= (2x-1)(9x^2-1)$
$= (2x-1)(3x+1)(3x-1)$

c. $32x^3 - 16x^2 - 2x + 1$
$$= 16x^2(2x-1) - 1(2x-1)$$
$$= (2x-1)(16x^2-1)$$
$$= (2x-1)(4x+1)(4x-1)$$

30. a.
$$x^2 = -x + 12$$
$$x^2 + x - 12 = 0$$
$$(x+4)(x-3) = 0$$
$$x + 4 = 0 \quad \text{or } x - 3 = 0$$
$$x = -4 \text{ or} \quad x = 3$$

b.
$$x^2 = -x + 20$$
$$x^2 + x - 20 = 0$$
$$(x+5)(x-4) = 0$$
$$x + 5 = 0 \quad \text{or } x - 4 = 0$$
$$x = -5 \text{ or} \quad x = 4$$

c.
$$x^2 = -2x + 24$$
$$x^2 + 2x - 24 = 0$$
$$(x+6)(x-4) = 0$$
$$x + 6 = 0 \quad \text{or } x - 4 = 0$$
$$x = -6 \text{ or} \quad x = 4$$

31. a.
$$6x^2 + x = 1$$
$$6x^2 + x - 1 = 0$$
$$(2x+1)(3x-1) = 0$$
$$2x + 1 = 0 \quad \text{or } 3x - 1 = 0$$
$$x = -\frac{1}{2} \text{ or} \quad x = \frac{1}{3}$$

b.
$$8x^2 + 2x = 1$$
$$8x^2 + 2x - 1 = 0$$
$$(2x+1)(4x-1) = 0$$
$$2x + 1 = 0 \quad \text{or } 4x - 1 = 0$$
$$x = -\frac{1}{2} \text{ or} \quad x = \frac{1}{4}$$

c.
$$10x^2 + 3x = 1$$
$$10x^2 + 3x - 1 = 0$$
$$(2x+1)(5x-1) = 0$$
$$2x + 1 = 0 \quad \text{or } 5x - 1 = 0$$
$$x = -\frac{1}{2} \text{ or} \quad x = \frac{1}{5}$$

32. a.
$$x^3 + 2x^2 - x - 2 = 0$$
$$x^2(x+2) - 1(x+2) = 0$$
$$(x+2)(x^2-1) = 0$$
$$(x+2)(x+1)(x-1) = 0$$
$$x + 2 = 0 \quad \text{or } x + 1 = 0 \quad \text{or } x - 1 = 0$$
$$x = -2 \text{ or} \quad x = -1 \text{ or} \quad x = 1$$

b.
$$x^3 + 4x^2 - x - 4 = 0$$
$$x^2(x+4) - 1(x+4) = 0$$
$$(x+4)(x^2-1) = 0$$
$$(x+4)(x+1)(x-1) = 0$$
$$x + 4 = 0 \quad \text{or } x + 1 = 0 \quad \text{or } x - 1 = 0$$
$$x = -4 \text{ or} \quad x = -1 \text{ or} \quad x = 1$$

c.
$$x^3 + 2x^2 - 9x - 18 = 0$$
$$x^2(x+2) - 9(x+2) = 0$$
$$(x+2)(x^2-9) = 0$$
$$(x+2)(x+3)(x-3) = 0$$
$$x + 2 = 0 \quad \text{or } x + 3 = 0 \quad \text{or } x - 3 = 0$$
$$x = -2 \text{ or} \quad x = -3 \text{ or} \quad x = 3$$

33. **a.** Using the Pythagorean Theorem:

$$x^2 + (x+7)^2 = (x+8)^2$$

$$x^2 + x^2 + 14x + 49 = x^2 + 16x + 64$$

$$x^2 - 2x - 15 = 0$$

$$(x-5)(x+3) = 0$$

$$x - 5 = 0 \text{ or } x + 3 = 0$$

$$x = 5 \text{ or } \quad x = -3$$

$$x + 7 = 12$$

$$x + 8 = 13$$

The lengths of the sides are 5, 12, and 13 units.

b. Using the Pythagorean Theorem:

$$x^2 + (x+3)^2 = (x+6)^2$$

$$x^2 + x^2 + 6x + 9 = x^2 + 12x + 36$$

$$x^2 - 6x - 27 = 0$$

$$(x-9)(x+3) = 0$$

$$x - 9 = 0 \quad \text{or } x + 3 = 0$$

$$x = 9 \text{ or } \quad x = -3$$

$$x + 3 = 12$$

$$x + 6 = 15$$

The lengths of the sides are 9, 12, and 15 units.

c. Using the Pythagorean Theorem:

$$x^2 + (x+4)^2 = (x+8)^2$$

$$x^2 + x^2 + 8x + 16 = x^2 + 16x + 64$$

$$x^2 - 8x - 48 = 0$$

$$(x-12)(x+4) = 0$$

$$x - 12 = 0 \text{ or } x + 4 = 0$$

$$x = 12 \text{ or } \quad x = -4$$

$$x + 4 = 16$$

$$x + 8 = 20$$

The lengths of the sides are 12, 16, and 20 units.

Cumulative Review Chapters 1–5

1. $\{2, 4, 6\}$

2. $0.19 = \dfrac{19}{100}$

3. $\cdot \dfrac{5}{9} \div \cdot \dfrac{1}{27} \cdot \ = \cdot \dfrac{5}{9} \div \cdot \dfrac{27}{1} \cdot \ = \dfrac{5}{\cancel{9}} \cdot \dfrac{\cancel{9} \cdot 3}{1} = 15$

4. $(7.5 - 10^3) - (8 - 10^{-7}) = 60 - 10^{-4}$

$$= 6 - 10^1 - 10^{-4}$$

$$= 6 - 10^{-3}$$

5. $-4^3 + \dfrac{(4-12)}{2} + 6 \div 3 = -4^3 + \dfrac{-8}{2} + 6 \div 3$

$$= -64 + \dfrac{-8}{2} + 6 \div 3$$

$$= -64 - 4 + 2 = -66$$

6. $\dfrac{3}{7} y \cdot 4 = 2$

$$\dfrac{3}{7} y = 6$$

$$\dfrac{7}{3} \cdot \dfrac{3}{7} y = \dfrac{7}{3} \cdot 6$$

$$y = 14$$

7.
$$\left|\frac{4}{3}x+6\right|+8=15$$
$$\left|\frac{4}{3}x+6\right|=7$$
$$\frac{4}{3}x+6=7 \quad \text{or} \quad \frac{4}{3}x+6=\cdot 7$$
$$\frac{4}{3}x=1 \quad \text{or} \quad \frac{4}{3}x=\cdot 13$$
$$x=1\cdot\frac{3}{4} \quad \text{or} \quad x=\cdot 13\cdot\frac{3}{4}$$
$$x=\frac{3}{4} \quad \text{or} \quad x=\cdot\frac{39}{4}$$

8.
$$\frac{x}{8}\mid\frac{x}{3}<\frac{x\mid 8}{8}$$
$$24\left|\frac{x}{8}\mid\frac{x}{3}\right|<24\left|\frac{x\mid 8}{8}\right|$$
$$3x\mid 8x<3x\mid 24$$
$$\mid 8x<\mid 24$$
$$x>3$$

9. $\left\{x\,\middle|\,x<\geq 2 \text{ or } x\geq 3\right\}$

10.
$$\mid 2x\leq 3\mid\leq 1$$
$$\leq 1\leq 2x\leq 3\leq 1$$
$$2\leq 2x\leq 4$$
$$1\leq x\leq 2$$

11.
$$B=\frac{2}{5}(A-11)$$
$$5B=2(A-11)$$
$$5B=2A-22$$
$$5B+22=2A$$
$$\frac{5B+22}{2}=A$$

12.

	Rate	Time	Distance
Freight	45	$T+2$	$45(T+2)$
Passenger	55	T	$55T$

Distance of freight = Distance of passenger
$$45(T+2)=55T$$
$$45T+90=55T$$
$$90=10T$$
$$9\,\text{hrs}=T$$
Since $D=RT=55(9)=495\,\text{mi}$

The passenger train overtakes the freight train 495 miles from the station.

13. $y=\mid 3x\mid 5$

If $x=0$, $y=\mid 5$

If $y=0$, $0=\mid 3x\mid 5$, $3x=\mid 5$ and $x=\mid\frac{5}{3}$

y-intercept $(0,-5)$ and x-intercept $\left|\mid\frac{5}{3},0\right|$

14. $4x=-16$

$x=-4$

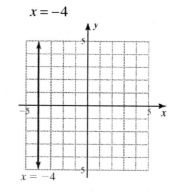

15.
$$m = \frac{-4-(-7)}{3-(-3)} = \frac{3}{6} = \frac{1}{2}$$

Parallel, since the slope is $\frac{1}{2}$.

16.
$$m = \frac{4-(-1)}{7-3} = \frac{5}{4}$$
$$y-4 = \frac{5}{4}(x-7)$$
$$4(y-4) = 5(x-7)$$
$$4y-16 = 5x-35$$
$$4y-5x = -19$$
$$5x-4y = 19$$

17. $y = -3x-1$

18. $x \geq 3$

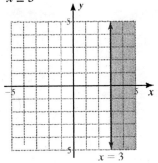

19.   To eliminate x, multiply the second equation by -3, add the result to the first equation, and solve:
$$3x+4y = -13 \rightarrow \quad 3x+4y = -13$$
$$x+y = -4 \rightarrow \underline{-3x-3y = 12}$$
$$y = -1$$
$$x-1 = -4 \rightarrow x = -3$$

Solution: $(-3, -1)$.

20.
$$3x-2y+z = 5$$
$$2x-y-3z = 0$$
$$3x-y-z = 4$$

Multiply the first equation by 3, add to the second equation to eliminate z:
$$9x-6y+3z = 15$$
$$\underline{2x-y-3z = 0}$$
$$11x-7y = 15$$

Add the first equation to the third equation to eliminate z:
$$6x-3y = 9$$

Multiply the first result by 3 and the second result by -7 to eliminate y and solve:
$$33x-21y = 45$$
$$\underline{-42x+21y = -63}$$
$$-9x = -18 \rightarrow x = 2$$

Substitute and solve for the other variables:
$$6(2)-3y = 9 \qquad 3(2)-2(1)+z = 5$$
$$-3y = -3 \qquad 6-2+z = 5$$
$$y = 1 \qquad z = 1$$

The solution is $(2, 1, 1)$.

21.
$$\begin{vmatrix} -1 & 2 & 4 \\ 3 & 0 & 0 \\ 3 & 3 & -4 \end{vmatrix} = -1\begin{vmatrix} 0 & 0 \\ 3 & -4 \end{vmatrix} - 2\begin{vmatrix} 3 & 0 \\ 3 & -4 \end{vmatrix} + 4\begin{vmatrix} 3 & 0 \\ 3 & 3 \end{vmatrix} = -1(0-0) - 2(-12-0) + 4(9-0)$$
$$= 0 + 24 + 36 = 60$$

22.
$$D = \begin{vmatrix} 2 & 1 & -3 \\ 1 & -2 & -2 \\ 3 & 1 & 3 \end{vmatrix} = 2\begin{vmatrix} -2 & -2 \\ 1 & 3 \end{vmatrix} - 1\begin{vmatrix} 1 & -2 \\ 3 & 3 \end{vmatrix} + (-3)\begin{vmatrix} 1 & -2 \\ 3 & 1 \end{vmatrix} = 2(-6+2) - 1(3+6) - 3(1+6)$$
$$= -8 - 9 - 21 = -38$$

$$D_z = \begin{vmatrix} 2 & 1 & -1 \\ 1 & -2 & 7 \\ 3 & 1 & 0 \end{vmatrix} = 2\begin{vmatrix} -2 & 7 \\ 1 & 0 \end{vmatrix} - 1\begin{vmatrix} 1 & 7 \\ 3 & 0 \end{vmatrix} + (-1)\begin{vmatrix} 1 & -2 \\ 3 & 1 \end{vmatrix} = 2(0-7) - 1(0-21) - 1(1+6)$$
$$= -14 + 21 - 7 = 0$$

$$z = \frac{D_z}{D} = \frac{0}{-38} = 0$$

23. Let n = the number of nickels
Let d = the number of dimes
$$5n + 10d = 375$$
$$n = d + 12$$
Substitute the second equation into the first equation and solve:
$$5(d+12) + 10d = 375$$
$$5d + 60 + 10d = 375$$
$$15d = 315$$
$$d = 21$$
$$n = 21 + 12 = 33$$
Bert has 33 nickels and 21 dimes.

24. Binomial

25. $R(x) = x^2 - 5x - 5$
$$R(-4) = (-4)^2 - 5(-4) - 5 = 16 + 20 - 5 = 31$$

26. $\left(7x^3 + 3x^2 + 7\right) - \left(8x^3 - 4x^2 + 7x - 7\right) = 7x^3 + 3x^2 + 7 - 8x^3 + 4x^2 - 7x + 7 = -x^3 + 7x^2 - 7x + 14$

27. $(2x \cdot 7y)^2 = (2x)^2 \cdot 2 \cdot 2x \cdot 7y + (7y)^2$
$$= 4x^2 \cdot 28xy + 49y^2$$

28. $625x^4 - y^4 = \left(25x^2 + y^2\right)\left(25x^2 - y^2\right)$
$$= \left(25x^2 + y^2\right)(5x + y)(5x - y)$$

29. $8c^3 + 125 = (2c)^3 + 5^3$
$$= (2c+5)\left((2c)^2 \cdot 2c \cdot 5 + 5^2\right)$$
$$= (2c+5)\left(4c^2 \cdot 10c + 25\right)$$

30.
$$2x^2 + 9x = -9$$
$$2x^2 + 9x + 9 = 0$$
$$(2x+3)(x+3) = 0$$
$$2x + 3 = 0 \quad \text{or} \quad x + 3 = 0$$
$$x = -\frac{3}{2} \quad \text{or} \quad x = -3$$

Chapter 6   Rational Expressions

6.1   Rational Expressions

Problems 6.1

1. **a.** Set the denominator equal to zero and solve:
$$2x = 0$$
$$x = 0$$
The expression is undefined when $x = 0$.

b. Set the denominator equal to zero and solve:
$$2x + 3 = 0$$
$$2x = -3$$
$$x = -\frac{3}{2}$$
The expression is undefined when $x = -\frac{3}{2}$.

c. Set the denominator equal to zero and solve:
$$x^2 - 6x - 7 = 0$$
$$(x - 7)(x + 1) = 0$$
$$x - 7 = 0 \text{ or } x + 1 = 0$$
$$x = 7 \text{ or } \quad x = -1$$
The expression is undefined when $x = 7$ or $x = -1$.

2. **a.** Multiply numerator and denominator by $4x^3$:
$$\frac{3}{7x} = \frac{3}{7x} \cdot \frac{4x^3}{4x^3} = \frac{12x^3}{28x^4}$$

b. Multiply numerator and denominator by 8:
$$\frac{4 \cdot x}{5y^2} = \frac{4 \cdot x}{5y^2} \cdot \frac{8}{8} = \frac{32 \cdot 8x}{40y^2}$$

c. Multiply numerator and denominator by $x - 3$:
$$\frac{x+1}{2x+1} = \frac{x+1}{2x+1} \cdot \frac{x-3}{x-3} = \frac{(x+1)(x-3)}{(2x+1)(x-3)}$$
$$= \frac{x^2 - 2x - 3}{2x^2 - 5x - 3}$$

3. **a.** $-\dfrac{7}{y} = \dfrac{-7}{y}$

b. $-\dfrac{x}{4} = \dfrac{x}{4}$

c. $-\dfrac{3a-b}{8} = \dfrac{-(3a-b)}{8} = \dfrac{-3a+b}{8} = \dfrac{b-3a}{8}$

4. **a.** $\dfrac{3x^5y^7}{6x^2y^3} = \dfrac{3x^2y^3 \cdot x^3y^4}{3x^2y^3 \cdot 2} = \dfrac{x^3y^4}{2}$

b. $\dfrac{2x^2 + 4xy}{x^2 - 4y^2} = \dfrac{2x(x+2y)}{(x+2y)(x-2y)} = \dfrac{2x}{x-2y}$

c. $\dfrac{3y + xy}{y^2} = \dfrac{y(3+x)}{y \cdot y} = \dfrac{3+x}{y}$

5.   a. $\dfrac{x^3-8}{2-x}=\dfrac{\cancel{(x-2)}\left(x^2+2x+4\right)}{-1\cancel{(x-2)}}$

$=-\left(x^2+2x+4\right)=-x^2-2x-4$

b. $\dfrac{1-9x^2}{27x^3-1}=\dfrac{(1+3x)(1-3x)}{(3x-1)\left(9x^2+3x+1\right)}$

$=\dfrac{(1+3x)(-1)\cancel{(3x-1)}}{\cancel{(3x-1)}\left(9x^2+3x+1\right)}$

$=\dfrac{-(1+3x)}{9x^2+3x+1}$

Exercises 6.1

1.   Set the denominator equal to zero and solve:

$x+3=0$

$x=-3$

The expression is undefined when $x=-3$.

3.   Set the denominator equal to zero and solve:

$5m+10=0$

$5m=-10$

$m=-2$

The expression is undefined when $m=-2$.

5.   Set the denominator equal to zero and solve:

$m^2-m-2=0$

$(m-2)(m+1)=0$

$m-2=0$ or $m+1=0$

$m=2$ or $m=-1$

The expression is undefined when
$m=2$ or $m=-1$.

7.   Set the denominator equal to zero and solve:

$p^2-9=0$

$(p+3)(p-3)=0$

$p+3=0$ or $p-3=0$

$p=-3$ or $p=3$

The expression is undefined when
$p=-3$ or $p=3$.

9.   Set the denominator equal to zero and solve:

$2a^2-11a-6=0$

$(2a+1)(a-6)=0$

$2a+1=0$ or $a-6=0$

$2a=-1$ or $a=6$

$a=-\dfrac{1}{2}$ or $a=6$

The expression is undefined when
$a=-\frac{1}{2}$ or $a=6$.

11.   Set the denominator equal to zero and solve:

$4v^2+9=0$

$4v^2=-9$

This is expression is false, thus the expression
is defined for all values of v.

13.   Multiply numerator and denominator by
$2y^2$:

$\dfrac{2x}{3y}=\dfrac{2x}{3y}\cdot\dfrac{2y^2}{2y^2}=\dfrac{4xy^2}{6y^3}$

15.   Multiply numerator and denominator by $x-y$:

$\dfrac{x}{x+y}=\dfrac{x}{x+y}\cdot\dfrac{x-y}{x-y}=\dfrac{x(x-y)}{(x+y)(x-y)}=\dfrac{x^2-xy}{x^2-y^2}$

17. Multiply numerator and denominator by $y+x$:

$$\frac{\cdot\,x}{y\cdot\,x}=\frac{\cdot\,x}{y\cdot\,x}\cdot\frac{y+x}{y+x}=\frac{\cdot\,x(y+x)}{(y\cdot\,x)(y+x)}$$

$$=\frac{\cdot\,xy\cdot\,x^2}{y^2\cdot\,x^2}$$

19. Multiply numerator and denominator by $2x+3y$:

$$\frac{\cdot\,x}{2x\cdot\,3y}=\frac{\cdot\,x}{2x\cdot\,3y}\cdot\frac{2x+3y}{2x+3y}$$

$$=\frac{\cdot\,x(2x+3y)}{(2x\cdot\,3y)(2x+3y)}=\frac{\cdot\,2x^2\cdot\,3xy}{4x^2\cdot\,9y^2}$$

21. Multiply numerator and denominator by $x-2$:

$$\frac{4x}{x+1}=\frac{4x}{x+1}\cdot\frac{x-2}{x-2}$$

$$=\frac{4x(x-2)}{(x+1)(x-2)}=\frac{4x^2-8x}{x^2-x-2}$$

23. Multiply numerator and denominator by $x-2$:

$$\frac{\cdot\,5x}{x+3}=\frac{\cdot\,5x}{x+3}\cdot\frac{x\cdot\,2}{x\cdot\,2}$$

$$=\frac{\cdot\,5x(x\cdot\,2)}{(x+3)(x\cdot\,2)}=\frac{\cdot\,5x^2+10x}{x^2+x\cdot\,6}$$

25. Multiply numerator and denominator by x^2-xy+y^2:

$$\frac{3}{x+y}=\frac{3}{x+y}\cdot\frac{x^2-xy+y^2}{x^2-xy+y^2}$$

$$=\frac{3\left(x^2-xy+y^2\right)}{(x+y)\left(x^2-xy+y^2\right)}=\frac{3x^2-3xy+3y^2}{x^3+y^3}$$

27. Multiply numerator and denominator by x^2+xy+y^2:

$$\frac{x}{x\cdot\,y}=\frac{x}{x\cdot\,y}\cdot\frac{x^2+xy+y^2}{x^2+xy+y^2}$$

$$=\frac{x\left(x^2+xy+y^2\right)}{(x\cdot\,y)\left(x^2+xy+y^2\right)}=\frac{x^3+x^2y+xy^2}{x^3\cdot\,y^3}$$

29. Multiply numerator and denominator by $x+y$:

$$\frac{x}{x^2\cdot\,xy+y^2}=\frac{x}{x^2\cdot\,xy+y^2}\cdot\frac{x+y}{x+y}$$

$$=\frac{x(x+y)}{\left(x^2\cdot\,xy+y^2\right)(x+y)}=\frac{x^2+xy}{x^3+y^3}$$

31. $-\dfrac{y}{-2}=\dfrac{y}{2}$

33. $-\dfrac{x}{x-5}=\dfrac{x}{-(x-5)}=\dfrac{x}{-x+5}=\dfrac{x}{5-x}$

35. $-\dfrac{-2x}{-5y}=\dfrac{-2x}{-(-5y)}=\dfrac{-2x}{5y}$

37. $\dfrac{-(x+y)}{-(x-y)}=\dfrac{x+y}{x-y}$

39. $\dfrac{-1}{-(x-2)}=\dfrac{1}{x-2}$

41. $\dfrac{x^4y^2}{xy^5}=\dfrac{\cancel{xy^2}\cdot\,x^3}{\cancel{xy^2}\cdot\,y^3}=\dfrac{x^3}{y^3}$

43. $\dfrac{3x\cdot\,3y}{x\cdot\,y}=\dfrac{3\cdot\cancel{(x\cdot\,y)}}{\cancel{x\cdot\,y}}=3$

45. $\dfrac{3x-2y}{9x^2-4y^2}=\dfrac{\cancel{3x-2y}}{(3x+2y)\cancel{(3x-2y)}}=\dfrac{1}{3x+2y}$

47. $\dfrac{(x-y)^3}{x^2-y^2}=\dfrac{\cancel{(x-y)}(x-y)^2}{(x+y)\cancel{(x-y)}}=\dfrac{(x-y)^2}{x+y}$

49. $\dfrac{ay^2\cdot\,ay}{ay}=\dfrac{\cancel{ay}\cdot(y\cdot\,1)}{\cancel{ay}}=y\cdot\,1$

51. $\dfrac{x^2+2xy+y^2}{x^2-y^2}=\dfrac{\cancel{(x+y)}(x+y)}{\cancel{(x+y)}(x-y)}=\dfrac{x+y}{x-y}$

53. $\dfrac{y^2-8y+15}{y^2+3y-18}=\dfrac{(y-5)\,(y-3)}{(y+6)\,(y-3)}=\dfrac{y-5}{y+6}$

55. $\dfrac{2\cdot y}{y\cdot 2}=\dfrac{\cdot 1\cdot (y-2)}{y-2}=\cdot 1$

57. $\dfrac{9-x^2}{x-3}=\dfrac{-\left(x^2-9\right)}{x-3}$

$\qquad =\dfrac{-(x+3)\,(x-3)}{x-3}=-(x+3)$

59. $\dfrac{y^3\cdot 8}{2\cdot y}=\dfrac{(y-2)\left(y^2+2y+4\right)}{\cdot 1\cdot (y-2)}=\cdot\left(y^2+2y+4\right)$

61. $\dfrac{3x\cdot 2y}{2y\cdot 3x}=\dfrac{\cdot 1\cdot (2y-3x)}{2y-3x}=\cdot 1$

63. $\dfrac{x^2+4x\cdot 5}{1\cdot x}=\dfrac{(x+5)\,(x-1)}{\cdot 1\cdot (x-1)}=\cdot (x+5)$

65. $\dfrac{x^2\cdot 6x+8}{4\cdot x}=\dfrac{(x\cdot 2)\,(x-4)}{\cdot 1\cdot (x-4)}=\cdot (x\cdot 2)=2\cdot x$

67. $\dfrac{2\cdot x}{x^2+4x\cdot 12}=\dfrac{\cdot 1\cdot (x-2)}{(x+6)\,(x-2)}=\dfrac{\cdot 1}{x+6}$

69. $-\dfrac{3-x}{x^2-5x+6}=\dfrac{x-3}{(x-2)\,(x-3)}=\dfrac{1}{x-2}$

71. $\dfrac{4x^3}{2x}=\dfrac{2x\cdot 2x^2}{2x}=2x^2$

73. $\dfrac{\cdot 16x^4}{8x^2}=\dfrac{\cdot 2x^2\cdot 8x^2}{8x^2}=\cdot 2x^2$

75. $x^2+x-12=(x+4)(x-3)$

77. $9x^2-4y^2=(3x)^2-(2y)^2=(3x+2y)(3x-2y)$

79. $x^3-1=(x-1)\left(x^2+x+1\right)$

81. $8x^3+1=(2x)^3+1^3=(2x+1)\left(4x^2-2x+1\right)$

83. $\dfrac{3}{7}\cdot\dfrac{14}{9}=\dfrac{3}{7}\cdot\dfrac{7\cdot 2}{3\cdot 3}=\dfrac{2}{3}$

85. $\cdot\dfrac{4}{9}\cdot\cdot\dfrac{27}{8}\cdot=\cdot\dfrac{4}{9}\cdot\cdot\dfrac{9\cdot 3\cdot}{4\cdot 2}\cdot=\dfrac{3}{2}$

87. $\cdot\dfrac{4}{5}\div\cdot\dfrac{8}{15}\cdot=\cdot\dfrac{4}{5}\cdot\dfrac{15}{8}\cdot=\cdot\dfrac{4}{8}\cdot\dfrac{8\cdot 3\cdot}{4\cdot 2}\cdot=\cdot\dfrac{3}{2}$

89. a. $P=\dfrac{2100p}{100-p}=\dfrac{2100(20)}{100-20}=\dfrac{42000}{80}=525$

The price is \$525 million.

b. $P=\dfrac{2100p}{100-p}=\dfrac{2100(40)}{100-40}=\dfrac{84000}{60}=1400$

The price is \$1400 million or \$1.4 billion.

c. $P=\dfrac{2100p}{100-p}=\dfrac{2100(60)}{100-60}=\dfrac{126000}{40}=3150$

The price is \$3150 million or \$3.15 billion.

d. No, $p=100$ causes division by zero. As p increases to 100, price increases without bound.

91. a. $\dfrac{5x^2-5}{x+1}=\dfrac{5\left(x^2-1\right)}{x+1}$

$\qquad =\dfrac{5\,(x+1)\,(x-1)}{x+1}=5(x-1)$

b. $P=5(x\cdot 1)=5(10\cdot 1)=5\cdot 9=\$45$

The profit on 10 sunglasses is \$45.

c. Maximum profit occurs when $x=100$.

$P=5(x\cdot 1)=5(100\cdot 1)=5\cdot 99=\$495$

The maximum profit is \$495.

93. Sample answer: y is a term in the denominator, not a factor, so it cannot be reduced with the y in the numerator.

95. Sample answer: The statement is true as long as the denominator is not zero. If the value of x is -2, -3, or 2, the denominator has a value of zero, and the fraction cannot be simplified.

97. Sample answer: Factoring a -1 out of the denominator produces factors that are identical and thus the expression simplifies to -1. Thus, any fraction where the numerator and denominator differ only in sign can be simplified to -1.

99.
$$\frac{x^5 y^7}{xy^3} = \frac{\cancel{xy^3} \cdot x^4 y^4}{\cancel{xy^3}} = x^4 y^4$$

101.
$$\frac{x^2 - y^2}{y^3 - x^3} = \frac{-\left(y^2 - x^2\right)}{y^3 - x^3}$$
$$= \frac{-\cancel{\left(y-x\right)}(y+x)}{\cancel{\left(y-x\right)}\left(y^2 + xy + x^2\right)} = \frac{-(y+x)}{y^2 + xy + x^2}$$

103.
$$\frac{4y^2 - xy^2}{y^2} = \frac{\cancel{y^2}(4-x)}{\cancel{y^2}} = 4 - x$$

105. Set the denominator equal to zero and solve:
$$x^2 - 1 = 0$$
$$(x+1)(x-1) = 0$$
$$x+1 = 0 \quad \text{or } x - 1 = 0$$
$$x = -1 \text{ or} \quad x = 1$$
The expression is undefined when $x = -1$ or $x = 1$.

107.
$$-\frac{-4}{x} = \frac{4}{x}$$

109. Multiply numerator and denominator by 2:
$$\frac{7}{8} = \frac{7}{8} \cdot \frac{2}{2} = \frac{14}{16}$$

111. Multiply numerator and denominator by $x - 3$:
$$\frac{4x+1}{x+2} = \frac{4x+1}{x+2} \cdot \frac{x-3}{x-3} = \frac{4x^2 - 11x - 3}{x^2 - x - 6}$$

6.2 Multiplication and Division of Rational Expressions

Problems 6.2

1. a.
$$\frac{x \cdot 2}{3x \cdot 2y} \cdot \frac{9x^2 \cdot 4y^2}{2x^2 \cdot 3x \cdot 2} = \frac{\cancel{x \cdot 2}}{3x \cdot 2y} \cdot \frac{(3x+2y)\cancel{(3x-2y)}}{(2x+1)\cancel{(x \cdot 2)}} = \frac{3x+2y}{2x+1}$$

b.
$$\frac{x^2 \cdot 9}{16x^2 \cdot 25} \cdot \frac{4x^2 + 5x}{3x+9} = \frac{\cancel{(x+3)}(x \cdot 3)}{\cancel{(4x+5)}(4x \cdot 5)} \cdot \frac{x\cancel{(4x+5)}}{3\cancel{(x+3)}} = \frac{x(x \cdot 3)}{3(4x \cdot 5)} = \frac{x^2 \cdot 3x}{12x \cdot 15}$$

2. a.
$$\frac{3 \cdot x}{2x+4} \cdot \frac{x^2+5x+6}{x^2 \cdot 9} = \frac{\cdot 1 \cdot \cancel{(x-3)}}{2\cancel{(x+2)}} \cdot \frac{\cancel{(x+2)}\cancel{(x+3)}}{\cancel{(x+3)}\cancel{(x-3)}} = \cdot \frac{1}{2}$$

b.
$$\frac{x^2+4x+3}{x^2+5x+6} \cdot \frac{x^2-2x-8}{x^2-2x-3} = \frac{\cancel{(x+3)}\cancel{(x+1)}}{\cancel{(x+2)}\cancel{(x+3)}} \cdot \frac{(x-4)\cancel{(x+2)}}{(x-3)\cancel{(x+1)}} = \frac{x-4}{x-3}$$

3. a.
$$\frac{x^5 y}{z} \div \frac{x^3 y}{z^6} = \frac{x^5 y}{z} \cdot \frac{z^6}{x^3 y} = \frac{\cancel{x^3}y \cdot x^2}{\cancel{z}} \cdot \frac{\cancel{z} \cdot z^5}{\cancel{x^3 y}} = x^2 z^5$$

b.
$$\frac{3x^2 \cdot 3}{5x+10} \div \frac{x+1}{5} = \frac{3x^2 \cdot 3}{5x+10} \cdot \frac{5}{x+1} = \frac{3\cancel{(x+1)}(x \cdot 1)}{\cancel{5}(x+2)} \cdot \frac{\cancel{5}}{\cancel{x+1}} = \frac{3(x \cdot 1)}{x+2} = \frac{3x \cdot 3}{x+2}$$

c.
$$\frac{x+2}{x \cdot 2} \div (x^2+4x+4) = \frac{x+2}{x \cdot 2} \cdot \frac{1}{x^2+4x+4} = \frac{\cancel{x+2}}{x \cdot 2} \cdot \frac{1}{\cancel{(x+2)}(x+2)} = \frac{1}{(x \cdot 2)(x+2)} = \frac{1}{x^2 \cdot 4}$$

4.
$$\frac{2 \cdot x}{x+3} \div \frac{x^3 \cdot 8}{x \cdot 5} \cdot \frac{x^3+27}{x \cdot 5} = \frac{2 \cdot x}{x+3} \cdot \frac{x \cdot 5}{x^3 \cdot 8} \cdot \frac{x^3+27}{x \cdot 5}$$

$$= \frac{\cdot \cancel{(x-2)}}{x+3} \cdot \frac{\cancel{x-5}}{\cancel{(x-2)}(x^2+2x+4)} \cdot \frac{\cancel{(x+3)}(x^2 \cdot 3x+9)}{\cancel{x-5}}$$

$$= \frac{\cdot (x^2 \cdot 3x+9)}{x^2+2x+4} = \frac{x^2+3x \cdot 9}{x^2+2x+4}$$

5.
$$\frac{x^3+x^2 \cdot x \cdot 1}{x+4} \div \frac{x^3+1}{x^2 \cdot 16} \cdot \frac{1}{x^2 \cdot 1} = \frac{x^2(x+1) \cdot 1(x+1)}{x+4} \cdot \frac{x^2 \cdot 16}{x^3+1} \cdot \frac{1}{x^2 \cdot 1}$$

$$= \frac{\cancel{(x+1)}\cancel{(x^2-1)}}{\cancel{x+4}} \cdot \frac{\cancel{(x+4)}(x \cdot 4)}{\cancel{(x+1)}(x^2 \cdot x+1)} \cdot \frac{1}{\cancel{(x^2-1)}} = \frac{x \cdot 4}{x^2 \cdot x+1}$$

Exercises 6.2

1.
$$\frac{3}{4} \cdot \frac{2}{5} = \frac{3}{2 \cdot \cancel{2}} \cdot \frac{\cancel{2}}{5} = \frac{3}{10}$$

3.
$$\frac{14x^2}{15} \cdot \frac{5}{7x} = \frac{\cancel{7}x \cdot 2x}{3 \cdot \cancel{5}} \cdot \frac{\cancel{5}}{\cancel{7}x} = \frac{2x}{3}$$

5.
$$\frac{\cdot 2xy^4}{9z^5} \cdot \frac{\cdot 3z}{7x^3 y^3} = \frac{\cdot 2 \cdot \cancel{x} \cdot y \cdot \cancel{y^3}}{3 \cdot \cancel{3z} \cdot z^4} \cdot \frac{\cdot 1 \cdot \cancel{3z}}{7 \cdot \cancel{x} \cdot x^2 \cdot \cancel{y}} = \frac{2y}{21x^2 z^4}$$

7.
$$\frac{10x+50}{6x+6} \cdot \frac{12}{5x+25} = \frac{2 \cdot \cancel{5} \cdot \cancel{(x+5)}}{\cancel{6}(x+1)} \cdot \frac{\cancel{6} \cdot 2}{\cancel{5}\cancel{(x+5)}} = \frac{4}{x+1}$$

9.
$$\frac{6y+3}{2y^2 \cdot 3y \cdot 2} \cdot \frac{y^2 \cdot 4}{3y+6} = \frac{\cancel{3}\,(2y+1)}{(2y+1)\,(y-2)} \cdot \frac{(y+2)\,(y-2)}{\cancel{3}\,(y+2)} = 1$$

11.
$$\frac{y \cdot x}{x^2+2xy} \cdot \frac{5x+10y}{x^2 \cdot y^2} = \frac{\cdot 1\,(x-y)}{x\,(x+2y)} \cdot \frac{5\,(x+2y)}{(x+y)\,(x-y)} = \frac{\cdot 5}{x(x+y)}$$

13.
$$\frac{3y^2 \cdot 17y+10}{y^2 \cdot 4y \cdot 5} \cdot \frac{y^2+3y+2}{y^2+y \cdot 2} = \frac{(3y \cdot 2)\,(y-5)}{(y-5)\,(y+1)} \cdot \frac{(y+2)\,(y+1)}{(y+2)(y \cdot 1)} = \frac{3y \cdot 2}{y \cdot 1}$$

15.
$$\frac{y^2+2y \cdot 8}{y^2+7y+12} \cdot \frac{y^2+2y \cdot 3}{y^2 \cdot 3y+2} = \frac{(y+4)\,(y-2)}{(y+3)\,(y+4)} \cdot \frac{(y-1)\,(y+3)}{(y-1)\,(y-2)} = 1$$

17.
$$\frac{x^3 \cdot 8}{4 \cdot x^2} \cdot \frac{x^2+x \cdot 2}{x^2+2x+4} = \frac{(x-2)\,(x^2+2x+4)}{\cdot \,(x+2)\,(x-2)} \cdot \frac{(x+2)\,(x \cdot 1)}{x^2+2x+4} = \cdot\,(x \cdot 1) = 1 \cdot x$$

19.
$$\frac{a^3+b^3}{a^3 \cdot b^3} \cdot \frac{a^2+ab+b^2}{a^2 \cdot ab+b^2} = \frac{(a+b)\,(a^2 \cdot ab+b^2)}{(a \cdot b)\,(a^2+ab+b^2)} \cdot \frac{a^2+ab+b^2}{a^2 \cdot ab+b^2} = \frac{a+b}{a \cdot b}$$

21.
$$\frac{3}{5} \div \frac{10}{9} = \frac{3}{5} \cdot \frac{9}{10} = \frac{27}{50}$$

23.
$$\frac{4}{5x^2} \div \frac{12}{25x^3} = \frac{4}{5x^2} \cdot \frac{25x^3}{12} = \frac{\cancel{4}}{5x^2} \cdot \frac{5x^2 \cdot 5x}{\cancel{4} \cdot 3} = \frac{5x}{3}$$

25.
$$\frac{24a^2b}{7c^2d} \div \frac{8ab}{21cd^2} = \frac{24a^2b}{7c^2d} \cdot \frac{21cd^2}{8ab} = \frac{8ab \cdot 3a}{7cd \cdot c} \cdot \frac{7cd \cdot 3d}{8ab} = \frac{9ad}{c}$$

27.
$$\frac{3x \cdot 3}{x} \div \frac{x^2 \cdot 1}{x^2} = \frac{3x \cdot 3}{x} \cdot \frac{x^2}{x^2 \cdot 1} = \frac{3\,(x-1)}{\cancel{x}} \cdot \frac{\cancel{x} \cdot x}{(x+1)\,(x-1)} = \frac{3x}{x+1}$$

29.
$$\frac{y^2 \cdot 25}{y^2 \cdot 4} \div \frac{3y \cdot 15}{4y \cdot 8} = \frac{y^2 \cdot 25}{y^2 \cdot 4} \cdot \frac{4y \cdot 8}{3y \cdot 15} = \frac{(y+5)\,(y-5)}{(y+2)\,(y-2)} \cdot \frac{4\,(y-2)}{3\,(y-5)} = \frac{4\,(y+5)}{3\,(y+2)} = \frac{4y+20}{3y+6}$$

31.
$$\frac{a^3+b^3}{a^3 \cdot b^3} \div \frac{a^2 \cdot ab+b^2}{a^2+ab+b^2} = \frac{a^3+b^3}{a^3 \cdot b^3} \cdot \frac{a^2+ab+b^2}{a^2 \cdot ab+b^2} = \frac{(a+b)\,(a^2 \cdot ab+b^2)}{(a \cdot b)\,(a^2+ab+b^2)} \cdot \frac{a^2+ab+b^2}{a^2 \cdot ab+b^2} = \frac{a+b}{a \cdot b}$$

33.
$$\frac{8a^3 \cdot 1}{6u^4w^3} \div \frac{1 \cdot 2a}{3u^2w} = \frac{8a^3 \cdot 1}{6u^4w^3} \cdot \frac{3u^2w}{1 \cdot 2a} = \frac{(2a-1)\,(4a^2+2a+1)}{3u^2w \cdot 2u^2w^2} \cdot \frac{3u^2w}{\cdot 1 \cdot (2a-1)} = \frac{\cdot\,(4a^2+2a+1)}{2u^2w^2}$$

35.
$$\frac{x - x^3}{2x^2+6x} \div \frac{5x^2 - 5x}{2x+6} = \frac{-x(x^2-1)}{2x^2+6x}\cdot\frac{2x+6}{5x^2-5x} = \frac{-1\cdot\cancel{x}(x+1)\cancel{(x-1)}}{\cancel{2}\cdot\cancel{x}\,(x+3)}\cdot\frac{\cancel{2}\,(x+3)}{5\cdot x\cancel{(x-1)}} = \frac{-(x+1)}{5x} = \frac{-x-1}{5x}$$

37.
$$\frac{y^2+y-12}{y^2-8y+15} \div \frac{3y^2+7y-20}{2y^2-7y-15} = \frac{y^2+y-12}{y^2-8y+15}\cdot\frac{2y^2-7y-15}{3y^2+7y-20}$$
$$= \frac{(y+4)\cancel{(y-3)}}{\cancel{(y-5)}\cancel{(y-3)}}\cdot\frac{(2y+3)\cancel{(y-5)}}{(3y-5)\cancel{(y+4)}} = \frac{2y+3}{3y-5}$$

39.
$$\frac{4x^2-12x+9}{25-4x^2} \div \frac{6x^2-5x-6}{6x^2+19x+10} = \frac{4x^2-12x+9}{25-4x^2}\cdot\frac{6x^2+19x+10}{6x^2-5x-6}$$
$$= \frac{\cancel{(2x-3)}(2x-3)}{-\cancel{(2x+5)}(2x-5)}\cdot\frac{\cancel{(2x+5)}\cancel{(3x+2)}}{\cancel{(3x+2)}\cancel{(2x-3)}} = \frac{2x-3}{5-2x}$$

41.
$$\frac{x^2+2x-3}{x-5} \div \frac{x^2+6x+9}{x^2-2x-15}\cdot\frac{1}{x^2-1} = \frac{x^2+2x-3}{x-5}\cdot\frac{x^2-2x-15}{x^2+6x+9}\cdot\frac{1}{x^2-1}$$
$$= \frac{(x+3)\cancel{(x-1)}}{\cancel{x-5}}\cdot\frac{\cancel{(x-5)}\cancel{(x+3)}}{\cancel{(x+3)}\cancel{(x+3)}}\cdot\frac{1}{(x+1)\cancel{(x-1)}} = \frac{1}{x+1}$$

43.
$$\frac{x^2-1}{x^2+3x-10} \div \frac{x^2-3x-4}{x^2-25}\cdot\frac{x-2}{x-5} = \frac{x^2-1}{x^2+3x-10}\cdot\frac{x^2-25}{x^2-3x-4}\cdot\frac{x-2}{x-5}$$
$$= \frac{\cancel{(x+1)}(x-1)}{\cancel{(x+5)}\cancel{(x-2)}}\cdot\frac{\cancel{(x+5)}\cancel{(x-5)}}{(x-4)\cancel{(x+1)}}\cdot\frac{\cancel{x-2}}{\cancel{x-5}} = \frac{x-1}{x-4}$$

45.
$$\frac{x-3}{3-x}\cdot\frac{x^2+3x-4}{x^2+7x+12} \div \frac{x^2+x-2}{x^2+5x+6} = \frac{x-3}{3-x}\cdot\frac{x^2+3x-4}{x^2+7x+12}\cdot\frac{x^2+5x+6}{x^2+x-2}$$
$$= \frac{-\cancel{(3-x)}}{\cancel{3-x}}\cdot\frac{\cancel{(x+4)}\cancel{(x-1)}}{\cancel{(x+4)}\cancel{(x+3)}}\cdot\frac{\cancel{(x+2)}\cancel{(x+3)}}{\cancel{(x+2)}\cancel{(x-1)}} = -1$$

47.
$$\frac{x^2-y^2}{x^2-2xy} \div \frac{x^2+xy-2y^2}{x^2-4y^2}\cdot\frac{x^2}{(x+y)^2} = \frac{x^2-y^2}{x^2-2xy}\cdot\frac{x^2-4y^2}{x^2+xy-2y^2}\cdot\frac{x^2}{(x+y)^2}$$
$$= \frac{\cancel{(x+y)}\cancel{(x-y)}}{\cancel{x}\cancel{(x-2y)}}\cdot\frac{\cancel{(x+2y)}\cancel{(x-2y)}}{\cancel{(x+2y)}\cancel{(x-y)}}\cdot\frac{\cancel{x}\cdot x}{\cancel{(x+y)}(x+y)} = \frac{x}{x+y}$$

49.
$$\frac{x^2+2xy-3y^2}{y^2-7y+10} \div \frac{x^2-3xy+2y^2}{y^2-3y-10}\cdot\frac{x^2-4y^2}{x^2-9y^2} = \frac{x^2+2xy-3y^2}{y^2-7y+10}\cdot\frac{y^2-3y-10}{x^2-3xy+2y^2}\cdot\frac{x^2-4y^2}{x^2-9y^2}$$
$$= \frac{(x+3y)\cancel{(x-y)}}{\cancel{(y-5)}(y-2)}\cdot\frac{\cancel{(y-5)}(y+2)}{\cancel{(x-2y)}\cancel{(x-y)}}\cdot\frac{(x+2y)\cancel{(x-2y)}}{\cancel{(x+3y)}(x-3y)}$$
$$= \frac{(y+2)(x+2y)}{(y-2)(x-3y)} = \frac{xy+2x+2y^2+4y}{xy-2x-3y^2+6y}$$

51.
$$\frac{x^3+x^2\cdot x\cdot 1}{x+2}\div\frac{x^3+1}{x+3}\cdot\frac{x+2}{x\cdot 1}=\frac{x^2(x+1)\cdot(x+1)}{x+2}\cdot\frac{x+3}{x^3+1}\cdot\frac{x+2}{x\cdot 1}$$

$$=\frac{\cancel{(x+1)}(x^2\cdot 1)}{\cancel{x+2}}\cdot\frac{x+3}{\cancel{(x+1)}(x^2\cdot x+1)}\cdot\frac{\cancel{x+2}}{x\cdot 1}=\frac{(x+1)\cancel{(x-1)}(x+3)}{\cancel{(x-1)}(x^2\cdot x+1)}$$

$$=\frac{(x+1)(x+3)}{x^2\cdot x+1}=\frac{x^2+4x+3}{x^2\cdot x+1}$$

53.
$$\frac{x\cdot 2}{x^2\cdot 9}\div\frac{x^3\cdot 8}{x+3}\cdot\frac{x\cdot 3}{x}=\frac{x\cdot 2}{x^2\cdot 9}\cdot\frac{x+3}{x^3\cdot 8}\cdot\frac{x\cdot 3}{x}=\frac{\cancel{x-2}}{\cancel{(x+3)}\cancel{(x-3)}}\cdot\frac{\cancel{x+3}}{\cancel{(x-2)}(x^2+2x+4)}\cdot\frac{\cancel{x-3}}{x}$$

$$=\frac{1}{x(x^2+2x+4)}=\frac{1}{x^3+2x^2+4x}$$

55.
$$\frac{3x+9}{4}\cdot\frac{600}{x^2+3x}=\frac{3\cancel{(x+3)}}{\cancel{4}}\cdot\frac{\cancel{4}\cdot 150}{x\cancel{(x+3)}}=\frac{450}{x}$$

57.
$$I=E\div R=\frac{4t}{t+3}\div\frac{t^2+9}{t^2+6t+9}=\frac{4t}{t+3}\cdot\frac{t^2+6t+9}{t^2+9}=\frac{4t}{\cancel{t+3}}\cdot\frac{\cancel{(t+3)}(t+3)}{t^2+9}=\frac{4t(t+3)}{t^2+9}=\frac{4t^2+12t}{t^2+9}$$

59. Multiply numerator and denominator by $x+2$:
$$\frac{x\cdot 1}{x\cdot 3}=\frac{x\cdot 1}{x\cdot 3}\cdot\frac{x+2}{x+2}=\frac{x^2+x\cdot 2}{x^2\cdot x\cdot 6}$$

61. Multiply numerator and denominator by $x-2$:
$$\frac{1}{x^2+2x+4}=\frac{1}{x^2+2x+4}\cdot\frac{x-2}{x-2}=\frac{x-2}{x^3-8}$$

63.
$$R\frac{R_T}{R-R_T}=\frac{RR_T}{R-R_T}$$

65.
$$R=\frac{3000}{x}\cdot(20+3x)=\frac{60,000+9000x}{x}$$

67.
$$A=\frac{\vdots\,2w\cdot L\cdot\,\cdot w\cdot}{2\because3}=\frac{2w^2\cdot Lw}{6}$$

69. Sample answer: Use the reciprocal of the rational expression following the division sign. Factor each of the expressions in the numerators and denominators of the rational expressions. Reduce if possible, and multiply the numerator factors and the denominator factors to find the solution.

71.
$$\frac{2\cdot x}{x+3}\div\frac{x^3\cdot 8}{x\cdot 5}\cdot\frac{x^3+27}{x\cdot 5}=\frac{2\cdot x}{x+3}\cdot\frac{x\cdot 5}{x^3\cdot 8}\cdot\frac{x^3+27}{x\cdot 5}$$

$$=\frac{\cdot 1\cancel{(x-2)}}{\cancel{x+3}}\cdot\frac{\cancel{x-5}}{\cancel{(x-2)}(x^2+2x+4)}\cdot\frac{\cancel{(x+3)}(x^2\cdot 3x+9)}{\cancel{x-5}}=\frac{\cdot(x^2\cdot 3x+9)}{x^2+2x+4}$$

73.
$$\frac{3x^2\cdot 3}{5x+10}\div\frac{x+1}{5}=\frac{3x^2\cdot 3}{5x+10}\cdot\frac{5}{x+1}=\frac{3\cancel{(x+1)}(x\cdot 1)}{\cancel{5}(x+2)}\cdot\frac{\cancel{5}}{\cancel{x+1}}=\frac{3(x\cdot 1)}{x+2}=\frac{3x\cdot 3}{x+2}$$

75.
$$\frac{3 \cdot x}{x+2} \cdot \frac{x^2+5x+6}{x^2 \cdot 9} = \frac{\cdot 1 (x-3)}{x+2} \cdot \frac{(x+2)(x+3)}{(x+3)(x-3)} = \cdot 1$$

77.
$$\frac{x^3+x^2 \cdot x \cdot 1}{x+4} \div \frac{x^3+1}{x^2 \cdot 16} \cdot \frac{1}{x^2 \cdot 1} = \frac{x^2(x+1) \cdot (x+1)}{x+4} \cdot \frac{x^2 \cdot 16}{x^3+1} \cdot \frac{1}{x^2 \cdot 1}$$

$$= \frac{(x+1)(x^2 \cdot 1)}{x+4} \cdot \frac{(x+4)(x \cdot 4)}{(x+1)(x^2 \cdot x+1)} \cdot \frac{1}{(x+1)(x \cdot 1)}$$

$$= \frac{(x+1)(x \cdot 1)(x \cdot 4)}{(x^2 \cdot x+1)(x+1)(x \cdot 1)} = \frac{x \cdot 4}{x^2 \cdot x+1}$$

79.
$$\frac{x^2 \cdot 9}{9x^2 \cdot 4y^2} \cdot \frac{3x^2 \cdot 2xy}{3x+9} = \frac{(x+3)(x \cdot 3)}{(3x+2y)(3x \cdot 2y)} \cdot \frac{x(3x \cdot 2y)}{3(x+3)} = \frac{x(x \cdot 3)}{3(3x+2y)} = \frac{x^2 \cdot 3x}{9x+6y}$$

6.3   Addition and Subtraction of Rational Expressions

Problems 6.3

1.   a.
$$\frac{4x}{5(x+3)} + \frac{6x}{5(x+3)} = \frac{10x}{5(x+3)} = \frac{5 \cdot 2x}{5(x+3)} = \frac{2x}{x+3}$$

b.
$$\frac{11x}{24(7+x)} + \frac{x}{24(7+x)} = \frac{12x}{24(7+x)} = \frac{12 \cdot x}{12 \cdot 2(7+x)} = \frac{x}{2(7+x)} = \frac{x}{14+2x}$$

c.
$$\frac{3x}{9x^2-16} - \frac{4}{9x^2-16} = \frac{3x-4}{9x^2-16} = \frac{3x-4}{(3x+4)(3x-4)} = \frac{1}{3x+4}$$

d.
$$\frac{5x}{x+8} - \frac{4x-8}{x+8} = \frac{5x-(4x-8)}{x+8} = \frac{5x-4x+8}{x+8} = \frac{x+8}{x+8} = 1$$

2.
$$\frac{5x}{x+1} - \frac{3x}{x+3} = \frac{5x(x+3)}{(x+1)(x+3)} - \frac{3x(x+1)}{(x+3)(x+1)} = \frac{5x^2+15x-3x^2-3x}{(x+3)(x+1)} = \frac{2x^2+12x}{(x+3)(x+1)} = \frac{2x(x+6)}{(x+3)(x+1)}$$

3.   a.
$$\frac{x+1}{(x+3)(x-1)} + \frac{x+4}{x^2-1} = \frac{x+1}{(x+3)(x-1)} + \frac{x+4}{(x+1)(x-1)} = \frac{(x+1)(x+1)}{(x+3)(x-1)(x+1)} + \frac{(x+4)(x+3)}{(x+1)(x-1)(x+3)}$$

$$= \frac{(x^2+2x+1)+(x^2+7x+12)}{(x+3)(x-1)(x+1)} = \frac{2x^2+9x+13}{(x+3)(x-1)(x+1)}$$

b.
$$\frac{x-3}{x^2-x-2}-\frac{x+3}{x^2-4}=\frac{x-3}{(x+1)(x-2)}-\frac{x+3}{(x+2)(x-2)}=\frac{(x-3)(x+2)}{(x+1)(x-2)(x+2)}-\frac{(x+3)(x+1)}{(x+2)(x-2)(x+1)}$$

$$=\frac{(x^2-x-6)-(x^2+4x+3)}{(x+1)(x-2)(x+2)}=\frac{x^2-x-6-x^2-4x-3}{(x+1)(x-2)(x+2)}=\frac{-5x-9}{(x+1)(x-2)(x+2)}$$

4.   a.
$$\frac{2x-2y}{x^2-2xy+y^2}+\frac{x-y}{x^2-y^2}=\frac{2\cancel{(x-y)}}{\cancel{(x-y)}(x-y)}+\frac{\cancel{x-y}}{(x+y)\cancel{(x-y)}}=\frac{2}{(x-y)}+\frac{1}{(x+y)}$$

$$=\frac{2(x+y)}{(x-y)(x+y)}+\frac{1(x-y)}{(x+y)(x-y)}=\frac{2x+2y+x-y}{(x-y)(x+y)}$$

$$=\frac{3x+y}{(x-y)(x+y)}=\frac{3x+y}{x^2-y^2}$$

b.
$$\frac{x}{(x+3)(x-3)}+\frac{3}{(3-x)(x+3)}=\frac{x}{(x+3)(x-3)}+\frac{-3}{(x-3)(x+3)}=\frac{\cancel{x-3}}{(x+3)\cancel{(x-3)}}=\frac{1}{x+3}$$

Exercises 6.3

1. $\dfrac{x}{5}+\dfrac{2x}{5}=\dfrac{3x}{5}$

3. $\dfrac{7x}{3}-\dfrac{2x}{3}=\dfrac{5x}{3}$

5. $\dfrac{3}{5x+10}+\dfrac{2x}{5(x+2)}=\dfrac{3}{5(x+2)}+\dfrac{2x}{5(x+2)}=\dfrac{2x+3}{5(x+2)}$

7. $\dfrac{2x+1}{2(x+1)}-\dfrac{x-1}{2x+2}=\dfrac{2x+1}{2(x+1)}-\dfrac{x-1}{2(x+1)}=\dfrac{2x+1-(x-1)}{2(x+1)}=\dfrac{2x+1-x+1}{2(x+1)}=\dfrac{x+2}{2(x+1)}$

9. $\dfrac{2x+1}{3(x-1)}+\dfrac{x+3}{3x-3}-\dfrac{x-1}{3(x-1)}=\dfrac{2x+1}{3(x-1)}+\dfrac{x+3}{3(x-1)}-\dfrac{x-1}{3(x-1)}=\dfrac{2x+1+x+3-x+1}{3(x-1)}=\dfrac{2x+5}{3(x-1)}$

11. $\dfrac{x}{x^2+3x-4}+\dfrac{x}{x^2-16}=\dfrac{x}{(x+4)(x-1)}+\dfrac{x}{(x+4)(x-4)}=\dfrac{x(x-4)}{(x+4)(x-1)(x-4)}+\dfrac{x(x-1)}{(x+4)(x-4)(x-1)}$

$$=\dfrac{x^2-4x+x^2-x}{(x+4)(x-4)(x-1)}=\dfrac{2x^2-5x}{(x+4)(x-4)(x-1)}$$

13. $\dfrac{3x}{x^2+3x-10}+\dfrac{2x}{x^2+x-6}=\dfrac{3x}{(x+5)(x-2)}+\dfrac{2x}{(x+3)(x-2)}=\dfrac{3x(x+3)}{(x+5)(x-2)(x+3)}+\dfrac{2x(x+5)}{(x+3)(x-2)(x+5)}$

$$=\dfrac{3x^2+9x+2x^2+10x}{(x+5)(x-2)(x+3)}=\dfrac{5x^2+19x}{(x+5)(x-2)(x+3)}$$

15. $\dfrac{1}{x^2-y^2}+\dfrac{5}{(x+y)^2}=\dfrac{1}{(x+y)(x-y)}+\dfrac{5}{(x+y)^2}=\dfrac{1(x+y)}{(x+y)(x-y)(x+y)}+\dfrac{5(x-y)}{(x+y)^2(x-y)}$

$$=\dfrac{x+y+5x-5y}{(x+y)^2(x-y)}=\dfrac{6x-4y}{(x+y)^2(x-y)}$$

17. $\dfrac{2}{x-5}-\dfrac{3x}{x^2-25}=\dfrac{2}{x-5}-\dfrac{3x}{(x+5)(x-5)}=\dfrac{2(x+5)}{(x-5)(x+5)}-\dfrac{3x}{(x+5)(x-5)}=\dfrac{2x+10-3x}{(x+5)(x-5)}$

$$=\dfrac{10-x}{(x+5)(x-5)}$$

19. $\dfrac{x-1}{x^2+3x+2}-\dfrac{x+7}{x^2+5x+6}=\dfrac{x-1}{(x+2)(x+1)}-\dfrac{x+7}{(x+3)(x+2)}=\dfrac{(x-1)(x+3)}{(x+2)(x+1)(x+3)}-\dfrac{(x+7)(x+1)}{(x+3)(x+2)(x+1)}$

$$=\dfrac{(x^2+2x-3)-(x^2+8x+7)}{(x+2)(x+1)(x+3)}=\dfrac{x^2+2x-3-x^2-8x-7}{(x+2)(x+1)(x+3)}$$

$$=\dfrac{-6x-10}{(x+2)(x+1)(x+3)}$$

21. $\dfrac{x+2}{x^2-4}+\dfrac{x+3}{x^2-9}=\dfrac{\cancel{x+2}}{\cancel{(x+2)}(x-2)}+\dfrac{\cancel{x+3}}{\cancel{(x+3)}(x-3)}=\dfrac{1}{x-2}+\dfrac{1}{x-3}=\dfrac{1(x-3)}{(x-2)(x-3)}+\dfrac{1(x-2)}{(x-3)(x-2)}$

$$=\dfrac{x-3+x-2}{(x-2)(x-3)}=\dfrac{2x-5}{(x-2)(x-3)}$$

23. $\dfrac{x-3}{x^2-9}+\dfrac{x+3}{x^2+6x+9}=\dfrac{\cancel{x-3}}{\cancel{(x-3)}(x+3)}+\dfrac{\cancel{x+3}}{\cancel{(x+3)}(x+3)}=\dfrac{1}{x+3}+\dfrac{1}{x+3}=\dfrac{2}{x+3}$

25. $\dfrac{a+3}{a^2+5a+6}+\dfrac{a+2}{a^2+6a+8}=\dfrac{\cancel{a+3}}{\cancel{(a+3)}(a+2)}+\dfrac{\cancel{a+2}}{\cancel{(a+2)}(a+4)}=\dfrac{1}{a+2}+\dfrac{1}{a+4}$

$$=\dfrac{1(a+4)}{(a+2)(a+4)}+\dfrac{1(a+2)}{(a+4)(a+2)}=\dfrac{a+4+a+2}{(a+4)(a+2)}=\dfrac{2a+6}{(a+4)(a+2)}$$

27. $\dfrac{3a+3}{a^2+5a+4}-\dfrac{a-3}{a^2+a-12}=\dfrac{3\cancel{(a+1)}}{\cancel{(a+1)}(a+4)}-\dfrac{\cancel{a-3}}{\cancel{(a-3)}(a+4)}=\dfrac{3}{a+4}-\dfrac{1}{a+4}=\dfrac{2}{a+4}$

29. $\dfrac{5a-15}{a^2+2a-15}-\dfrac{a^2+5a}{a^2+8a+15}=\dfrac{5\cancel{(a-3)}}{\cancel{(a-3)}(a+5)}-\dfrac{a\cancel{(a+5)}}{\cancel{(a+5)}(a+3)}=\dfrac{5}{a+5}-\dfrac{a}{a+3}$

$$=\dfrac{5(a+3)}{(a+5)(a+3)}-\dfrac{a(a+5)}{(a+3)(a+5)}=\dfrac{5a+15-a^2-5a}{(a+5)(a+3)}=\dfrac{15-a^2}{(a+5)(a+3)}$$

31. $\dfrac{y}{y^2-1}+\dfrac{y}{y-1}=\dfrac{y}{(y+1)(y-1)}+\dfrac{y}{y-1}=\dfrac{y}{(y+1)(y-1)}+\dfrac{y(y+1)}{(y-1)(y+1)}=\dfrac{y+y^2+y}{(y+1)(y-1)}=\dfrac{y^2+2y}{(y+1)(y-1)}$

33. $\dfrac{3y+1}{y^2-16}-\dfrac{2y-1}{y-4}=\dfrac{3y+1}{(y+4)(y-4)}-\dfrac{2y-1}{y-4}=\dfrac{3y+1}{(y+4)(y-4)}-\dfrac{(2y-1)(y+4)}{(y-4)(y+4)}=\dfrac{3y+1-\left(2y^2+7y-4\right)}{(y+4)(y-4)}$

$=\dfrac{3y+1-2y^2-7y+4}{(y+4)(y-4)}=\dfrac{-2y^2-4y+5}{(y+4)(y-4)}$

35. $\dfrac{5x+2y}{5x-2y}+\dfrac{5x-2y}{5x+2y}=\dfrac{(5x+2y)(5x+2y)}{(5x-2y)(5x+2y)}+\dfrac{(5x-2y)(5x-2y)}{(5x+2y)(5x-2y)}=\dfrac{25x^2+20xy+4y^2+25x^2-20xy+4y^2}{(5x-2y)(5x+2y)}$

$=\dfrac{50x^2+8y^2}{(5x-2y)(5x+2y)}$

37. $\dfrac{3x-y}{2x-y}-\dfrac{2x+y}{3x+y}=\dfrac{(3x-y)(3x+y)}{(2x-y)(3x+y)}-\dfrac{(2x+y)(2x-y)}{(3x+y)(2x-y)}=\dfrac{\left(9x^2-y^2\right)-\left(4x^2-y^2\right)}{(2x-y)(3x+y)}=\dfrac{9x^2-y^2-4x^2+y^2}{(2x-y)(3x+y)}$

$=\dfrac{5x^2}{(2x-y)(3x+y)}$

39. $\dfrac{x+3}{x^2-x-2}+\dfrac{x-1}{x^2+2x+1}=\dfrac{x+3}{(x-2)(x+1)}+\dfrac{x-1}{(x+1)^2}=\dfrac{(x+3)(x+1)}{(x-2)(x+1)(x+1)}+\dfrac{(x-1)(x-2)}{(x+1)^2(x-2)}$

$=\dfrac{x^2+4x+3+x^2-3x+2}{(x+1)^2(x-2)}=\dfrac{2x^2+x+5}{(x+1)^2(x-2)}$

41. $\dfrac{x+1}{x^2-x-2}-\dfrac{x}{x^2-5x+4}=\dfrac{\cancel{x+1}}{(x-2)\cancel{(x+1)}}-\dfrac{x}{(x-1)(x-4)}=\dfrac{1(x-1)(x-4)}{(x-2)(x-1)(x-4)}-\dfrac{x(x-2)}{(x-1)(x-4)(x-2)}$

$=\dfrac{x^2-5x+4-x^2+2x}{(x-2)(x-1)(x-4)}=\dfrac{4-3x}{(x-2)(x-1)(x-4)}$

43. $\dfrac{2}{5+x}+\dfrac{5x}{x^2-25}+\dfrac{7}{5-x}=\dfrac{2}{x+5}+\dfrac{5x}{(x+5)(x-5)}+\dfrac{-7}{x-5}=\dfrac{2(x-5)}{(x+5)(x-5)}+\dfrac{5x}{(x+5)(x-5)}+\dfrac{-7(x+5)}{(x-5)(x+5)}$

$=\dfrac{2x-10+5x-7x-35}{(x+5)(x-5)}=\dfrac{-45}{(x+5)(x-5)}$

45. $\dfrac{x}{(x-y)(2-x)}-\dfrac{y}{(y-x)(2-x)}+\dfrac{y}{(x-y)(x-2)}=\dfrac{-x}{(x-y)(x-2)}-\dfrac{y}{(x-y)(x-2)}+\dfrac{y}{(x-y)(x-2)}$

$=\dfrac{-x-y+y}{(x-y)(x-2)}=\dfrac{-x}{(x-y)(x-2)}=\dfrac{x}{(x-y)(2-x)}$

47.
$$\frac{4a^2-9b^2}{4a^2-12ab+9b^2}+\frac{12a+18b}{4a^2+12ab+9b^2}-\frac{2a+3b}{2a+3b}=\frac{(2a+3b)\,\cancel{(2a-3b)}}{(2a-3b)\,\cancel{(2a-3b)}}+\frac{6\,\cancel{(2a+3b)}}{(2a+3b)\,\cancel{(2a+3b)}}-\frac{2a+3b}{2a+3b}$$

$$=\frac{(2a+3b)(2a+3b)}{(2a-3b)(2a+3b)}+\frac{6(2a-3b)}{(2a+3b)(2a-3b)}-\frac{(2a+3b)(2a-3b)}{(2a+3b)(2a-3b)}$$

$$=\frac{4a^2+12ab+9b^2+12a-18b-4a^2+9b^2}{(2a-3b)(2a+3b)}=\frac{18b^2+12ab+12a-18b}{(2a-3b)(2a+3b)}$$

49.
$$\frac{x+5}{x^3+125}+\frac{x-5}{x^2-25}-\frac{1}{x+5}=\frac{\cancel{x+5}}{\cancel{(x+5)}(x^2-5x+25)}+\frac{\cancel{x-5}}{(x+5)\cancel{(x-5)}}-\frac{1}{x+5}$$

$$=\frac{1}{x^2-5x+25}+\frac{1}{x+5}-\frac{1}{x+5}=\frac{1}{x^2-5x+25}$$

51.
$$\cdot\,\frac{w_0x^3}{6L}+\frac{w_0Lx}{2}\cdot\frac{w_0L^2}{3}=\cdot\,\frac{w_0x^3}{6L}+\frac{w_0Lx}{2}\cdot\frac{3L}{3L}\cdot\frac{w_0L^2}{3}\cdot\frac{2L}{2L}=\frac{\cdot\,w_0x^3+3w_0L^2x\cdot\,2w_0L^3}{6L}$$

53.
$$\frac{p^2}{2mr^2}\cdot\frac{gmM}{r}=\frac{p^2}{2mr^2}\cdot\frac{gmM}{r}\cdot\frac{2mr}{2mr}=\frac{p^2\cdot\,2gm^2Mr}{2mr^2}$$

55.
$$9\,\vdots\,2+\frac{2}{9}\,\vdots\,=9\cdot2+9\cdot\frac{2}{9}=18+2=20$$

57.
$$12xy\,\vdots\,\frac{2}{y}+\frac{3}{2x}\,\vdots\,=12x\cancel{y}\cdot\frac{2}{\cancel{y}}+12\cancel{x}y\cdot\frac{3}{2\cancel{x}}$$
$$=24x+18y$$

59.
$$x^2\,\vdots\,1\cdot\frac{1}{x^2}\,\vdots\,=x^2\cdot1\cdot\cancel{x^2}\cdot\frac{1}{\cancel{x^2}}=x^2\cdot1$$

61.
$$P(x+h)=(x+h)^2=x^2+2xh+h^2$$

63.
$$\frac{P(x+h)-P(x)}{h}=\frac{2xh+h^2}{h}=\frac{\cancel{h}(2x+h)}{\cancel{h}}$$
$$=2x+h$$

65. Sample answer: Find the factors of each denominator. Choose the most occurrences of each factor in the denominators. Find the product of these factors.

67. Sample answer: Any common denominator will do the job. The advantage of having the LCD is that there will be less reducing that needs to be done in the final answer.

69.
$$\frac{x-3}{x^2-x-2}-\frac{x+3}{x^2-4}=\frac{x-3}{(x-2)(x+1)}-\frac{x+3}{(x+2)(x-2)}=\frac{(x-3)(x+2)}{(x-2)(x+1)(x+2)}-\frac{(x+3)(x+1)}{(x+2)(x-2)(x+1)}$$

$$=\frac{(x^2-x-6)-(x^2+4x+3)}{(x+2)(x-2)(x+1)}=\frac{x^2-x-6-x^2-4x-3}{(x+2)(x-2)(x+1)}=\frac{-5x-9}{(x+2)(x-2)(x+1)}$$

71.
$$\frac{x}{(x+3)(x-3)}+\frac{3}{(3-x)(x+3)}=\frac{x}{(x+3)(x-3)}+\frac{-3}{(x-3)(x+3)}=\frac{\cancel{x-3}}{(x+3)\cancel{(x-3)}}=\frac{1}{x+3}$$

73.
$$\frac{4x}{5(x-2)}+\frac{6x}{5(x-2)}=\frac{10x}{5(x-2)}=\frac{\cancel{5}\cdot 2x}{\cancel{5}(x-2)}=\frac{2x}{x-2}$$

75.
$$\frac{x}{x^2-9}-\frac{3}{x^2-9}=\frac{x-3}{x^2-9}=\frac{\cancel{x-3}}{(x+3)\cancel{(x-3)}}=\frac{1}{x+3}$$

6.4 Complex Fractions

Problems 6.4

1. Multiply the numerator and denominator by the LCD which is $4ab$:

$$\frac{\dfrac{2}{b}\cdot\dfrac{3}{a}}{\dfrac{1}{2b}\cdot\dfrac{3}{4a}}=\frac{4ab\cdot\dfrac{2}{b}\cdot\dfrac{3}{a}\cdot}{4ab\cdot\dfrac{1}{2b}\cdot\dfrac{3}{4a}\cdot}=\frac{4a\cancel{b}\cdot\dfrac{2}{\cancel{b}}\cdot 4\cancel{a}b\cdot\dfrac{3}{\cancel{a}}}{\cancel{2}\cdot 2a\cancel{b}\cdot\dfrac{1}{\cancel{2b}}\cdot \cancel{4}\cancel{a}b\cdot\dfrac{3}{\cancel{4a}}}=\frac{8a\cdot 12b}{2a\cdot 3b}=\frac{4(2a-3b)}{2a-3b}=4$$

2. Multiply the numerator and denominator by the LCD which is $4x^2$:

$$\frac{2x^2+\dfrac{1}{4x}}{4\cdot\dfrac{1}{x^2}}=\frac{4x^2\cdot 2x^2+\dfrac{1}{4x}\cdot}{4x^2\cdot 4\cdot\dfrac{1}{x^2}\cdot}=\frac{4x^2\cdot 2x^2+\cancel{4x}\cdot x\cdot\dfrac{1}{\cancel{4x}}}{4x^2\cdot 4\cdot 4\cancel{x^2}\cdot\dfrac{1}{\cancel{x^2}}}=\frac{8x^4+x}{16x^2-4}=\frac{x(8x^3+1)}{4(4x^2-1)}$$

$$=\frac{x\cancel{(2x+1)}(4x^2-2x+1)}{4\cancel{(2x+1)}(2x-1)}=\frac{x(4x^2-2x+1)}{4(2x-1)}=\frac{4x^3-2x^2+x}{8x-4}$$

3. Multiply the numerator and denominator by the LCD which is $(x+3)(x-3)$:

$$\frac{\dfrac{x}{x+3}+x}{1-\dfrac{7}{x^2-9}}=\frac{\dfrac{x}{x+3}+x}{1-\dfrac{7}{(x+3)(x-3)}}=\frac{(x+3)(x-3)\left|\dfrac{x}{x+3}+x\right|}{(x+3)(x-3)\left|1-\dfrac{7}{(x+3)(x-3)}\right|}=\frac{x(x-3)+x(x+3)(x-3)}{(x+3)(x-3)-7}$$

$$=\frac{x^2-3x+x^3-9x}{x^2-9-7}=\frac{x^3+x^2-12x}{x^2-16}=\frac{x(x^2+x-12)}{(x+4)(x-4)}=\frac{x\cancel{(x+4)}(x-3)}{\cancel{(x+4)}(x-4)}=\frac{x(x-3)}{x-4}=\frac{x^2-3x}{x-4}$$

4.
$$2+\frac{a}{2+\dfrac{2}{2+a}}=2+\frac{a}{\dfrac{2(2+a)}{2+a}+\dfrac{2}{2+a}}=2+\frac{a}{\dfrac{4+2a}{2+a}+\dfrac{2}{2+a}}=2+\frac{a}{\dfrac{6+2a}{2+a}}=2+a\cdot\frac{2+a}{6+2a}$$

$$=\frac{2(6+2a)}{6+2a}+\frac{2a+a^2}{6+2a}=\frac{12+4a+2a+a^2}{6+2a}=\frac{a^2+6a+12}{2a+6}=\frac{a^2+6a+12}{2(a+3)}$$

5. Multiply the numerator and denominator by the LCD which is $x-2$:

$$\frac{x+(x-2)^{-1}}{1-x(2-x)^{-1}} = \frac{x+\dfrac{1}{x-2}}{1-\dfrac{x}{2-x}} = \frac{x+\dfrac{1}{x-2}}{1-\dfrac{x}{-1(x-2)}} = \frac{(x-2)\left[x+\dfrac{1}{x-2}\right]}{(x-2)\left[1-\dfrac{x}{-1(x-2)}\right]} = \frac{x(x-2)+1}{x-2+x} = \frac{x^2-2x+1}{2x-2}$$

$$= \frac{(x-1)(x-1)}{2(x-1)} = \frac{x-1}{2}$$

Exercises 6.4

1. Multiply the numerator and denominator by the LCD which is 4:

$$\frac{50\cdot 5\frac{1}{2}}{7\frac{3}{4}+\frac{1}{2}} = \frac{50\cdot\dfrac{11}{2}}{\dfrac{31}{4}+\dfrac{1}{2}} = \frac{4\cdot 50\cdot\dfrac{11}{2}}{4\cdot\dfrac{31}{4}+4\cdot\dfrac{1}{2}} = \frac{4\cdot 50\cdot 2\cdot\cancel{2}\cdot\dfrac{11}{\cancel{2}}}{\cancel{4}\cdot\dfrac{31}{\cancel{4}}+2\cdot\cancel{2}\cdot\dfrac{1}{\cancel{2}}} = \frac{200\cdot 22}{31+2} = \frac{178}{33}$$

3.
$$\frac{\dfrac{a}{b}}{\dfrac{c}{b}} = \frac{a}{\cancel{b}}\cdot\frac{\cancel{b}}{c} = \frac{a}{c}$$

5.
$$\frac{\dfrac{x}{y}}{\dfrac{x^2}{z}} = \frac{\cancel{x}}{y}\cdot\frac{z}{\cancel{x}\cdot x} = \frac{z}{xy}$$

7.
$$\frac{\dfrac{3x}{5y}}{\dfrac{3x}{2z}} = \frac{\cancel{3x}}{5y}\cdot\frac{2z}{\cancel{3x}} = \frac{2z}{5y}$$

9. Multiply the numerator and denominator by the LCD which is 2:

$$\frac{\dfrac{1}{2}}{2-\dfrac{1}{2}} = \frac{2\left[\dfrac{1}{2}\right]}{2\left[2-\dfrac{1}{2}\right]} = \frac{1}{4-1} = \frac{1}{3}$$

11. Multiply the numerator and denominator by the LCD which is b:

$$\frac{a-\dfrac{a}{b}}{1+\dfrac{a}{b}} = \frac{b\left[a-\dfrac{a}{b}\right]}{b\left[1+\dfrac{a}{b}\right]} = \frac{ab-a}{b+a}$$

13. Multiply the numerator and denominator by the LCD which is x:

$$\frac{\dfrac{1}{x}+2}{2-\dfrac{1}{x}} = \frac{x\left[\dfrac{1}{x}+2\right]}{x\left[2-\dfrac{1}{x}\right]} = \frac{1+2x}{2x-1}$$

15. Multiply the numerator and denominator by the LCD which is 6:

$$\frac{\dfrac{2}{3}+x}{x-\dfrac{1}{2}} = \frac{6\left[\dfrac{2}{3}+x\right]}{6\left[x-\dfrac{1}{2}\right]} = \frac{4+6x}{6x-3}$$

17. Multiply the numerator and denominator by the LCD which is x^2:

$$\frac{y+\dfrac{2}{x}}{y^2-\dfrac{4}{x^2}} = \frac{x^2\left|y+\dfrac{2}{x}\right|}{x^2\left|y^2-\dfrac{4}{x^2}\right|} = \frac{x^2y+2x}{x^2y^2-4} = \frac{x(xy+2)}{(xy+2)(xy-2)} = \frac{x}{xy-2}$$

19. Multiply the numerator and denominator by the LCD which is x^2y^2:

$$\frac{\dfrac{x}{y^2}-\dfrac{y}{x^2}}{x^2+xy+y^2} = \frac{x^2y^2\left|\dfrac{x}{y^2}-\dfrac{y}{x^2}\right|}{x^2y^2\left(x^2+xy+y^2\right)} = \frac{x^3-y^3}{x^2y^2\left(x^2+xy+y^2\right)} = \frac{(x-y)\left(x^2+xy+y^2\right)}{x^2y^2\left(x^2+xy+y^2\right)} = \frac{x-y}{x^2y^2}$$

21.
$$3\cdot\frac{3}{3\cdot\dfrac{1}{2}} = 3\cdot\frac{3}{\dfrac{6}{2}\cdot\dfrac{1}{2}} = 3\cdot\frac{3}{\dfrac{5}{2}} = 3\cdot 3\cdot\frac{2}{5} = 3\cdot\frac{6}{5} = \frac{15}{5}\cdot\frac{6}{5} = \frac{9}{5}$$

23.
$$a\cdot\frac{a}{a+\dfrac{1}{2}} = a\cdot\frac{a}{\dfrac{2a}{2}+\dfrac{1}{2}} = a\cdot\frac{a}{\dfrac{2a+1}{2}} = a\cdot a\cdot\frac{2}{2a+1} = a\cdot\frac{2a}{2a+1} = \frac{a(2a+1)}{2a+1}\cdot\frac{2a}{2a+1} = \frac{2a^2+a\cdot 2a}{2a+1}$$
$$= \frac{2a^2\cdot a}{2a+1}$$

25.
$$x\cdot\frac{x}{1\cdot\dfrac{x}{1\cdot x}} = x\cdot\frac{x}{\dfrac{1\cdot x}{1\cdot x}\cdot\dfrac{x}{1\cdot x}} = x\cdot\frac{x}{\dfrac{1\cdot x\cdot x}{1\cdot x}} = x\cdot\frac{x}{\dfrac{1\cdot 2x}{1\cdot x}} = x\cdot x\cdot\frac{1\cdot x}{1\cdot 2x} = x\cdot\frac{x\cdot x^2}{1\cdot 2x}$$
$$= \frac{x(1\cdot 2x)}{1\cdot 2x}\cdot\frac{x\cdot x^2}{1\cdot 2x} = \frac{x\cdot 2x^2\cdot x+x^2}{1\cdot 2x} = \frac{\cdot x^2}{1\cdot 2x} = \frac{x^2}{2x\cdot 1}$$

27. Multiply the numerator and denominator by the LCD which is $(x+1)(x-1)$:

$$\frac{\dfrac{x-1}{x+1}+\dfrac{x+1}{x-1}}{\dfrac{x-1}{x+1}-\dfrac{x+1}{x-1}} = \frac{(x+1)(x-1)\left|\dfrac{x-1}{x+1}+\dfrac{x+1}{x-1}\right|}{(x+1)(x-1)\left|\dfrac{x-1}{x+1}-\dfrac{x+1}{x-1}\right|} = \frac{(x-1)^2+(x+1)^2}{(x-1)^2-(x+1)^2} = \frac{x^2-2x+1+x^2+2x+1}{x^2-2x+1-x^2-2x-1} = \frac{2x^2+2}{-4x}$$
$$= \frac{\cancel{2}\left(x^2+1\right)}{\cancel{2}\left(\,-2x\right)} = \frac{-\left(x^2+1\right)}{2x}$$

29. Multiply the numerator and denominator by the LCD which is $(x+y)(x-y)$:

$$\frac{(x-y)^{-1}+(x+y)^{-1}}{(x-y)^{-1}-(x+y)^{-1}} = \frac{\dfrac{1}{x-y}+\dfrac{1}{x+y}}{\dfrac{1}{x-y}-\dfrac{1}{x+y}} = \frac{(x+y)(x-y)\left|\dfrac{1}{x-y}+\dfrac{1}{x+y}\right|}{(x+y)(x-y)\left|\dfrac{1}{x-y}-\dfrac{1}{x+y}\right|} = \frac{(x+y)+(x-y)}{(x+y)-(x-y)}$$
$$= \frac{x+y+x-y}{x+y-x+y} = \frac{\cancel{2}x}{\cancel{2}y} = \frac{x}{y}$$

31. Multiply the numerator and denominator by the LCD which is $x-2$:

$$\frac{x(x-2)^{-1}-x}{x(2-x)^{-1}+x} = \frac{\dfrac{x}{x-2}-x}{\dfrac{x}{2-x}+x} = \frac{(x-2)\left|\dfrac{x}{x-2}-x\right|}{(x-2)\left|\dfrac{x}{-1(x-2)}+x\right|} = \frac{x-x(x-2)}{-x+x(x-2)} = \frac{x-x^2+2x}{-x+x^2-2x} = \frac{3x-x^2}{x^2-3x} = -1$$

33. Multiply the numerator and denominator by the LCD which is x^2:

$$\frac{\dfrac{1}{x^2}+\dfrac{3}{x}-4}{\dfrac{1}{x^2}+\dfrac{5}{x}+4} = \frac{x^2\left|\dfrac{1}{x^2}+\dfrac{3}{x}-4\right|}{x^2\left|\dfrac{1}{x^2}+\dfrac{5}{x}+4\right|} = \frac{1+3x-4x^2}{1+5x+4x^2} = \frac{(1-4x)(1-x)}{(1-4x)(1+x)} = \frac{1-x}{1+x}$$

35. Multiply the numerator and denominator by the LCD which is $y+3$:

$$\frac{y+3-\dfrac{16}{y+3}}{y-6+\dfrac{20}{y+3}} = \frac{(y+3)\left|y+3-\dfrac{16}{y+3}\right|}{(y+3)\left|y-6+\dfrac{20}{y+3}\right|} = \frac{(y+3)^2-16}{(y+3)(y-6)+20} = \frac{y^2+6y+9-16}{y^2-3y-18+20} = \frac{y^2+6y-7}{y^2-3y+2}$$

$$= \frac{(y-1)(y+7)}{(y-1)(y-2)} = \frac{y+7}{y-2}$$

37. Multiply the numerator and denominator by the LCD which is $x(3x+1)$:

$$\frac{\dfrac{8x}{3x+1}-\dfrac{3x-1}{x}}{\dfrac{x}{3x+1}-\dfrac{2x-2}{x}} = \frac{x(3x+1)\left|\dfrac{8x}{3x+1}-\dfrac{3x-1}{x}\right|}{x(3x+1)\left|\dfrac{x}{3x+1}-\dfrac{2x-2}{x}\right|} = \frac{8x^2-(3x+1)(3x-1)}{x^2-(3x+1)(2x-2)} = \frac{8x^2-(9x^2-1)}{x^2-(6x^2-4x-2)}$$

$$= \frac{8x^2-9x^2+1}{x^2-6x^2+4x+2} = \frac{-x^2+1}{-5x^2+4x+2} = \frac{-1(x^2-1)}{-1(5x^2-4x-2)} = \frac{x^2-1}{5x^2-4x-2}$$

39. Multiply the numerator and denominator by the LCD which is cd:

$$\frac{\dfrac{c}{d}-\dfrac{d}{c}}{\dfrac{c}{d}+2+\dfrac{d}{c}} = \frac{cd\left|\dfrac{c}{d}-\dfrac{d}{c}\right|}{cd\left|\dfrac{c}{d}+2+\dfrac{d}{c}\right|} = \frac{c^2-d^2}{c^2-2cd+d^2} = \frac{(c-d)(c+d)}{(c-d)(c-d)} = \frac{c+d}{c-d}$$

41. Multiply the numerator and denominator by the LCD which is $\left(a^2+b^2\right)(a+b)(a-b)$:

$$\frac{\dfrac{a^2\cdot b^2}{a^2+b^2}\cdot\dfrac{a^2+b^2}{a^2\cdot b^2}}{\dfrac{a\cdot b}{a+b}\cdot\dfrac{a+b}{a\cdot b}}=\frac{\left(a^2+b^2\right)(a+b)(a\cdot b)\cdot\dfrac{a^2\cdot b^2}{a^2+b^2}\cdot\dfrac{a^2+b^2}{(a+b)(a\cdot b)}\cdot}{\left(a^2+b^2\right)(a+b)(a\cdot b)\cdot\dfrac{a\cdot b}{a+b}\cdot\dfrac{a+b}{a\cdot b}\cdot}$$

$$=\frac{\left(a^2\cdot b^2\right)(a+b)(a\cdot b)\cdot\left(a^2+b^2\right)\left(a^2+b^2\right)}{\left(a^2+b^2\right)\cdot(a\cdot b)(a\cdot b)\cdot(a+b)(a+b)\cdot}$$

$$=\frac{\left(a^4\cdot 2a^2b^2+b^4\right)\cdot\left(a^4+2a^2b^2+b^4\right)}{\left(a^2+b^2\right)\cdot\left(a^2\cdot 2ab+b^2\right)\cdot\left(a^2+2ab+b^2\right)\cdot}$$

$$=\frac{a^4\cdot 2a^2b^2+b^4\cdot a^4\cdot 2a^2b^2\cdot b^4}{\left(a^2+b^2\right)\cdot a^2\cdot 2ab+b^2\cdot a^2\cdot 2ab\cdot b^2\cdot}$$

$$=\frac{\cdot 4a^2b^2}{\left(a^2+b^2\right)[\cdot 4ab]}=\frac{\cdot\cancel{4ab}\cdot ab}{\cdot\cancel{4ab}\left(a^2+b^2\right)}=\frac{ab}{a^2+b^2}$$

43. Multiply the numerator and denominator by the LCD which is R_1R_2:

$$R=\frac{1}{\dfrac{1}{R_1}+\dfrac{1}{R_2}}=\frac{R_1R_2\cdot 1}{R_1R_2\left(\dfrac{1}{R_1}+\dfrac{1}{R_2}\right)}=\frac{R_1R_2}{R_2+R_1}$$

45. Multiply the numerator and denominator by the LCD which is c:

$$f=f_{\text{static}}\sqrt{\frac{1+\dfrac{v}{c}}{1\cdot\dfrac{v}{c}}}=f_{\text{static}}\sqrt{\frac{\cdot 1+\dfrac{v\cdot}{c\cdot}\cdot c}{\cdot 1\cdot\dfrac{v\cdot}{c\cdot}\cdot c}}=f_{\text{static}}\sqrt{\frac{c+v}{c\cdot v}}$$

47. $\dfrac{8x^4}{2x^3}=\dfrac{\cancel{2x^3}\cdot 4x}{\cancel{2x^3}}=4x$

49. $\dfrac{\cdot 30x^6}{6x^4}=\dfrac{\cancel{6x^4}\cdot\left(\cdot 5x^2\right)}{\cancel{6x^4}}=\cdot 5x^2$

51. $\dfrac{\cdot 30x}{10x^3}=\dfrac{\cancel{10x}\cdot(\cdot 3)}{\cancel{10x}\cdot x^2}=\dfrac{\cdot 3}{x^2}$

53. $x^2(5x+5)=5x^3+5x^2$

55. $3x^4(3x-5)=9x^5-15x^4$

57. $6x^2+7x-3=6x^2+9x-2x-3$
$$=3x(2x+3)-(2x+3)=(2x+3)(3x-1)$$

59. $\left(6x^3+25x^2+2x-8\right)-\left(6x^3+24x^2\right)=6x^3+25x^2+2x-8-6x^3-24x^2=x^2+2x-8$

61. Multiply the numerator and denominator by the LCD which is $12N$:

$$APR = \dfrac{\dfrac{24(NM - P)}{N}}{P + \dfrac{NM}{12}} = \dfrac{12N\left|\dfrac{24(NM - P)}{N}\right|}{12N\left|P + \dfrac{NM}{12}\right|} = \dfrac{288(NM - P)}{N(12P + NM)}$$

63. Multiply the numerator and denominator by the LCD which is 6:

$$\dfrac{1}{4 + \dfrac{1}{6}} = \dfrac{6\left|1\right|}{6\left|4 + \dfrac{1}{6}\right|} = \dfrac{6}{24 + 1} = \dfrac{6}{25} \text{ yr}$$

65. Multiply the numerator and denominator by the LCD which is 43:

$$11 + \dfrac{1}{1 + \dfrac{7}{43}} = 11 + \dfrac{43\left|1\right|}{43\left|1 + \dfrac{7}{43}\right|} = 11 + \dfrac{43}{43 + 7} = 11\dfrac{43}{50} \text{ yr}$$

67. Sample answer: A complex fraction contains fractions in the numerator or denominator or both.

69. Sample answer: Method I is usually the preferred method when there is a single numerator and a single denominator in the complex fraction. There are no complex fractions within the complex fraction. (See exercise 73.)

71.

$$2 + \dfrac{a}{2 + \dfrac{2}{2 + a}} = 2 + \dfrac{a}{2 \cdot \dfrac{(2+a)}{(2+a)} + \dfrac{2}{2+a}} = 2 + \dfrac{a}{\dfrac{4 + 2a + 2}{2+a}} = 2 + \dfrac{a}{\dfrac{2a + 6}{2+a}} = 2 + a \cdot \dfrac{2+a}{2a+6}$$

$$= 2 + \dfrac{2a + a^2}{2a + 6} = 2\dfrac{(2a+6)}{(2a+6)} + \dfrac{2a + a^2}{2a+6} = \dfrac{4a + 12 + 2a + a^2}{2a+6} = \dfrac{a^2 + 6a + 12}{2(a+3)}$$

73. Multiply the numerator and denominator by the LCD which is $(x+2)(x-2)$:

$$\dfrac{\dfrac{x}{x+2} + x}{1 - \dfrac{5}{x^2 - 4}} = \dfrac{(x+2)(x-2)\left|\dfrac{x}{x+2} + x\right|}{(x+2)(x-2)\left|1 - \dfrac{5}{(x+2)(x-2)}\right|} = \dfrac{x(x-2) + x(x+2)(x-2)}{(x+2)(x-2) - 5} = \dfrac{x^2 - 2x + x^3 - 4x}{x^2 - 4 - 5}$$

$$= \dfrac{x^3 + x^2 - 6x}{x^2 - 9} = \dfrac{x\cancel{(x+3)}(x-2)}{\cancel{(x+3)}(x-3)} = \dfrac{x(x-2)}{x-3}$$

75. Multiply the numerator and denominator by the LCD which is $4ab$:

$$\dfrac{\dfrac{2}{b} - \dfrac{3}{a}}{\dfrac{1}{2b} + \dfrac{3}{4a}} = \dfrac{4ab\left|\dfrac{2}{b} - \dfrac{3}{a}\right|}{4ab\left|\dfrac{1}{2b} + \dfrac{3}{4a}\right|} = \dfrac{8a - 12b}{2a + 3b} = \dfrac{4(2a - 3b)}{2a + 3b}$$

77. Multiply the numerator and denominator by the LCD which is x^3:

$$\frac{x-\frac{1}{x^3}}{x-\frac{1}{x^2}} = \frac{x^3\left|x-\frac{1}{x^3}\right|}{x^3\left|x-\frac{1}{x^2}\right|} = \frac{x^4-1}{x^4-x} = \frac{(x^2+1)(x^2-1)}{x(x^3-1)} = \frac{(x^2+1)(x+1)(x-1)}{x(x-1)(x^2+x+1)} = \frac{(x^2+1)(x+1)}{x(x^2+x+1)}$$

79. Multiply the numerator and denominator by the LCD which is $x-4$:

$$\frac{x(x-4)^{-1}+x}{x(4-x)^{-1}-x} = \frac{\frac{x}{x-4}+x}{\frac{x}{4-x}-x} = \frac{(x-4)\left|\frac{x}{x-4}+x\right|}{(x-4)\left|\frac{x}{-(x-4)}-x\right|} = \frac{x+x(x-4)}{-x-x(x-4)} = \frac{x+x^2-4x}{-x-x^2+4x} = \frac{x^2-3x}{-x^2+3x}$$

$$= \frac{x^2-3x}{-(x^2-3x)} = -1$$

6.5   Division of Polynomials and Synthetic Division

Problems 6.5

1. **a.** $\dfrac{24x^7-18x^5-12x^3}{-6x^3} = \dfrac{24x^7}{-6x^3} - \dfrac{18x^5}{-6x^3} - \dfrac{12x^3}{-6x^3} = -4x^4+3x^2+2$

b. $\dfrac{16x^4-4x^3+8x^2-16x+40}{8x^2} = \dfrac{16x^4}{8x^2} - \dfrac{4x^3}{8x^2} + \dfrac{8x^2}{8x^2} - \dfrac{16x}{8x^2} + \dfrac{40}{8x^2} = 2x^2 - \dfrac{x}{2} + 1 - \dfrac{2}{x} + \dfrac{5}{x^2}$

2.

$$
\begin{array}{r}
x+6 \\
x^2-2x+3\overline{)\ x^3+4x^2-15x} \\
\underline{-(x^3-2x^2+\ 3x)} \\
6x^2-18x \\
\underline{-(6x^2-12x+18)} \\
-6x-18
\end{array}
$$

Quotient: $x+6$ R: $-6x-18$

3.

$$
\begin{array}{r}
2x^2+2x+3 \\
3x-3\overline{)\ 6x^3+0x^2+3x-9} \\
\underline{-(6x^3-6x^2)} \\
6x^2+3x \\
\underline{-(6x^2-6x)} \\
9x-9 \\
\underline{-(9x-9)} \\
0
\end{array}
$$

Quotient: $2x^2+2x+3$

4.

$$
\begin{array}{r}
2x^2 - 7x + 3 \\
x+3{\overline{\smash{\big)}\,2x^3 -\ x^2 - 18x + 9}} \\
\underline{-(2x^3 + 6x^2)} \\
-7x^2 - 18x \\
\underline{-(-7x^2 - 21x)} \\
3x + 9 \\
\underline{-(3x + 9)} \\
0
\end{array}
$$

The factorization is:

$$(x+3)(2x^2 - 7x + 3) = (x+3)(2x-1)(x-3)$$

5.

$$
\begin{array}{r}
2{\overline{\smash{\big)}\,3 - 2\ \ 0 -\ 9 +\ 1}} \\
\underline{+6 + 8 + 16 + 14} \\
3 + 4 + 8 +\ 7 + 15
\end{array}
$$

Quotient: $3x^3 + 4x^2 + 8x + 7$ R: 15

$$3x^3 + 4x^2 + 8x + 7 + \dfrac{15}{x-2}$$

6.

$$
\begin{array}{r}
1{\overline{\smash{\big)}\,2 + 1 - 35 - 16 + 48}} \\
\underline{+ 2 +\ \ 3 - 32 - 48} \\
2 + 3 - 32 - 48 \quad\ \ 0
\end{array}
$$

Since the remainder is zero, 1 is a solution.

Exercises 6.5

1. $\dfrac{3x^3 + 9x^2 - 6x}{3x} = \dfrac{3x^3}{3x} + \dfrac{9x^2}{3x} - \dfrac{6x}{3x} = x^2 + 3x - 2$

3. $\dfrac{10x^3 - 5x^2 + 15x}{-5x} = \dfrac{10x^3}{-5x} - \dfrac{5x^2}{-5x} + \dfrac{15x}{-5x} = -2x^2 + x - 3$

5. $\dfrac{8y^4 - 32y^3 + 12y^2}{-4y^2} = \dfrac{8y^4}{-4y^2} - \dfrac{32y^3}{-4y^2} + \dfrac{12y^2}{-4y^2} = -2y^2 + 8y - 3$

7. $\dfrac{10x^5 + 8x^4 - 16x^3 + 6x^2}{2x^3} = \dfrac{10x^5}{2x^3} + \dfrac{8x^4}{2x^3} - \dfrac{16x^3}{2x^3} + \dfrac{6x^2}{2x^3} = 5x^2 + 4x - 8 + \dfrac{3}{x}$

9. $\dfrac{15x^3y^2 - 10x^2y + 15x}{5x^2y} = \dfrac{15x^3y^2}{5x^2y} - \dfrac{10x^2y}{5x^2y} + \dfrac{15x}{5x^2y} = 3xy - 2 + \dfrac{3}{xy}$

11.

$$
\begin{array}{r}
x + 3 \\
x+2{\overline{\smash{\big)}\,x^2 + 5x + 6}} \\
\underline{-(x^2 + 2x)} \\
3x + 6 \\
\underline{-(3x + 6)} \\
0
\end{array}
$$

Quotient: $x + 3$

13.

$$
\begin{array}{r}
y + 5 \\
y-2{\overline{\smash{\big)}\,y^2 + 3y - 10}} \\
\underline{-(y^2 - 2y)} \\
5y - 10 \\
\underline{-(5y - 10)} \\
0
\end{array}
$$

Quotient: $y + 5$

15.

$$2x+2 \overline{)\begin{array}{r} x^2-x-1 \\ 2x^3+0x^2-4x-2 \end{array}}$$

$$\begin{array}{r} -(2x^3+2x^2) \\ \hline -2x^2-4x \\ -(-2x^2-2x) \\ \hline -2x-2 \\ -(-2x-2) \\ \hline 0 \end{array}$$

Quotient: x^2-x-1

17.

$$3x-1 \overline{)\begin{array}{r} x^2+5x+6 \\ 3x^3+14x^2+13x-6 \end{array}}$$

$$\begin{array}{r} -(3x^3-x^2) \\ \hline 15x^2+13x \\ -(15x^2-5x) \\ \hline 18x-6 \\ -(18x-6) \\ \hline 0 \end{array}$$

Quotient: x^2+5x+6

19.

$$2x-3 \overline{)\begin{array}{r} x^2-2x-8 \\ 2x^3-7x^2-10x+24 \end{array}}$$

$$\begin{array}{r} -(2x^3-3x^2) \\ \hline -4x^2-10x \\ -(-4x^2+6x) \\ \hline -16x+24 \\ -(-16x+24) \\ \hline 0 \end{array}$$

Quotient: x^2-2x-8

21.

$$2x-1 \overline{)\begin{array}{r} x^2+4x+3 \\ 2x^3+7x^2+2x-2 \end{array}}$$

$$\begin{array}{r} -(2x^3-x^2) \\ \hline 8x^2+2x \\ -(8x^2-4x) \\ \hline 6x-2 \\ -(6x-3) \\ \hline 1 \end{array}$$

Quotient: x^2+4x+3 R: 1

23.

$$y^2+y+1 \overline{)\begin{array}{r} y^2-y-1 \\ y^4+0y^3-y^2-2y-1 \end{array}}$$

$$\begin{array}{r} -(y^4+y^3+y^2) \\ \hline -y^3-2y^2-2y \\ -(-y^3-y^2-y) \\ \hline -y^2-y-1 \\ -(-y^2-y-1) \\ \hline 0 \end{array}$$

Quotient: y^2-y-1

25.

$$2x-3 \overline{)\begin{array}{r} 4x^2+3x+7 \\ 8x^3-6x^2+5x-9 \end{array}}$$

$$\begin{array}{r} -(8x^3-12x^2) \\ \hline 6x^2+5x \\ -(6x^2-9x) \\ \hline 14x-9 \\ -(14x-21) \\ \hline 12 \end{array}$$

Quotient: $4x^2+3x+7$ R: 12

27.

$$x+2 \overline{)\begin{array}{r} x^2-2x+4 \\ x^3+0x^2+0x+8 \end{array}}$$

$$\begin{array}{r} -(x^3+2x^2) \\ \hline -2x^2+0x \\ -(-2x^2-4x) \\ \hline 4x+8 \\ -(4x+8) \\ \hline 0 \end{array}$$

Quotient: x^2-2x+4

29.

$$2y-4 \overline{)\begin{array}{r} 4y^2+8y+16 \\ 8y^3+0y^2+0y-64 \end{array}}$$

$$\begin{array}{r} -(8y^3-16y^2) \\ \hline 16y^2+0y \\ -(16y^2-32y) \\ \hline 32y-64 \\ -(32y-64) \\ \hline 0 \end{array}$$

Quotient: $4y^2+8y+16$

31.

$$a^2 + 2a + 1 \overline{\smash{)}\, \begin{array}{l} a^2 - 2a - 1 \\ a^4 + 0a^3 - 4a^2 - 4a - 1 \end{array}}$$

$$\underline{-(a^4 + 2a^3 + \ a^2)}$$
$$-2a^3 - 5a^2 - 4a$$
$$\underline{-(-2a^3 - 4a^2 - 2a)}$$
$$-a^2 - 2a - 1$$
$$\underline{-(-a^2 - 2a - 1)}$$
$$0$$

Quotient: $a^2 - 2a - 1$

33.

$$x^2 - 2x + 5 \overline{\smash{)}\, \begin{array}{l} x^3 + 2x^2 - x \\ x^5 + 0x^4 + 0x^3 + 12x^2 - 5x \end{array}}$$

$$\underline{-(x^5 - 2x^4 + 5x^3)}$$
$$2x^4 - 5x^3 + 12x^2$$
$$\underline{-(2x^4 - 4x^3 + 10x^2)}$$
$$-x^3 + 2x^2 - 5x$$
$$\underline{-(-x^3 + 2x^2 - 5x)}$$
$$0$$

Quotient: $x^3 + 2x^2 - x$

35.

$$2x + 5 \overline{\smash{)}\, \begin{array}{l} 2x^3 - 3x^2 + x - 4 \\ 4x^4 + 4x^3 - 13x^2 - 3x - 21 \end{array}}$$

$$\underline{-(4x^4 + 10x^3)}$$
$$-6x^3 - 13x^2$$
$$\underline{-(-6x^3 - 15x^2)}$$
$$2x^2 - 3x$$
$$\underline{-(2x^2 + 5x)}$$
$$-8x - 21$$
$$\underline{-(-8x - 20)}$$
$$-1$$

Quotient: $2x^3 - 3x^2 + x - 4$ R: -1

37.

$$x + 1 \overline{\smash{)}\, \begin{array}{l} x^2 - 5x + 6 \\ x^3 - 4x^2 + \ x + 6 \end{array}}$$

$$\underline{-(x^3 + \ x^2)}$$
$$-5x^2 + \ x$$
$$\underline{-(-5x^2 - 5x)}$$
$$6x + 6$$
$$\underline{-(6x + 6)}$$
$$0$$

The factorization is:
$$(x+1)(x^2 - 5x + 6) = (x+1)(x-3)(x-2)$$

39.

$$x^2 - 4x + 4 \overline{\smash{)}\, \begin{array}{l} x^2 - 1 \\ x^4 - 4x^3 + 3x^2 + 4x - 4 \end{array}}$$

$$\underline{-(x^4 - 4x^3 + 4x^2)}$$
$$-x^2 + 4x - 4$$
$$\underline{-(-x^2 + 4x - 4)}$$
$$0$$

The factorization is:
$$(x^2 - 4x + 4)(x^2 - 1)$$
$$= (x-2)(x-2)(x+1)(x-1)$$

41.

$$x^2 - 3x + 7 \overline{\smash{)}\, \begin{array}{l} x^2 + 9x + 20 \\ x^4 + 6x^3 + \ 0x^2 + 3x + 140 \end{array}}$$

$$\underline{-(x^4 - 3x^3 + \ 7x^2)}$$
$$9x^3 - 7x^2 + 3x$$
$$\underline{-(9x^3 - 27x^2 + 63x)}$$
$$20x^2 - 60x + 140$$
$$\underline{-(20x^2 - 60x + 140)}$$
$$0$$

The factorization is:
$$(x^2 - 3x + 7)(x^2 + 9x + 20)$$
$$= (x^2 - 3x + 7)(x+5)(x+4)$$

43.

$$3 \overline{\smash{)}\, 1 \quad 0 - 8 - 3}$$
$$\underline{ + 3 + 9 + 3}$$
$$1 + 3 + 1 + 0$$

Quotient: $v^2 + 3v + 1$

45.

$$2 \overline{\smash{)}\, 1 + 4 - 7 + 5}$$
$$\underline{ + 2 + 12 + 10}$$
$$1 + 6 + \ 5 + 15$$

Quotient: $x^2 + 6x + 5$ R: 15

47.
$$-6\overline{)1\quad 0 - 32 + 24}$$
$$\underline{\quad -6 + 36 - 24}$$
$$1 - 6 + \;\; 4 \quad\;\; 0$$
Quotient: $z^2 - 6z + 4$

49.
$$4\overline{)3\quad 0 - 41 - 13 - \;\;8}$$
$$\underline{\quad +12 + 48 + 28 + 60}$$
$$3 + 12 + \;\;7 + 15 + 52$$
Quotient: $3y^3 + 12y^2 + 7y + 15$ R: 52

51.
$$6\overline{)2 - 13 + 6 + 5 - 30}$$
$$\underline{\quad +12 - \;6 + 0 + 30}$$
$$2 - \;\;1 + 0 + 5 + \;\;0$$
Quotient: $2y^3 - y^2 + 5$

53.
$$4\overline{)1 + 6 - \;\;6 - 136}$$
$$\underline{\quad + 4 + 40 + 136}$$
$$1 + 10 + 34 + \;\;0$$
Since the remainder is zero, 4 is a solution.

55.
$$-4\overline{)5 + 18 - 1 + 28}$$
$$\underline{\quad -20 + 8 - 28}$$
$$5 - \;\;2 + 7 + \;\;0$$
Since the remainder is zero, –4 is a solution.

57.
$$5\overline{)3 - 14 - 7 + 21 - 55}$$
$$\underline{\quad +15 + 5 - 10 + 55}$$
$$3 + \;\;1 - 2 + 11 + \;\;0$$
Since the remainder is zero, 5 is a solution.

59.
$$-1\overline{)1 + 1 + 2 + 5 - \;\;2 - 5}$$
$$\underline{\quad -1 + 0 - 2 - 3 + 5}$$
$$1 + 0 + 2 + 3 - 5 + 0$$
Since the remainder is zero, –1 is a solution.

61.
$$\overline{C} = \frac{C}{x} = \frac{500 + 4x}{x} = \frac{500}{x} + \frac{4x}{x} = \frac{500}{x} + 4$$

63.
$$4x + 8 = 6x$$
$$8 = 2x$$
$$4 = x$$

65.
$$x(x+2) - (x-3)(x-4) = 4x + 3$$
$$x^2 + 2x - (x^2 - 7x + 12) = 4x + 3$$
$$x^2 + 2x - x^2 + 7x - 12 = 4x + 3$$
$$9x - 12 = 4x + 3$$
$$5x = 15$$
$$x = 3$$

67. $a = 1;\; b = 1, 2, 3, 6$

Factors: $x + 1,\; x + 2,\; x + 3,\; x + 6$

69. Answers will vary.

71. Sample answer: Perform synthetic division by k and determine whether the remainder is zero. The alternate method would be to evaluate the function at k and determine whether the value is zero.

73.
$$3\overline{)1 - 3\quad 0 + 3}$$
$$\underline{\quad + 3 + 0 + 0}$$
$$1 + 0 + 0 + 3$$
Quotient: x^2 R: 3

75.
$$2\overline{)2 + 5 - \;\;8 - 20}$$
$$\underline{\quad + 4 + 18 + 20}$$
$$2 + 9 + 10 + \;\;0$$
Since the remainder is zero, 2 is a solution.

77.

$$x+3 \overline{)\,2x^3 - x^2 - 18x + 9\,}$$

quotient: $2x^2 - 7x + 3$

$\quad -(2x^3 + 6x^2)$

$\qquad\qquad -7x^2 - 18x$

$\qquad\qquad -(-7x^2 - 21x)$

$\qquad\qquad\qquad 3x + 9$

$\qquad\qquad\qquad -(3x + 9)$

$\qquad\qquad\qquad\qquad 0$

The factorization is:

$$(x+3)(2x^2 - 7x + 3) = (x+3)(2x-1)(x-3)$$

79.

$$x-2 \overline{)\,6x^3 + 0x^2 + 3x - 9\,}$$

quotient: $6x^2 + 12x + 27$

$\quad -(6x^3 - 12x^2)$

$\qquad\qquad 12x^2 + 3x$

$\qquad\qquad -(12x^2 - 24x)$

$\qquad\qquad\qquad 27x - 9$

$\qquad\qquad\qquad -(27x - 54)$

$\qquad\qquad\qquad\qquad 45$

Quotient: $6x^2 + 12x + 27$ R: 45

81.

$$\frac{24x^5 - 18x^4 + 12x^3}{6x^2} = \frac{24x^5}{6x^2} - \frac{18x^4}{6x^2} + \frac{12x^3}{6x^2} = 4x^3 - 3x^2 + 2x$$

6.6   Equations Involving Rational Expressions

Problems 6.6

1. **a.** Multiply both sides by the LCD $x(2x+1)$:

$$\frac{10}{2x+1} = \frac{5}{x}$$

$$\frac{x(2x+1)}{1}\cdot\frac{10}{2x+1} = \frac{x(2x+1)}{1}\cdot\frac{5}{x}$$

$$10x = 10x + 5$$

$$20x = 5$$

$$x = \frac{5}{20} = \frac{1}{4}$$

b. Multiply both sides by the LCD $(x+4)(x-3)$:

$$\frac{6}{x+4} = \frac{5}{x-3}$$

$$\frac{(x+4)(x-3)}{1}\cdot\frac{6}{x+4} = \frac{(x+4)(x-3)}{1}\cdot\frac{5}{x-3}$$

$$6x - 18 = 5x + 20$$

$$x = 38$$

2. **a.** Multiply both sides by the LCD $(x-6)(x-4)(x-2)$:

$$\frac{1}{x-6} + \frac{1}{x-4} = \frac{6}{(x-6)(x-2)}$$

$$\frac{(x-6)(x-4)(x-2)}{1}\cdot\left(\frac{1}{x-6} + \frac{1}{x-4}\right) = \frac{(x-6)(x-4)(x-2)}{1}\cdot\frac{6}{(x-6)(x-2)}$$

$$(x-4)(x-2) + (x-6)(x-2) = 6(x-4)$$

$$(x^2 - 6x + 8) + (x^2 - 8x + 12) = 6x - 24$$

$$x^2 - 6x + 8 + x^2 + 8x - 12 = 6x - 24$$

$$2x - 4 = 6x - 24$$

$$20 = 4x$$

$$5 = x$$

b. Multiply both sides by the LCD $(x-4)(x+3)$:

$$\frac{x}{x-4}-\frac{x-5}{x+3}=\frac{12x+16}{x^2-x-12}$$

$$\frac{(x-4)(x+3)}{1}\cdot\left|\frac{x}{x-4}-\frac{x-5}{x+3}\right|=\frac{(x-4)(x+3)}{1}\cdot\frac{12x+16}{(x-4)(x+3)}$$

$$x(x+3)-(x-4)(x-5)=12x+16$$

$$\left(x^2+3x\right)-\left(x^2-9x+20\right)=12x+16$$

$$x^2+3x-x^2+9x-20=12x+16$$

$$12x-20=12x+16$$

$$-20=16$$

This is a contradiction, so there is no solution.

3. Multiply both sides by the LCD $(x+1)(x-1)$:

$$3-\frac{4}{x^2-1}=\frac{-2}{x-1}$$

$$\frac{(x+1)(x-1)}{1}\cdot\left|3-\frac{4}{(x+1)(x-1)}\right|=\frac{(x+1)(x-1)}{1}\cdot\frac{-2}{x-1}$$

$$3(x+1)(x-1)-4=-2(x+1)$$

$$3x^2-3-4=-2x-2$$

$$3x^2+2x-5=0$$

$$(3x+5)(x-1)=0$$

$$3x+5=0 \quad \text{or } x-1=0$$

$$3x=-5 \quad \text{or} \quad x=1$$

$$x=-\frac{5}{3} \quad \text{or} \quad x=1$$

1 makes the denominator $x-1=0$ and thus is extraneous. The solution is $-\frac{5}{3}$.

4. Multiply both sides by the LCD $x(x+8)(x-8)$:

$$\frac{x-7}{x^2-8x}=\frac{2}{x^2-64}$$

$$\frac{x(x+8)(x-8)}{1}\cdot\left|\frac{x-7}{x(x-8)}\right|=\frac{x(x+8)(x-8)}{1}\cdot\frac{2}{(x+8)(x-8)}$$

$$(x+8)(x-7)=2x$$

$$x^2+x-56=2x$$

$$x^2-x-56=0$$

$$(x-8)(x+7)=0$$

$$x-8=0 \text{ or } x+7=0$$

$$x=8 \text{ or} \quad x=-7$$

8 makes the denominator $x^2-64=0$ and thus is extraneous. The solution is –7.

5. Multiply both sides by the LCD $(x+1)(x-4)$:

$$2(x+1)^{-1}+4(x-4)^{-1}=1$$

$$\frac{2}{x+1}+\frac{4}{x-4}=1$$

$$\frac{(x+1)(x-4)}{1}\left[\frac{2}{x+1}+\frac{4}{x-4}\right]=\frac{(x+1)(x-4)}{1}$$

$$2(x-4)+4(x+1)=(x+1)(x-4)$$

$$2x-8+4x+4=x^2-3x-4$$

$$6x-4=x^2-3x-4$$

$$0=x^2-9x$$

$$0=x(x-9)$$

$$x=0 \text{ or } x-9=0$$

$$x=0 \text{ or } \quad x=9$$

The solutions are 0 and 9.

6.

$$\frac{1\text{ in}}{25\text{ mi}}=\frac{3.5\text{ in}}{x}$$

$$1x=25(3.5)$$

$$x=87.5\text{ mi}$$

The distance between the cities is 87.5 mi.

Exercises 6.6

1. Multiply both sides by the LCD 6:

$$\frac{x}{3}+\frac{x}{6}=3$$

$$6\left[\frac{x}{3}+\frac{x}{6}\right]=6\cdot 3$$

$$2x+x=18$$

$$3x=18$$

$$x=6$$

3. Multiply both sides by the LCD 10:

$$\frac{x}{5}-\frac{3x}{10}=\frac{1}{2}$$

$$10\left[\frac{x}{5}-\frac{3x}{10}\right]=10\cdot\frac{1}{2}$$

$$2x-3x=5$$

$$-x=5$$

$$x=-5$$

5. Multiply both sides by the LCD $3y$:

$$\frac{1}{y}+\frac{4}{3y}=7$$

$$3y\left[\frac{1}{y}+\frac{4}{3y}\right]=3y\cdot 7$$

$$3+4=21y$$

$$7=21y$$

$$\frac{1}{3}=y$$

7. Multiply both sides by the LCD $(y-8)(y-2)$:

$$\frac{2}{y-8}=\frac{1}{y-2}$$

$$(y-8)(y-2)\cdot\frac{2}{y-8}=(y-8)(y-2)\cdot\frac{1}{y-2}$$

$$2(y-2)=1(y-8)$$

$$2y-4=y-8$$

$$y=-4$$

9. Multiply both sides by the LCD $(3z+4)(5z-6)$:

$$\frac{3}{3z+4} = \frac{2}{5z-6}$$

$$(3z+4)(5z-6)\cdot\frac{3}{3z+4} = (3z+4)(5z-6)\cdot\frac{2}{5z-6}$$

$$3(5z-6) = 2(3z+4)$$

$$15z-18 = 6z+8$$

$$9z = 26$$

$$z = \frac{26}{9}$$

11. Multiply both sides by the LCD $(2x+1)(3x-1)$:

$$\frac{-2}{2x+1} = \frac{3}{3x-1}$$

$$(2x+1)(3x-1)\cdot\frac{-2}{2x+1} = (2x+1)(3x-1)\cdot\frac{3}{3x-1}$$

$$-2(3x-1) = 3(2x+1)$$

$$-6x+2 = 6x+3$$

$$-12x = 1$$

$$x = -\frac{1}{12}$$

13. Multiply both sides by the LCD $(x+1)(2x-1)$:

$$\frac{-1}{x+1} = \frac{-2}{2x-1}$$

$$(x+1)(2x-1)\cdot\frac{-1}{x+1} = (x+1)(2x-1)\cdot\frac{-2}{2x-1}$$

$$-1(2x-1) = -2(x+1)$$

$$-2x+1 = -2x-2$$

$$1 = -2$$

This is a contradiction, so there is no solution.

15. Multiply both sides by the LCD $2(3x+1)$:

$$\frac{2}{3x+1} = \frac{5}{6x+2}$$

$$2(3x+1)\cdot\frac{2}{3x+1} = 2(3x+1)\cdot\frac{5}{2(3x+1)}$$

$$2\cdot2 = 5$$

$$4 = 5$$

This is a contradiction, so there is no solution.

17. Multiply both sides by the LCD $(x+2)(x-2)$:

$$\frac{2}{x^2-4}+\frac{5}{x+2}=\frac{7}{x-2}$$

$$(x+2)(x-2)\left[\frac{2}{(x+2)(x-2)}+\frac{5}{x+2}\right]=(x+2)(x-2)\frac{7}{x-2}$$

$$2+5(x-2)=7(x+2)$$

$$2+5x-10=7x+14$$

$$5x-8=7x+14$$

$$-2x=22$$

$$x=-11$$

19. Multiply both sides by the LCD $(t-1)(t-2)$:

$$\frac{t+2}{t^2-3t+2}=\frac{3}{t-1}-\frac{1}{t-2}$$

$$(t-1)(t-2)\frac{t+2}{(t-1)(t-2)}=(t-1)(t-2)\left[\frac{3}{t-1}-\frac{1}{t-2}\right]$$

$$t+2=3(t-2)-1(t-1)$$

$$t+2=3t-6-t+1$$

$$t+2=2t-5$$

$$7=t$$

21. Multiply both sides by the LCD $(x+1)(x-1)$:

$$\frac{x^2}{x^2-1}=1+\frac{1}{x+1}$$

$$(x+1)(x-1)\frac{x^2}{(x+1)(x-1)}=(x+1)(x-1)\left[1+\frac{1}{x+1}\right]$$

$$x^2=(x+1)(x-1)+1(x-1)$$

$$x^2=x^2-1+x-1$$

$$2=x$$

23. Multiply both sides by the LCD $(x-3)(x-1)(x+1)$:

$$\frac{1}{x^2-4x+3}+\frac{1}{x^2-2x-3}=\frac{1}{x^2-1}$$

$$(x-3)(x-1)(x+1)\left[\frac{1}{(x-3)(x-1)}+\frac{1}{(x-3)(x+1)}\right]=(x-3)(x-1)(x+1)\frac{1}{(x+1)(x-1)}$$

$$x+1+x-1=x-3$$

$$2x=x-3$$

$$x=-3$$

25. Multiply both sides by the LCD $(3x+1)(x+1)(x+2)$:

$$\frac{x+2}{3x^2+4x+1}=\frac{x+1}{3x^2+7x+2}$$

$$(3x+1)(x+1)(x+2)\frac{x+2}{(3x+1)(x+1)}=(3x+1)(x+1)(x+2)\frac{x+1}{(3x+1)(x+2)}$$

$$(x+2)^2=(x+1)^2$$

$$x^2+4x+4=x^2+2x+1$$

$$2x=-3$$

$$x=-\frac{3}{2}$$

27. Multiply both sides by the LCD $(2z-1)(z+3)$:

$$\frac{2z+13}{2z^2+5z-3}+\frac{3}{z+3}=\frac{4}{2z-1}$$

$$(2z-1)(z+3)\left[\frac{2z+13}{(2z-1)(z+3)}+\frac{3}{z+3}\right]=(2z-1)(z+3)\frac{4}{2z-1}$$

$$2z+13+3(2z-1)=4(z+3)$$

$$2z+13+6z-3=4z+12$$

$$8z+10=4z+12$$

$$4z=2$$

$$z=\frac{1}{2}$$

$\frac{1}{2}$ makes the denominator $2z-1=0$ and thus is extraneous. There is no solution.

29. Multiply both sides by the LCD $(5x+1)(x-1)$:

$$\frac{3-x}{5x^2-4x-1}+\frac{2}{5x+1}=\frac{1}{x-1}$$

$$(5x+1)(x-1)\left[\frac{3-x}{(5x+1)(x-1)}+\frac{2}{5x+1}\right]=(5x+1)(x-1)\frac{1}{x-1}$$

$$3-x+2(x-1)=1(5x+1)$$

$$3-x+2x-2=5x+1$$

$$x+1=5x+1$$

$$-4x=0$$

$$x=0$$

31. Multiply both sides by the LCD x:

$$4x^{-1} + 2 = 7$$

$$\frac{4}{x} = 5$$

$$x \cdot \frac{4}{x} = x \cdot 5$$

$$4 = 5x$$

$$\frac{4}{5} = x$$

33. Multiply both sides by the LCD $x(x+1)$:

$$4x^{-1} + 6x^{-1} = 15(x+1)^{-1}$$

$$\frac{4}{x} + \frac{6}{x} = \frac{15}{x+1}$$

$$x(x+1) \cdot \frac{10}{x} = x(x+1) \cdot \frac{15}{x+1}$$

$$10(x+1) = 15x$$

$$10x + 10 = 15x$$

$$10 = 5x$$

$$2 = x$$

35. Multiply both sides by the LCD $(x-8)(x-2)$:

$$2(x \cdot 8)^{-1} = (x \cdot 2)^{-1}$$

$$\frac{2}{x \cdot 8} = \frac{1}{x \cdot 2}$$

$$(x \cdot 8)(x \cdot 2) \cdot \frac{2}{x \cdot 8} = (x \cdot 8)(x \cdot 2) \cdot \frac{1}{x \cdot 2}$$

$$2(x \cdot 2) = 1(x \cdot 8)$$

$$2x \cdot 4 = x \cdot 8$$

$$x = \cdot 4$$

37.

$$\frac{2}{y-8} = \frac{1}{y-2}$$

$$2(y-2) = 1(y-8)$$

$$2y - 4 = y - 8$$

$$y = -4$$

39.

$$\frac{3}{3z+4} = \frac{2}{5z-6}$$

$$3(5z-6) = 2(3z+4)$$

$$15z - 18 = 6z + 8$$

$$9z = 26$$

$$z = \frac{26}{9}$$

41.

$$\frac{-2}{2x+1} = \frac{3}{3x-1}$$

$$-2(3x-1) = 3(2x+1)$$

$$-6x + 2 = 6x + 3$$

$$-12x = 1$$

$$x = -\frac{1}{12}$$

43.

$$\frac{-1}{x+1} = \frac{-2}{2x-1}$$

$$-1(2x-1) = -2(x+1)$$

$$-2x + 1 = -2x - 2$$

$$1 = -2$$

This is a contradiction, so there is no solution.

45.
$$\frac{2}{3x+1} = \frac{5}{6x+2}$$
$$2(6x+2) = 5(3x+1)$$
$$12x+4 = 15x+5$$
$$-3x = 1$$
$$x = -\frac{1}{3}$$
$-\frac{1}{3}$ makes the denominator $3x+1=0$ and thus is extraneous. There is no solution.

47. Let t = the time to process 20 rolls of film
$$\frac{2 \text{ hr}}{9 \text{ rolls}} = \frac{t}{20 \text{ rolls}}$$
$$9t = 2 \cdot 20$$
$$9t = 40$$
$$t = \frac{40}{9} = 4\frac{4}{9} \text{ hr}$$

49. **a.** Let t = the driving hours for the trip
$$\frac{7 \text{ hr}}{425 \text{ mi}} = \frac{t}{1000 \text{ mi}}$$
$$425t = 7 \cdot 1000$$
$$425t = 7000$$
$$t \approx 16.5 \text{ hr}$$

b. Let t = the number of days for the trip
$$\frac{1 \text{ day}}{425 \text{ mi}} = \frac{t}{1000 \text{ mi}}$$
$$425t = 1 \cdot 1000$$
$$425t = 1000$$
$$t \approx 2.4 \text{ days}$$

51. Let x = the number of consecutive hits
$$\frac{5+x}{20+x} = 0.350$$
$$5+x = 0.350(20+x)$$
$$5+x = 7+0.350x$$
$$0.650x = 2$$
$$x \approx 3$$
The softball player needs approximately 3 consecutive hits.

53. Let x = first odd integer
$x+2$ = second consecutive odd integer
$x+4$ = third consecutive odd integer
$$x+x+2+x+4 = 69$$
$$3x+6 = 69$$
$$3x = 63$$
$$x = 21$$
$$x+2 = 23$$
$$x+4 = 25$$
The odd integers are 21, 23, and 25.

55.

Percent of Salt	Amount to be mixed	Amount of Pure Salt
20% or 0.20	x	$0.20x$
15% or 0.15	40	0.15(40)
18% or 0.18	$x+40$	0.18(x + 40)

$$0.20x + 0.15(40) = 0.18(x+40)$$
$$0.20x + 6 = 0.18x + 7.2$$
$$0.02x = 1.2$$
$$x = 60 \text{ gallons}$$
60 gallons of 20% salt solution must be added.

57. Solve for h:
$$A = \frac{h(b_1+b_2)}{2}$$
$$2A = h(b_1+b_2)$$
$$\frac{2A}{b_1+b_2} = h$$

59. Solve for Q_1:

$$\frac{Q_1}{Q_2 - Q_1} = P$$
$$Q_1 = P(Q_2 - Q_1)$$
$$Q_1 = PQ_2 - PQ_1$$
$$Q_1 + PQ_1 = PQ_2$$
$$Q_1(1 + P) = PQ_2$$
$$Q_1 = \frac{PQ_2}{1 + P}$$

61. Solve for f:

$$\frac{1}{f} = \frac{1}{a} + \frac{1}{b}$$
$$fab\left[\frac{1}{f}\right] = fab\left[\frac{1}{a} + \frac{1}{b}\right]$$
$$ab = fb + fa$$
$$ab = f(b + a)$$
$$\frac{ab}{b + a} = f$$

63. Sample answer: An extraneous root is a solution to the rational equation that causes division by zero when it is substituted back into the equation.

65. Sample answer: The solutions need to be checked because there is the possibility of division by zero because there is a variable in the denominator.

67. Multiply both sides by the LCD $x(x + 8)(x - 8)$:

$$\frac{x - 7}{x^2 - 8x} = \frac{2}{x^2 - 64}$$
$$\frac{x(x + 8)(x - 8)}{1}\left[\frac{x - 7}{x(x - 8)}\right] = \frac{x(x + 8)(x - 8)}{1}\left[\frac{2}{(x + 8)(x - 8)}\right]$$
$$(x + 8)(x - 7) = 2x$$
$$x^2 + x - 56 = 2x$$
$$x^2 - x - 56 = 0$$
$$(x - 8)(x + 7) = 0$$
$$x - 8 = 0 \text{ or } x + 7 = 0$$
$$x = 8 \text{ or } \quad x = -7$$

8 makes the denominator $x^2 - 64 = 0$ and thus is extraneous. The solution is -7.

69.
$$\frac{3}{x} = \frac{5}{x + 2}$$
$$3(x + 2) = 5x$$
$$3x + 6 = 5x$$
$$6 = 2x$$
$$3 = x$$

71.   Multiply both sides by the LCD $(x+3)(x-2)$:

$$4(x+3)^{-1}+3x(x-2)^{-1}=-2$$

$$\frac{4}{x+3}+\frac{3x}{x-2}=-2$$

$$(x+3)(x-2)\left[\frac{4}{x+3}+\frac{3x}{x-2}\right]=(x+3)(x-2)(-2)$$

$$4(x-2)+3x(x+3)=-2\left(x^2+x-6\right)$$

$$4x-8+3x^2+9x=-2x^2-2x+12$$

$$5x^2+15x-20=0$$

$$x^2+3x-4=0$$

$$(x+4)(x-1)=0$$

$$x+4=0 \quad \text{or} \quad x-1=0$$

$$x=-4 \text{ or} \quad x=1$$

73.   Multiply both sides by the LCD $(x-6)(x-4)(x-2)$:

$$\frac{1}{x-6}-\frac{1}{x-4}=\frac{6}{(x-6)(x-2)}$$

$$\frac{(x-6)(x-4)(x-2)}{1}\left[\frac{1}{x-6}-\frac{1}{x-4}\right]=\frac{(x-6)(x-4)(x-2)}{1}\left[\frac{6}{(x-6)(x-2)}\right]$$

$$(x-4)(x-2)-(x-6)(x-2)=6(x-4)$$

$$\left(x^2-6x+8\right)-\left(x^2-8x+12\right)=6x-24$$

$$x^2-6x+8-x^2+8x-12=6x-24$$

$$2x-4=6x-24$$

$$20=4x$$

$$5=x$$

6.7   Applications: Problem Solving

Problems 6.7

1.   Let $x=$ the first odd number
Let $x+2=$ the consecutive odd number

$$\frac{1}{x}+\frac{1}{x+2}=-\frac{8}{15}$$

$$15x(x+2)\left[\frac{1}{x}+\frac{1}{x+2}\right]=15x(x+2)\left[-\frac{8}{15}\right]$$

$$15(x+2)+15x=-8\,x(x+2)$$

$$15x+30+15x=-8x^2-16x$$

$$8x^2+46x+30=0$$

$$4x^2+23x+15=0$$

$$(4x+3)(x+5)=0$$

$$4x+3=0 \quad \text{or} \quad x+5=0$$

$$4x=-3 \quad \text{or} \quad x=-5$$

$$x=-\frac{3}{4} \text{ or} \quad x=-5$$

The integers are -5 and -3.

2. Let t = the time when working together

$$\frac{1}{5}+\frac{1}{10}=\frac{1}{t}$$

$$10t\left(\frac{1}{5}+\frac{1}{10}\right)=10t\cdot\frac{1}{t}$$

$$2t+t=10$$

$$3t=10$$

$$t=\frac{10}{3}=3\frac{1}{3}\ \text{hr}$$

Working together they can do the job in $3\frac{1}{3}$ hr or 3 hr 20 min.

3. Let t = the time with both pipes open

$$\frac{1}{6}+\frac{1}{7}=\frac{1}{t}$$

$$42t\left(\frac{1}{6}+\frac{1}{7}\right)=42t\cdot\frac{1}{t}$$

$$7t+6t=42$$

$$t=42\ \text{hr}$$

With both pipes open the pool will be filled in 42 hr.

4. Let R = the speed of the boat in still water
$R + 18$ = the speed downstream
$R - 18$ = the speed upstream

$$\frac{36}{R+18}=\text{the time downstream}$$

$$\frac{12}{R-18}=\text{the time upstream}$$

$$\frac{36}{R+18}=\frac{12}{R-18}$$

$$36(R-18)=12(R+18)$$

$$36R-648=12R+216$$

$$24R=864$$

$$R=36$$

The speed of the boat in still water is 36 mi/hr.

5. Solve for $\cos(2u)$:

$$\sin^2 u=\frac{1-\cos(2u)}{2}$$

$$2\sin^2 u=1-\cos(2u)$$

$$2\sin^2 u-1=-\cos(2u)$$

$$\cos(2u)=1-2\sin^2 u$$

Exercises 6.7

1. Let x = the integer

$$x+\frac{1}{x}=\frac{65}{8}$$

$$8x\cdot x+\frac{1}{x}=8x\cdot\frac{65}{8}$$

$$8x^2+8=65x$$

$$8x^2-65x+8=0$$

$$(8x-1)(x-8)=0$$

$$8x-1=0\ \text{ or } x-8=0$$

$$8x=1\ \text{ or }\quad x=8$$

$$x=\frac{1}{8}\ \text{ or }\quad x=8$$

The integer is 8.

3. Let x = the number
$2x$ = twice the number

$$\frac{1}{x}+\frac{1}{2x}=\frac{3}{10}$$

$$10x\left(\frac{1}{x}+\frac{1}{2x}\right)=10x\left(\frac{3}{10}\right)$$

$$10+5=3x$$

$$15=3x$$

$$5=x$$

$$2x=10$$

The numbers are 5 and 10.

5. Let x = the first even integer
Let $x + 2$ = the consecutive even integer

$$\frac{1}{x} + \frac{1}{x+2} = \frac{7}{24}$$

$$24x(x+2)\frac{1}{x} + \frac{1}{x+2} = 24x(x+2)\frac{7}{24}$$

$$24(x+2) + 24x = 7x(x+2)$$

$$24x + 48 + 24x = 7x^2 + 14x$$

$$0 = 7x^2 - 34x - 48$$

$$(7x+8)(x-6) = 0$$

$$7x + 8 = 0 \quad \text{or } x - 6 = 0$$

$$7x = -8 \quad \text{or} \quad x = 6$$

$$x = -\frac{8}{7} \quad \text{or} \quad x = 6$$

The integers are 6 and 8.

7. Let x = the numerator
Let $x + 5$ = the denominator

$$\frac{x+3}{x+5+3} = \frac{1}{2}$$

$$\frac{x+3}{x+8} = \frac{1}{2}$$

$$2(x+3) = x+8$$

$$2x+6 = x+8$$

$$x = 2$$

$$x+5 = 7$$

The fraction is $\frac{2}{7}$.

9. Let x = the increase in current liabilities

$$CR = \frac{90,000}{40,000+x} = \frac{3}{2}$$

$$3(40,000+x) = 2(90,000)$$

$$120,000 + 3x = 180,000$$

$$3x = 60,000$$

$$x = 20,000$$

Increase the current liabilities by \$20,000.

11. Let t = the time when working together

$$\frac{1}{3} + \frac{1}{5} = \frac{1}{t}$$

$$15t \left[\frac{1}{3} + \frac{1}{5}\right] = 15t \left[\frac{1}{t}\right]$$

$$5t + 3t = 15$$

$$8t = 15$$

$$t = \frac{15}{8} = 1\frac{7}{8} \text{ hr}$$

Working together they do the job in $1\frac{7}{8}$ hr.

13. Let t = the time when working together

$$\frac{1}{9} + \frac{1}{10} = \frac{1}{t}$$

$$90t \left[\frac{1}{9} + \frac{1}{10}\right] = 90t \left[\frac{1}{t}\right]$$

$$10t + 9t = 90$$

$$19t = 90$$

$$t = \frac{90}{19} = 4\frac{14}{19} \text{ hr}$$

Working together they do the job in $4\frac{14}{19}$ hr.

15. Let t = the time for the first press
Let $2t$ = the time for the second press

$$\frac{1}{t} + \frac{1}{2t} = \frac{1}{2}$$

$$2t \left[\frac{1}{t} + \frac{1}{2t}\right] = 2t \left[\frac{1}{2}\right]$$

$$2 + 1 = t$$

$$3 = t$$

$$t = 3 \text{ hr}$$

$$2t = 6 \text{ hr}$$

The first press takes 3 hr and the second takes 6 hr.

17. Let t = the time with both pipes open

$$\frac{1}{9} + \frac{1}{21} = \frac{1}{t}$$

$$63t\left(\frac{1}{9} + \frac{1}{21}\right) = 63t\left(\frac{1}{t}\right)$$

$$7t + 3t = 63$$

$$4t = 63$$

$$t = \frac{63}{4} = 15\frac{3}{4} \text{ hr}$$

It takes $15\frac{3}{4}$ hr to fill the tank.

19. Let t = the time with both pipes open

$$\frac{1}{7} + \frac{1}{21} = \frac{1}{t}$$

$$21t\left(\frac{1}{7} + \frac{1}{21}\right) = 21t\left(\frac{1}{t}\right)$$

$$3t + t = 21$$

$$4t = 21$$

$$t = \frac{21}{4} = 5\frac{1}{4} \text{ hr}$$

It takes $5\frac{1}{4}$ hr to fill the tank.

21. Let t = the time with both engines

$$\frac{1}{60} + \frac{1}{90} = \frac{1}{t}$$

$$180t\left(\frac{1}{60} + \frac{1}{90}\right) = 180t\left(\frac{1}{t}\right)$$

$$3t + 2t = 180$$

$$5t = 180$$

$$t = 36 \text{ sec}$$

Both engines will burn for 36 sec.

23. Let t = the time with both pipes open

$$\frac{1}{6} - \frac{1}{9} = \frac{1}{t}$$

$$18t\left(\frac{1}{6} - \frac{1}{9}\right) = 18t\left(\frac{1}{t}\right)$$

$$3t - 2t = 18$$

$$t = 18 \text{ hr}$$

It takes 18 hr to empty the tank.

25. Let R = the plane's speed in still air
$R + 30$ = the speed with the wind
$R - 30$ = the speed against the wind

$$\frac{360}{R + 30} = \text{the time with the wind}$$

$$\frac{240}{R - 30} = \text{the time against the wind}$$

$$\frac{360}{R + 30} = \frac{240}{R - 30}$$

$$360(R - 30) = 240(R + 30)$$

$$360R - 10800 = 240R + 7200$$

$$120R = 18,000$$

$$R = 150$$

The plane's speed in still air is 150 mi/hr.

27. Let R = the wind velocity
$120 + R$ = the speed with the wind
$120 - R$ = the speed against the wind

$$\frac{450}{120 + R} = \text{the time with the wind}$$

$$\frac{270}{120 - R} = \text{the time against the wind}$$

$$\frac{450}{120 + R} = \frac{270}{120 - R}$$

$$450(120 - R) = 270(120 + R)$$

$$54,000 - 450R = 32,400 + 270R$$

$$-720R = -21,600$$

$$R = 30$$

The wind velocity is 30 mi/hr.

29. Let R = the speed of the automobile
$R + 100$ = the speed of the plane

$$\frac{200}{R} = \text{the time for the automobile}$$

$$\frac{1000}{R + 100} = \text{the time for the plane}$$

$$\frac{200}{R} = \frac{1000}{R + 100}$$

$$200(R + 100) = 1000R$$

$$200R + 20,000 = 1000R$$

$$20,000 = 800R$$

$$25 = R$$

$$R + 100 = 125$$

The auto's speed is 25 mi/hr and the plane's speed is 125 mi/hr.

31.
$$\frac{x^{-5}}{x^3} = x^{-5-3} = x^{-8} = \frac{1}{x^8}$$

33.
$$\left(x^{-4}\right)^{-5} = x^{20}$$

35.
$$\left(-2xy^2\right)^3 = \left(-2\right)^3 x^3 \left(y^2\right)^3 = -8x^3 y^6$$

37.
$$x^{-9} \cdot x^7 = x^{-9+7} = x^{-2} = \frac{1}{x^2}$$

39.
$$\left|\frac{a^{|4}}{b^3}\right|^2 = \frac{\left(a^{|4}\right)^2}{\left(b^3\right)^2} = \frac{a^{|8}}{b^6} = \frac{1}{a^8 b^6}$$

41. Solve for F:
$$\frac{1}{F} = \frac{1}{f_1} + \frac{1}{f_2}$$
$$Ff_1 f_2 \left|\frac{1}{F}\right. = Ff_1 f_2 \left|\left|\frac{1}{f_1} + \frac{1}{f_2}\right|\right.$$
$$f_1 f_2 = Ff_2 + Ff_1$$
$$f_1 f_2 = F\left(f_2 + f_1\right)$$
$$\frac{f_1 f_2}{f_2 + f_1} = F$$

43. Solve for R:
$$i = \frac{2E}{R + 2r}$$
$$i(R + 2r) = 2E$$
$$iR + 2ir = 2E$$
$$iR = 2E - 2ir$$
$$R = \frac{2E - 2ir}{i}$$

45.
$$d = \frac{24 \cdot 3}{5 + 1} = \frac{72}{6} = 12$$
Adult dosage is 12 tablets per day.

47. Solve for $\cos(2u)$:
$$\cos^2 u = \frac{1 + \cos(2u)}{2}$$
$$2\cos^2 u = 1 + \cos(2u)$$
$$2\cos^2 u - 1 = \cos(2u)$$

49. Sample answer: The sum should equal 1 because they are completing one job together.

51. Let R = the speed of the boat in still water
$R + 18$ = the speed downstream
$R - 18$ = the speed upstream
$\dfrac{36}{R + 18}$ = the time downstream
$\dfrac{12}{R - 18}$ = the time upstream
$$\frac{36}{R + 18} = \frac{12}{R - 18}$$
$$36(R - 18) = 12(R + 18)$$
$$36R - 648 = 12R + 216$$
$$24R = 864$$
$$R = 36$$
The speed of the boat in still water is 36 mi/hr.

53. Let x = the first even integer
Let $x + 2$ = the consecutive even integer
$$\frac{1}{x} + \frac{1}{x + 2} = \frac{5}{12}$$
$$12x(x+2)\frac{1}{x} + \frac{1}{x+2} = 12x(x+2)\frac{5}{12}$$
$$12(x+2) + 12x = 5 x(x+2)$$
$$12x + 24 + 12x = 5x^2 + 10x$$
$$5x^2 - 14x - 24 = 0$$
$$(5x + 6)(x - 4) = 0$$
$$5x + 6 = 0 \quad \text{or} \quad x - 4 = 0$$
$$5x = -6 \quad \text{or} \quad x = 4$$
$$x = -\frac{6}{5} \quad \text{or} \quad x = 4$$
The integers are 4 and 6.

6.8   Variation

Problems 6.8

1. **a.** $L = kt$
    b. $6 \text{ in} = k \cdot 2 \text{ mo}$

$$k = \frac{6 \text{ in}}{2 \text{ mo}} = 3 \text{ in/mo}$$

2. **a.** $P = \dfrac{k}{r}$
    b. $\$100 = \dfrac{k}{0.08}$

$$k = \$100(0.08) = \$8$$

3. **a.** $f = \dfrac{k}{a}$

   b. $6 = \dfrac{k}{\frac{1}{2}}$

$$k = 6 \cdot \tfrac{1}{2} = 3$$

   c. $f = \dfrac{3}{a}; \quad 18 = \dfrac{3}{a}$

$$18a = 3$$

$$a = \frac{3}{18} = \frac{1}{6}$$

4. **a.** $F = kAV^2$
    b. $\quad 2.2 = k \cdot 1 \cdot 20^2$

$$2.2 = 400k$$

$$0.0055 = k$$

   c. $F = 0.0055(2)(60)^2$

$$= 0.0055(7200) = 39.6 \text{ lb}$$

5. $g = kA$

$$375 = \frac{50}{31} A$$

$$A = \frac{31}{50} \approx 375 \approx 233 \text{ mi}^2$$

Exercises 6.8

1. $T = ks$

3. $W = kh^3$

5. $W = kB$

7. $R = \dfrac{k}{D^2}$

9. $I = kPr$

11. $A = ksv^2$

13. $V = kdw$

15. $I = \dfrac{ki}{d^2}$

17. $R = \dfrac{kL}{A}$

19. $W = \dfrac{k}{d^2}$

21. **a.** $I = km$

    b. $26.40 = k(480)$

$$k = \frac{26.40}{480} = 0.055$$

    c. $I = 0.055m = 0.055(750) = \$41.25$

23. **a.** $d = ks^2$

    b. $54 = k(30)^2$

$$k = \frac{54}{30^2} = \frac{54}{900} = 0.06$$

    c. $d = 0.06s^2 = 0.06(60)^2 = 216 \text{ ft}$

25. **a.** $S = \dfrac{k}{y}$

    b. $50 = \dfrac{k}{3}$

$$k = 50 \cdot 3 = 150$$

$$S = \frac{150}{y} = \frac{150}{5} = 30 \text{ new songs}$$

27. **a.** $W = \dfrac{k}{d^2}$

    b. $121 = \dfrac{k}{3960^2}$

$$k = 121(3960)^2$$

    c. $W = \dfrac{121(3960)^2}{4840^2} = 81 \text{ lb}$

29. **a.** $d = ks$

    b. $501 = k(28.41)$

$$k = \frac{501}{28.41} \approx 17.63$$

    c. k represents the number of hours needed to travel d distance at speed s.

31. **a.** $C = k(F - 37)$

$$80 = k(57 - 37) = 20k$$

$$k = \frac{80}{20} = 4$$

$$C = 4(F - 37)$$

    b. $C = 4(90 - 37) = 4(53) = 212 \text{ chirps}$

33. **a.** $I = kn$

    b. $31.4 = k(23)$

$$k = \frac{31.4}{23} \approx 1.365$$

    c. $I = 1.365(35) = 47.775$

        $\text{Conc} = 319.9 + 47.775 = 367.675 \text{ ppm}$

35. $I = ktP$

$$100 = k \cdot \tfrac{1}{4} \cdot 8000 = 2000k$$

$$k = \frac{100}{2000} = 0.05$$

$$I = 0.05 \cdot \tfrac{5}{12} \cdot 10{,}000 = \$208.33$$

37. $C = \dfrac{kw}{L}$

$$98 = \frac{k \cdot 15}{21}$$

$$k = \frac{98 \cdot 21}{15} = 137.2$$

$$C = \frac{137.2(15)}{20} = 102.9$$

39. $2x - y = 2$

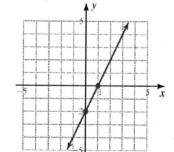

41. $y = -x - 3$

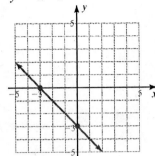

43.

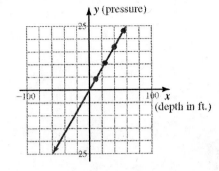

45. $p = kd$

47. The slope and k are equal.

49. Sample answer: In a direct variation, as one quantity increases, the other quantity also increases. In an inverse variation, as one quantity increases, the other quantity decreases.

51. Sample answer: Another way of expressing the idea is: "y varies inversely as x".

53. **a.** $F = kAV^2$

    b. $\quad 1.8 = k \cdot 1 \cdot 20^2$

$\quad\quad\quad 1.8 = 400k$

$\quad\quad 0.0045 = k$

    c. $F = 0.0045(2)(60)^2$

$\quad\quad\quad = 0.0045(7200) = 32.4 \text{ lb}$

55. **a.** $P = \dfrac{k}{r}$

    b. $\$100 = \dfrac{k}{0.10}$

$\quad\quad k = \$100(0.10) = \$10$

Review Exercises

1. **a.** Set the denominator equal to zero and solve:

$$x - 4 = 0$$
$$x = 4$$

The expression is undefined when $x = 4$.

    b. Set the denominator equal to zero and solve:

$$x^2 + 10x + 9 = 0$$
$$(x + 9)(x + 1) = 0$$
$$x + 9 = 0 \quad \text{or} \; x + 1 = 0$$
$$x = -9 \text{ or} \quad x = -1$$

The expression is undefined when $x = -9$ or $x = -1$.

    c. Set the denominator equal to zero and solve:

$$x^2 + 25 = 0$$
$$x^2 = -25$$

This is expression is false, thus the expression is defined for all values of x.

2. **a.** Multiply numerator and denominator by $4y^3$:

$$\frac{2x^2}{9y^4} = \frac{2x^2}{9y^4} \cdot \frac{4y^3}{4y^3} = \frac{8x^2y^3}{36y^7}$$

b. Multiply numerator and denominator by $5y^4$:

$$\frac{2x^2}{9y^4} = \frac{2x^2}{9y^4} \cdot \frac{5y^4}{5y^4} = \frac{10x^2y^4}{45y^8}$$

c. Multiply numerator and denominator by $x+5$:

$$\frac{2x+1}{x+1} = \frac{2x+1}{x+1} \cdot \frac{x+5}{x+5} = \frac{2x^2+11x+5}{x^2+6x+5}$$

d. Multiply numerator and denominator by $x+6$:

$$\frac{2x+1}{x+1} = \frac{2x+1}{x+1} \cdot \frac{x+6}{x+6} = \frac{2x^2+13x+6}{x^2+7x+6}$$

3. **a.** $-\dfrac{-6}{y} = \dfrac{6}{y}$

b. $\dfrac{-7}{-y} = \dfrac{7}{y}$

c. $-\dfrac{-8}{-y} = \dfrac{-8}{y}$

4. **a.** $-\dfrac{x-y}{6} = \dfrac{y-x}{6}$

b. $-\dfrac{7}{x-y} = \dfrac{7}{y-x}$

c. $-\dfrac{x-y}{8} = \dfrac{y-x}{8}$

5. **a.** $\dfrac{x^4y^7}{xy^2} = \dfrac{\cancel{xy^2} \cdot x^3y^5}{\cancel{xy^2}} = x^3y^5$

b. $\dfrac{x^4y^8}{xy^2} = \dfrac{\cancel{xy^2} \cdot x^3y^6}{\cancel{xy^2}} = x^3y^6$

c. $\dfrac{x^4y^9}{xy^2} = \dfrac{\cancel{xy^2} \cdot x^3y^7}{\cancel{xy^2}} = x^3y^7$

6. **a.** $\dfrac{xy^2+y^3}{x^2-y^2} = \dfrac{y^2\cancel{(x+y)}}{\cancel{(x+y)}(x-y)} = \dfrac{y^2}{x-y}$

b. $\dfrac{xy^3+y^4}{x^2-y^2} = \dfrac{y^3\cancel{(x+y)}}{\cancel{(x+y)}(x-y)} = \dfrac{y^3}{x-y}$

c. $\dfrac{xy^4+y^5}{x^2-y^2} = \dfrac{y^4\cancel{(x+y)}}{\cancel{(x+y)}(x-y)} = \dfrac{y^4}{x-y}$

7. **a.** $\dfrac{4y^2-x^2}{x^3+8y^3} = \dfrac{\cancel{(2y+x)}(2y-x)}{\cancel{(x+2y)}(x^2-2xy+4y^2)}$

$$= \dfrac{2y-x}{x^2-2xy+4y^2}$$

b. $\dfrac{4y^2-x^2}{x^3-8y^3} = \dfrac{(2y+x)(2y-x)}{(x-2y)(x^2+2xy+4y^2)}$

$$= \dfrac{-(x+2y)\cancel{(x-2y)}}{\cancel{(x-2y)}(x^2+2xy+4y^2)}$$

$$= \dfrac{-(x+2y)}{x^2+2xy+4y^2}$$

c. $\dfrac{9y^2-x^2}{x^3-27y^3} = \dfrac{(3y+x)(3y-x)}{(x-3y)(x^2+3xy+9y^2)}$

$$= \dfrac{-(x+3y)\cancel{(x-3y)}}{\cancel{(x-3y)}(x^2+3xy+9y^2)}$$

$$= \dfrac{-(x+3y)}{x^2+3xy+9y^2}$$

8. **a.** $\dfrac{x \cdot 2}{3x \cdot 2y} \cdot \dfrac{9x^2 \cdot 4y^2}{3x^2 \cdot 5x \cdot 2} = \dfrac{\cancel{x \cdot 2}}{3x \cancel{\cdot 2y}} \cdot \dfrac{(3x+2y)\cancel{(3x-2y)}}{(3x+1)\cancel{(x-2)}} = \dfrac{3x+2y}{3x+1}$

b.

$$\frac{x\cdot 2}{3x\cdot 2y}\cdot\frac{9x^2\cdot 4y^2}{4x^2\cdot 11x+6}=\frac{\cancel{x\cdot 2}}{\cancel{3x\cdot 2y}}\cdot\frac{(3x+2y)\cancel{(3x-2y)}}{(4x\cdot 3)\cancel{(x\cdot 2)}}=\frac{3x+2y}{4x\cdot 3}$$

c.

$$\frac{x\cdot 2}{3x\cdot 2y}\cdot\frac{9x^2\cdot 4y^2}{5x^2\cdot 8x\cdot 4}=\frac{\cancel{x\cdot 2}}{\cancel{3x-2y}}\cdot\frac{(3x+2y)\cancel{(3x-2y)}}{(5x+2)\cancel{(x\cdot 2)}}=\frac{3x+2y}{5x+2}$$

9. a.

$$\frac{x+4}{x\cdot 2}\div(x^2+8x+16)=\frac{x+4}{x\cdot 2}\cdot\frac{1}{x^2+8x+16}=\frac{\cancel{x+4}}{x\cdot 2}\cdot\frac{1}{\cancel{(x+4)}(x+4)}=\frac{1}{(x\cdot 2)(x+4)}$$

b.

$$\frac{x+5}{x\cdot 2}\div(x^2+10x+25)=\frac{x+5}{x\cdot 2}\cdot\frac{1}{x^2+10x+25}=\frac{\cancel{x+5}}{x\cdot 2}\cdot\frac{1}{\cancel{(x+5)}(x+5)}=\frac{1}{(x\cdot 2)(x+5)}$$

c.

$$\frac{x+6}{x\cdot 2}\div(x^2+12x+36)=\frac{x+6}{x\cdot 2}\cdot\frac{1}{x^2+12x+36}=\frac{\cancel{x+6}}{x\cdot 2}\cdot\frac{1}{\cancel{(x+6)}(x+6)}=\frac{1}{(x\cdot 2)(x+6)}$$

10. a.

$$\frac{2\cdot x}{x+3}\div\frac{x^3\cdot 8}{x+6}\cdot\frac{x^3+27}{x+6}=\frac{\cdot(x\cdot 2)}{x+3}\cdot\frac{x+6}{x^3\cdot 8}\cdot\frac{x^3+27}{x+6}$$

$$=\frac{\cdot\cancel{(x\cdot 2)}}{\cancel{x+3}}\cdot\frac{\cancel{x+6}}{\cancel{(x\cdot 2)}(x^2+2x+4)}\cdot\frac{\cancel{(x+3)}(x^2\cdot 3x+9)}{\cancel{x+6}}=\frac{\cdot(x^2\cdot 3x+9)}{x^2+2x+4}$$

b.

$$\frac{4\cdot x}{x+3}\div\frac{x^3\cdot 64}{x+7}\cdot\frac{x^3+27}{x+7}=\frac{\cdot(x\cdot 4)}{x+3}\cdot\frac{x+7}{x^3\cdot 64}\cdot\frac{x^3+27}{x+7}$$

$$=\frac{\cdot\cancel{(x\cdot 4)}}{\cancel{x+3}}\cdot\frac{\cancel{x+7}}{\cancel{(x\cdot 4)}(x^2+4x+16)}\cdot\frac{\cancel{(x+3)}(x^2\cdot 3x+9)}{\cancel{x+7}}=\frac{\cdot(x^2\cdot 3x+9)}{x^2+4x+16}$$

c.

$$\frac{2\cdot x}{x+5}\div\frac{x^3\cdot 8}{x+8}\cdot\frac{x^3+125}{x+8}=\frac{\cdot(x\cdot 2)}{x+5}\cdot\frac{x+8}{x^3\cdot 8}\cdot\frac{x^3+125}{x+8}$$

$$=\frac{\cdot\cancel{(x\cdot 2)}}{\cancel{x+5}}\cdot\frac{\cancel{x+8}}{\cancel{(x\cdot 2)}(x^2+2x+4)}\cdot\frac{\cancel{(x+5)}(x^2\cdot 5x+25)}{\cancel{x+8}}=\frac{\cdot(x^2\cdot 5x+25)}{x^2+2x+4}$$

11. a.

$$\frac{x}{x^2-4}+\frac{2}{x^2-4}=\frac{x+2}{x^2-4}=\frac{\cancel{x+2}}{\cancel{(x+2)}(x-2)}=\frac{1}{x-2}$$

b.

$$\frac{x}{x^2-9}+\frac{3}{x^2-9}=\frac{x+3}{x^2-9}=\frac{\cancel{x+3}}{\cancel{(x+3)}(x-3)}=\frac{1}{x-3}$$

c.

$$\frac{x}{x^2-16}+\frac{4}{x^2-16}=\frac{x+4}{x^2-16}=\frac{\cancel{x+4}}{\cancel{(x+4)}(x-4)}=\frac{1}{x-4}$$

12. a.
$$\frac{x}{x^2-9}-\frac{3}{x^2-9}=\frac{x-3}{x^2-9}=\frac{\cancel{x-3}}{\cancel{(x-3)}(x+3)}=\frac{1}{x+3}$$

b.
$$\frac{x}{x^2-16}-\frac{4}{x^2-16}=\frac{x-4}{x^2-16}=\frac{\cancel{x-4}}{\cancel{(x-4)}(x+4)}=\frac{1}{x+4}$$

c.
$$\frac{x}{x^2-25}-\frac{5}{x^2-25}=\frac{x-5}{x^2-25}=\frac{\cancel{x-5}}{\cancel{(x-5)}(x+5)}=\frac{1}{x+5}$$

13. a.
$$\frac{x+1}{x^2+x-2}+\frac{x+5}{x^2-1}=\frac{x+1}{(x+2)(x-1)}+\frac{x+5}{(x+1)(x-1)}=\frac{(x+1)(x+1)}{(x+2)(x-1)(x+1)}+\frac{(x+5)(x+2)}{(x+1)(x-1)(x+2)}$$
$$=\frac{x^2+2x+1+x^2+7x+10}{(x+1)(x-1)(x+2)}=\frac{2x^2+9x+11}{(x+1)(x-1)(x+2)}$$

b.
$$\frac{x+1}{x^2+x-2}+\frac{x+6}{x^2-1}=\frac{x+1}{(x+2)(x-1)}+\frac{x+6}{(x+1)(x-1)}=\frac{(x+1)(x+1)}{(x+2)(x-1)(x+1)}+\frac{(x+6)(x+2)}{(x+1)(x-1)(x+2)}$$
$$=\frac{x^2+2x+1+x^2+8x+12}{(x+1)(x-1)(x+2)}=\frac{2x^2+10x+13}{(x+1)(x-1)(x+2)}$$

c.
$$\frac{x+1}{x^2+x-2}+\frac{x+7}{x^2-1}=\frac{x+1}{(x+2)(x-1)}+\frac{x+7}{(x+1)(x-1)}=\frac{(x+1)(x+1)}{(x+2)(x-1)(x+1)}+\frac{(x+7)(x+2)}{(x+1)(x-1)(x+2)}$$
$$=\frac{x^2+2x+1+x^2+9x+14}{(x+1)(x-1)(x+2)}=\frac{2x^2+11x+15}{(x+1)(x-1)(x+2)}$$

14. a.
$$\frac{x-4}{x^2-x-6}-\frac{x+1}{x^2-9}=\frac{x-4}{(x-3)(x+2)}-\frac{x+1}{(x+3)(x-3)}=\frac{(x-4)(x+3)}{(x-3)(x+2)(x+3)}-\frac{(x+1)(x+2)}{(x+3)(x-3)(x+2)}$$
$$=\frac{x^2-x-12-x^2-3x-2}{(x-3)(x+2)(x+3)}=\frac{-4x-14}{(x-3)(x+2)(x+3)}$$

b.
$$\frac{x-3}{x^2-x-6}-\frac{x+1}{x^2-9}=\frac{x-3}{(x-3)(x+2)}-\frac{x+1}{(x+3)(x-3)}=\frac{(x-3)(x+3)}{(x-3)(x+2)(x+3)}-\frac{(x+1)(x+2)}{(x+3)(x-3)(x+2)}$$
$$=\frac{x^2-9-x^2-3x-2}{(x-3)(x+2)(x+3)}=\frac{-3x-11}{(x-3)(x+2)(x+3)}$$

c.
$$\frac{x-1}{x^2-x-6}-\frac{x+1}{x^2-9}=\frac{x-1}{(x-3)(x+2)}-\frac{x+1}{(x+3)(x-3)}=\frac{(x-1)(x+3)}{(x-3)(x+2)(x+3)}-\frac{(x+1)(x+2)}{(x+3)(x-3)(x+2)}$$
$$=\frac{x^2+2x-3-x^2-3x-2}{(x-3)(x+2)(x+3)}=\frac{-x-5}{(x-3)(x+2)(x+3)}$$

15. **a.** Multiply the numerator and denominator by the LCD which is x^5:

$$\dfrac{\frac{1}{x}+\frac{1}{x^4}}{\frac{1}{x}-\frac{1}{x^5}}=\dfrac{x^5\left|\frac{1}{x}+\frac{1}{x^4}\right|}{x^5\left|\frac{1}{x}-\frac{1}{x^5}\right|}=\dfrac{x^4+x}{x^4-1}=\dfrac{x\left(x^3+1\right)}{\left(x^2+1\right)\left(x^2-1\right)}=\dfrac{x\cancel{(x+1)}\left(x^2-x+1\right)}{\left(x^2+1\right)\cancel{(x+1)}(x-1)}=\dfrac{x\left(x^2-x+1\right)}{\left(x^2+1\right)(x-1)}$$

b. Multiply the numerator and denominator by the LCD which is x^6:

$$\dfrac{\frac{1}{x^2}+\frac{1}{x^5}}{\frac{1}{x^2}-\frac{1}{x^6}}=\dfrac{x^6\left|\frac{1}{x^2}+\frac{1}{x^5}\right|}{x^6\left|\frac{1}{x^2}-\frac{1}{x^6}\right|}=\dfrac{x^4+x}{x^4-1}=\dfrac{x\left(x^3+1\right)}{\left(x^2+1\right)\left(x^2-1\right)}=\dfrac{x\cancel{(x+1)}\left(x^2-x+1\right)}{\left(x^2+1\right)\cancel{(x+1)}(x-1)}=\dfrac{x\left(x^2-x+1\right)}{\left(x^2+1\right)(x-1)}$$

c. Multiply the numerator and denominator by the LCD which is x^7:

$$\dfrac{\frac{1}{x^3}+\frac{1}{x^6}}{\frac{1}{x^3}-\frac{1}{x^7}}=\dfrac{x^7\left|\frac{1}{x^3}+\frac{1}{x^6}\right|}{x^7\left|\frac{1}{x^3}-\frac{1}{x^7}\right|}=\dfrac{x^4+x}{x^4-1}=\dfrac{x\left(x^3+1\right)}{\left(x^2+1\right)\left(x^2-1\right)}=\dfrac{x\cancel{(x+1)}\left(x^2-x+1\right)}{\left(x^2+1\right)\cancel{(x+1)}(x-1)}=\dfrac{x\left(x^2-x+1\right)}{\left(x^2+1\right)(x-1)}$$

16. **a.**

$$4+\dfrac{a}{4+\dfrac{4}{4+a}}=4+\dfrac{a}{\dfrac{4(4+a)}{4+a}+\dfrac{4}{4+a}}=4+\dfrac{a}{\dfrac{16+4a+4}{4+a}}=4+a\cdot\dfrac{4+a}{4a+20}=4+\dfrac{4a+a^2}{4a+20}$$

$$=\dfrac{4(4a+20)}{4a+20}+\dfrac{4a+a^2}{4a+20}=\dfrac{16a+80+4a+a^2}{4a+20}=\dfrac{a^2+20a+80}{4a+20}$$

b.

$$5+\dfrac{a}{5+\dfrac{5}{5+a}}=5+\dfrac{a}{\dfrac{5(5+a)}{5+a}+\dfrac{5}{5+a}}=5+\dfrac{a}{\dfrac{25+5a+5}{5+a}}=5+a\cdot\dfrac{5+a}{5a+30}=5+\dfrac{5a+a^2}{5a+30}$$

$$=\dfrac{5(5a+30)}{5a+30}+\dfrac{5a+a^2}{5a+30}=\dfrac{25a+150+5a+a^2}{5a+30}=\dfrac{a^2+30a+150}{5a+30}$$

c.

$$6+\dfrac{a}{6+\dfrac{6}{6+a}}=6+\dfrac{a}{\dfrac{6(6+a)}{6+a}+\dfrac{6}{6+a}}=6+\dfrac{a}{\dfrac{36+6a+6}{6+a}}=6+a\cdot\dfrac{6+a}{6a+42}=6+\dfrac{6a+a^2}{6a+42}$$

$$=\dfrac{6(6a+42)}{6a+42}+\dfrac{6a+a^2}{6a+42}=\dfrac{36a+252+6a+a^2}{6a+42}=\dfrac{a^2+42a+252}{6a+42}$$

17. **a.** $\dfrac{18x^5-12x^3+6x^2}{6x^2}=\dfrac{18x^5}{6x^2}-\dfrac{12x^3}{6x^2}+\dfrac{6x^2}{6x^2}=3x^3-2x+1$

b. $\dfrac{18x^5-12x^3+6x^2}{6x^3}=\dfrac{18x^5}{6x^3}-\dfrac{12x^3}{6x^3}+\dfrac{6x^2}{6x^3}=3x^2-2+\dfrac{1}{x}$

c. $\dfrac{18x^5-12x^3+6x^2}{6x^4}=\dfrac{18x^5}{6x^4}-\dfrac{12x^3}{6x^4}+\dfrac{6x^2}{6x^4}=3x-\dfrac{2}{x}+\dfrac{1}{x^2}$

18. a.

$$2x+2 \overline{\smash{)}\ 2x^3 + 0x^2 - 4x - 8} \quad \overset{x^2 - x - 1}{}$$

$$\underline{-(2x^3 + 2x^2)}$$
$$-2x^2 - 4x$$
$$\underline{-(-2x^2 - 2x)}$$
$$-2x - 8$$
$$\underline{-(-2x - 2)}$$
$$-6$$

Quotient: $x^2 - x - 1$ R: -6

b.

$$2x+2 \overline{\smash{)}\ 2x^3 + 0x^2 - 4x - 9} \quad \overset{x^2 - x - 1}{}$$

$$\underline{-(2x^3 + 2x^2)}$$
$$-2x^2 - 4x$$
$$\underline{-(-2x^2 - 2x)}$$
$$-2x - 9$$
$$\underline{-(-2x - 2)}$$
$$-7$$

Quotient: $x^2 - x - 1$ R: -7

c.

$$2x+2 \overline{\smash{)}\ 2x^3 + 0x^2 - 4x - 10} \quad \overset{x^2 - x - 1}{}$$

$$\underline{-(2x^3 + 2x^2)}$$
$$-2x^2 - 4x$$
$$\underline{-(-2x^2 - 2x)}$$
$$-2x - 10$$
$$\underline{-(-2x - 2)}$$
$$-8$$

Quotient: $x^2 - x - 1$ R: -8

19. a.

$$x-1 \overline{\smash{)}\ x^3 - 6x^2 + 11x - 6} \quad \overset{x^2 - 5x + 6}{}$$

$$\underline{-(x^3 - x^2)}$$
$$-5x^2 + 11x$$
$$\underline{-(-5x^2 + 5x)}$$
$$6x - 6$$
$$\underline{-(6x - 6)}$$
$$0$$

The factorization is:
$$(x-1)(x^2 - 5x + 6) = (x-1)(x-3)(x-2)$$

b.

$$x-2 \overline{\smash{)}\ x^3 - 6x^2 + 11x - 6} \quad \overset{x^2 - 4x + 3}{}$$

$$\underline{-(x^3 - 2x^2)}$$
$$-4x^2 + 11x$$
$$\underline{-(-4x^2 + 8x)}$$
$$3x - 6$$
$$\underline{-(3x - 6)}$$
$$0$$

The factorization is:
$$(x-2)(x^2 - 4x + 3) = (x-2)(x-3)(x-1)$$

c.

$$x-3 \overline{\smash{)}\ x^3 - 6x^2 + 11x - 6} \quad \overset{x^2 - 3x + 2}{}$$

$$\underline{-(x^3 - 3x^2)}$$
$$-3x^2 + 11x$$
$$\underline{-(-3x^2 + 9x)}$$
$$2x - 6$$
$$\underline{-(2x - 6)}$$
$$0$$

The factorization is:
$$(x-3)(x^2 - 3x + 2) = (x-3)(x-2)(x-1)$$

20. a.

$$-1 \overline{\smash{)}\ 1 + 10 + 35 + 50 + 28}$$
$$\underline{\ -1 -\ 9 - 26 - 24}$$
$$1 +\ 9 + 26 + 24 +\ 4$$

Quotient: $x^3 + 9x^2 + 26x + 24$ R: 4

b.

$$-2 \overline{\smash{)}\ 1 + 10 + 35 + 50 + 28}$$
$$\underline{\ -2 - 16 - 38 - 24}$$
$$1 +\ 8 + 19 + 12 +\ 4$$

Quotient: $x^3 + 8x^2 + 19x + 12$ R: 4

257

c.
$$-3)\overline{1 + 10 + 35 + 50 + 28}$$
$$\underline{-3 - 21 - 42 - 24}$$
$$1 + 7 + 14 + 8 + 4$$

Quotient: $x^3+7x^2+14x+8$ R: 4

b.
$$-2)\overline{1 + 10 + 35 + 50 + 24}$$
$$\underline{-2 - 16 - 38 - 24}$$
$$1 + 8 + 19 + 12 + 0$$

Since the remainder is zero, -2 is a solution.

21. a.
$$-1)\overline{1 + 10 + 35 + 50 + 24}$$
$$\underline{-1 - 9 - 26 - 24}$$
$$1 + 9 + 26 + 24 + 0$$

Since the remainder is zero, -1 is a solution.

c.
$$-3)\overline{1 + 10 + 35 + 50 + 24}$$
$$\underline{-3 - 21 - 42 - 24}$$
$$1 + 7 + 14 + 8 + 0$$

Since the remainder is zero, -3 is a solution.

22. a. Multiply both sides by the LCD $(x+4)(x-4)$:

$$\frac{x}{x+4}-\frac{x}{x-4}=\frac{x^2+16}{x^2-16}$$

$$(x+4)(x-4)\left[\frac{x}{x+4}-\frac{x}{x-4}\right]=(x+4)(x-4)\cdot\frac{x^2+16}{(x+4)(x-4)}$$

$$x(x-4)-x(x+4)=x^2+16$$

$$x^2-4x-x^2-4x=x^2+16$$

$$-8x=x^2+16$$

$$0=x^2+8x+16$$

$$0=(x+4)(x+4)$$

$$x+4=0 \quad \text{or} \quad x+4=0$$

$$x=-4 \quad \text{or} \quad x=-4$$

-4 makes the denominator $x+4=0$ and thus is extraneous. There is no solution.

b. Multiply both sides by the LCD $(x+5)(x-5)$:

$$1+\frac{4}{x-5}=\frac{40}{x^2-25}$$

$$(x+5)(x-5)\left[1+\frac{4}{x-5}\right]=(x+5)(x-5)\cdot\frac{40}{(x+5)(x-5)}$$

$$(x+5)(x-5)+4(x+5)=40$$

$$x^2-25+4x+20=40$$

$$x^2+4x-45=0$$

$$(x+9)(x-5)=0$$

$$x+9=0 \quad \text{or} \quad x-5=0$$

$$x=-9 \quad \text{or} \quad x=5$$

5 makes the denominator $x-5=0$ and thus is extraneous. The solution is -9.

c. Multiply both sides by the LCD $(x+6)(x-6)$:

$$1+\frac{5}{x-6}=\frac{60}{x^2-36}$$

$$(x+6)(x-6)\left[1+\frac{5}{x-6}\right]=(x+6)(x-6)\left[\frac{60}{(x+6)(x-6)}\right]$$

$$(x+6)(x-6)+5(x+6)=60$$

$$x^2-36+5x+30=60$$

$$x^2+5x-66=0$$

$$(x+11)(x-6)=0$$

$$x+11=0 \quad \text{or } x-6=0$$

$$x=-11 \text{ or} \quad x=6$$

6 makes the denominator $x-6=0$ and thus is extraneous. The solution is -11.

23. a.

$$\frac{\frac{1}{4}\ \text{in}}{2\ \text{ft}}=\frac{4\ \text{in}}{x} \qquad \frac{\frac{1}{4}\ \text{in}}{2\ \text{ft}}=\frac{6\ \text{in}}{x}$$

$$\frac{1}{4}x=2\cdot4 \qquad\qquad \frac{1}{4}x=2\cdot6$$

$$\frac{1}{4}x=8 \qquad\qquad \frac{1}{4}x=12$$

$$x=32\ \text{ft} \qquad\qquad x=48\ \text{ft}$$

The dimensions are 32 ft by 48 ft.

b.

$$\frac{75}{100}=\frac{x}{10}$$

$$100x=75\cdot10$$

$$100x=750$$

$$x=7.5$$

He is expected to make 7 or 8 free throws.

24. a. Let $x =$ the first even integer
Let $x + 2 =$ the consecutive even integer

$$\frac{1}{x}+\frac{1}{x+2}=\frac{11}{60}$$

$$60x(x+2)\frac{1}{x}+\frac{1}{x+2}=60x(x+2)\frac{11}{60}$$

$$60(x+2)+60x=11x(x+2)$$

$$60x+120+60x=11x^2+22x$$

$$0=11x^2-98x-120$$

$$(11x+12)(x-10)=0$$

$$11x+12=0 \quad \text{or } x-10=0$$

$$11x=-12 \quad \text{or} \quad x=10$$

$$x=-\frac{12}{11} \quad \text{or} \quad x=10$$

The integers are 10 and 12.

b. Let $x =$ the first even integer
Let $x + 2 =$ the consecutive even integer

$$\frac{1}{x}+\frac{1}{x+2}=\frac{13}{84}$$

$$84x(x+2)\frac{1}{x}+\frac{1}{x+2}=84x(x+2)\frac{13}{84}$$

$$84(x+2)+84x=13x(x+2)$$

$$84x+168+84x=13x^2+26x$$

$$0=13x^2-142x-168$$

$$(13x+14)(x-12)=0$$

$$13x+14=0 \quad \text{or } x-12=0$$

$$13x=-14 \quad \text{or} \quad x=12$$

$$x=-\frac{14}{13} \quad \text{or} \quad x=12$$

The integers are 12 and 14.

c. Let $x =$ the first even integer

Let $x + 2 =$ the consecutive even integer

$$\frac{1}{x} + \frac{1}{x+2} = \frac{15}{112}$$

$$112x(x+2)\left[\frac{1}{x} + \frac{1}{x+2}\right] = 112x(x+2)\left[\frac{15}{112}\right]$$

$$112(x+2) + 112x = 15 \cdot x(x+2)$$

$$112x + 224 + 112x = 15x^2 + 30x$$

$$0 = 15x^2 - 194x - 224$$

$$(15x + 16)(x - 14) = 0$$

$$15x + 16 = 0 \quad \text{or} \quad x - 14 = 0$$

$$15x = -16 \quad \text{or} \quad x = 14$$

$$x = -\frac{16}{15} \quad \text{or} \quad x = 14$$

The integers are 14 and 16.

25. a. Let $t =$ the time when working together

$$\frac{1}{4} + \frac{1}{5} = \frac{1}{t}$$

$$20t\left[\frac{1}{4} + \frac{1}{5}\right] = 20t\left[\frac{1}{t}\right]$$

$$5t + 4t = 20$$

$$9t = 20$$

$$t = \frac{20}{9} = 2\frac{2}{9} \text{ hr}$$

Working together they do the job in $2\frac{2}{9}$ hr.

b. Let $t =$ the time when working together

$$\frac{1}{4} + \frac{1}{6} = \frac{1}{t}$$

$$12t\left[\frac{1}{4} + \frac{1}{6}\right] = 12t\left[\frac{1}{t}\right]$$

$$3t + 2t = 12$$

$$5t = 12$$

$$t = \frac{12}{5} = 2\frac{2}{5} \text{ hr}$$

Working together they do the job in $2\frac{2}{5}$ hr.

c. Let $t =$ the time when working together

$$\frac{1}{4} + \frac{1}{7} = \frac{1}{t}$$

$$28t\left[\frac{1}{4} + \frac{1}{7}\right] = 28t\left[\frac{1}{t}\right]$$

$$7t + 4t = 28$$

$$11t = 28$$

$$t = \frac{28}{11} = 2\frac{6}{11} \text{ hr}$$

Working together they do the job in $2\frac{6}{11}$ hr.

26. a. Let $R =$ the plane's speed in still air

$R + 25 =$ the speed with the wind

$R - 25 =$ the speed against the wind

$\dfrac{1200}{R+25} =$ the time with the wind

$\dfrac{960}{R-25} =$ the time against the wind

$$\frac{1200}{R+25} = \frac{960}{R-25}$$

$$1200(R - 25) = 960(R + 25)$$

$$1200R - 30,000 = 960R + 24,000$$

$$240R = 54,000$$

$$R = 225$$

The plane's speed in still air is 225 mi/hr.

b. Let R = the plane's speed in still air
$R + 25$ = the speed with the wind
$R - 25$ = the speed against the wind
$\dfrac{1200}{R+25}$ = the time with the wind
$\dfrac{1000}{R-25}$ = the time against the wind

$$\frac{1200}{R+25} = \frac{1000}{R-25}$$
$$1200(R-25) = 1000(R+25)$$
$$1200R - 30,000 = 1000R + 25,000$$
$$200R = 55,000$$
$$R = 275$$

The plane's speed in still air is 275 mi/hr.

c. Let R = the plane's speed in still air
$R + 25$ = the speed with the wind
$R - 25$ = the speed against the wind
$\dfrac{1200}{R+25}$ = the time with the wind
$\dfrac{1040}{R-25}$ = the time against the wind

$$\frac{1200}{R+25} = \frac{1040}{R-25}$$
$$1200(R-25) = 1040(R+25)$$
$$1200R - 30,000 = 1040R + 26,000$$
$$160R = 56,000$$
$$R = 350$$

The plane's speed in still air is 350 mi/hr.

27. a. Solve for a:

$$A = \frac{a+2b+3c}{2}$$
$$2A = a + 2b + 3c$$
$$2A - 2b - 3c = a$$

b. Solve for b:

$$A = \frac{a+2b+3c}{2}$$
$$2A = a + 2b + 3c$$
$$2A - a - 3c = 2b$$
$$\frac{2A - a - 3c}{2} = b$$
$$b = A - \frac{a}{2} - \frac{3c}{2}$$

c. Solve for c:

$$A = \frac{a+2b+3c}{2}$$
$$2A = a + 2b + 3c$$
$$2A - a - 2b = 3c$$
$$\frac{2A - a - 2b}{3} = c$$
$$c = \frac{2A}{3} - \frac{a}{3} - \frac{2b}{3}$$

28. a. $P = kT$

b. $3 = k \cdot 360$

$$k = \frac{3}{360} = \frac{1}{120}$$

29. a. $P = \dfrac{k}{V}$

b. $1600 = \dfrac{k}{2}$

$$k = 1600 \cdot 2 = 3200$$

30. $g = kxt^2$

Cumulative Review Chapters 1-6

1. $-8 - 6 = -14$

2. Associative law of multiplication

3. $\left[(5x^2 - 1) + (x+2)\right] - \left[(x-6) + (8x^2 - 8)\right]$

$= \left[5x^2 + x + 1\right] - \left[8x^2 + x - 14\right]$

$= 5x^2 + x + 1 - 8x^2 - x + 14$

$= -3x^2 + 15$

4.
$$\frac{x+6}{5} \cdot \frac{x \cdot 6}{7} = 2$$

$$35 \cdot \frac{x+6}{5} \cdot \frac{x \cdot 6}{7} = 35 \cdot 2$$

$$7(x+6) \cdot 5(x \cdot 6) = 70$$

$$7x + 42 \cdot 5x + 30 = 70$$

$$2x + 72 = 70$$

$$2x = -2$$

$$x = -1$$

5. $x - -4$

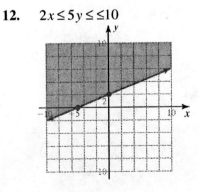

6. $x + 6 - 7$ and $-4x < 16$

$x - 1$ and $x > -4$

7. $H = 2.75h + 73.49$; $H = 136.74$

$136.74 = 2.75h + 73.49$

$63.25 = 2.75h$

$23 = h$

8. Let $x =$ first odd integer

$x + 2 =$ second consecutive odd integer

$x + 4 =$ third consecutive odd integer

$x + x + 2 + x + 4 = 93$

$3x + 6 = 93$

$3x = 87$

$x = 29$

$x + 2 = 31$

$x + 4 = 33$

The odd integers are 29, 31, and 33.

9. $m = \dfrac{4-2}{-2-4} = \dfrac{2}{-6} = -\dfrac{1}{3}$

10. $y - 1 = -5(x - 1)$

$y - 1 = -5x + 5$

$5x + y = 6$

11. $6x - 3y = 36$

$-3y = -6x + 36$

$y = 2x - 12$

$m = 2;$ y-intercept: $(0, -12)$

12. $2x \le 5y \le \le 10$

13. $|x+4|>2$

$x+4>2 \quad$ or $x+4<-2$

$x>-2$ or $\quad x<-6$

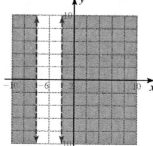

14. $y+x=-1$

$2y=-2x-4$

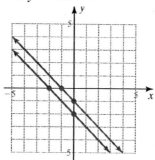

No solution

15. $x-4y=10 \qquad \rightarrow \quad x=4y+10$

$2x=8y+18$

Substitute the first equation into the second and solve:

$2(4y+10)=8y+18$

$8y+20=8y+18$

$20=18$

This is a contradiction. There is no solution and the system is inconsistent. The solution set is $\varnothing$.

16. To eliminate x, multiply the second equation by -3, add the result to the first equation, and solve:

$3x+4y=18 - \quad 3x+4y= \ 18$

$x+ \ y= \ 5 - \ -3x-3y=-15$

$y=3$

$x+3=5 \rightarrow x=2$

Solution: $(2, 3)$. The system is consistent.

17. $\begin{vmatrix} -4 & -4 \\ 1 & -5 \end{vmatrix} = (-4)(-5)-(1)(-4)=20+4=24$

18.

$D=\begin{vmatrix} 1 & 2 & 1 \\ 1 & 1 & -1 \\ 3 & 5 & 1 \end{vmatrix}=1\begin{vmatrix} 1 & -1 \\ 5 & 1 \end{vmatrix}-2\begin{vmatrix} 1 & -1 \\ 3 & 1 \end{vmatrix}+1\begin{vmatrix} 1 & 1 \\ 3 & 5 \end{vmatrix}=1(1+5)-2(1+3)+1(5-3)=6-8+2=0$

$D_x=\begin{vmatrix} 5 & 2 & 1 \\ 6 & 1 & -1 \\ 7 & 5 & 1 \end{vmatrix}=5\begin{vmatrix} 1 & -1 \\ 5 & 1 \end{vmatrix}-2\begin{vmatrix} 6 & -1 \\ 7 & 1 \end{vmatrix}+1\begin{vmatrix} 6 & 1 \\ 7 & 5 \end{vmatrix}=5(1+5)-2(6+7)+1(30-7)$

$=30-26+23=27$

There is no solution because the value of D is zero and one of the other determinants is not zero.

19. $\begin{vmatrix} -3 & -3 & 2 \\ -2 & 2 & -3 \\ 0 & -2 & -3 \end{vmatrix}=2\begin{vmatrix} -2 & 2 \\ 0 & -2 \end{vmatrix}-(-3)\begin{vmatrix} -3 & -3 \\ 0 & -2 \end{vmatrix}+(-3)\begin{vmatrix} -3 & -3 \\ -2 & 2 \end{vmatrix}=2(4-0)+3(6-0)-3(-6-6)$

$=8+18+36=62$

20. Let x = the height of the building
Let y = the height of the flagpole
$$x + y = 210$$
$$x = 9y$$
Substitute the second equation into the first equation and solve:
$$9y + y = 210$$
$$10y = 210$$
$$y = 21$$
The height of the flagpole is 21 ft.

21. $\{(x, y)\mid y = 4 + x\}$

$D = \{\text{all real numbers}\}$

$R = \{\text{all real numbers}\}$

22. $\left(2x^3 - 7x^2 - 2x - 3\right) + \left(3 - 7x + 6x^2 - 7x^3\right) = -5x^3 - x^2 - 9x$

23. $\left(2n - 1\right)\left(3n - 5\right) = 6n^2 - 10n - 3n + 5 = 6n^2 - 13n + 5$

24. $12t^2 - 7t - 10 = 12t^2 - 15t + 8t - 10 = 3t\left(4t - 5\right) + 2\left(4t - 5\right) = \left(4t - 5\right)\left(3t + 2\right)$

25. $x^3 + 3x^2 - 16x - 48 = 0$
$$x^2\left(x + 3\right) - 16\left(x + 3\right) = 0$$
$$\left(x + 3\right)\left(x^2 - 16\right) = 0$$
$$\left(x + 3\right)\left(x + 4\right)\left(x - 4\right) = 0$$
$$x + 3 = 0 \quad \text{or } x + 4 = 0 \quad \text{or } x - 4 = 0$$
$$x = -3 \text{ or} \quad x = -4 \text{ or} \quad x = 4$$

26. $\dfrac{1 \cdot x}{x + 3} \div \dfrac{x^3 \cdot 1}{x + 2} \cdot \dfrac{x^3 + 27}{x + 2} = \dfrac{\cdot \left(x \cdot 1\right)}{x + 3} \cdot \dfrac{x + 2}{x^3 \cdot 1} \cdot \dfrac{x^3 + 27}{x + 2}$

$$= \dfrac{\cdot \left(\cancel{x \cdot 1}\right)}{\cancel{x + 3}} \cdot \dfrac{\cancel{x + 2}}{\left(\cancel{x \cdot 1}\right)\left(x^2 + x + 1\right)} \cdot \dfrac{\left(\cancel{x + 3}\right)\left(x^2 \cdot 3x + 9\right)}{\cancel{x + 2}} = \dfrac{\cdot \left(x^2 \cdot 3x + 9\right)}{x^2 + x + 1}$$

27. $\dfrac{x - 1}{x^2 - 5x + 6} + \dfrac{x + 3}{x^2 - 4} = \dfrac{x - 1}{\left(x - 2\right)\left(x - 3\right)} + \dfrac{x + 3}{\left(x + 2\right)\left(x - 2\right)} = \dfrac{\left(x - 1\right)\left(x + 2\right)}{\left(x - 2\right)\left(x - 3\right)\left(x + 2\right)} + \dfrac{\left(x + 3\right)\left(x - 3\right)}{\left(x + 2\right)\left(x - 2\right)\left(x - 3\right)}$

$$= \dfrac{x^2 + x - 2 + x^2 - 9}{\left(x - 2\right)\left(x - 3\right)\left(x + 2\right)} = \dfrac{2x^2 + x - 11}{\left(x - 2\right)\left(x - 3\right)\left(x + 2\right)}$$

28. Multiply both sides by the LCD $(x+2)(x-2)$:

$$1+\frac{3}{x+2}=\frac{-12}{x^2-4}$$

$$(x+2)(x-2)\left[1+\frac{3}{x+2}\right]=(x+2)(x-2)\left[\frac{-12}{(x+2)(x-2)}\right]$$

$$(x+2)(x-2)+3(x-2)=-12$$

$$x^2-4+3x-6=-12$$

$$x^2+3x+2=0$$

$$(x+2)(x+1)=0$$

$$x+2=0 \quad \text{or } x+1=0$$

$$x=-2 \text{ or } \quad x=-1$$

-2 makes the denominator $x+2=0$ and thus is extraneous. The solution is -1.

29. Let t = the time when working together

$$\frac{1}{3}+\frac{1}{5}=\frac{1}{t}$$

$$15t\left[\frac{1}{3}+\frac{1}{5}\right]=15t\left[\frac{1}{t}\right]$$

$$5t+3t=15$$

$$8t=15$$

$$t=\frac{15}{8}=1\frac{7}{8}\ \text{hr}$$

Working together they do the job in $1\frac{7}{8}$ hr.

30. $P=kT$

$$8=k\cdot480$$

$$k=\frac{8}{480}=\frac{1}{60}$$

Chapter 7   Rational Exponents and Radicals

7.1   Rational Exponents and Radicals

Problems 7.1

1. **a.** $\sqrt{\neq 25}$ is not a real number because

$n^2 \neq \neq 25$

   b. $\sqrt[3]{-125} = -5$ because $(-5)^3 = -125$

   c. $\sqrt[3]{\left|\frac{1}{27}\right|} = \left|\frac{1}{3}\right|$ because $\left|\left|\frac{1}{3}\right|\right|^3 = \left|\frac{1}{27}\right|$

2. **a.** $49^{1/2} = \sqrt{49} = 7$

   b. $(-216)^{1/3} = \sqrt[3]{-216} = -6$

   c. $\left|\frac{1}{81}\right|^{1/4} = \sqrt[4]{\frac{1}{81}} = \frac{1}{3}$

3. **a.** $27^{2/3} = \left(\sqrt[3]{27}\right)^2 = 3^2 = 9$

   b. $(-64)^{2/3} = \left(\sqrt[3]{-64}\right)^2 = (-4)^2 = 16$

   c. $(-36)^{3/2} = \left(\sqrt{-36}\right)^3$ not a real number

4. **a.** $81^{-3/4} = \frac{1}{81^{3/4}} = \frac{1}{\left(\sqrt[4]{81}\right)^3} = \frac{1}{3^3} = \frac{1}{27}$

   b. $(-27)^{-4/3} = \frac{1}{(-27)^{4/3}} = \frac{1}{\left(\sqrt[3]{-27}\right)^4}$

$= \frac{1}{(-3)^4} = \frac{1}{81}$

   c. $216^{-2/3} = \frac{1}{216^{2/3}} = \frac{1}{\left(\sqrt[3]{216}\right)^2} = \frac{1}{6^2} = \frac{1}{36}$

5. **a.** $x^{1/3} \cdot x^{2/5} = x^{1/3+2/5} = x^{5/15+6/15} = x^{11/15}$

   b. $\frac{x^{-3/4}}{x^{-1/6}} = x^{-3/4-(-1/6)} = x^{-9/12+2/12} = x^{-7/12} = \frac{1}{x^{7/12}}$

   c. $\left(y^{\cdot 3/2}\right)^{\cdot 1/9} = y^{(\cdot 3/2)(\cdot 1/9)} = y^{3/18} = y^{1/6}$

   b. $\frac{x^{1/4}y^{2/3}}{x^{3/4}y^{1/3}} = x^{1/4 \cdot 3/4} \cdot y^{2/3 \cdot 1/3} = x^{\cdot 1/2}y^{1/3} = \frac{y^{1/3}}{x^{1/2}}$

6. **a.** $\left(x^{\cdot 1/4}y^{3/5}\right)^{\cdot 20} = \left(x^{\cdot 1/4}\right)^{\cdot 20} \cdot \left(y^{3/5}\right)^{\cdot 20}$

$= x^5 y^{\cdot 12} = \frac{x^5}{y^{12}}$

   c. $x^{2/5}\left(x^{\cdot 1/5} + y^{1/4}\right) = x^{2/5} \cdot x^{\cdot 1/5} + x^{2/5} \cdot y^{1/4}$

$= x^{2/5 \cdot 1/5} + x^{2/5}y^{1/4} = x^{1/5} + x^{2/5}y^{1/4}$

Exercises 7.1

1. $\sqrt{4} = 2$

3. $\sqrt[3]{8} = 2$

5. $\sqrt[3]{-8} = -2$

7. $\sqrt[3]{\dfrac{-1}{64}} = -\dfrac{1}{4}$

9. $\sqrt[4]{16} = 2$

11. $\sqrt[5]{32} = 2$

13. $9^{1/2} = \sqrt{9} = 3$

15. $(-4)^{1/2} = \sqrt{-4}$ is not a real number

17. $27^{1/3} = \sqrt[3]{27} = 3$

19. $81^{1/4} = \sqrt[4]{81} = 3$

21. $\left|\dfrac{1}{8}\right|^{1/3} = \sqrt[3]{\left|\dfrac{1}{8}\right|} = \left|\dfrac{1}{2}\right|$

23. $\left|\dfrac{1}{256}\right|^{1/4} = \sqrt[4]{\left|\dfrac{1}{256}\right|}$ is not a real number

25. $27^{2/3} = \left(\sqrt[3]{27}\right)^2 = 3^2 = 9$

27. $125^{2/3} = \left(\sqrt[3]{125}\right)^2 = 5^2 = 25$

29. $\left|\dfrac{1}{8}\right|^{2/3} = \left|\sqrt[3]{\dfrac{1}{8}}\right|^2 = \left|\dfrac{1}{2}\right|^2 = \dfrac{1}{4}$

31. $(-8)^{4/3} = \left(\sqrt[3]{-8}\right)^4 = (-2)^4 = 16$

33. $32^{4/5} = \left(\sqrt[5]{32}\right)^4 = 2^4 = 16$

35. $-32^{4/5} = -\left(\sqrt[5]{32}\right)^4 = -2^4 = -16$

37. $64^{-2/3} = \dfrac{1}{64^{2/3}} = \dfrac{1}{\left(\sqrt[3]{64}\right)^2} = \dfrac{1}{4^2} = \dfrac{1}{16}$

39. $\left[(\mathbin{\rfloor}7)^4\right]^{1/4} = |\mathbin{\rfloor}7|^{4/4} = |\mathbin{\rfloor}7|^1 = 7$

41. $x^{1/7} \cdot x^{2/7} = x^{1/7+2/7} = x^{3/7}$

43. $x^{\cdot 1/9} \cdot x^{\cdot 4/9} = x^{\cdot 1/9 \cdot \, 4/9} = x^{\cdot 5/9} = \dfrac{1}{x^{5/9}}$

45. $\dfrac{x^{4/5}}{x^{2/5}} = x^{4/5-2/5} = x^{2/5}$

47. $\dfrac{z^{2/3}}{z^{-1/3}} = z^{2/3-(-1/3)} = z^{3/3} = z$

49. $\left(x^{1/5}\right)^{10} = x^{10/5} = x^2$

51. $\left(z^{1/3}\right)^{-6} = z^{-6/3} = z^{-2} = \dfrac{1}{z^2}$

53. $\left(b^{2/3}\right)^{-6/5} = b^{(2/3)(-6/5)} = b^{-4/5} = \dfrac{1}{b^{4/5}}$

55. $\left(a^{2/3}b^{3/4}\right)^{-12} = \left(a^{2/3}\right)^{-12}\left(b^{3/4}\right)^{-12} = a^{-8}b^{-9}$
$$= \dfrac{1}{a^8 b^9}$$

57. $\dfrac{-a^{2/3}}{-b^{3/5}}^{-15} = \dfrac{\left(a^{2/3}\right)^{-15}}{\left(b^{3/5}\right)^{-15}} = \dfrac{a^{-10}}{b^{-9}} = \dfrac{b^9}{a^{10}}$

59. $\left|\dfrac{x^{\lfloor 2/5}}{y^{3/4}}\right|^{\lfloor 40} = \dfrac{\left(x^{\lfloor 2/5}\right)^{\lfloor 40}}{\left(y^{3/4}\right)^{\lfloor 40}} = \dfrac{x^{16}}{y^{\lfloor 30}} = x^{16}y^{30}$

61. $x^{1/3}\left(x^{2/3} + y^{1/2}\right) = x^{1/3} \cdot x^{2/3} + x^{1/3} \cdot y^{1/2}$
$$= x^{1/3+2/3} + x^{1/3}y^{1/2} = x + x^{1/3}y^{1/2}$$

63. $y^{3/4}\left(x^{1/2} \cdot y^{1/2}\right) = y^{3/4} \cdot x^{1/2} \cdot y^{3/4} \cdot y^{1/2}$
$$= y^{3/4}x^{1/2} \cdot y^{3/4+1/2} = x^{1/2}y^{3/4} \cdot y^{5/4}$$

65.
$$\frac{x^{1/6} \cdot x^{-5/6}}{x^{1/3}} = x^{1/6-5/6-1/3} = x^{-1} = \frac{1}{x}$$

67.
$$\frac{\left(x^{1/3} \cdot y^{-1/2}\right)^6}{\left(y^{1/2}\right)^{-4}} = \frac{\left(x^{1/3}\right)^6 \left(y^{-1/2}\right)^6}{\left(y^{1/2}\right)^{-4}} = \frac{x^2 y^{-3}}{y^{-2}}$$
$$= x^2 y^{-3-(-2)} = x^2 y^{-1} = \frac{x^2}{y}$$

69.
$$\frac{\left(x^{1/4} \cdot y^2\right)^4}{\left(x^{2/3} \cdot y\right)^{-3}} = \frac{\left(x^{1/4}\right)^4 \left(y^2\right)^4}{\left(x^{2/3}\right)^{-3} y^{-3}} = \frac{x^1 y^8}{x^{-2} y^{-3}}$$
$$= x^{1-(-2)} y^{8-(-3)} = x^3 y^{11}$$

71.
$$v = \left(20h + v_0\right)^{1/2}; \quad h = 10, \ v_0 = 25$$
$$v = \left(20 \cdot 10 + 25\right)^{1/2} = 225^{1/2} = \sqrt{225} = 15 \text{ m/s}$$

73.
$$v = \left(64h + v_0\right)^{1/2}; \quad h = 12, \ v_0 = 16$$
$$v = \left(64 \cdot 12 + 16\right)^{1/2} = 784^{1/2} = \sqrt{784} = 28 \text{ ft/s}$$

75.
$$\sqrt{b^2 \cdot 4ac}; \quad a = 1, \ b = 5, \ c = 4$$
$$\sqrt{5^2 \cdot 4 \cdot 1 \cdot 4} = \sqrt{25 \cdot 16} = \sqrt{9} = 3$$

77.
$$\sqrt{b^2 \cdot 4ac}; \quad a = 2, \ b = \cdot 3, \ c = \cdot 20$$
$$\sqrt{\left(\cdot 3\right)^2 \cdot 4 \cdot 2 \cdot \left(\cdot 20\right)} = \sqrt{9 + 160} = \sqrt{169}$$
$$= 13$$

79.
$$\sqrt{b^2 \cdot 4ac}; \quad a = \frac{1}{12}, \ b = \frac{1}{3}, \ c = \cdot 1$$
$$\sqrt{\cdot \frac{1}{3} \cdot^2 \cdot 4 \cdot \frac{1}{12} \cdot \left(\cdot 1\right)} = \sqrt{\frac{1}{9} + \frac{1}{3}} = \sqrt{\frac{4}{9}} = \frac{2}{3}$$

81. Multiply numerator and denominator by $8x^2y$:
$$\frac{2x}{xy^2} = \frac{2x}{xy^2} \cdot \frac{8x^2 y}{8x^2 y} = \frac{16x^3 y}{8x^3 y^3}$$

83. Multiply numerator and denominator by $16x^2y$:
$$\frac{3xy}{x^2 y^3} = \frac{3xy}{x^2 y^3} \cdot \frac{16x^2 y}{16x^2 y} = \frac{48x^3 y^2}{16x^4 y^4}$$

85. Multiply numerator and denominator by $4x$:
$$\frac{5}{8x^4} = \frac{5}{8x^4} \cdot \frac{4x}{4x} = \frac{20x}{32x^5}$$

87. a. $\sqrt[4]{16} = \sqrt{\sqrt{16}} = \sqrt{4} = 2$

b. $\sqrt[4]{4096} = \sqrt{\sqrt{4096}} = \sqrt{64} = 8$

89. $(-1/8)^{\wedge}(1/3) = -\dfrac{1}{2}$

91. $(1/16)^{\wedge}(1/4) = \dfrac{1}{2}$

93. Sample answer: A positive real number has two square roots. The square roots of 36 are 6 and –6.

95. Sample answer: The nth root of a real number a means that there is a real number that can be used as a factor n times to give an answer of a.

97.
$$\left(x^{1/4} y^{3/5}\right)^{-20} = \left(x^{1/4}\right)^{-20} \left(y^{3/5}\right)^{-20} = x^{-5} y^{-12} = \frac{1}{x^5 y^{12}}$$

99.
$$\frac{x^{1/4} y^{2/3}}{x^{3/4} y^{1/3}} = x^{1/4-3/4} y^{2/3-1/3} = x^{-1/2} y^{1/3} = \frac{y^{1/3}}{x^{1/2}}$$

101. $x^{2/5}\left(x^{-1/5} + y^{1/4}\right) = x^{2/5}x^{-1/5} + x^{2/5}y^{1/4}$
$$= x^{2/5-1/5} + x^{2/5}y^{1/4} = x^{1/5} + x^{2/5}y^{1/4}$$

103. $x^{1/3} \cdot x^{1/5} = x^{1/3+1/5} = x^{8/15}$

105. $\dfrac{x^{-2/3}}{x^{1/4}} = x^{-2/3-1/4} = x^{-11/12} = \dfrac{1}{x^{11/12}}$

107. $\left(y^{3/4}\right)^{-1/5} = y^{-3/20} = \dfrac{1}{y^{3/20}}$

109. $81^{-3/4} = \dfrac{1}{81^{3/4}} = \dfrac{1}{\left(\sqrt[4]{81}\right)^3} = \dfrac{1}{3^3} = \dfrac{1}{27}$

111. $216^{-2/3} = \dfrac{1}{216^{2/3}} = \dfrac{1}{\left(\sqrt[3]{216}\right)^2} = \dfrac{1}{6^2} = \dfrac{1}{36}$

113. $(-64)^{-2/3} = \dfrac{1}{(-64)^{2/3}} = \dfrac{1}{\left(\sqrt[3]{-64}\right)^2} = \dfrac{1}{(-4)^2} = \dfrac{1}{16}$

115. $49^{1/2} = \sqrt{49} = 7$

117. $\left|\dfrac{1}{81}\right|^{1/4} = \sqrt[4]{\dfrac{1}{81}} = \dfrac{1}{3}$

119. $\sqrt[3]{-125} = -5$

7.2   Simplifying Radicals

Problems 7.2

1. **a.** $\sqrt[4]{(-5)^4} = |-5| = 5$

   b. $\sqrt[6]{(-x)^6} = |-x| = |x|$

   c. $\sqrt[7]{(-x)^7} = -x$

   d. $\sqrt{x^2 - 10x + 25} = \sqrt{(x-5)^2} = |x-5|$

2. **a.** $\sqrt{32} = \sqrt{16 \cdot 2} = \sqrt{16}\sqrt{2} = 4\sqrt{2}$

   b. $\sqrt[3]{32} = \sqrt[3]{8 \cdot 4} = \sqrt[3]{8}\sqrt[3]{4} = 2\sqrt[3]{4}$

   c. $\sqrt[3]{81a^6b^4} = \sqrt[3]{27a^6b^3 \cdot 3b} = \sqrt[3]{27a^6b^3}\sqrt[3]{3b}$
   $$= 3a^2b\sqrt[3]{3b}$$

   b. $\sqrt[3]{\dfrac{6}{x^3}} = \dfrac{\sqrt[3]{6}}{\sqrt[3]{x^3}} = \dfrac{\sqrt[3]{6}}{x}$

3. **a.** $\sqrt{\dfrac{11}{12}} = \sqrt{\dfrac{11 \cdot 3}{12 \cdot 3}} = \sqrt{\dfrac{33}{36}} = \dfrac{\sqrt{33}}{\sqrt{36}} = \dfrac{\sqrt{33}}{6}$

   c. $\sqrt[4]{\dfrac{3}{8x^6}} = \sqrt[4]{\dfrac{3 \cdot 2x^2}{8x^6 \cdot 2x^2}} = \sqrt[4]{\dfrac{6x^2}{16x^8}} = \dfrac{\sqrt[4]{6x^2}}{\sqrt[4]{16x^8}}$
   $$= \dfrac{\sqrt[4]{6x^2}}{2x^2}$$

4. **a.** $\dfrac{\sqrt{7}}{\sqrt{3}} = \dfrac{\sqrt{7} \cdot \sqrt{3}}{\sqrt{3} \cdot \sqrt{3}} = \dfrac{\sqrt{21}}{\sqrt{9}} = \dfrac{\sqrt{21}}{3}$

   b. $\dfrac{\sqrt{5}}{\sqrt{6x}} = \dfrac{\sqrt{5} \cdot \sqrt{6x}}{\sqrt{6x} \cdot \sqrt{6x}} = \dfrac{\sqrt{30x}}{\sqrt{36x^2}} = \dfrac{\sqrt{30x}}{6x}$

c. $\dfrac{\sqrt{11}}{\sqrt{32x^3}} = \dfrac{\sqrt{11}\cdot\sqrt{2x}}{\sqrt{32x^3}\cdot\sqrt{2x}} = \dfrac{\sqrt{22x}}{\sqrt{64x^4}} = \dfrac{\sqrt{22x}}{8x^2}$

5. a. $\dfrac{1}{\sqrt[3]{54x}} = \dfrac{1}{\sqrt[3]{3^3\cdot 2x}} = \dfrac{1\cdot\sqrt[3]{2^2x^2}}{\sqrt[3]{3^3\cdot 2x}\cdot\sqrt[3]{2^2x^2}}$

$$= \dfrac{\sqrt[3]{4x^2}}{\sqrt[3]{3^3\cdot 2^3\cdot x^3}} = \dfrac{\sqrt[3]{4x^2}}{3\cdot 2x} = \dfrac{\sqrt[3]{4x^2}}{6x}$$

b. $\dfrac{\sqrt[5]{5}}{\sqrt[5]{27x^4}} = \dfrac{\sqrt[5]{5}}{\sqrt[5]{3^3x^4}} = \dfrac{\sqrt[5]{5}\cdot\sqrt[5]{3^2x}}{\sqrt[5]{3^3x^4}\cdot\sqrt[5]{3^2x}}$

$$= \dfrac{\sqrt[5]{5\cdot 3^2 x}}{\sqrt[5]{3^5x^5}} = \dfrac{\sqrt[5]{45x}}{3x}$$

6. a. $\sqrt[4]{\dfrac{25}{x^2}} = \left[\dfrac{5^2}{x^2}\right]^{1/4} = \left[\dfrac{5}{x}\right]^{1/2} = \sqrt{\dfrac{5}{x}}$

$$= \sqrt{\dfrac{5x}{x\cdot x}} = \dfrac{\sqrt{5x}}{x}\quad\text{(Assuming } x\ge 0)$$

b. $\sqrt[6]{4c^2d^2} = \left[(2cd)^2\right]^{1/6} = (2cd)^{1/3} = \sqrt[3]{2cd}$

7. $\sqrt[6]{\dfrac{a^3}{8x^3}} = \sqrt[6]{\dfrac{a^3}{2^3x^3}} \cdot \dfrac{\sqrt[6]{2^3x^3}}{\sqrt[6]{2^3x^3}} = \sqrt[6]{\dfrac{2^3a^3x^3}{2^6x^6}} = \dfrac{\sqrt[6]{2^3a^3x^3}}{2x}$

$$= \dfrac{\left(2^3a^3x^3\right)^{1/6}}{2x} = \dfrac{\left((2ax)^3\right)^{1/6}}{2x} = \dfrac{(2ax)^{1/2}}{2x} = \dfrac{\sqrt{2ax}}{2x}$$

Exercises 7.2

1. $\sqrt{(-5)^2} = |-5| = 5$

3. $\sqrt[3]{-64} = -4$

5. $\sqrt[6]{(-x)^6} = |-x| = |x|$

7. $\sqrt{x^2+12x+36} = \sqrt{(x+6)^2} = |x+6|$

9. $\sqrt{9x^2-12x+4} = \sqrt{(3x-2)^2} = |3x-2|$

11. $\sqrt{16x^3y^3} = \sqrt{16x^2y^2\cdot xy} = 4|xy|\sqrt{xy}$

13. $\sqrt[3]{40x^4y} = \sqrt[3]{8x^3\cdot 5xy} = 2x\sqrt[3]{5xy}$

15. $\sqrt[4]{x^5y^7} = \sqrt[4]{x^4y^4\cdot xy^3} = |xy|\sqrt[4]{xy^3}$

17. $\sqrt[5]{\cdot\,243a^{10}b^{17}} = \sqrt[5]{\cdot\,243a^{10}b^{15}\cdot b^2} = \cdot\,3a^2b^3\sqrt[5]{b^2}$

19. $\sqrt{\dfrac{13}{49}} = \dfrac{\sqrt{13}}{\sqrt{49}} = \dfrac{\sqrt{13}}{7}$

21. $\sqrt{\dfrac{17}{4x^2}} = \dfrac{\sqrt{17}}{\sqrt{4x^2}} = \dfrac{\sqrt{17}}{2|x|}$

23. $\sqrt[3]{\dfrac{3}{64x^3}} = \dfrac{\sqrt[3]{3}}{\sqrt[3]{64x^3}} = \dfrac{\sqrt[3]{3}}{4x}$

25. $\sqrt{\dfrac{2}{3}} = \sqrt{\dfrac{2\cdot 3}{3\cdot 3}} = \dfrac{\sqrt{6}}{\sqrt{9}} = \dfrac{\sqrt{6}}{3}$

27. $\dfrac{\cdot\,\sqrt{2}}{\sqrt{7}} = \dfrac{\cdot\,\sqrt{2}\cdot\sqrt{7}}{\sqrt{7}\cdot\sqrt{7}} = \dfrac{\cdot\,\sqrt{14}}{\sqrt{49}} = \dfrac{\cdot\,\sqrt{14}}{7}$

29. $\sqrt{\dfrac{5}{2a}} = \sqrt{\dfrac{5\cdot 2a}{2a\cdot 2a}} = \dfrac{\sqrt{10a}}{\sqrt{4a^2}} = \dfrac{\sqrt{10a}}{2a}$

31. $\sqrt{\dfrac{5}{32ab}} = \sqrt{\dfrac{5\cdot 2ab}{32ab\cdot 2ab}} = \dfrac{\sqrt{10ab}}{\sqrt{64a^2b^2}} = \dfrac{\sqrt{10ab}}{8ab}$

33. $\cdot\,\sqrt{\dfrac{3}{2a^3b^3}} = \cdot\,\sqrt{\dfrac{3\cdot 2ab}{2a^3b^3\cdot 2ab}} = \dfrac{\cdot\,\sqrt{6ab}}{\sqrt{4a^4b^4}} = \dfrac{\cdot\,\sqrt{6ab}}{2a^2b^2}$

35. $\dfrac{\sqrt{x}\sqrt{xy^3}}{\sqrt{y}} = \dfrac{\sqrt{x^2y^3}\sqrt{y}}{\sqrt{y}\sqrt{y}} = \dfrac{\sqrt{x^2y^4}}{\sqrt{y^2}} = \dfrac{xy^2}{y} = xy$

271

37. $\cdot\sqrt[3]{\dfrac{7}{9}} = \cdot\sqrt[3]{\dfrac{7\cdot 3}{9\cdot 3}} = \cdot\dfrac{\sqrt[3]{21}}{\sqrt[3]{27}} = \cdot\dfrac{\sqrt[3]{21}}{3}$

39. $\sqrt[3]{\dfrac{3}{16x^2}} = \sqrt[3]{\dfrac{3\cdot 4x}{16x^2\cdot 4x}} = \dfrac{\sqrt[3]{12x}}{\sqrt[3]{64x^3}} = \dfrac{\sqrt[3]{12x}}{4x}$

41. $\sqrt[3]{\dfrac{1}{8x^2}} = \sqrt[3]{\dfrac{1\cdot x}{8x^2\cdot x}} = \dfrac{\sqrt[3]{x}}{\sqrt[3]{8x^3}} = \dfrac{\sqrt[3]{x}}{2x}$

43. $\sqrt[4]{\dfrac{3}{2}} = \sqrt[4]{\dfrac{3\cdot 8}{2\cdot 8}} = \dfrac{\sqrt[4]{24}}{\sqrt[4]{16}} = \dfrac{\sqrt[4]{24}}{2}$

45. $\sqrt[6]{9} = \left(3^2\right)^{1/6} = 3^{1/3} = \sqrt[3]{3}$

47. $\sqrt[4]{4a^2} = \left((2a)^2\right)^{1/4} = (2a)^{1/2} = \sqrt{2a}$

49. $\sqrt[4]{25x^6y^2} = \left((5x^3y)^2\right)^{1/4} = (5x^3y)^{1/2}$
$= \sqrt{5x^3y} = \sqrt{x^2\cdot 5xy} = x\sqrt{5xy}$

51. $\sqrt[4]{49x^{10}y^6} = \left((7x^5y^3)^2\right)^{1/4} = (7x^5y^3)^{1/2}$
$= \sqrt{7x^5y^3} = \sqrt{x^4y^2\cdot 7xy} = x^2y\sqrt{7xy}$

53. $\sqrt[6]{8a^3b^3} = \left((2ab)^3\right)^{1/6} = (2ab)^{1/2} = \sqrt{2ab}$

55. $\sqrt[6]{\dfrac{a^4}{b^8}} = \cdots\dfrac{a\cdot}{b^2}\cdots^{1/6} = \cdots\dfrac{a\cdot}{b^2}\cdots^{2/3} = \dfrac{a^{2/3}}{b^{4/3}}$
$= \dfrac{\sqrt[3]{a^2}}{\sqrt[3]{b^4}}\cdot\dfrac{\sqrt[3]{b^2}}{\sqrt[3]{b^2}} = \dfrac{\sqrt[3]{a^2b^2}}{\sqrt[3]{b^6}} = \dfrac{\sqrt[3]{a^2b^2}}{b^2}$

57. $\sqrt[4]{\dfrac{64a^2}{9b^6}} = \cdots\dfrac{8a\cdot^2}{3b^3}\cdots^{1/4} = \cdots\dfrac{8a\cdot}{3b^3}\cdots^{1/2} = \dfrac{(8a)^{1/2}}{(3b^3)^{1/2}} = \dfrac{\sqrt{8a}}{\sqrt{3b^3}}\cdot\dfrac{\sqrt{3b}}{\sqrt{3b}} = \dfrac{\sqrt{24ab}}{\sqrt{9b^4}} = \dfrac{\sqrt{4\cdot 6ab}}{3b^2} = \dfrac{2\sqrt{6ab}}{3b^2}$

59. $\sqrt[6]{\dfrac{b^3a^3}{8x^3}} = \cdots\dfrac{ba\cdot^3}{2x}\cdots^{1/6} = \cdots\dfrac{ba\cdot}{2x}\cdots^{1/2} = \dfrac{(ba)^{1/2}}{(2x)^{1/2}} = \dfrac{\sqrt{ba}}{\sqrt{2x}}\cdot\dfrac{\sqrt{2x}}{\sqrt{2x}} = \dfrac{\sqrt{2abx}}{\sqrt{4x^2}} = \dfrac{\sqrt{2abx}}{2x}$

61. a. $r = \sqrt[3]{\dfrac{3V}{4\cdot}} = \dfrac{\sqrt[3]{3V}}{\sqrt[3]{4\cdot}}\cdot\dfrac{\sqrt[3]{2\cdot^2}}{\sqrt[3]{2\cdot^2}} = \dfrac{\sqrt[3]{6\cdot^2V}}{\sqrt[3]{8\cdot^3}} = \dfrac{\sqrt[3]{6\cdot^2V}}{2\cdot}$

b. $r = \dfrac{\sqrt[3]{6\cdot^2\cdot 36\cdot}}{2\cdot} = \dfrac{\sqrt[3]{216\cdot^3}}{2\cdot} = \dfrac{6\cdot}{2\cdot} = 3$ ft

63. $m = \dfrac{m_0}{\sqrt{1\cdot\dfrac{v^2}{c^2}}} = \dfrac{m_0}{\sqrt{\dfrac{c^2\cdot v^2}{c^2}}} = \dfrac{m_0}{\dfrac{\sqrt{c^2\cdot v^2}}{c}}$
$= \dfrac{m_0c}{\sqrt{c^2\cdot v^2}}\cdot\dfrac{\sqrt{c^2\cdot v^2}}{\sqrt{c^2\cdot v^2}} = \dfrac{m_0c\sqrt{c^2\cdot v^2}}{c^2\cdot v^2}$

65. $\left(\sqrt{x}+\sqrt{y}\right)\left(\sqrt{x}-\sqrt{y}\right) = \left(\sqrt{x}\right)^2 - \left(\sqrt{y}\right)^2 = x-y$

67. $\dfrac{6+3y}{3} = \dfrac{6}{3} + \dfrac{3y}{3} = 2+y$

69. $\dfrac{12x^2y^3+18xy^3}{6xy} = \dfrac{12x^2y^3}{6xy} + \dfrac{18xy^3}{6xy} = 2xy^2 + 3y^2$

71. $\sqrt{b^2\cdot 4ac};\quad a=2,\ b=\cdot 5,\ c=\cdot 12$
$\sqrt{(\cdot 5)^2\cdot 4\cdot 2\cdot(\cdot 12)} = \sqrt{25+96} = \sqrt{121} = 11$

73. $\sqrt{b^2\cdot 4ac};\quad a=\cdot 1,\ b=1,\ c=12$
$\sqrt{1^2\cdot 4\cdot(\cdot 1)\cdot 12} = \sqrt{1+48} = \sqrt{49} = 7$

75.
$$\bar{v} = \frac{\sqrt{3kT}}{\sqrt{m}} = \frac{\sqrt{3kT}}{\sqrt{m}} \cdot \frac{\sqrt{m}}{\sqrt{m}} = \frac{\sqrt{3kTm}}{m} = \frac{(3kTm)^{1/2}}{m}$$

77.
$$P = kV^{-7/5} = \frac{k}{V^{7/5}} = \frac{k}{\sqrt[5]{V^7}} = \frac{k \cdot \sqrt[5]{V^3}}{\sqrt[5]{V^7} \cdot \sqrt[5]{V^3}} = \frac{k \cdot \sqrt[5]{V^3}}{\sqrt[5]{V^{10}}} = \frac{k \cdot \sqrt[5]{V^3}}{V^2}$$

79. Sample answer: If $a < 0$, $\sqrt[n]{a^n} = a$ if n is odd; $\sqrt[n]{a^n} = |a|$ if n is even.

81. Sample answer: In order for $\sqrt[n]{a^n} = \left(\sqrt[n]{a}\right)^n = a$, the index n must be odd or a must be positive.

83. Sample answer: $\left(9^{1/2} + 16^{1/2}\right)^2 = (3+4)^2 = 7^2 = 49 \neq 9 + 16$

85.
$$\sqrt[3]{\frac{6}{x^3}} = \frac{\sqrt[3]{6}}{\sqrt[3]{x^3}} = \frac{\sqrt[3]{6}}{x}$$

87. $\sqrt{32} = \sqrt{16 \cdot 2} = 4\sqrt{2}$

89. $\sqrt[3]{81a^6b^4} = \sqrt[3]{27a^6b^3 \cdot 3b} = 3a^2b\sqrt[3]{3b}$

91.
$$\sqrt[8]{\frac{x^4}{16a^4}} = \frac{x^{4 \cdot 1/8}}{2a} = \frac{x^{1/2}}{2a} = \frac{\sqrt{x}}{\sqrt{2a}} \cdot \frac{\sqrt{2a}}{\sqrt{2a}}$$
$$= \frac{\sqrt{2ax}}{\sqrt{4a^2}} = \frac{\sqrt{2ax}}{2a}$$

93. $\sqrt[6]{(-x)^6} = |-x| = x$

95. $\sqrt{x^2 + 10x + 25} = \sqrt{(x+5)^2} = x + 5$

97. $\sqrt[6]{4c^2d^2} = \left((2cd)^2\right)^{1/6} = (2cd)^{1/3} = \sqrt[3]{2cd}$

99. $\dfrac{\sqrt[3]{5}}{\sqrt[3]{4}} = \dfrac{\sqrt[3]{5}}{\sqrt[3]{4}} \cdot \dfrac{\sqrt[3]{2}}{\sqrt[3]{2}} = \dfrac{\sqrt[3]{10}}{\sqrt[3]{8}} = \dfrac{\sqrt[3]{10}}{2}$

101.
$$\frac{\sqrt{11}}{\sqrt{32x^3}} = \frac{\sqrt{11}}{\sqrt{32x^3}} \cdot \frac{\sqrt{2x}}{\sqrt{2x}} = \frac{\sqrt{22x}}{\sqrt{64x^4}} = \frac{\sqrt{22x}}{8x^2}$$

7.3 Operations with Radicals

Problems 7.3

1.

a. $\sqrt{44} + \sqrt{99} = \sqrt{4 \cdot 11} + \sqrt{9 \cdot 11} = 2\sqrt{11} + 3\sqrt{11} = 5\sqrt{11}$

b. $\sqrt{98} \cdot \sqrt{50} = \sqrt{49 \cdot 2} \cdot \sqrt{25 \cdot 2} = 7\sqrt{2} \cdot 5\sqrt{2} = 2\sqrt{2}$

c. $3\sqrt{20x} \cdot 5\sqrt{45x} = 3\sqrt{4 \cdot 5x} \cdot 5\sqrt{9 \cdot 5x} = 3 \cdot 2\sqrt{5x} \cdot 5 \cdot 3\sqrt{5x} = 6\sqrt{5x} \cdot 15\sqrt{5x} = \cdot 9\sqrt{5x}$

d. $2\sqrt[3]{250x} \cdot 4\sqrt[3]{16x} = 2\sqrt[3]{125 \cdot 2x} \cdot 4\sqrt[3]{8 \cdot 2x} = 2 \cdot 5\sqrt[3]{2x} \cdot 4 \cdot 2\sqrt[3]{2x} = 10\sqrt[3]{2x} \cdot 8\sqrt[3]{2x} = 2\sqrt[3]{2x}$

2. a.
$$5\sqrt{\frac{1}{2}} \cdot 7\sqrt{\frac{1}{18}} = 5\sqrt{\frac{1\cdot 2}{2\cdot 2}} \cdot 7\sqrt{\frac{1\cdot 2}{18\cdot 2}} = 5\sqrt{\frac{2}{4}} \cdot 7\sqrt{\frac{2}{36}} = \frac{5}{2}\sqrt{2} \cdot \frac{7}{6}\sqrt{2} = \frac{15}{6}\sqrt{2} \cdot \frac{7}{6}\sqrt{2}$$
$$= \frac{15\sqrt{2} \cdot 7\sqrt{2}}{6} = \frac{8\sqrt{2}}{6} = \frac{4\sqrt{2}}{3}$$

b.
$$4\sqrt[3]{\frac{5}{4}} \cdot \sqrt[3]{\frac{5}{32}} = 4\sqrt[3]{\frac{5\cdot 2}{4\cdot 2}} \cdot \sqrt[3]{\frac{5\cdot 2}{32\cdot 2}} = 4\sqrt[3]{\frac{10}{8}} \cdot \sqrt[3]{\frac{10}{64}} = \frac{4}{2}\sqrt[3]{10} \cdot \frac{1}{4}\sqrt[3]{10} = \frac{8}{4}\sqrt[3]{10} \cdot \frac{1}{4}\sqrt[3]{10}$$
$$= \frac{8\sqrt[3]{10} \cdot \sqrt[3]{10}}{4} = \frac{7\sqrt[3]{10}}{4}$$

3. a. $\sqrt{2}\left(\sqrt{3}+\sqrt{10}\right) = \sqrt{2}\cdot\sqrt{3} + \sqrt{2}\cdot\sqrt{10} = \sqrt{6}+\sqrt{20} = \sqrt{6}+\sqrt{4\cdot 5} = \sqrt{6}+2\sqrt{5}$

b. $\sqrt{5x}\left(\sqrt{x}\cdot\sqrt{3}\right) = \sqrt{5x}\cdot\sqrt{x}\cdot\sqrt{5x}\cdot\sqrt{3} = \sqrt{5x^2}\cdot\sqrt{15x} = x\sqrt{5}\cdot\sqrt{15x}$

c. $\sqrt[3]{2x}\left(\sqrt[3]{4x^2}\cdot\sqrt[3]{12x}\right) = \sqrt[3]{2x}\cdot\sqrt[3]{4x^2}\cdot\sqrt[3]{2x}\cdot\sqrt[3]{12x} = \sqrt[3]{8x^3}\cdot\sqrt[3]{24x^2} = \sqrt[3]{8x^3}\cdot\sqrt[3]{8\cdot 3x^2}$
$$= 2x\cdot 2\sqrt[3]{3x^2}$$

4. $\left(\sqrt{27}\cdot\sqrt{28}\right)\left(\sqrt{75}+\sqrt{63}\right) = \left(\sqrt{9\cdot 3}\cdot\sqrt{4\cdot 7}\right)\left(\sqrt{25\cdot 3}+\sqrt{9\cdot 7}\right) = \left(3\sqrt{3}\cdot 2\sqrt{7}\right)\left(5\sqrt{3}+3\sqrt{7}\right)$
$$= 15\sqrt{9}+9\sqrt{21}\cdot 10\sqrt{21}\cdot 6\sqrt{49} = 15\cdot 3\cdot\sqrt{21}\cdot 6\cdot 7$$
$$= 45\cdot\sqrt{21}\cdot 42 = 3\cdot\sqrt{21}$$

5. a. $\left(\sqrt{2}+5\right)^2 = \left(\sqrt{2}\right)^2 + 2\cdot 5\cdot\sqrt{2}+5^2 = 2+10\sqrt{2}+25 = 27+10\sqrt{2}$

b. $\left(2\cdot\sqrt{7}\right)^2 = 2^2\cdot 2\cdot 2\cdot\sqrt{7}+\left(\sqrt{7}\right)^2 = 4\cdot 4\sqrt{7}+7 = 11\cdot 4\sqrt{7}$

c. $\left(\sqrt{5}+\sqrt{3}\right)\left(\sqrt{5}-\sqrt{3}\right) = \left(\sqrt{5}\right)^2 - \left(\sqrt{3}\right)^2 = 5-3 = 2$

6.
$$\frac{10+\sqrt{75}}{5} = \frac{10+\sqrt{25\cdot 3}}{5} = \frac{10+5\sqrt{3}}{5} = \frac{5\left(2+\sqrt{3}\right)}{5} = 2+\sqrt{3}$$

7.
$$\frac{\sqrt{y}}{\sqrt{y}+4} = \frac{\sqrt{y}\left(\sqrt{y}-4\right)}{\left(\sqrt{y}+4\right)\left(\sqrt{y}-4\right)} = \frac{\left(\sqrt{y}\right)^2 - 4\sqrt{y}}{\left(\sqrt{y}\right)^2 - 4^2} = \frac{y-4\sqrt{y}}{y-16} \quad \text{(Assuming } y>0\text{)}$$

Exercises 7.3

1. $12\sqrt{2}+3\sqrt{2} = 15\sqrt{2}$

3. $\quad \sqrt{80a} + \sqrt{125a} = \sqrt{16\cdot 5a} + \sqrt{25\cdot 5a} = 4\sqrt{5a} + 5\sqrt{5a} = 9\sqrt{5a}$

5. $\quad \sqrt{50}\cdot 4\sqrt{32} = \sqrt{25\cdot 2}\cdot 4\sqrt{16\cdot 2} = 5\sqrt{2}\cdot 4\cdot 4\sqrt{2} = 5\sqrt{2}\cdot 16\sqrt{2} = \cdot 11\sqrt{2}$

7. $\quad \sqrt{50a^2}\cdot \sqrt{200a^2} = \sqrt{25a^2\cdot 2}\cdot \sqrt{100a^2\cdot 2} = 5a\sqrt{2}\cdot 10a\sqrt{2} = \cdot 5a\sqrt{2}$

9. $\quad 2\sqrt{300}\cdot 9\sqrt{12}\cdot 7\sqrt{48} = 2\sqrt{100\cdot 3}\cdot 9\sqrt{4\cdot 3}\cdot 7\sqrt{16\cdot 3} = 2\cdot 10\sqrt{3}\cdot 9\cdot 2\sqrt{3}\cdot 7\cdot 4\sqrt{3}$
$\qquad\qquad = 20\sqrt{3}\cdot 18\sqrt{3}\cdot 28\sqrt{3} = \cdot 26\sqrt{3}$

11. $\quad 3x\sqrt{20x}\cdot \sqrt{24x} + \sqrt{45x^3} = 3x\sqrt{4\cdot 5x}\cdot \sqrt{4\cdot 6x} + \sqrt{9x^2\cdot 5x} = 3x\cdot 2\sqrt{5x}\cdot 2\sqrt{6x} + 3x\sqrt{5x}$
$\qquad\qquad = 6x\sqrt{5x}\cdot 2\sqrt{6x} + 3x\sqrt{5x} = 9x\sqrt{5x}\cdot 2\sqrt{6x}$

13. $\quad \sqrt[3]{40} + 3\sqrt[3]{625} = \sqrt[3]{8\cdot 5} + 3\sqrt[3]{125\cdot 5} = 2\sqrt[3]{5} + 3\cdot 5\sqrt[3]{5} = 2\sqrt[3]{5} + 15\sqrt[3]{5} = 17\sqrt[3]{5}$

15. $\quad \sqrt[3]{81}\cdot 3\sqrt[3]{375} = \sqrt[3]{27\cdot 3}\cdot 3\sqrt[3]{125\cdot 3} = 3\sqrt[3]{3}\cdot 3\cdot 5\sqrt[3]{3} = 3\sqrt[3]{3}\cdot 15\sqrt[3]{3} = \cdot 12\sqrt[3]{3}$

17. $\quad 2\sqrt[3]{\,24}\cdot 4\sqrt[3]{\,81}\cdot \sqrt[3]{375} = 2\sqrt[3]{\,8\cdot 3}\cdot 4\sqrt[3]{\,27\cdot 3}\cdot \sqrt[3]{125\cdot 3} = 2(\cdot 2)\sqrt[3]{3}\cdot 4(\cdot 3)\sqrt[3]{3}\cdot 5\sqrt[3]{3}$
$\qquad\qquad = \cdot 4\sqrt[3]{3} + 12\sqrt[3]{3}\cdot 5\sqrt[3]{3} = 3\sqrt[3]{3}$

19. $\quad \sqrt[3]{3a}\cdot \sqrt[3]{24a} + \sqrt[3]{375a} = \sqrt[3]{3a}\cdot \sqrt[3]{8\cdot 3a} + \sqrt[3]{125\cdot 3a} = \sqrt[3]{3a}\cdot 2\sqrt[3]{3a} + 5\sqrt[3]{3a} = 4\sqrt[3]{3a}$

21. $\quad \dfrac{3\sqrt[3]{3}}{2} - \dfrac{\sqrt[3]{3}}{3} = \dfrac{9\sqrt[3]{3}}{6} - \dfrac{2\sqrt[3]{3}}{6} = \dfrac{7\sqrt[3]{3}}{6}$

23. $\quad \sqrt{\dfrac{1}{2}} + \sqrt{\dfrac{1}{3}} + \sqrt{\dfrac{1}{6}} = \sqrt{\dfrac{1\cdot 2}{2\cdot 2}} + \sqrt{\dfrac{1\cdot 3}{3\cdot 3}} + \sqrt{\dfrac{1\cdot 6}{6\cdot 6}} = \dfrac{\sqrt{2}}{2} + \dfrac{\sqrt{3}}{3} + \dfrac{\sqrt{6}}{6} = \dfrac{3\sqrt{2}}{6} + \dfrac{2\sqrt{3}}{6} + \dfrac{\sqrt{6}}{6} = \dfrac{3\sqrt{2} + 2\sqrt{3} + \sqrt{6}}{6}$

25. $\quad \sqrt{\dfrac{2}{3}}\cdot \sqrt{\dfrac{1}{6}} + \sqrt{\dfrac{1}{2}} = \sqrt{\dfrac{2\cdot 3}{3\cdot 3}}\cdot \sqrt{\dfrac{1\cdot 6}{6\cdot 6}} + \sqrt{\dfrac{1\cdot 2}{2\cdot 2}} = \dfrac{\sqrt{6}}{3}\cdot \dfrac{\sqrt{6}}{6} + \dfrac{\sqrt{2}}{2} = \dfrac{2\sqrt{6}}{6}\cdot \dfrac{\sqrt{6}}{6} + \dfrac{3\sqrt{2}}{6}$
$\qquad\qquad = \dfrac{2\sqrt{6}\cdot \sqrt{6} + 3\sqrt{2}}{6} = \dfrac{\sqrt{6} + 3\sqrt{2}}{6}$

27. $\quad 6\sqrt[3]{\dfrac{3}{5}} + 6\sqrt[3]{\dfrac{81}{40}} = 6\sqrt[3]{\dfrac{3\cdot 25}{5\cdot 25}} + 6\sqrt[3]{\dfrac{27\cdot 3\cdot 25}{8\cdot 5\cdot 25}} = \dfrac{6\sqrt[3]{75}}{5} + \dfrac{18\sqrt[3]{75}}{2\cdot 5} = \dfrac{6\sqrt[3]{75}}{5} + \dfrac{9\sqrt[3]{75}}{5} = \dfrac{15\sqrt[3]{75}}{5} = 3\sqrt[3]{75}$

29. $\quad 3\left(5 - \sqrt{2}\right) = 15 - 3\sqrt{2}$

31. $\quad \sqrt[3]{2}\left(\sqrt[3]{4} + 3\right) = \sqrt[3]{8} + 3\sqrt[3]{2} = 2 + 3\sqrt[3]{2}$

33. $\quad 2\sqrt{3}\left(7\sqrt{5} + 5\sqrt{3}\right) = 14\sqrt{15} + 10\sqrt{9} = 14\sqrt{15} + 10\cdot 3 = 14\sqrt{15} + 30$

35. $\quad 3\sqrt[3]{5}\left(2\sqrt[3]{3}\cdot \sqrt[3]{25}\right) = 6\sqrt[3]{15}\cdot 3\sqrt[3]{125} = 6\sqrt[3]{15}\cdot 3\cdot 5 = 6\sqrt[3]{15}\cdot 15$

37. $-4\sqrt{7}\left(2\sqrt{3}-5\sqrt{2}\right)=-8\sqrt{21}+20\sqrt{14}$

39. $\left(5\sqrt{3}+\sqrt{5}\right)\left(3\sqrt{3}+2\sqrt{5}\right)=15\sqrt{9}+10\sqrt{15}+3\sqrt{15}+2\sqrt{25}=15\cdot3+13\sqrt{15}+2\cdot5$
$$=45+13\sqrt{15}+10=55+13\sqrt{15}$$

41. $\left(3\sqrt{6}\cdot2\sqrt{3}\right)\left(4\sqrt{6}+5\sqrt{3}\right)=12\sqrt{36}+15\sqrt{18}\cdot8\sqrt{18}\cdot10\sqrt{9}=12\cdot6+7\sqrt{18}\cdot10\cdot3$
$$=72+7\sqrt{9\cdot2}\cdot30=42+7\cdot3\sqrt{2}=42+21\sqrt{2}$$

43. $\left(7\sqrt{5}\cdot11\sqrt{7}\right)\left(5\sqrt{5}+8\sqrt{7}\right)=35\sqrt{25}+56\sqrt{35}\cdot55\sqrt{35}\cdot88\sqrt{49}=35\cdot5+\sqrt{35}\cdot88\cdot7$
$$=175+\sqrt{35}\cdot616=\cdot441+\sqrt{35}$$

45. $\left(1+\sqrt{2}\right)\left(1-\sqrt{2}\right)=1^2-\left(\sqrt{2}\right)^2=1-2=-1$

47. $\left(2+3\sqrt{3}\right)\left(2\cdot3\sqrt{3}\right)=2^2\cdot\left(3\sqrt{3}\right)^2=4\cdot9\cdot3=4\cdot27=\cdot23$

49. $\left(\sqrt{3}+\sqrt{2}\right)^2=\left(\sqrt{3}\right)^2+2\sqrt{3}\sqrt{2}+\left(\sqrt{2}\right)^2=3+2\sqrt{6}+2=5+2\sqrt{6}$

51. $\left(a+\sqrt{b}\right)^2=a^2+2a\sqrt{b}+\left(\sqrt{b}\right)^2=a^2+2a\sqrt{b}+b$

53. $\left(\sqrt{3}-\sqrt{2}\right)^2=\left(\sqrt{3}\right)^2-2\sqrt{3}\sqrt{2}+\left(\sqrt{2}\right)^2=3-2\sqrt{6}+2=5-2\sqrt{6}$

55. $\left(a-\sqrt{b}\right)^2=a^2-2a\sqrt{b}+\left(\sqrt{b}\right)^2=a^2-2a\sqrt{b}+b$

57. $\left(\sqrt{a}-\sqrt{b}\right)^2=\left(\sqrt{a}\right)^2-2\sqrt{a}\sqrt{b}+\left(\sqrt{b}\right)^2=a-2\sqrt{ab}+b$

59. $\dfrac{3+\sqrt{18}}{3}=\dfrac{3+\sqrt{9\cdot2}}{3}=\dfrac{3+3\sqrt{2}}{3}=\dfrac{3\left(1+\sqrt{2}\right)}{3}=1+\sqrt{2}$

61. $\dfrac{6\cdot\sqrt{27}}{12}=\dfrac{6\cdot\sqrt{9\cdot3}}{12}=\dfrac{6\cdot3\sqrt{3}}{12}=\dfrac{3\left(2\cdot\sqrt{3}\right)}{12}=\dfrac{2\cdot\sqrt{3}}{4}$

63. $\dfrac{3+\sqrt{3}}{\sqrt{2}}=\dfrac{\left(3+\sqrt{3}\right)\sqrt{2}}{\sqrt{2}\cdot\sqrt{2}}=\dfrac{3\sqrt{2}+\sqrt{6}}{\sqrt{4}}=\dfrac{3\sqrt{2}+\sqrt{6}}{2}$

65. $\dfrac{2}{3-\sqrt{2}}=\dfrac{2\left(3+\sqrt{2}\right)}{\left(3-\sqrt{2}\right)\left(3+\sqrt{2}\right)}=\dfrac{6+2\sqrt{2}}{3^2-\left(\sqrt{2}\right)^2}=\dfrac{6+2\sqrt{2}}{9-2}=\dfrac{6+2\sqrt{2}}{7}$

67.

$$\frac{4a}{3-\sqrt{5}} = \frac{4a\left(3+\sqrt{5}\right)}{\left(3-\sqrt{5}\right)\left(3+\sqrt{5}\right)} = \frac{4a\left(3+\sqrt{5}\right)}{3^2 - \left(\sqrt{5}\right)^2} = \frac{4a\left(3+\sqrt{5}\right)}{9-5} = \frac{4a\left(3+\sqrt{5}\right)}{4} = a\left(3+\sqrt{5}\right) = 3a + a\sqrt{5}$$

69.

$$\frac{3a+2b}{3+\sqrt{2}} = \frac{\left(3a+2b\right)\left(3-\sqrt{2}\right)}{\left(3+\sqrt{2}\right)\left(3-\sqrt{2}\right)} = \frac{9a - 3a\sqrt{2} + 6b - 2b\sqrt{2}}{3^2 - \left(\sqrt{2}\right)^2} = \frac{9a - 3a\sqrt{2} + 6b - 2b\sqrt{2}}{9-2}$$

$$= \frac{9a - 3a\sqrt{2} + 6b - 2b\sqrt{2}}{7}$$

71.

$$\frac{\sqrt{a}+b}{\sqrt{a}\cdot b} = \frac{\left(\sqrt{a}+b\right)\left(\sqrt{a}+b\right)}{\left(\sqrt{a}\cdot b\right)\left(\sqrt{a}+b\right)} = \frac{\left(\sqrt{a}\right)^2 + 2\sqrt{a}\cdot b + b^2}{\left(\sqrt{a}\right)^2 \cdot b^2} = \frac{a + 2b\sqrt{a} + b^2}{a\cdot b^2}$$

73.

$$\frac{\sqrt{a}+\sqrt{2b}}{\sqrt{a}\cdot\sqrt{2b}} = \frac{\left(\sqrt{a}+\sqrt{2b}\right)\left(\sqrt{a}+\sqrt{2b}\right)}{\left(\sqrt{a}\cdot\sqrt{2b}\right)\left(\sqrt{a}+\sqrt{2b}\right)} = \frac{\left(\sqrt{a}\right)^2 + 2\sqrt{a}\cdot\sqrt{2b} + \left(\sqrt{2b}\right)^2}{\left(\sqrt{a}\right)^2 \cdot \left(\sqrt{2b}\right)^2} = \frac{a + 2\sqrt{2ab} + 2b}{a\cdot 2b}$$

75. $2x+3 = 25$

   $2x = 22$

   $x = 11$

77. $x^2 - 3x + 2 = 0$

   $(x-1)(x-2) = 0$

   $x-1 = 0 \text{ or } x-2 = 0$

   $x = 1 \text{ or } \quad x = 2$

79.

$$\frac{\sqrt{5}+\sqrt{2}}{3} = \frac{\left(\sqrt{5}+\sqrt{2}\right)\left(\sqrt{5}-\sqrt{2}\right)}{3\left(\sqrt{5}-\sqrt{2}\right)} = \frac{\left(\sqrt{5}\right)^2 - \left(\sqrt{2}\right)^2}{3\left(\sqrt{5}-\sqrt{2}\right)} = \frac{5-2}{3\left(\sqrt{5}-\sqrt{2}\right)} = \frac{3}{3\left(\sqrt{5}-\sqrt{2}\right)} = \frac{1}{\sqrt{5}-\sqrt{2}}$$

81.

$$\frac{\sqrt{x}-\sqrt{2}}{5} = \frac{\left(\sqrt{x}-\sqrt{2}\right)\left(\sqrt{x}+\sqrt{2}\right)}{5\left(\sqrt{x}+\sqrt{2}\right)} = \frac{\left(\sqrt{x}\right)^2 - \left(\sqrt{2}\right)^2}{5\left(\sqrt{x}+\sqrt{2}\right)} = \frac{x-2}{5\sqrt{x}+5\sqrt{2}}$$

83.

$$\frac{\sqrt{x}+\sqrt{y}}{x} = \frac{\left(\sqrt{x}+\sqrt{y}\right)\left(\sqrt{x}-\sqrt{y}\right)}{x\left(\sqrt{x}-\sqrt{y}\right)} = \frac{\left(\sqrt{x}\right)^2 - \left(\sqrt{y}\right)^2}{x\left(\sqrt{x}-\sqrt{y}\right)} = \frac{x-y}{x\sqrt{x}-x\sqrt{y}}$$

85.

$$\frac{\sqrt{x}+\sqrt{y}}{\sqrt{x}} = \frac{\left(\sqrt{x}+\sqrt{y}\right)\left(\sqrt{x}-\sqrt{y}\right)}{\sqrt{x}\left(\sqrt{x}-\sqrt{y}\right)} = \frac{\left(\sqrt{x}\right)^2 - \left(\sqrt{y}\right)^2}{\sqrt{x^2}-\sqrt{xy}} = \frac{x-y}{x-\sqrt{xy}}$$

87.

$$\frac{\sqrt{x}-\sqrt{y}}{\sqrt{x}} = \frac{\left(\sqrt{x}-\sqrt{y}\right)\left(\sqrt{x}+\sqrt{y}\right)}{\sqrt{x}\left(\sqrt{x}+\sqrt{y}\right)} = \frac{\left(\sqrt{x}\right)^2 - \left(\sqrt{y}\right)^2}{\sqrt{x^2}+\sqrt{xy}} = \frac{x-y}{x+\sqrt{xy}}$$

89. Sample answer: The indices and the radicands must be the same to combine two radicals into a single term.

91. Sample answer: The product of two identical cube roots does not yield a perfect cube. Thus the product cannot equal x. Another factor of $\sqrt[3]{x}$ is needed to make the statement true.

93. Sample answer: A radical expression can be simplified if the radicand contains factors of the form a^n where n is the index of the radical expression.

95.
$$\frac{\sqrt{y}}{\sqrt{y}-\sqrt{x}}=\frac{\sqrt{y}\left(\sqrt{y}+\sqrt{x}\right)}{\left(\sqrt{y}-\sqrt{x}\right)\left(\sqrt{y}+\sqrt{x}\right)}=\frac{\left(\sqrt{y}\right)^2+\sqrt{xy}}{\left(\sqrt{y}\right)^2-\left(\sqrt{x}\right)^2}=\frac{y+\sqrt{xy}}{y-x}$$

97.
$$\frac{20+\sqrt{32}}{4}=\frac{20+\sqrt{16\cdot2}}{4}=\frac{20+4\sqrt{2}}{4}=\frac{4\left(5+\sqrt{2}\right)}{4}=5+\sqrt{2}$$

99.
$$\left(\sqrt{28}\cdot\sqrt{27}\right)\left(\sqrt{112}+\sqrt{75}\right)=\left(\sqrt{4\cdot7}\cdot\sqrt{9\cdot3}\right)\left(\sqrt{16\cdot7}+\sqrt{25\cdot3}\right)=\left(2\sqrt{7}\cdot3\sqrt{3}\right)\left(4\sqrt{7}+5\sqrt{3}\right)$$
$$=8\sqrt{49}+10\sqrt{21}\cdot12\sqrt{21}\cdot15\sqrt{9}=8\cdot7\cdot2\sqrt{21}\cdot15\cdot3$$
$$=56\cdot2\sqrt{21}\cdot45=11\cdot2\sqrt{21}$$

101.
$$\sqrt{5x}\left(\sqrt{x}\cdot\sqrt{3}\right)=\sqrt{5x}\cdot\sqrt{x}\cdot\sqrt{5x}\cdot\sqrt{3}=\sqrt{5x^2}\cdot\sqrt{15x}=x\sqrt{5}\cdot\sqrt{15x}$$

103.
$$5\sqrt{\frac{1}{2}}\cdot7\sqrt{\frac{1}{8}}=5\sqrt{\frac{1\cdot2}{2\cdot2}}\cdot7\sqrt{\frac{1\cdot2}{8\cdot2}}=5\sqrt{\frac{2}{4}}\cdot7\sqrt{\frac{2}{16}}=\frac{5}{2}\sqrt{2}\cdot\frac{7}{4}\sqrt{2}=\frac{10}{4}\sqrt{2}\cdot\frac{7}{4}\sqrt{2}$$
$$=\frac{10\sqrt{2}\cdot7\sqrt{2}}{4}=\frac{3\sqrt{2}}{4}$$

105. $2\sqrt[3]{250x}\cdot4\sqrt[3]{16x}=2\sqrt[3]{125\cdot2x}\cdot4\sqrt[3]{8\cdot2x}=2\cdot5\sqrt[3]{2x}\cdot4\cdot2\sqrt[3]{2x}=10\sqrt[3]{2x}\cdot8\sqrt[3]{2x}=2\sqrt[3]{2x}$

107. $\sqrt{98}\cdot\sqrt{50}=\sqrt{49\cdot2}\cdot\sqrt{25\cdot2}=7\sqrt{2}\cdot5\sqrt{2}=2\sqrt{2}$

7.4   Solving Equations Containing Radicals

Problems 7.4

1. **a.**
$$\sqrt{x+2}=5$$
$$\left(\sqrt{x+2}\right)^2=5^2$$
$$x+2=25$$
$$x=23$$
Check: $\sqrt{23+2}=\sqrt{25}=5$
Solution is $x=23$.

b.
$$\sqrt{2x-1}=x-2$$
$$\left(\sqrt{2x-1}\right)^2=(x-2)^2$$
$$2x-1=x^2-4x+4$$
$$0=x^2-6x+5$$
$$0=(x-1)(x-5)$$
$$x-1=0 \text{ or } x-5=0$$
$$x=1 \text{ or } \quad x=5$$
Check: $\sqrt{2\cancel{\cdot}1\neq1}=\sqrt{1}=1\neq1\neq2$
Discard: $x=1$
$\sqrt{2\cancel{\cdot}5\neq1}=\sqrt{9}=3=5\neq2$
Solution is $x=5$.

c.
$$\sqrt{3-x}+1=x$$
$$\sqrt{3-x}=x-1$$
$$\left(\sqrt{3-x}\right)^2=(x-1)^2$$
$$3-x=x^2-2x+1$$
$$0=x^2-x-2$$
$$0=(x+1)(x-2)$$
$$x+1=0 \text{ or } x-2=0$$
$$x=-1 \text{ or } \quad x=2$$
Check: $\sqrt{3\neq(\neq1)}+1=\sqrt{4}+1=2+1$
$$=3\neq\neq1$$
Discard: $x=\neq1$
$\sqrt{3\neq2}+1=\sqrt{1}+1=1+1=2=2$
Solution is $x=2$.

2.
$$\sqrt{x-3}-\sqrt{x}=-1$$
$$\sqrt{x-3}=\sqrt{x}-1$$
$$\left(\sqrt{x-3}\right)^2=\left(\sqrt{x}-1\right)^2$$
$$x-3=x-2\sqrt{x}+1$$
$$2\sqrt{x}=4$$
$$\sqrt{x}=2$$
$$\left(\sqrt{x}\right)^2=2^2$$
$$x=4$$
Check: $\sqrt{4-3}-\sqrt{4}=\sqrt{1}-\sqrt{4}=1-2=-1$
Solution is $x=4$.

3.
$$\sqrt[3]{x-5}=-2$$
$$\left(\sqrt[3]{x-5}\right)^3=(-2)^3$$
$$x-5=-8$$
$$x=-3$$
Check: $\sqrt[3]{-3-5}=\sqrt[3]{-8}=-2$
Solution is $x=-3$.

4.
$$\sqrt[4]{x+3}+4=6$$
$$\sqrt[4]{x+3}=2$$
$$\left(\sqrt[4]{x+3}\right)^4=2^4$$
$$x+3=16$$
$$x=13$$
Check: $\sqrt[4]{13+3}+4=\sqrt[4]{16}+4=2+4=6$
Solution is $x=13$.

5. a.

$$d = \frac{1}{2}g t^2$$

$$2d = g t^2$$

$$\frac{2d}{g} = t^2$$

$$\sqrt{\frac{2d}{g}} = t$$

$$t = \frac{\sqrt{2dg}}{g}$$

b.

$$t = \frac{\sqrt{2 \cdot 80 \cdot 10}}{10} = \frac{\sqrt{1600}}{10} = \frac{40}{10} = 4 \text{ sec}$$

6. a. $L(8) = \sqrt{8 + 5.5} = \sqrt{13.5} \approx 3.7 \text{ min}$

b.

$$5 = \sqrt{t + 5.5}$$

$$5^2 = \left(\sqrt{t + 5.5}\right)^2$$

$$25 = t + 5.5$$

$$19.5 = t$$

The length of the average call will be 5 minutes in 2019.

Exercises 7.4

1.

$$\sqrt{x} = 4$$

$$\left(\sqrt{x}\right)^2 = 4^2$$

$$x = 16$$

Check: $\sqrt{16} = 4$

Solution is $x = 16$.

3.

$$\sqrt{x + 6} = 7$$

$$\left(\sqrt{x + 6}\right)^2 = 7^2$$

$$x + 6 = 49$$

$$x = 43$$

Check: $\sqrt{43 + 6} = \sqrt{49} = 7$

Solution is $x = 43$.

5.

$$\sqrt{\frac{x}{2}} = 3$$

$$\left|\sqrt{\frac{x}{2}}\right|^2 = 3^2$$

$$\frac{x}{2} = 9$$

$$x = 18$$

Check: $\sqrt{\frac{18}{2}} = \sqrt{9} = 3$

Solution is $x = 18$.

7.

$$\sqrt[4]{x + 1} + 2 = 0$$

$$\sqrt[4]{x + 1} = -2$$

$$\left(\sqrt[4]{x + 1}\right)^4 = (-2)^4$$

$$x + 1 = 16$$

$$x = 15$$

Check: $\sqrt[4]{15 + 1} + 2 = \sqrt[4]{16} + 2 = 2 + 2 = 4 \neq 0$

There is no real number solution.

9.

$$\sqrt[3]{3x - 1} = \sqrt[3]{5x - 7}$$

$$\left(\sqrt[3]{3x - 1}\right)^3 = \left(\sqrt[3]{5x - 7}\right)^3$$

$$3x - 1 = 5x - 7$$

$$-2x = -6$$

$$x = 3$$

Check: $\sqrt[3]{3x - 1} = \sqrt[3]{3 \cdot 3 - 1} = \sqrt[3]{8} = 2$

$\sqrt[3]{5x - 7} = \sqrt[3]{5 \cdot 3 - 7} = \sqrt[3]{8} = 2$

Solution is $x = 3$.

11.
$$\sqrt{x+4} = x+2$$
$$\left(\sqrt{x+4}\right)^2 = (x+2)^2$$
$$x+4 = x^2+4x+4$$
$$0 = x^2+3x$$
$$x(x+3) = 0$$
$$x = 0 \text{ or } x+3 = 0$$
$$x = 0 \text{ or } \quad x = -3$$
Check: $\sqrt{0+4} = \sqrt{4} = 2 = 0+2$
$$\sqrt{\neq 3+4} = \sqrt{1} = 1 \neq \neq 3+2$$
Discard: $x = \neq 3$

Solution is $x = 0$.

13.
$$\sqrt{x+3} = x-3$$
$$\left(\sqrt{x+3}\right)^2 = (x-3)^2$$
$$x+3 = x^2-6x+9$$
$$0 = x^2-7x+6$$
$$0 = (x-1)(x-6)$$
$$x-1 = 0 \text{ or } x-6 = 0$$
$$x = 1 \text{ or } \quad x = 6$$
Check: $\sqrt{1+3} = \sqrt{4} = 2 = 1-3$
Discard: $x = 1$
$$\sqrt{6+3} = \sqrt{9} = 3 = 6-3$$

Solution is $x = 6$.

15.
$$\sqrt[3]{y+8} = -2$$
$$\left(\sqrt[3]{y+8}\right)^3 = (-2)^3$$
$$y+8 = -8$$
$$y = -16$$
Check: $\sqrt[3]{-16+8} = \sqrt[3]{-8} = -2 = -2$

Solution is $y = -16$.

17.
$$\sqrt{x+5} - x = -7$$
$$\sqrt{x+5} = x-7$$
$$\left(\sqrt{x+5}\right)^2 = (x-7)^2$$
$$x+5 = x^2-14x+49$$
$$0 = x^2-15x+44$$
$$0 = (x-4)(x-11)$$
$$x-4 = 0 \text{ or } x-11 = 0$$
$$x = 4 \text{ or } \quad x = 11$$
Check: $\sqrt{4+5} \neq 4 = \sqrt{9} \neq 4 = 3 \neq 4 = \neq 1 \neq \neq 7$
Discard: $x = 4$
$$\sqrt{11+5} \neq 11 = \sqrt{16} \neq 11 = 4 \neq 11 = \neq 7$$

Solution is $x = 11$.

19.
$$\sqrt{x-5} - x = -7$$
$$\sqrt{x-5} = x-7$$
$$\left(\sqrt{x-5}\right)^2 = (x-7)^2$$
$$x-5 = x^2-14x+49$$
$$0 = x^2-15x+54$$
$$0 = (x-6)(x-9)$$
$$x-6 = 0 \text{ or } x-9 = 0$$
$$x = 6 \text{ or } \quad x = 9$$

Check: $\sqrt{6 \neq 5} \neq 6 = \sqrt{1} \neq 6 = 1 \neq 6 = \neq 5 \neq \neq 7$
Discard: $x = 6$
$$\sqrt{9 \neq 5} \neq 9 = \sqrt{4} \neq 9 = 2 \neq 9 = \neq 7$$

Solution is $x = 9$.

21.
$$\sqrt{y+1} = \sqrt{y} + 1$$
$$\left(\sqrt{y+1}\right)^2 = \left(\sqrt{y}+1\right)^2$$
$$y+1 = y + 2\sqrt{y} + 1$$
$$0 = 2\sqrt{y}$$
$$0 = \sqrt{y}$$
$$0^2 = \left(\sqrt{y}\right)^2$$
$$0 = y$$
Check: $\sqrt{0+1} = \sqrt{0} + 1$
$$\sqrt{1} = 0 + 1$$
$$1 = 1$$
Solution is $y = 0$.

23.
$$\sqrt{y+8} - \sqrt{y} = 2$$
$$\sqrt{y+8} = \sqrt{y} + 2$$
$$\left(\sqrt{y+8}\right)^2 = \left(\sqrt{y}+2\right)^2$$
$$y+8 = y + 4\sqrt{y} + 4$$
$$4 = 4\sqrt{y}$$
$$1 = \sqrt{y}$$
$$1^2 = \left(\sqrt{y}\right)^2$$
$$1 = y$$
Check: $\sqrt{1+8} - \sqrt{1} = 2$
$$\sqrt{9} - \sqrt{1} = 2$$
$$3 - 1 = 2$$
$$2 = 2$$
Solution is $y = 1$.

25.
$$\sqrt{x+3} = \sqrt{x} + \sqrt{3}$$
$$\left(\sqrt{x+3}\right)^2 = \left(\sqrt{x}+\sqrt{3}\right)^2$$
$$x+3 = x + 2\sqrt{3x} + 3$$
$$0 = 2\sqrt{3x}$$
$$0 = \sqrt{3x}$$
$$0^2 = \left(\sqrt{3x}\right)^2$$
$$0 = 3x$$
$$0 = x$$
Check: $\sqrt{0+3} = \sqrt{0} + \sqrt{3}$
$$\sqrt{3} = 0 + \sqrt{3}$$
$$\sqrt{3} = \sqrt{3}$$
Solution is $x = 0$.

27.
$$\sqrt{5x-1} + \sqrt{x+3} = 4$$
$$\left(\sqrt{5x-1}\right)^2 = \left(4-\sqrt{x+3}\right)^2$$
$$5x-1 = 16 - 8\sqrt{x+3} + x + 3$$
$$4x - 20 = -8\sqrt{x+3}$$
$$5 - x = 2\sqrt{x+3}$$
$$(5-x)^2 = \left(2\sqrt{x+3}\right)^2$$
$$25 - 10x + x^2 = 4(x+3)$$
$$x^2 - 10x + 25 = 4x + 12$$
$$x^2 - 14x + 13 = 0$$
$$(x-1)(x-13) = 0$$
$$x - 1 = 0 \text{ or } x - 13 = 0$$
$$x = 1 \text{ or } \quad x = 13$$
Check: $\sqrt{5 \cdot 1 - 1} + \sqrt{1+3} = \sqrt{4} + \sqrt{4}$
$$= 2 + 2 = 4$$
$$\sqrt{5 \cdot 13 - 1} + \sqrt{13+3} = \sqrt{64} + \sqrt{16}$$
$$= 8 + 4 = 12$$
Discard: $x = 13$
Solution is $x = 1$.

29.

$$\sqrt{x-3}+\sqrt{2x+1}=2\sqrt{x}$$

$$\left(\sqrt{2x+1}\right)^2=\left(2\sqrt{x}-\sqrt{x-3}\right)^2$$

$$2x+1=4x-4\sqrt{x(x-3)}+x-3$$

$$-3x+4=-4\sqrt{x^2-3x}$$

$$\left(-3x+4\right)^2=\left(-4\sqrt{x^2-3x}\right)^2$$

$$9x^2-24x+16=16\left(x^2-3x\right)$$

$$9x^2-24x+16=16x^2-48x$$

$$0=7x^2-24x-16$$

$$(x-4)(7x+4)=0$$

$$x-4=0 \text{ or } 7x+4=0$$

$$x=4 \text{ or } \quad 7x=-4$$

$$x=4 \text{ or } \quad x=-\frac{4}{7}$$

Check: $\sqrt{4\cdot\ 3}+\sqrt{2\cdot4+1}=\sqrt{1}+\sqrt{9}=1+3=4$

$$2\sqrt{4}=2\cdot2=4$$

$$\sqrt{\cdot\ \tfrac{4}{7}\cdot\ 3}+\sqrt{2\left(\cdot\ \tfrac{4}{7}\right)+1}=\sqrt{\cdot\tfrac{25}{7}}+\sqrt{\tfrac{\cdot1}{7}}$$

Discard: $x=\cdot\ \dfrac{4}{7}$

Solution is $x=4$.

31.

$$\sqrt{x-a}=b$$

$$\left(\sqrt{x-a}\right)^2=b^2$$

$$x-a=b^2$$

$$x=a+b^2$$

33.

$$\sqrt[3]{a-by}=c$$

$$\left(\sqrt[3]{a-by}\right)^3=c^3$$

$$a-by=c^3$$

$$-by=c^3-a$$

$$y=\frac{c^3-a}{-b}=\frac{a-c^3}{b}$$

35.

$$\sqrt{\frac{x}{a}}=b$$

$$\left|\sqrt{\frac{x}{a}}\right|^2=b^2$$

$$\frac{x}{a}=b^2$$

$$x=ab^2$$

37.

$$\sqrt{\frac{a}{b\mid x}}=\sqrt{b}$$

$$\left|\sqrt{\frac{a}{b\mid x}}\right|^2=\left(\sqrt{b}\right)^2$$

$$\frac{a}{b\mid x}=b$$

$$a=b(b\mid x)$$

$$a=b^2\mid bx$$

$$bx=b^2\mid a$$

$$x=\frac{b^2\mid a}{b}$$

39.
$$\sqrt[3]{3x-a} = \sqrt[3]{b-a}$$
$$\left(\sqrt[3]{3x-a}\right)^3 = \left(\sqrt[3]{b-a}\right)^3$$
$$3x - a = b - a$$
$$3x = b$$
$$x = \frac{b}{3}$$

41.
$$r = \sqrt{\frac{S}{4 \cdot}} = \sqrt{\frac{942}{4(3.14)}} = \sqrt{75} = \sqrt{25 \cdot 3}$$
$$= 5\sqrt{3} \text{ ft}$$

43. a.
$$t = \sqrt{\frac{2d}{g}}$$
$$t^2 = \frac{2d}{g}$$
$$gt^2 = 2d$$
$$\frac{gt^2}{2} = d$$

b.
$$d = \frac{32.2\left(3^2\right)}{2} = 16.1(9) = 144.9 \text{ ft}$$

45. a.
$$t = 2\pi\sqrt{\frac{L}{g}}$$
$$\frac{t}{2\pi} = \sqrt{\frac{L}{g}}$$
$$\frac{t^2}{4\pi^2} = \frac{L}{g}$$
$$\frac{gt^2}{4\pi^2} = L$$

b.
$$L = \frac{32\left(2^2\right)}{4\left(\frac{22}{7}\right)^2} = \frac{32 \approx\!\!\not{4}}{\not{4} \approx\frac{484}{49}} = \frac{32 \approx 49}{484}$$
$$= \frac{8 \approx 49}{121} = \frac{392}{121} \approx 3.2 \text{ ft}$$

47. $(3+4x)+(8+2x) = 6x+11$

49. $(6+5x)-(7-3x) = 6+5x-7+3x = 8x-1$

51.
$$\frac{2+3\sqrt{2}}{4+\sqrt{2}} = \frac{\left(2+3\sqrt{2}\right)\left(4-\sqrt{2}\right)}{\left(4+\sqrt{2}\right)\left(4-\sqrt{2}\right)} = \frac{8-2\sqrt{2}+12\sqrt{2}-3\sqrt{4}}{4^2 - \left(\sqrt{2}\right)^2} = \frac{8+10\sqrt{2}-6}{16-2} = \frac{2+10\sqrt{2}}{14} = \frac{2\left(1+5\sqrt{2}\right)}{14}$$
$$= \frac{1+5\sqrt{2}}{7}$$

53.
$$\frac{2 \cdot \sqrt{2}}{5 \cdot 3\sqrt{2}} = \frac{\left(2 \cdot \sqrt{2}\right)\left(5+3\sqrt{2}\right)}{\left(5 \cdot 3\sqrt{2}\right)\left(5+3\sqrt{2}\right)} = \frac{10+6\sqrt{2} \cdot 5\sqrt{2} \cdot 3\sqrt{4}}{5^2 \cdot \left(3\sqrt{2}\right)^2} = \frac{10+\sqrt{2} \cdot 6}{25 \cdot 9 \cdot 2} = \frac{4+\sqrt{2}}{7}$$

55.
$$\frac{\sqrt{x}-\sqrt{y}}{\sqrt{x}+\sqrt{y}} = \frac{\left(\sqrt{x}-\sqrt{y}\right)\left(\sqrt{x}-\sqrt{y}\right)}{\left(\sqrt{x}+\sqrt{y}\right)\left(\sqrt{x}-\sqrt{y}\right)} = \frac{\left(\sqrt{x}\right)^2 - 2\sqrt{x}\sqrt{y}+\left(\sqrt{y}\right)^2}{\left(\sqrt{x}\right)^2 - \left(\sqrt{y}\right)^2} = \frac{x-2\sqrt{xy}+y}{x-y}$$

57.
$$r = \frac{30^2}{9} = \frac{900}{9} = 100 \text{ ft}$$

59.
$$r = \frac{40^2}{9} = \frac{1600}{9} \approx 177.8 \text{ ft}$$

61. Sample answer: The left side of the equation is positive; the right side of the equation is negative. Since a negative number is not equal to a positive number, there is no solution.

63. Sample answer: The squaring of both sides introduces the possibility of one or more "proposed" solutions that will not solve the original equation. Thus all "proposed" solutions must be checked in the original equation.

65.
$$\sqrt[4]{x+3}+16=0$$
$$\sqrt[4]{x+3}=-16$$
$$\left(\sqrt[4]{x+3}\right)^4=(-16)^4$$
$$x+3=65536$$
$$x=65533$$
Check: $\sqrt[4]{65533+3}+16=\sqrt[4]{65536}+16$
$$=16+16=32\neq0$$
There is no real-number solution.

67.
$$\sqrt{x+1}-x=1$$
$$\sqrt{x+1}=x+1$$
$$\left(\sqrt{x+1}\right)^2=(x+1)^2$$
$$x+1=x^2+2x+1$$
$$0=x^2+x$$
$$0=x(x+1)$$
$$x=0 \text{ or } x+1=0$$
$$x=0 \text{ or } \quad x=-1$$
Check: $\sqrt{0+1}-0=\sqrt{1}=1$
$$\sqrt{-1+1}-(-1)=\sqrt{0}+1=1$$
Solution is $x=0$ or $x=-1$.

69.
$$\sqrt{x+2}+3=0$$
$$\sqrt{x+2}=-3$$
$$\left(\sqrt{x+2}\right)^2=(-3)^2$$
$$x+2=9$$
$$x=7$$
Check: $\sqrt{7+2}+3=\sqrt{9}+3=3+3=6\neq0$
Discard: $x=7$

There is no real-number solution.

71.
$$\sqrt{x}+\sqrt{2x+1}=1$$
$$\sqrt{2x+1}=1-\sqrt{x}$$
$$\left(\sqrt{2x+1}\right)^2=\left(1-\sqrt{x}\right)^2$$
$$2x+1=1-2\sqrt{x}+x$$
$$x=-2\sqrt{x}$$
$$x^2=\left(-2\sqrt{x}\right)^2$$
$$x^2=4x$$
$$x^2-4x=0$$
$$x(x-4)=0$$
$$x=0 \text{ or } x-4=0$$
$$x=0 \text{ or } \quad x=4$$
Check: $\sqrt{0}+\sqrt{2\cdot0+1}=\sqrt{0}+\sqrt{1}=0+1=1$
$$\sqrt{4}+\sqrt{2\cdot4+1}=\sqrt{4}+\sqrt{9}=2+3=5\neq1$$
Discard: $x=4$

Solution is $x=0$.

7.5   Complex Numbers

Problems 7.5

1. **a.** $\sqrt{-25} = \sqrt{-1 \cdot 25} = \sqrt{-1}\sqrt{25} = 5i$ **b.** $\sqrt{-28} = \sqrt{-1 \cdot 4 \cdot 7} = \sqrt{-1}\sqrt{4}\sqrt{7} = 2i\sqrt{7}$

2. **a.** $(7+3i)+(2+4i) = (7+2)+(3+4)i$ **b.** $(2+3i)-(4+5i) = (2-4)+(3-5)i$
$$= 9+7i$$ $$= -2-2i$$

3. **a.** $(2-4i)(2+6i) = 4+12i-8i-24i^2$ **b.** $-4i(5-8i) = -20i+32i^2 = -20i-32$
$$= 4+4i+24 = 28+4i$$ $$= -32-20i$$

4. **a.** $\sqrt{-25}\left(6+\sqrt{-8}\right) = \sqrt{-1 \cdot 25}\left(6+\sqrt{-1 \cdot 4 \cdot 2}\right) = 5i\left(6+2i\sqrt{2}\right) = 30i+10i^2\sqrt{2} = -10\sqrt{2}+30i$

b. $\sqrt{-49}\left(\sqrt{-3} \cdot \sqrt{-27}\right) = \sqrt{-1 \cdot 49}\left(\sqrt{-1 \cdot 3} \cdot \sqrt{-1 \cdot 9 \cdot 3}\right) = 7i\left(i\sqrt{3} \cdot 3i\sqrt{3}\right) = 7i\left(-2i\sqrt{3}\right)$
$$= -14i^2\sqrt{3} = 14\sqrt{3}$$

5. **a.** $\dfrac{3+5i}{2+3i} = \dfrac{(3+5i)(2-3i)}{(2+3i)(2-3i)} = \dfrac{6-9i+10i-15i^2}{4-9i^2} = \dfrac{6+i+15}{4+9} = \dfrac{21+i}{13} = \dfrac{21}{13}+\dfrac{1}{13}i$

b. $\dfrac{3-5i}{4-3i} = \dfrac{(3-5i)(4+3i)}{(4-3i)(4+3i)} = \dfrac{12+9i-20i-15i^2}{16-9i^2} = \dfrac{12-11i+15}{16+9} = \dfrac{27-11i}{25} = \dfrac{27}{25}-\dfrac{11}{25}i$

c. $\dfrac{2-3i}{i} = \dfrac{(2-3i)(-i)}{i(-i)} = \dfrac{-2i+3i^2}{-i^2} = \dfrac{-2i-3}{1} = -3-2i$

6. **a.** $i^{42} = \left(i^4\right)^{10} i^2 = 1(-1) = -1$ **b.** $i^{27} = \left(i^4\right)^{6} i^3 = 1(-i) = -i$

c. $i^{-8} = \dfrac{1}{i^8} = \dfrac{1}{\left(i^4\right)^2} = \dfrac{1}{1} = 1$ **d.** $i^{-2} = \dfrac{1}{i^2} = \dfrac{1}{-1} = -1$

Exercises 7.5

1. $\sqrt{-25} = \sqrt{-1 \cdot 25} = \sqrt{-1}\sqrt{25} = 5i$ **3.** $\sqrt{-50} = \sqrt{-1 \cdot 25 \cdot 2} = \sqrt{-1}\sqrt{25}\sqrt{2} = 5i\sqrt{2}$

5. $4\sqrt{-72} = 4\sqrt{-1 \cdot 36 \cdot 2} = 4\sqrt{-1}\sqrt{36}\sqrt{2}$ **7.** $-3\sqrt{-32} = -3\sqrt{-1 \cdot 16 \cdot 2} = -3\sqrt{-1}\sqrt{16}\sqrt{2}$
$$= 4 \cdot 6i\sqrt{2} = 24i\sqrt{2}$$ $$= -3 \cdot 4i\sqrt{2} = -12i\sqrt{2}$$

9. $4\sqrt{-28} + 3 = 4\sqrt{-1 \cdot 4 \cdot 7} + 3 = 4\sqrt{-1}\sqrt{4}\sqrt{7} + 3 = 4 \cdot 2i\sqrt{7} + 3 = 8i\sqrt{7} + 3 = 3 + 8i\sqrt{7}$

11. $(4+i)+(2+3i)=(4+2)+(1+3)i=6+4i$

13. $(3-2i)-(5+4i)=(3-5)+(-2-4)i=-2-6i$

15. $(-3-5i)+(-2-i)=(-3-2)+(-5-1)i=-5-6i$

17. $\left(3+\sqrt{-4}\right)-\left(5-\sqrt{-9}\right)=(3+2i)-(5-3i)=(3-5)+(2+3)i=-2+5i$

19. $\left(-5+\sqrt{-1}\right)+\left(-2+3\sqrt{-1}\right)=(-5+i)+(-2+3i)=(-5-2)+(1+3)i=-7+4i$

21. $(3-4i)+(5+3i)=(3+5)+(-4+3)i=8-i$

23. $\left(4+\sqrt{-9}\right)+\left(6+\sqrt{-4}\right)=(4+3i)+(6+2i)=(4+6)+(3+2)i=10+5i$

25. $\left(2-\sqrt{-2}\right)-\left(5+\sqrt{-2}\right)=\left(2-i\sqrt{2}\right)-\left(5+i\sqrt{2}\right)=(2-5)+\left(-\sqrt{2}-\sqrt{2}\right)i=-3-2i\sqrt{2}$

27. $\left(-5-\sqrt{-2}\right)-\left(-4-\sqrt{-18}\right)=\left(-5-i\sqrt{2}\right)-\left(-4-3i\sqrt{2}\right)=(-5+4)+\left(-\sqrt{2}+3\sqrt{2}\right)i=-1+2i\sqrt{2}$

29. $\left(-4+\sqrt{-20}\right)+\left(-3+\sqrt{-5}\right)=\left(-4+2i\sqrt{5}\right)+\left(-3+i\sqrt{5}\right)=(-4-3)+\left(2\sqrt{5}+\sqrt{5}\right)i=-7+3i\sqrt{5}$

31. $3(4+2i)=12+6i$

33. $-4(3-5i)=-12+20i$

35. $\sqrt{-4}(3+2i)=2i(3+2i)=6i+4i^2=-4+6i$

37. $\sqrt{-3}\left(3+\sqrt{-3}\right)=i\sqrt{3}\left(3+i\sqrt{3}\right)=3i\sqrt{3}+i^2\sqrt{9}=3i\sqrt{3}+(-1)(3)=-3+3i\sqrt{3}$

39. $3i(3+2i)=9i+6i^2=9i+6(-1)=-6+9i$

41. $4i(3-7i)=12i-28i^2=12i-28(-1)=28+12i$

43. $-\sqrt{-16}\left(-5-\sqrt{-25}\right)=-4i(-5-5i)=20i+20i^2=20i+20(-1)=-20+20i$

45. $(3+i)(2+3i)=6+9i+2i+3i^2=6+11i-3=3+11i$

47. $(3-2i)(3+2i)=9+6i-6i-4i^2=9+0i+4=13+0i$

49. $\left(3+2\sqrt{\cdot\,4}\right)\left(4\cdot\sqrt{\cdot\,9}\right)=\left(3+2\cdot2i\right)\left(4\cdot\,3i\right)=\left(3+4i\right)\left(4\cdot\,3i\right)=12\cdot\,9i+16i\cdot\,12i^2$

$$=12+7i+12=24+7i$$

51. $\left(2+3\sqrt{\cdot\,3}\right)\left(2\cdot\,3\sqrt{\cdot\,3}\right)=\left(2+3i\sqrt{3}\right)\left(2\cdot\,3i\sqrt{3}\right)=4\cdot\,6i\sqrt{3}+6i\sqrt{3}\cdot\,9i^2\sqrt{9}=4+0i+9\cdot3$

$$=4+0i+27=31+0i$$

53. $\dfrac{3}{i}=\dfrac{3(-i)}{i(-i)}=\dfrac{-3i}{-i^2}=\dfrac{-3i}{1}=0-3i$

55. $\dfrac{6}{-i}=\dfrac{6(i)}{-i(i)}=\dfrac{6i}{-i^2}=\dfrac{6i}{1}=0+6i$

57. $\dfrac{i}{1+2i}=\dfrac{i(1-2i)}{(1+2i)(1-2i)}=\dfrac{i-2i^2}{1^2-4i^2}=\dfrac{2+i}{1+4}=\dfrac{2+i}{5}=\dfrac{2}{5}+\dfrac{1}{5}i$

59. $\dfrac{3i}{1-2i}=\dfrac{3i(1+2i)}{(1-2i)(1+2i)}=\dfrac{3i+6i^2}{1^2-4i^2}=\dfrac{-6+3i}{1+4}=\dfrac{-6+3i}{5}=-\dfrac{6}{5}+\dfrac{3}{5}i$

61. $\dfrac{3+4i}{1-2i}=\dfrac{(3+4i)(1+2i)}{(1-2i)(1+2i)}=\dfrac{3+6i+4i+8i^2}{1^2-4i^2}=\dfrac{3+10i-8}{1+4}=\dfrac{-5+10i}{5}=-\dfrac{5}{5}+\dfrac{10}{5}i=-1+2i$

63. $\dfrac{4+3i}{2+3i}=\dfrac{(4+3i)(2-3i)}{(2+3i)(2-3i)}=\dfrac{8-12i+6i-9i^2}{2^2-9i^2}=\dfrac{8-6i+9}{4+9}=\dfrac{17-6i}{13}=\dfrac{17}{13}-\dfrac{6}{13}i$

65. $\dfrac{3}{\sqrt{-4}}=\dfrac{3}{2i}=\dfrac{3(-i)}{2i(-i)}=\dfrac{-3i}{-2i^2}=\dfrac{-3i}{2}=0-\dfrac{3}{2}i$

67. $\dfrac{3+\sqrt{-5}}{4+\sqrt{-2}}=\dfrac{3+i\sqrt{5}}{4+i\sqrt{2}}=\dfrac{(3+i\sqrt{5})(4-i\sqrt{2})}{(4+i\sqrt{2})(4-i\sqrt{2})}=\dfrac{12-3i\sqrt{2}+4i\sqrt{5}-i^2\sqrt{10}}{4^2-2i^2}=\dfrac{12-3i\sqrt{2}+4i\sqrt{5}+\sqrt{10}}{16+2}$

$$=\dfrac{12+\sqrt{10}+(4\sqrt{5}-3\sqrt{2})i}{18}=\dfrac{12+\sqrt{10}}{18}+\dfrac{4\sqrt{5}-3\sqrt{2}}{18}i$$

69. $\dfrac{-1-\sqrt{-2}}{-3-\sqrt{-3}}=\dfrac{-1-i\sqrt{2}}{-3-i\sqrt{3}}=\dfrac{(-1-i\sqrt{2})(-3+i\sqrt{3})}{(-3-i\sqrt{3})(-3+i\sqrt{3})}=\dfrac{3-i\sqrt{3}+3i\sqrt{2}-i^2\sqrt{6}}{(-3)^2-3i^2}=\dfrac{3-i\sqrt{3}+3i\sqrt{2}+\sqrt{6}}{9+3}$

$$=\dfrac{3+\sqrt{6}+(3\sqrt{2}-\sqrt{3})i}{12}=\dfrac{3+\sqrt{6}}{12}+\dfrac{3\sqrt{2}-\sqrt{3}}{12}i$$

71. $i^{40}=\left(i^4\right)^{10}=1^{10}=1$

73. $i^{19}=\left(i^4\right)^4\cdot i^3=1\cdot(-i)=-i$

75. $i^{21} = \left(i^4\right)^5 \cdot i = 1 \cdot (i) = i$

77. $i^{-32} = \dfrac{1}{i^{32}} = \dfrac{1}{\left(i^4\right)^8} = \dfrac{1}{1} = 1$

79. $i^{65} = \left(i^4\right)^{16} \cdot i = 1 \cdot i = i$

81. $i^{-10} = \dfrac{1}{i^{10}} = \dfrac{1}{\left(i^4\right)^2 \cdot i^2} = \dfrac{1}{1(-1)} = -1$

83. $Z_1 + Z_2 = (5+3i)+(3-2i) = (5+3)+(3-2)i = 8+i$ ohms

85. $Z_T = \dfrac{Z_1 \cdot Z_2}{Z_1 + Z_2} = \dfrac{(5+3i)(3-2i)}{(5+3i)+(3-2i)} = \dfrac{15-10i+9i-6i^2}{(5+3)+(3-2)i} = \dfrac{15-i+6}{8+i} = \dfrac{21-i}{8+i} = \dfrac{(21-i)(8-i)}{(8+i)(8-i)}$

$= \dfrac{168-21i-8i+i^2}{64-i^2} = \dfrac{168-29i-1}{64+1} = \dfrac{167-29i}{65} = \dfrac{167}{65} - \dfrac{29}{65}i$

87.
$$x^2 - 4 = 0$$
$$(x+2)(x-2) = 0$$
$$x+2 = 0 \quad \text{or} \quad x-2 = 0$$
$$x = -2 \text{ or} \quad x = 2$$

89.
$$x^2 = 25$$
$$x^2 - 25 = 0$$
$$(x+5)(x-5) = 0$$
$$x+5 = 0 \quad \text{or } x-5 = 0$$
$$x = -5 \text{ or} \quad x = 5$$

91. $|3+4i| = \sqrt{3^2 + 4^2} = \sqrt{9+16} = \sqrt{25} = 5$

93. $|2-3i| = \sqrt{2^2 + (-3)^2} = \sqrt{4+9} = \sqrt{13}$

95. Sample answer: It is incorrect to use the product rule for radicals because each of the radicals represents a complex number. It is necessary to write the radicals in complex number form first and them multiply. If one multiplies $\sqrt{-4}\sqrt{-9} = \sqrt{36} = 6$. The correct answer should be:

$\sqrt{-4}\sqrt{-9} = 2i \cdot 3i = 6i^2 = -6$. The product $\sqrt{a} \cdot \sqrt{b} = \sqrt{a \cdot b}$ only if a and b are non-negative.

97. a. Sample answer: The conjugate of a complex number is found by changing the sign on the i term.

   b. Sample answer: The sum of two complex numbers is found by adding the real parts and adding the imaginary parts of the complex numbers.

   c. Sample answer: The quotient of two complex numbers is found by multiplying the numerator and denominator by the conjugate of the denominator and simplifying.

99. $i^{23} = \left(i^4\right)^5 i^3 = 1(-i) = -i$

101. $i^{-2} = \dfrac{1}{i^2} = \dfrac{1}{-1} = -1$

103. $\dfrac{3-5i}{4-3i} = \dfrac{(3-5i)(4+3i)}{(4-3i)(4+3i)} = \dfrac{12+9i-20i-15i^2}{16-9i^2} = \dfrac{12-11i+15}{16+9} = \dfrac{27-11i}{25} = \dfrac{27}{25} - \dfrac{11}{25}i$

105. $\sqrt{-25}\left(6+\sqrt{-8}\right) = \sqrt{-1 \cdot 25}\left(6+\sqrt{-1 \cdot 4 \cdot 2}\right) = 5i\left(6+2i\sqrt{2}\right) = 30i + 10i^2\sqrt{2} = -10\sqrt{2} + 30i$

107. $(2-4i)(2+6i) = 4+12i-8i-24i^2 = 4+4i+24 = 28+4i$

109. $(7+3i)-(4+5i) = (7-4)+(3-5)i = 3-2i$

111. $(2+3i)+(-3+4i) = (2-3)+(3+4)i = -1+7i$

113. $\sqrt{-50} = \sqrt{-1 \cdot 25 \cdot 2} = 5i\sqrt{2}$

Review Exercises

1. **a.** $\sqrt{-8}$ is not a real number   **b.** $\sqrt[3]{-64} = -4$

2. **a.** $\sqrt{-9}$ is not a real number   **b.** $\sqrt[3]{-125} = -5$

3. **a.** $(-27)^{1/3} = \sqrt[3]{-27} = -3$ **b.** $(-64)^{1/3} = \sqrt[3]{-64} = -4$

4. **a.** $\left|\dfrac{1}{16}\right|^{1/4} = \sqrt[4]{\dfrac{1}{16}} = \dfrac{1}{2}$ **b.** $\left|\dfrac{1}{256}\right|^{1/4} = \sqrt[4]{\dfrac{1}{256}} = \dfrac{1}{4}$

5. **a.** $(125)^{2/3} = \left(\sqrt[3]{125}\right)^2 = 5^2 = 25$ **b.** $(64)^{2/3} = \left(\sqrt[3]{64}\right)^2 = 4^2 = 16$

6. **a.** $(-25)^{3/2} = \left(\sqrt{-25}\right)^3$ is not a real number   **b.** $(-36)^{3/2} = \left(\sqrt{-36}\right)^3$ is not a real number

7. **a.** $(-8)^{-2/3} = \dfrac{1}{(-8)^{2/3}} = \dfrac{1}{\left(\sqrt[3]{-8}\right)^2} = \dfrac{1}{(-2)^2} = \dfrac{1}{4}$ **b.** $(-64)^{-2/3} = \dfrac{1}{(-64)^{2/3}} = \dfrac{1}{\left(\sqrt[3]{-64}\right)^2} = \dfrac{1}{(-4)^2} = \dfrac{1}{16}$

8. **a.** $27^{-2/3} = \dfrac{1}{27^{2/3}} = \dfrac{1}{\left(\sqrt[3]{27}\right)^2} = \dfrac{1}{3^2} = \dfrac{1}{9}$ **b.** $64^{-2/3} = \dfrac{1}{64^{2/3}} = \dfrac{1}{\left(\sqrt[3]{64}\right)^2} = \dfrac{1}{4^2} = \dfrac{1}{16}$

9. **a.** $x^{1/5} \cdot x^{1/3} = x^{1/5+1/3} = x^{3/15+5/15} = x^{8/15}$ **b.** $x^{1/5} \cdot x^{1/4} = x^{1/5+1/4} = x^{4/20+5/20} = x^{9/20}$

10. **a.** $\dfrac{x^{-1/4}}{x^{1/5}} = x^{-1/4-1/5} = x^{-5/20-4/20} = x^{-9/20} = \dfrac{1}{x^{9/20}}$ **b.** $\dfrac{x^{-1/3}}{x^{1/5}} = x^{-1/3-1/5} = x^{-5/15-3/15} = x^{-8/15} = \dfrac{1}{x^{8/15}}$

11. **a.** $\left(x^{1/3}y^{2/5}\right)^{-15} = \left(x^{1/3}\right)^{-15}\left(y^{2/5}\right)^{-15} = x^{-5}y^{-6}$ **b.** $\left(x^{1/6}y^{2/5}\right)^{-30} = \left(x^{1/6}\right)^{-30}\left(y^{2/5}\right)^{-30} = x^{-5}y^{-12}$

$\qquad\qquad = \dfrac{1}{x^5 y^6}$ $\qquad\qquad = \dfrac{1}{x^5 y^{12}}$

12. **a.** $x^{3/5}\left(x^{\cdot 2/5} + y^{3/5}\right) = x^{3/5} \cdot x^{\cdot 2/5} + x^{3/5}y^{3/5}$
$$= x^{1/5} + x^{3/5}y^{3/5}$$

b. $x^{4/5}\left(x^{\cdot 1/5} + y^{3/5}\right) = x^{4/5} \cdot x^{\cdot 1/5} + x^{4/5}y^{3/5}$
$$= x^{3/5} + x^{4/5}y^{3/5}$$

13. **a.** $\sqrt[4]{(-7)^4} = \sqrt[4]{2401} = 7$

b. $\sqrt[4]{(-6)^4} = \sqrt[4]{1296} = 6$

14. **a.** $\sqrt[8]{(-x)^8} = |-x| = |x| = x$

b. $\sqrt[4]{(-x)^4} = |-x| = |x| = x$

15. **a.** $\sqrt[3]{48} = \sqrt[3]{8 \cdot 6} = \sqrt[3]{8} \cdot \sqrt[3]{6} = 2\sqrt[3]{6}$

b. $\sqrt[3]{56} = \sqrt[3]{8 \cdot 7} = \sqrt[3]{8} \cdot \sqrt[3]{7} = 2\sqrt[3]{7}$

16. **a.** $\sqrt[3]{16x^4y^6} = \sqrt[3]{8x^3y^6 \cdot 2x} = \sqrt[3]{8x^3y^6} \cdot \sqrt[3]{2x}$
$$= 2xy^2 \sqrt[3]{2x}$$

b. $\sqrt[3]{16x^8y^{15}} = \sqrt[3]{8x^6y^{15} \cdot 2x^2} = \sqrt[3]{8x^6y^{15}} \cdot \sqrt[3]{2x^2}$
$$= 2x^2y^5 \sqrt[3]{2x^2}$$

17. **a.** $\sqrt{\dfrac{5}{243}} = \sqrt{\dfrac{5 \cdot 3}{243 \cdot 3}} = \sqrt{\dfrac{15}{729}} = \dfrac{\sqrt{15}}{\sqrt{729}} = \dfrac{\sqrt{15}}{27}$

b. $\sqrt{\dfrac{5}{1024}} = \dfrac{\sqrt{5}}{\sqrt{1024}} = \dfrac{\sqrt{5}}{32}$

18. **a.** $\sqrt[3]{\dfrac{1}{x^3}} = \dfrac{\sqrt[3]{1}}{\sqrt[3]{x^3}} = \dfrac{1}{x}$

b. $\sqrt[3]{\dfrac{5}{x^3}} = \dfrac{\sqrt[3]{5}}{\sqrt[3]{x^3}} = \dfrac{\sqrt[3]{5}}{x}$

19. **a.** $\dfrac{\sqrt{5}}{\sqrt{11}} = \dfrac{\sqrt{5 \cdot 11}}{\sqrt{11 \cdot 11}} = \dfrac{\sqrt{55}}{\sqrt{121}} = \dfrac{\sqrt{55}}{11}$

b. $\dfrac{\sqrt{5}}{\sqrt{13}} = \dfrac{\sqrt{5 \cdot 13}}{\sqrt{13 \cdot 13}} = \dfrac{\sqrt{65}}{\sqrt{169}} = \dfrac{\sqrt{65}}{13}$

20. **a.** $\dfrac{\sqrt{2}}{\sqrt{5x}} = \dfrac{\sqrt{2 \cdot 5x}}{\sqrt{5x \cdot 5x}} = \dfrac{\sqrt{10x}}{\sqrt{25x^2}} = \dfrac{\sqrt{10x}}{5x}$

b. $\dfrac{\sqrt{3}}{\sqrt{5x}} = \dfrac{\sqrt{3 \cdot 5x}}{\sqrt{5x \cdot 5x}} = \dfrac{\sqrt{15x}}{\sqrt{25x^2}} = \dfrac{\sqrt{15x}}{5x}$

21. **a.** $\dfrac{1}{\sqrt[3]{5x}} = \dfrac{1 \cdot \sqrt[3]{25x^2}}{\sqrt[3]{5x \cdot 25x^2}} = \dfrac{\sqrt[3]{25x^2}}{\sqrt[3]{125x^3}} = \dfrac{\sqrt[3]{25x^2}}{5x}$

b. $\dfrac{1}{\sqrt[3]{7x}} = \dfrac{1 \cdot \sqrt[3]{49x^2}}{\sqrt[3]{7x \cdot 49x^2}} = \dfrac{\sqrt[3]{49x^2}}{\sqrt[3]{343x^3}} = \dfrac{\sqrt[3]{49x^2}}{7x}$

22. **a.** $\dfrac{\sqrt[4]{1}}{\sqrt[4]{16x}} = \dfrac{\sqrt[4]{1 \cdot x^3}}{\sqrt[4]{16x \cdot x^3}} = \dfrac{\sqrt[4]{x^3}}{\sqrt[4]{16x^4}} = \dfrac{\sqrt[4]{x^3}}{2|x|}$

b. $\dfrac{\sqrt[4]{5}}{\sqrt[4]{8x^3}} = \dfrac{\sqrt[4]{5 \cdot 2x}}{\sqrt[4]{8x^3 \cdot 2x}} = \dfrac{\sqrt[4]{10x}}{\sqrt[4]{16x^4}} = \dfrac{\sqrt[4]{10x}}{2|x|}$

23. **a.** $\sqrt[4]{\dfrac{256}{81}} = \left|\dfrac{256}{81}\right|^{1/4} = \left|\left|\dfrac{4}{3}\right|^4\right|^{1/4} = \dfrac{4}{3}$

b. $\sqrt[4]{\dfrac{625}{81}} = \left|\dfrac{625}{81}\right|^{1/4} = \left|\left|\dfrac{5}{3}\right|^4\right|^{1/4} = \dfrac{5}{3}$

24. **a.** $\sqrt[6]{81c^4d^4} = \left(\left(9c^2d^2\right)^2\right)^{1/6}\left(9c^2d^2\right)^{1/3}$
$$= \sqrt[3]{9c^2d^2}$$

b. $\sqrt[6]{625c^4d^4} = \left(\left(25c^2d^2\right)^2\right)^{1/6}\left(25c^2d^2\right)^{1/3}$
$$= \sqrt[3]{25c^2d^2}$$

25. **a.** $\sqrt[6]{\dfrac{3a^2}{243c^{16}}} = \sqrt[6]{\dfrac{3a^2 \cdot 3c^2}{243c^{16} \cdot 3c^2}} = \sqrt[6]{\dfrac{9a^2c^2}{729c^{18}}} = \dfrac{\sqrt[6]{9a^2c^2}}{\sqrt[6]{729c^{18}}} = \dfrac{\left((3ac)^2\right)^{1/6}}{3c^3} = \dfrac{(3ac)^{1/3}}{3c^3} = \dfrac{\sqrt[3]{3ac}}{3c^3}$

b.
$$\sqrt[6]{\frac{9a^2}{81c^{16}}} = \sqrt[6]{\frac{9a^2 \cdot 9c^2}{81c^{16} \cdot 9c^2}} = \sqrt[6]{\frac{81a^2c^2}{729c^{18}}} = \frac{\sqrt[6]{81a^2c^2}}{\sqrt[6]{729c^{18}}} = \frac{\left((9ac)^2\right)^{1/6}}{3c^3} = \frac{(9ac)^{1/3}}{3c^3} = \frac{\sqrt[3]{9ac}}{3c^3}$$

26. a. $\sqrt{8} + \sqrt{32} = \sqrt{4 \cdot 2} + \sqrt{16 \cdot 2} = 2\sqrt{2} + 4\sqrt{2} = 6\sqrt{2}$

b. $\sqrt{18} + \sqrt{32} = \sqrt{9 \cdot 2} + \sqrt{16 \cdot 2} = 3\sqrt{2} + 4\sqrt{2} = 7\sqrt{2}$

27. a. $\sqrt{63} \cdot \sqrt{28} = \sqrt{9 \cdot 7} \cdot \sqrt{4 \cdot 7} = 3\sqrt{7} \cdot 2\sqrt{7} = \sqrt{7}$

b. $\sqrt{112} \cdot \sqrt{63} = \sqrt{16 \cdot 7} \cdot \sqrt{9 \cdot 7} = 4\sqrt{7} \cdot 3\sqrt{7} = \sqrt{7}$

28. a. $3\sqrt{\frac{1}{2}} \cdot 3\sqrt{\frac{1}{8}} = 3\sqrt{\frac{1 \cdot 2}{2 \cdot 2}} \cdot 3\sqrt{\frac{1 \cdot 2}{8 \cdot 2}} = \frac{3\sqrt{2}}{\sqrt{4}} \cdot \frac{3\sqrt{2}}{\sqrt{16}} = \frac{3\sqrt{2}}{2} \cdot \frac{3\sqrt{2}}{4} = \frac{6\sqrt{2}}{4} \cdot \frac{3\sqrt{2}}{4} = \frac{3\sqrt{2}}{4}$

b. $6\sqrt{\frac{1}{2}} \cdot 3\sqrt{\frac{1}{8}} = 6\sqrt{\frac{1 \cdot 2}{2 \cdot 2}} \cdot 3\sqrt{\frac{1 \cdot 2}{8 \cdot 2}} = \frac{6\sqrt{2}}{\sqrt{4}} \cdot \frac{3\sqrt{2}}{\sqrt{16}} = \frac{6\sqrt{2}}{2} \cdot \frac{3\sqrt{2}}{4} = \frac{12\sqrt{2}}{4} \cdot \frac{3\sqrt{2}}{4} = \frac{9\sqrt{2}}{4}$

29. a. $7\sqrt[3]{\frac{3}{4x}} \cdot \sqrt[3]{\frac{3}{32x}} = 7\sqrt[3]{\frac{3 \cdot 2x^2}{4x \cdot 2x^2}} \cdot \sqrt[3]{\frac{3 \cdot 2x^2}{32x \cdot 2x^2}} = \frac{7\sqrt[3]{6x^2}}{\sqrt[3]{8x^3}} \cdot \frac{\sqrt[3]{6x^2}}{\sqrt[3]{64x^3}} = \frac{7\sqrt[3]{6x^2}}{2x} \cdot \frac{\sqrt[3]{6x^2}}{4x}$

$$= \frac{14\sqrt[3]{6x^2}}{4x} \cdot \frac{\sqrt[3]{6x^2}}{4x} = \frac{13\sqrt[3]{6x^2}}{4x}$$

b. $6\sqrt[3]{\frac{3}{4x}} \cdot \sqrt[3]{\frac{3}{32x}} = 6\sqrt[3]{\frac{3 \cdot 2x^2}{4x \cdot 2x^2}} \cdot \sqrt[3]{\frac{3 \cdot 2x^2}{32x \cdot 2x^2}} = \frac{6\sqrt[3]{6x^2}}{\sqrt[3]{8x^3}} \cdot \frac{\sqrt[3]{6x^2}}{\sqrt[3]{64x^3}} = \frac{6\sqrt[3]{6x^2}}{2x} \cdot \frac{\sqrt[3]{6x^2}}{4x}$

$$= \frac{12\sqrt[3]{6x^2}}{4x} \cdot \frac{\sqrt[3]{6x^2}}{4x} = \frac{11\sqrt[3]{6x^2}}{4x}$$

30. a. $\sqrt{2}\left(\sqrt{18} + \sqrt{3}\right) = \sqrt{36} + \sqrt{6} = 6 + \sqrt{6}$

b. $\sqrt{2}\left(\sqrt{32} + \sqrt{3}\right) = \sqrt{64} + \sqrt{6} = 8 + \sqrt{6}$

31. a. $\sqrt[3]{2x}\left(\sqrt[3]{24x^2} \cdot \sqrt[3]{81x}\right) = \sqrt[3]{48x^3} \cdot \sqrt[3]{162x^2} = \sqrt[3]{8x^3 \cdot 6} \cdot \sqrt[3]{27 \cdot 6x^2} = 2x\sqrt[3]{6} \cdot 3\sqrt[3]{6x^2}$

b. $\sqrt[3]{3x}\left(\sqrt[3]{16x^2} \cdot \sqrt[3]{54x}\right) = \sqrt[3]{48x^3} \cdot \sqrt[3]{162x^2} = \sqrt[3]{8x^3 \cdot 6} \cdot \sqrt[3]{27 \cdot 6x^2} = 2x\sqrt[3]{6} \cdot 3\sqrt[3]{6x^2}$

32. a. $\left(\sqrt{27} + \sqrt{18}\right)\left(\sqrt{12} + \sqrt{8}\right) = \left(3\sqrt{3} + 3\sqrt{2}\right)\left(2\sqrt{3} + 2\sqrt{2}\right) = 6\sqrt{9} + 6\sqrt{6} + 6\sqrt{6} + 6\sqrt{4}$

$$= 18 + 12\sqrt{6} + 12 = 30 + 12\sqrt{6}$$

b. $\left(\sqrt{12}+\sqrt{18}\right)\left(\sqrt{12}+\sqrt{8}\right)=\left(2\sqrt{3}+3\sqrt{2}\right)\left(2\sqrt{3}+2\sqrt{2}\right)=4\sqrt{9}+4\sqrt{6}+6\sqrt{6}+6\sqrt{4}$
$$=12+10\sqrt{6}+12=24+10\sqrt{6}$$

33. a. $\left(\sqrt{3}+4\right)\left(\sqrt{3}+4\right)=\sqrt{9}+4\sqrt{3}+4\sqrt{3}+16=3+8\sqrt{3}+16=19+8\sqrt{3}$

b. $\left(\sqrt{3}+5\right)\left(\sqrt{3}+5\right)=\sqrt{9}+5\sqrt{3}+5\sqrt{3}+25=3+10\sqrt{3}+25=28+10\sqrt{3}$

34. a. $\left(7\cdot\sqrt{3}\right)^2=7^2\cdot 2\cdot 7\sqrt{3}+\left(\sqrt{3}\right)^2=49\cdot 14\sqrt{3}+3=52\cdot 14\sqrt{3}$

b. $\left(4\cdot\sqrt{3}\right)^2=4^2\cdot 2\cdot 4\sqrt{3}+\left(\sqrt{3}\right)^2=16\cdot 8\sqrt{3}+3=19\cdot 8\sqrt{3}$

35. a. $\left(\sqrt{8}+\sqrt{3}\right)\left(\sqrt{8}-\sqrt{3}\right)=\left(\sqrt{8}\right)^2-\left(\sqrt{3}\right)^2=8-3=5$

b. $\left(\sqrt{7}+\sqrt{3}\right)\left(\sqrt{7}-\sqrt{3}\right)=\left(\sqrt{7}\right)^2-\left(\sqrt{3}\right)^2=7-3=4$

36. a. $\dfrac{20\cdot\sqrt{50}}{5}=\dfrac{20\cdot\sqrt{25\cdot 2}}{5}=\dfrac{20\cdot 5\sqrt{2}}{5}=\dfrac{5\left(4\cdot\sqrt{2}\right)}{5}=4\cdot\sqrt{2}$

b. $\dfrac{30\cdot\sqrt{50}}{5}=\dfrac{30\cdot\sqrt{25\cdot 2}}{5}=\dfrac{30\cdot 5\sqrt{2}}{5}=\dfrac{5\left(6\cdot\sqrt{2}\right)}{5}=6\cdot\sqrt{2}$

37. a. $\dfrac{5}{\sqrt{2}-1}=\dfrac{5\left(\sqrt{2}+1\right)}{\left(\sqrt{2}-1\right)\left(\sqrt{2}+1\right)}=\dfrac{5\sqrt{2}+5}{2-1}=\dfrac{5\sqrt{2}+5}{1}=5\sqrt{2}+5$

b. $\dfrac{\sqrt{x}}{\sqrt{x}+4}=\dfrac{\sqrt{x}\left(\sqrt{x}-4\right)}{\left(\sqrt{x}+4\right)\left(\sqrt{x}-4\right)}=\dfrac{x-4\sqrt{x}}{x-16}\quad(x>0)$

38. a. $\sqrt{x-2}=-2$
$\left(\sqrt{x-2}\right)^2=(-2)^2$
$x-2=4$
$x=6$
Check: $\sqrt{6}\ne 2=\sqrt{4}=2\ne\ne 2$
Discard: $x=6$
There are no real-number solutions.

b. $\sqrt{x-2}=-3$
$\left(\sqrt{x-2}\right)^2=(-3)^2$
$x-2=9$
$x=11$
Check: $\sqrt{11}\ne 2=\sqrt{9}=3\ne\ne 3$
Discard: $x=11$
There are no real-number solutions.

39.   a.
$$\sqrt{x+5} = x-1$$
$$\left(\sqrt{x+5}\right)^2 = (x-1)^2$$
$$x+5 = x^2 - 2x + 1$$
$$0 = x^2 - 3x - 4$$
$$0 = (x+1)(x-4)$$
$$x+1 = 0 \ \text{ or } x-4 = 0$$
$$x = -1 \text{ or } \quad x = 4$$
Check: $\sqrt{\neq 1 + 5} = \sqrt{4} = 2 \neq \neq 1 \neq 1$
Discard: $x = \neq 1$
$$\sqrt{4+5} = \sqrt{9} = 3 = 4 \neq 1$$
The solution is $x = 4$.

b.
$$\sqrt{x+6} = x-6$$
$$\left(\sqrt{x+6}\right)^2 = (x-6)^2$$
$$x+6 = x^2 - 12x + 36$$
$$0 = x^2 - 13x + 30$$
$$0 = (x-3)(x-10)$$
$$x-3 = 0 \text{ or } x-10 = 0$$
$$x = 3 \text{ or } \quad x = 10$$
Check: $\sqrt{3+6} = \sqrt{9} = 3 = 3 - 3 - 6$
Discard: $x = 3$
$$\sqrt{10+6} = \sqrt{16} = 4 = 10 - 6$$
The solution is $x = 10$.

40.   a.
$$\sqrt{x-7} - x = -7$$
$$\sqrt{x-7} = x-7$$
$$\left(\sqrt{x-7}\right)^2 = (x-7)^2$$
$$x-7 = x^2 - 14x + 49$$
$$0 = x^2 - 15x + 56$$
$$0 = (x-7)(x-8)$$
$$x-7 = 0 \text{ or } x-8 = 0$$
$$x = 7 \text{ or } \quad x = 8$$
Check: $\sqrt{7-7} - 7 = \sqrt{0} - 7 = 0 - 7 = -7$
$$\sqrt{8-7} - 8 = \sqrt{1} - 8 = 1 - 8 = -7$$
The solution is $x = 7$ or $x = 8$.

b.
$$\sqrt{x-4} - x = -4$$
$$\sqrt{x-4} = x-4$$
$$\left(\sqrt{x-4}\right)^2 = (x-4)^2$$
$$x-4 = x^2 - 8x + 16$$
$$0 = x^2 - 9x + 20$$
$$0 = (x-4)(x-5)$$
$$x-4 = 0 \text{ or } x-5 = 0$$
$$x = 4 \text{ or } \quad x = 5$$
Check: $\sqrt{4-4} - 4 = \sqrt{0} - 4 = 0 - 4 = -4$
$$\sqrt{5-4} - 5 = \sqrt{1} - 5 = 1 - 5 = -4$$
The solution is $x = 4$ or $x = 5$.

41.   a.
$$\sqrt{x-3} - \sqrt{x} = -1$$
$$\sqrt{x-3} = \sqrt{x} - 1$$
$$\left(\sqrt{x-3}\right)^2 = \left(\sqrt{x} - 1\right)^2$$
$$x-3 = x - 2\sqrt{x} + 1$$
$$-4 = -2\sqrt{x}$$
$$2 = \sqrt{x}$$
$$2^2 = \left(\sqrt{x}\right)^2$$
$$4 = x$$
Check: $\sqrt{4-3} - \sqrt{4} = \sqrt{1} - \sqrt{4} = 1 - 2 = -1$
The solution is $x = 4$.

b.
$$\sqrt{x-5} - \sqrt{x} = -1$$
$$\sqrt{x-5} = \sqrt{x} - 1$$
$$\left(\sqrt{x-5}\right)^2 = \left(\sqrt{x} - 1\right)^2$$
$$x-5 = x - 2\sqrt{x} + 1$$
$$-6 = -2\sqrt{x}$$
$$3 = \sqrt{x}$$
$$3^2 = \left(\sqrt{x}\right)^2$$
$$9 = x$$
Check: $\sqrt{9-5} - \sqrt{9} = \sqrt{4} - \sqrt{9} = 2 - 3 = -1$
The solution is $x = 9$.

42. a.

$$\sqrt[3]{x-5} = 3$$

$$\left(\sqrt[3]{x-5}\right)^3 = 3^3$$

$$x - 5 = 27$$

$$x = 32$$

Check: $\sqrt[3]{32-5} = \sqrt[3]{27} = 3$

The solution is $x = 32$.

b.

$$\sqrt[3]{x-3} = 4$$

$$\left(\sqrt[3]{x-3}\right)^3 = 4^3$$

$$x - 3 = 64$$

$$x = 67$$

Check: $\sqrt[3]{67-3} = \sqrt[3]{64} = 4$

The solution is $x = 67$.

43. a.

$$\sqrt[4]{x+1} + 1 = 0$$

$$\sqrt[4]{x+1} = -1$$

$$\left(\sqrt[4]{x+1}\right)^4 = (-1)^4$$

$$x + 1 = 1$$

$$x = 0$$

Check: $\sqrt[4]{0+1} + 1 = \sqrt[4]{1} + 1 = 1 + 1 = 2 \neq 0$

There is no real-number solution.

b.

$$\sqrt[4]{x-2} + 16 = 0$$

$$\sqrt[4]{x-2} = -16$$

$$\left(\sqrt[4]{x-2}\right)^4 = (-16)^4$$

$$x - 2 = 65536$$

$$x = 65538$$

Check: $\sqrt[4]{65538 \neq 2} + 16 = \sqrt[4]{65536} + 16$

$$= 16 + 16 = 32 \neq 0$$

There is no real-number solution.

44. a.

$$d = \sqrt{\frac{k}{I}}$$

$$d^2 = \frac{k}{I}$$

$$Id^2 = k$$

$$I = \frac{k}{d^2}$$

b.

$$d = \sqrt{\frac{k}{I}}$$

$$d^2 = \frac{k}{I}$$

$$Id^2 = k$$

45. a. $\sqrt{-100} = \sqrt{-1 \cdot 100} = 10i$

b. $\sqrt{-121} = \sqrt{-1 \cdot 121} = 11i$

46. a. $\sqrt{-72} = \sqrt{-1 \cdot 36 \cdot 2} = 6i\sqrt{2}$

b. $\sqrt{-50} = \sqrt{-1 \cdot 25 \cdot 2} = 5i\sqrt{2}$

47. a. $(3+5i)+(7-2i) = (3+7)+(5-2)i$
$$= 10 + 3i$$

b. $(4+7i)+(2-4i) = (4+2)+(7-4)i$
$$= 6 + 3i$$

48. a. $(3+5i)-(7-2i) = (3-7)+(5+2)i$
$$= -4 + 7i$$

b. $(4+7i)-(2-4i) = (4-2)+(7+4)i$
$$= 2 + 11i$$

49. a. $(3+2i)(5-3i) = 15 - 9i + 10i - 6i^2$
$$= 15 + i + 6 = 21 + i$$

b. $(4+5i)(2-3i) = 8 - 12i + 10i - 15i^2$
$$= 8 - 2i + 15 = 23 - 2i$$

50. a. $\sqrt{-16}\left(4 - \sqrt{-72}\right) = 4i\left(4 - 6i\sqrt{2}\right) = 16i - 24i^2\sqrt{2} = 16i + 24\sqrt{2} = 24\sqrt{2} + 16i$

b. $\sqrt{-36}\left(4 - \sqrt{-72}\right) = 6i\left(4 - 6i\sqrt{2}\right) = 24i - 36i^2\sqrt{2} = 24i + 36\sqrt{2} = 36\sqrt{2} + 24i$

51. a. $\dfrac{2+3i}{4+3i} = \dfrac{(2+3i)(4-3i)}{(4+3i)(4-3i)} = \dfrac{8-6i+12i-9i^2}{16-9i^2} = \dfrac{8+6i+9}{16+9} = \dfrac{17+6i}{25} = \dfrac{17}{25} + \dfrac{6}{25}i$

b. $\dfrac{3-5i}{4-3i} = \dfrac{(3-5i)(4+3i)}{(4-3i)(4+3i)} = \dfrac{12+9i-20i-15i^2}{16-9i^2} = \dfrac{12-11i+15}{16+9} = \dfrac{27-11i}{25} = \dfrac{27}{25} - \dfrac{11}{25}i$

52. a. $i^{38} = \left(i^4\right)^9 \cdot i^2 = 1(-1) = -1$ **b.** $i^{75} = \left(i^4\right)^{18} \cdot i^3 = 1(-i) = -i$

53. a. $i^{\,14} = \dfrac{1}{i^{14}} = \dfrac{1}{\left(i^4\right)^3 \cdot i^2} = \dfrac{1}{1(\cdot\,1)} = \dfrac{1}{\cdot\,1} = \cdot\,1$ **b.** $i^{\,27} = \dfrac{1}{i^{27}} = \dfrac{1}{\left(i^4\right)^6 \cdot i^3} = \dfrac{1}{1(\cdot\,i)} = \dfrac{1\cdot i}{\cdot\,i\cdot i} = \dfrac{i}{1} = i$

Cumulative Review Chapters 1–7

1. $\sqrt{19}$ is an irrational number and a real number.

2. $\left(2x^2 y^{-4}\right)^{-4} = 2^{-4}\left(x^2\right)^{-4}\left(y^{-4}\right)^{-4} = 2^{-4} x^{-8} y^{16}$

$$= \frac{y^{16}}{2^4 x^8} = \frac{y^{16}}{16x^8}$$

3. $\because\; 9(9+6) \cdot +7 = \left[\cdot\;9\cdot15\right] + 7 = \cdot\;135 + 7$

$$= \cdot\;128$$

4. $\dfrac{7}{6} \mid \dfrac{x}{18} = \dfrac{3(x+9)}{54}$

$18 \left|\dfrac{7}{6} \mid \dfrac{x}{18}\right| = 18 \left|\dfrac{3(x+9)}{54}\right|$

$21 \mid x = x + 9$

$\mid 2x = \mid 12$

$x = 6$

5. $\geq 7 \geq \geq 7x \geq 14 < 7$

$7 \geq \geq 7x < 21$

$\dfrac{7}{\geq 7} \geq \dfrac{\geq 7x}{\geq 7} > \dfrac{21}{\geq 7}$

$\geq 1 \geq x > \geq 3$

$\geq 3 < x \geq \geq 1$

6. Let $L = W + 30$

$100 = 2(W + 30) + 2W$

$100 = 2W + 60 + 2W$

$40 = 4W$

$10 \text{ ft} = W$

$L = 10 + 30 = 40 \text{ ft}$

The dimensions are 10 ft by 40 ft.

7. If $y = 0$, $0 = 8x + 6$, $\cdot\, 8x = 6$, $x = \cdot\, \dfrac{3}{4}$

If $x = 0$, $y = 8 \cdot 0 + 6 = 6$

The x-intercept is $\left(-\frac{3}{4},\, 0\right)$.

The y-intercept is $(0,\, 6)$.

8. Since the line is perpendicular, the slope is $\frac{1}{4}$.

$$\frac{y - (-2)}{-3 - 5} = \frac{1}{4}$$

$$\frac{y + 2}{-8} = \frac{1}{4}$$

$$y + 2 = -2$$

$$y = -4$$

9.
$$m = \frac{7 - 4}{8 - (-6)} = \frac{3}{14}$$

$$y - 7 = \frac{3}{14}(x - 8)$$

$$14(y - 7) = 3(x - 8)$$

$$14y - 98 = 3x - 24$$

$$14y - 3x = 74$$

$$3x - 14y = -74$$

10. $y - x - 2$

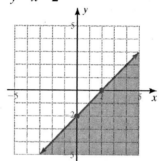

11. $|y| \le 2$

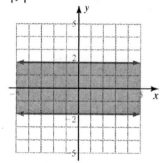

12. $x - 2y = 3$

$$x = -3 - y$$

Substitute the second equation into the first and solve:

$$-3 - y - 2y = 3$$

$$-3y = 6$$

$$y = -2$$

$$x = -3 - (-2) = -3 + 2 = -1$$

The solution is $(-1, -2)$.

13. To eliminate the fractions, multiply the first equation by 12, multiply the second equation by 36:

$$\frac{x}{4} + \frac{y}{3} = -6 \rightarrow 3x + 4y = -72$$

$$\frac{x}{12} + \frac{y}{9} = -2 \rightarrow 3x + 4y = -72$$

Both equations are the same so there are infinitely many solutions of the form $y = -\frac{3}{4}x - 18$.

14.
$$2x + y + z = -5$$
$$-4x + y + z = -3$$
$$3y + 3z = -2$$

Multiply the first equation by 2 and add to the second equation to eliminate x:

$$4x + 2y + 2z = -10$$
$$\underline{-4x + y + z = -3}$$
$$3y + 3z = -13$$

Multiply this result by -1, add to the third equation, and solve:

$$-3y - 3z = 13$$
$$\underline{3y + 3z = -2}$$
$$0 = 11$$

This result is a contradiction. There is no solution. The system is inconsistent.

15.

$$D = \begin{vmatrix} 3 & 2 & 3 \\ 2 & 1 & 1 \\ 1 & 2 & -2 \end{vmatrix} = 3\begin{vmatrix} 1 & 1 \\ 2 & -2 \end{vmatrix} - 2\begin{vmatrix} 2 & 1 \\ 1 & -2 \end{vmatrix} + 3\begin{vmatrix} 2 & 1 \\ 1 & 2 \end{vmatrix} = 3(-2-2) - 2(-4-1) + 3(4-1)$$

$$= -12 + 10 + 9 = 7$$

$$D_x = \begin{vmatrix} 29 & 2 & 3 \\ 15 & 1 & 1 \\ -1 & 2 & -2 \end{vmatrix} = 29\begin{vmatrix} 1 & 1 \\ 2 & -2 \end{vmatrix} - 2\begin{vmatrix} 15 & 1 \\ -1 & -2 \end{vmatrix} + 3\begin{vmatrix} 15 & 1 \\ -1 & 2 \end{vmatrix} = 29(-2-2) - 2(-30+1) + 3(30+1)$$

$$= -116 + 58 + 93 = 35$$

$$x = \frac{D_x}{D} = \frac{35}{7} = 5$$

16.   Let x = the amount invested at 7%
Let y = the amount invested at 9%
Let z = the amount invested at 11%

$$x + y + z = 90,000$$

$$0.07x + 0.09y + 0.11z = 8500$$

$$0.07x + 0.09y = 0.11z - 300$$

Substitute the third equation into the second equation and solve:

$$0.11z - 300 + 0.11z = 8500$$

$$0.22z = 8800$$

$$z = 40,000$$

Substitute this result into the first and third equations, solve the first equation for x and substitute into the second result:

$$x + y + 40,000 = 90,000 - \quad x + y = 50,000$$

$$x = 50,000 - y$$

$$0.07x + 0.09y = 0.11(40,000) - 300$$

$$0.07x + 0.09y = 4400 - 300 = 4100$$

$$0.07(50,000 - y) + 0.09y = 4100$$

$$3500 - 0.07y + 0.09y = 4100$$

$$0.02y = 600$$

$$y = 30,000$$

$$x = 50,000 - 30,000$$

$$x = 20,000$$

7% investment is $20,000; 9% investment is $30,000; 11% investment is $40,000.

17.   Range: $\{-5, -3, 0\}$

18. $f(x) = 4x \cdot 3$

$$f(2) = 4 \cdot 2 \cdot 3 = 8 \cdot 3 = 5$$

19. $(x-4)(x^2 + 4x + 2)$

$$= x^2(x-4) + 4x(x-4) + 2(x-4)$$

$$= x^3 - 4x^2 + 4x^2 - 16x + 2x - 8$$

$$= x^3 - 14x - 8$$

20. $9x^6 - 12x^5 + 15x^4 + 12x^3$

$$= 3x^3(3x^3 - 4x^2 + 5x + 4)$$

21. $9x^3y - 15x^2y^2 - 6xy^3 = 3xy(3x^2 - 5xy - 2y^2)$

$$= 3xy(3x + y)(x - 2y)$$

22.

$$6x^2 - 7x = -2$$
$$6x^2 - 7x + 2 = 0$$
$$(3x - 2)(2x - 1) = 0$$
$$3x - 2 = 0 \ \text{ or } \ 2x - 1 = 0$$
$$3x = 2 \ \text{ or } \quad 2x = 1$$
$$x = \frac{2}{3} \ \text{ or } \quad x = \frac{1}{2}$$

23.

$$\frac{x^2 - 9}{27 - x^3} = \frac{-(9 - x^2)}{(3 - x)(9 + 3x + x^2)}$$
$$= \frac{-(3-x)(3+x)}{(3-x)(9 + 3x + x^2)} = \frac{-(x+3)}{9 + 3x + x^2}$$

24.

$$
\begin{array}{r}
3x^2 + 4x + 1 \\
x + 4 \overline{\smash{\big)}\ 3x^3 + 16x^2 + 17x + 4} \\
\underline{-(3x^3 + 12x^2)} \\
4x^2 + 17x \\
\underline{-(4x^2 + 16x)} \\
x + 4 \\
\underline{-(x + 4)} \\
0
\end{array}
$$

The factorization is:
$$(x + 4)(3x^2 + 4x + 1) = (x + 4)(x + 1)(3x + 1)$$

25. Multiply both sides by the LCD $(x + 5)(x - 5)$:

$$\frac{x}{x + 5} - \frac{x}{x - 5} = \frac{x^2 + 25}{x^2 - 25}$$

$$(x + 5)(x - 5)\left[\frac{x}{x + 5} - \frac{x}{x - 5}\right] = (x + 5)(x - 5)\left[\frac{x^2 + 25}{(x + 5)(x - 5)}\right]$$

$$x(x - 5) - x(x + 5) = x^2 + 25$$

$$x^2 - 5x - x^2 - 5x = x^2 + 25$$

$$-10x = x^2 + 25$$

$$0 = x^2 + 10x + 25$$

$$0 = (x + 5)(x + 5)$$

$$x + 5 = 0 \quad \text{or } x + 5 = 0$$

$$x = -5 \ \text{or} \quad x = -5$$

-5 makes the denominator $x + 5 = 0$ and thus is extraneous. There is no solution.

26. Let x = the first even integer
Let $x + 2$ = the consecutive even integer

$$\frac{1}{x} + \frac{1}{x+2} = \frac{11}{60}$$

$$60x(x+2)\frac{1}{x} + \frac{1}{x+2} = 60x(x+2)\frac{11}{60}$$

$$60(x+2) + 60x = 11 \cdot x(x+2)$$

$$60x + 120 + 60x = 11x^2 + 22x$$

$$0 = 11x^2 - 98x - 120$$

$$(11x + 12)(x - 10) = 0$$

$$11x + 12 = 0 \quad \text{or } x - 10 = 0$$

$$11x = -12 \quad \text{or} \quad x = 10$$

$$x = -\frac{12}{11} \quad \text{or} \quad x = 10$$

The integers are 10 and 12.

27.

$$P = \frac{k}{V}$$

$$1720 = \frac{k}{4}$$

$$k = 1720 \cdot 4 = 6880$$

28. $9^{3/2} = \left(\sqrt{9}\right)^3 = 3^3 = 27$

29. $x^{2/5}\left(x^{-1/5} + y^{4/5}\right) = x^{2/5 - 1/5} + x^{2/5}y^{4/5}$

$$= x^{1/5} + x^{2/5}y^{4/5}$$

30. $\sqrt{175} + \sqrt{112} = \sqrt{25 \cdot 7} + \sqrt{16 \cdot 7}$

$$= 5\sqrt{7} + 4\sqrt{7} = 9\sqrt{7}$$

31. $\left(7 \cdot \sqrt{7}\right)^2 = 7^2 \cdot 2 \cdot 7 \cdot \sqrt{7} + \left(\sqrt{7}\right)^2$

$$= 49 \cdot 14\sqrt{7} + 7 = 56 \cdot 14\sqrt{7}$$

32.

$$\frac{\sqrt{x}}{\sqrt{x} - \sqrt{5}} = \frac{\sqrt{x}\left(\sqrt{x} + \sqrt{5}\right)}{\left(\sqrt{x} - \sqrt{5}\right)\left(\sqrt{x} + \sqrt{5}\right)}$$

$$= \frac{x + 5\sqrt{x}}{x - 5} \quad (x > 0)$$

33.

$$\sqrt{x + 34} = x + 14$$

$$\left(\sqrt{x + 34}\right)^2 = (x + 14)^2$$

$$x + 34 = x^2 + 28x + 196$$

$$0 = x^2 + 27x + 162$$

$$0 = (x + 18)(x + 9)$$

$$x + 18 = 0 \quad \text{or } x + 9 = 0$$

$$x = -18 \quad \text{or} \quad x = -9$$

Check: $\sqrt{\neq 18 + 34} = \sqrt{16} = 4 \neq \neq 18 + 14$

Discard: $x = \neq 18$

$$\sqrt{\neq 9 + 34} = \sqrt{25} = 5 = \neq 9 + 14$$

The solution is $x = \neq 9$.

34. $\frac{7 + 6i}{1 + 9i} = \frac{(7 + 6i)(1 - 9i)}{(1 + 9i)(1 - 9i)} = \frac{7 - 63i + 6i - 54i^2}{1 - 81i^2} = \frac{7 - 57i + 54}{1 + 81} = \frac{61 - 57i}{82} = \frac{61}{82} - \frac{57}{82}i$

35. $i^{49} = \left(i^4\right)^{12} \cdot i = 1 \cdot i = i$

300

Chapter 8   Quadratic Equations and Inequalities

8.1   Solving Quadratics by Completing the Square

Problems 8.1

1.   a.
$$9x^2 - 4 = 0$$
$$9x^2 = 4$$
$$x^2 = \frac{4}{9}$$
$$x = \pm\sqrt{\frac{4}{9}} = \pm\frac{2}{3}$$
The solution set is $\left\{\frac{2}{3}, -\frac{2}{3}\right\}$.

b.
$$3x^2 + 54 = 0$$
$$3x^2 = \cdot\ 54$$
$$x^2 = \cdot\ 18$$
$$x = \pm\sqrt{\cdot\ 18} = \pm\sqrt{\cdot\ 1\cdot 9\cdot 2} = \pm 3i\sqrt{2}$$
The solution set is $\left\{3i\sqrt{2}, -3i\sqrt{2}\right\}$.

c.
$$3x^2 - 16 = 0$$
$$3x^2 = 16$$
$$x^2 = \frac{16}{3}$$
$$x = \pm\sqrt{\frac{16}{3}} = \pm\frac{4}{\sqrt{3}}$$
$$= \pm\frac{4\sqrt{3}}{\sqrt{3}\sqrt{3}} = \pm\frac{4\sqrt{3}}{3}$$
The solution set is $\left\{\dfrac{4\sqrt{3}}{3},\ \text{J}\ \dfrac{4\sqrt{3}}{3}\right\}$.

2.   a.
$$(x-2)^2 = 24$$
$$\sqrt{(x-2)^2} = \pm\sqrt{24}$$
$$x - 2 = \pm 2\sqrt{6}$$
$$x = 2 \pm 2\sqrt{6}$$
The solution set is $\left\{2 + 2\sqrt{6},\ 2 - 2\sqrt{6}\right\}$.

b.
$$4(x+7)^2 = 3$$
$$(x+7)^2 = \frac{3}{4}$$
$$\sqrt{(x+7)^2} = \pm\sqrt{\frac{3}{4}}$$
$$x + 7 = \pm\frac{\sqrt{3}}{2}$$
$$x = -7 \pm \frac{\sqrt{3}}{2} = \frac{-14 \pm \sqrt{3}}{2}$$
The solution set is
$$\left\{\frac{\text{J}\ 14 + \sqrt{3}}{2}, \frac{\text{J}\ 14\ \text{J}\ \sqrt{3}}{2}\right\}.$$

c.
$$5(x-3)^2 + 36 = 0$$
$$5(x-3)^2 = -36$$
$$(x-3)^2 = -\frac{36}{5}$$
$$\sqrt{(x-3)^2} = \pm\sqrt{\frac{-36}{5}}$$
$$x - 3 = \pm\frac{6i}{\sqrt{5}}$$
$$x - 3 = \pm\frac{6i\sqrt{5}}{5}$$
$$x = 3 \pm \frac{6\sqrt{5}}{5}i$$
The solution set is $\left\{3 + \dfrac{6\sqrt{5}}{5}i,\ 3\ \text{J}\ \dfrac{6\sqrt{5}}{5}i\right\}$.

3. $x^2 + 6x + 9 = 24$

$(x+3)^2 = 24$

$\sqrt{(x+3)^2} = \pm\sqrt{24}$

$x + 3 = \pm 2\sqrt{6}$

$x = -3 \pm 2\sqrt{6}$

The solution set is $\left\{-3 + 2\sqrt{6}, -3 - 2\sqrt{6}\right\}$.

4. $x^2 + 12x - 8 = 0$

$x^2 + 12x = 8$

$x^2 + 12x + 6^2 = 8 + 6^2$

$(x+6)^2 = 44$

$\sqrt{(x+6)^2} = \pm\sqrt{44}$

$x + 6 = \pm 2\sqrt{11}$

$x = -6 \pm 2\sqrt{11}$

The solution set is $\left\{-6 + 2\sqrt{11}, -6 - 2\sqrt{11}\right\}$.

5. $5x^2 \cdot 5x \cdot 1 = 0$

$5x^2 \cdot 5x = 1$

$x^2 \cdot x = \dfrac{1}{5}$

$x^2 \cdot x + \left(\dfrac{1 \cdot}{2}\right)^2 = \dfrac{1}{5} + \left(\dfrac{1 \cdot}{2}\right)^2$

$\left(x \cdot \dfrac{1 \cdot}{2}\right)^2 = \dfrac{9}{20}$

$\sqrt{\left(x \cdot \dfrac{1 \cdot}{2}\right)^2} = \pm\sqrt{\dfrac{9}{20}}$

$x \cdot \dfrac{1}{2} = \pm\sqrt{\dfrac{9 \cdot 5}{20 \cdot 5}} = \pm\dfrac{3\sqrt{5}}{10}$

$x = \dfrac{1}{2} \pm \dfrac{3\sqrt{5}}{10} = \dfrac{5 \pm 3\sqrt{5}}{10}$

The solution set is $\left\{\dfrac{5 + 3\sqrt{5}}{10}, \dfrac{5 \, | \, 3\sqrt{5}}{10}\right\}$.

Exercises 8.1

1. $x^2 = 64$

$x = \pm\sqrt{64} = \pm 8$

The solution set is $\{8, -8\}$.

3. $x^2 = \cdot 121$

$x = \pm\sqrt{\cdot 121} = \pm\sqrt{\cdot 1 \cdot 121} = \pm 11i$

The solution set is $\{11i, -11i\}$.

5. $x^2 - 169 = 0$

$x^2 = 169$

$x = \pm\sqrt{169} = \pm 13$

The solution set is $\{13, -13\}$.

7. $x^2 + 4 = 0$

$x^2 = \cdot 4$

$x = \pm\sqrt{\cdot 4} = \pm\sqrt{\cdot 1 \cdot 4} = \pm 2i$

The solution set is $\{2i, -2i\}$.

9.
$$36x^2 - 49 = 0$$
$$36x^2 = 49$$
$$x^2 = \frac{49}{36}$$
$$x = \pm\sqrt{\frac{49}{36}} = \pm\frac{7}{6}$$
The solution set is $\left\{\frac{7}{6}, \ \frac{7}{6}\right\}$.

11.
$$4x^2 + 81 = 0$$
$$4x^2 = \cdot \ 81$$
$$x^2 = \cdot \ \frac{81}{4}$$
$$x = \pm\sqrt{\frac{\cdot \ 81}{4}} = \pm\sqrt{\frac{\cdot \ 1 \cdot 81}{4}} = \pm\frac{9}{2}i$$
The solution set is $\left\{\frac{9}{2}i, \ \frac{9}{2}i\right\}$.

13.
$$3x^2 \cdot 25 = 0$$
$$3x^2 = 25$$
$$x^2 = \frac{25}{3}$$
$$x = \pm\sqrt{\frac{25}{3}} = \pm\sqrt{\frac{25 \cdot 3}{3 \cdot 3}} = \pm\frac{5\sqrt{3}}{3}$$
The solution set is $\left\{\frac{5\sqrt{3}}{3}, \ \frac{5\sqrt{3}}{3}\right\}$.

15.
$$5x^2 + 36 = 0$$
$$5x^2 = \cdot \ 36$$
$$x^2 = \cdot \ \frac{36}{5}$$
$$x = \pm\sqrt{\frac{\cdot \ 36}{5}} = \pm\sqrt{\frac{\cdot \ 1 \cdot 36 \cdot 5}{5 \cdot 5}} = \pm\frac{6\sqrt{5}}{5}i$$
The solution set is $\left\{\frac{6\sqrt{5}}{5}i, \ \frac{6\sqrt{5}}{5}i\right\}$.

17.
$$3x^2 \cdot 100 = 0$$
$$3x^2 = 100$$
$$x^2 = \frac{100}{3}$$
$$x = \pm\sqrt{\frac{100}{3}} = \pm\sqrt{\frac{100 \cdot 3}{3 \cdot 3}} = \pm\frac{10\sqrt{3}}{3}$$
The solution set is $\left\{\frac{10\sqrt{3}}{3}, \ \frac{10\sqrt{3}}{3}\right\}$.

19.
$$13x^2 + 81 = 0$$
$$13x^2 = \cdot \ 81$$
$$x^2 = \cdot \ \frac{81}{13}$$
$$x = \pm\sqrt{\frac{\cdot \ 81}{13}} = \pm\sqrt{\frac{\cdot \ 1 \cdot 81 \cdot 13}{13 \cdot 13}} = \pm\frac{9\sqrt{13}}{13}i$$
The solution set is $\left\{\frac{9\sqrt{13}}{13}i, \ \frac{9\sqrt{13}}{13}i\right\}$.

21.
$$(x+5)^2 = 4$$
$$\sqrt{(x+5)^2} = \pm\sqrt{4}$$
$$x + 5 = \pm 2$$
$$x = -5 \pm 2$$
$$x = -3 \text{ or } x = -7$$
The solution set is $\{-3, \ -7\}$.

23.
$$x^2 + 4x + 4 = -25$$
$$(x+2)^2 = -25$$
$$\sqrt{(x+2)^2} = \pm\sqrt{-25}$$
$$x + 2 = \pm 5i$$
$$x = -2 \pm 5i$$
$$x = -2 + 5i \text{ or } x = -2 - 5i$$
The solution set is $\{-2 + 5i, \ -2 - 5i\}$.

25.
$$(x-6)^2 = 18$$
$$\sqrt{(x-6)^2} = \pm\sqrt{18} = \pm\sqrt{9\cdot 2}$$
$$x-6 = \pm 3\sqrt{2}$$
$$x = 6 \pm 3\sqrt{2}$$
$$x = 6+3\sqrt{2} \text{ or } x = 6-3\sqrt{2}$$
The solution set is $\left\{6+3\sqrt{2},\, 6-3\sqrt{2}\right\}$.

27.
$$x^2 - 2x + 1 = -28$$
$$(x-1)^2 = -28$$
$$\sqrt{(x-1)^2} = \pm\sqrt{-28} = \pm\sqrt{-1\cdot 4\cdot 7}$$
$$x-1 = \pm 2i\sqrt{7}$$
$$x = 1 \pm 2i\sqrt{7}$$
$$x = 1+2i\sqrt{7} \text{ or } x = 1-2i\sqrt{7}$$
The solution set is $\left\{1+2i\sqrt{7},\, 1-2i\sqrt{7}\right\}$.

29.
$$(x-1)^2 - 50 = 0$$
$$(x-1)^2 = 50$$
$$\sqrt{(x-1)^2} = \pm\sqrt{50} = \pm\sqrt{25\cdot 2}$$
$$x-1 = \pm 5\sqrt{2}$$
$$x = 1 \pm 5\sqrt{2}$$
$$x = 1+5\sqrt{2} \text{ or } x = 1-5\sqrt{2}$$
The solution set is $\left\{1+5\sqrt{2},\, 1-5\sqrt{2}\right\}$.

31.
$$(x-5)^2 - 32 = 0$$
$$(x-5)^2 = 32$$
$$\sqrt{(x-5)^2} = \pm\sqrt{32} = \pm\sqrt{16\cdot 2}$$
$$x-5 = \pm 4\sqrt{2}$$
$$x = 5 \pm 4\sqrt{2}$$
$$x = 5+4\sqrt{2} \text{ or } x = 5-4\sqrt{2}$$
The solution set is $\left\{5+4\sqrt{2},\, 5-4\sqrt{2}\right\}$.

33.
$$(x-9)^2 + 64 = 0$$
$$(x-9)^2 = -64$$
$$\sqrt{(x-9)^2} = \pm\sqrt{-64} = \pm\sqrt{-1\cdot 64}$$
$$x-9 = \pm 8i$$
$$x = 9 \pm 8i$$
$$x = 9+8i \text{ or } x = 9-8i$$
The solution set is $\left\{9+8i,\, 9-8i\right\}$.

35.
$$3x^2 + 6x + 3 = 96$$
$$x^2 + 2x + 1 = 32$$
$$(x+1)^2 = 32$$
$$\sqrt{(x+1)^2} = \pm\sqrt{32} = \pm\sqrt{16\cdot 2}$$
$$x+1 = \pm 4\sqrt{2}$$
$$x = -1 \pm 4\sqrt{2}$$
$$x = -1+4\sqrt{2} \text{ or } x = -1-4\sqrt{2}$$
The solution set is $\left\{-1+4\sqrt{2},\, -1-4\sqrt{2}\right\}$.

37.
$$7(x-2)^2 - 350 = 0$$
$$7(x-2)^2 = 350$$
$$(x-2)^2 = 50$$
$$\sqrt{(x-2)^2} = \pm\sqrt{50} = \pm\sqrt{25\cdot 2}$$
$$x-2 = \pm 5\sqrt{2}$$
$$x = 2 \pm 5\sqrt{2}$$
$$x = 2+5\sqrt{2} \text{ or } x = 2-5\sqrt{2}$$
The solution set is $\left\{2+5\sqrt{2},\, 2-5\sqrt{2}\right\}$.

39.
$$7(x-5)^2 + 189 = 0$$
$$7(x-5)^2 = -189$$
$$(x-5)^2 = -27$$
$$\sqrt{(x-5)^2} = \pm\sqrt{-27} = \pm\sqrt{-1\cdot 9\cdot 3}$$
$$x-5 = \pm 3i\sqrt{3}$$
$$x = 5 \pm 3i\sqrt{3}$$
$$x = 5+3i\sqrt{3} \text{ or } x = 5-3i\sqrt{3}$$
The solution set is $\left\{5+3i\sqrt{3},\, 5-3i\sqrt{3}\right\}$.

41.
$$x^2 + 6x + 5 = 0$$
$$x^2 + 6x = -5$$
$$x^2 + 6x + 3^2 = -5 + 3^2$$
$$(x+3)^2 = 4$$
$$\sqrt{(x+3)^2} = \pm\sqrt{4}$$
$$x + 3 = \pm 2$$
$$x = -3 \pm 2$$
$$x = -1 \text{ or } x = -5$$
The solution set is $\{-1, -5\}$.

43.
$$x^2 + 8x + 15 = 0$$
$$x^2 + 8x = -15$$
$$x^2 + 8x + 4^2 = -15 + 4^2$$
$$(x+4)^2 = 1$$
$$\sqrt{(x+4)^2} = \pm\sqrt{1}$$
$$x + 4 = \pm 1$$
$$x = -4 \pm 1$$
$$x = -3 \text{ or } x = -5$$
The solution set is $\{-3, -5\}$.

45.
$$x^2 + 6x + 10 = 0$$
$$x^2 + 6x = -10$$
$$x^2 + 6x + 3^2 = -10 + 3^2$$
$$(x+3)^2 = -1$$
$$\sqrt{(x+3)^2} = \pm\sqrt{-1}$$
$$x + 3 = \pm i$$
$$x = -3 \pm i$$
$$x = -3 + i \text{ or } x = -3 - i$$
The solution set is $\{-3+i, -3-i\}$.

47.
$$x^2 - 10x + 24 = 0$$
$$x^2 - 10x = -24$$
$$x^2 - 10x + 5^2 = -24 + 5^2$$
$$(x-5)^2 = 1$$
$$\sqrt{(x-5)^2} = \pm\sqrt{1}$$
$$x - 5 = \pm 1$$
$$x = 5 \pm 1$$
$$x = 6 \text{ or } x = 4$$
The solution set is $\{6, 4\}$.

49.
$$x^2 - 10x + 21 = 0$$
$$x^2 - 10x = -21$$
$$x^2 - 10x + 5^2 = -21 + 5^2$$
$$(x-5)^2 = 4$$
$$\sqrt{(x-5)^2} = \pm\sqrt{4}$$
$$x - 5 = \pm 2$$
$$x = 5 \pm 2$$
$$x = 7 \text{ or } x = 3$$
The solution set is $\{7, 3\}$.

51.
$$x^2 - 8x + 17 = 0$$
$$x^2 - 8x = -17$$
$$x^2 - 8x + 4^2 = -17 + 4^2$$
$$(x-4)^2 = -1$$
$$\sqrt{(x-4)^2} = \pm\sqrt{-1}$$
$$x - 4 = \pm i$$
$$x = 4 \pm i$$
$$x = 4 + i \text{ or } x = 4 - i$$
The solution set is $\{4+i, 4-i\}$.

53. $2x^2 + 4x + 3 = 0$

$2x^2 + 4x = \cdot\ 3$

$x^2 + 2x = \cdot\ \dfrac{3}{2}$

$x^2 + 2x + 1^2 = \cdot\ \dfrac{3}{2} + 1^2$

$(x+1)^2 = \cdot\ \dfrac{1}{2}$

$\sqrt{(x+1)^2} = \pm\sqrt{\dfrac{\cdot\ 1}{2}} = \pm\sqrt{\dfrac{\cdot\ 1 \cdot 2}{2 \cdot 2}} = \pm\dfrac{i\sqrt{2}}{2}$

$x + 1 = \pm\dfrac{i\sqrt{2}}{2}$

$x = \cdot\ 1 \pm \dfrac{i\sqrt{2}}{2}$

$x = \cdot\ 1 + \dfrac{\sqrt{2}}{2}i \text{ or } x = \cdot\ 1 \cdot\ \dfrac{\sqrt{2}}{2}i$

The solution set is $\left\} J\, 1 + \dfrac{\sqrt{2}}{2}i,\ J\, 1 J\ \dfrac{\sqrt{2}}{2}i \right\}$.

55. $3x^2 + 6x + 78 = 0$

$3x^2 + 6x = \cdot\ 78$

$x^2 + 2x = \cdot\ 26$

$x^2 + 2x + 1^2 = \cdot\ 26 + 1^2$

$(x+1)^2 = \cdot\ 25$

$\sqrt{(x+1)^2} = \pm\sqrt{\cdot\ 25} = \pm\sqrt{\cdot\ 1 \cdot 25}$

$x + 1 = \pm 5i$

$x = \cdot\ 1 \pm 5i$

$x = \cdot\ 1 + 5i \text{ or } x = \cdot\ 1 \cdot\ 5i$

The solution set is $\{-1 + 5i,\ -1 - 5i\}$.

57. $25y^2\ |\ 25y + 6 = 0$

$25y^2\ |\ 25y =\ |6$

$y^2\ |\ y =\ |\dfrac{6}{25}$

$y^2\ |\ y + \left|\dfrac{1}{2}\right|^2 =\ |\dfrac{6}{25} + \left|\dfrac{1}{2}\right|^2 =\ |\dfrac{24}{100} + \dfrac{25}{100}$

$\left|y\ |\ \dfrac{1}{2}\right|^2 = \dfrac{1}{100}$

$\sqrt{\left|y\ |\ \dfrac{1}{2}\right|^2} = \pm\sqrt{\dfrac{1}{100}}$

$y\ |\ \dfrac{1}{2} = \pm\dfrac{1}{10}$

$x = \dfrac{1}{2} \pm \dfrac{1}{10}$

$x = \dfrac{3}{5} \text{ or } x = \dfrac{2}{5}$

The solution set is $\left\}\dfrac{3}{5}, \dfrac{2}{5}\right\}$.

59. $4y^2\ |\ 4y + 5 = 0$

$4y^2\ |\ 4y =\ |5$

$y^2\ |\ y =\ |\dfrac{5}{4}$

$y^2\ |\ y + \left|\dfrac{1}{2}\right|^2 =\ |\dfrac{5}{4} + \left|\dfrac{1}{2}\right|^2 =\ |\dfrac{5}{4} + \dfrac{1}{4}$

$\left|y\ |\ \dfrac{1}{2}\right|^2 =\ |1$

$\sqrt{\left|y\ |\ \dfrac{1}{2}\right|^2} = \pm\sqrt{|1}$

$y\ |\ \dfrac{1}{2} = \pm i$

$x = \dfrac{1}{2} \pm i$

$x = \dfrac{1}{2} + i \text{ or } x = \dfrac{1}{2}\ |\ i$

The solution set is $\left\}\dfrac{1}{2} + i, \dfrac{1}{2} J\ i \right\}$.

61.
$$4x^2 - 7 = 4x$$
$$4x^2 - 4x = 7$$
$$x^2 - x = \frac{7}{4}$$
$$x^2 - x + \left(\frac{1}{2}\right)^2 = \frac{7}{4} + \left(\frac{1}{2}\right)^2$$
$$\left(x - \frac{1}{2}\right)^2 = 2$$
$$\sqrt{\left(x - \frac{1}{2}\right)^2} = \pm\sqrt{2}$$
$$x - \frac{1}{2} = \pm\sqrt{2}$$
$$x = \frac{1}{2} \pm \sqrt{2} = \frac{1 \pm 2\sqrt{2}}{2}$$
$$x = \frac{1 + 2\sqrt{2}}{2} \text{ or } x = \frac{1 - 2\sqrt{2}}{2}$$
The solution set is $\left\{\dfrac{1 + 2\sqrt{2}}{2}, \dfrac{1 - 2\sqrt{2}}{2}\right\}$.

63.
$$2x^2 + 1 = 4x$$
$$2x^2 - 4x = -1$$
$$x^2 - 2x = -\frac{1}{2}$$
$$x^2 - 2x + 1^2 = -\frac{1}{2} + 1^2$$
$$(x - 1)^2 = \frac{1}{2}$$
$$\sqrt{(x - 1)^2} = \pm\sqrt{\frac{1}{2}} = \pm\sqrt{\frac{1 \cdot 2}{2 \cdot 2}}$$
$$x - 1 = \pm\frac{\sqrt{2}}{2}$$
$$x = 1 \pm \frac{\sqrt{2}}{2} = \frac{2 \pm \sqrt{2}}{2}$$
$$x = \frac{2 + \sqrt{2}}{2} \text{ or } x = \frac{2 - \sqrt{2}}{2}$$
The solution set is $\left\{\dfrac{2 + \sqrt{2}}{2}, \dfrac{2 - \sqrt{2}}{2}\right\}$.

65.
$$(x + 3)(x - 2) = -4$$
$$x^2 + x - 6 = -4$$
$$x^2 + x = 2$$
$$x^2 + x + \left(\frac{1}{2}\right)^2 = 2 + \left(\frac{1}{2}\right)^2$$
$$\left(x + \frac{1}{2}\right)^2 = \frac{9}{4}$$
$$\sqrt{\left(x + \frac{1}{2}\right)^2} = \pm\sqrt{\frac{9}{4}}$$
$$x + \frac{1}{2} = \pm\frac{3}{2}$$
$$x = -\frac{1}{2} \pm \frac{3}{2}$$
$$x = 1 \text{ or } x = -2$$
The solution set is $\{1, -2\}$.

67.
$$2x(x + 5) - 1 = 0$$
$$2x^2 + 10x = 1$$
$$x^2 + 5x = \frac{1}{2}$$
$$x^2 + 5x + \left(\frac{5}{2}\right)^2 = \frac{1}{2} + \left(\frac{5}{2}\right)^2$$
$$\left(x + \frac{5}{2}\right)^2 = \frac{27}{4}$$
$$\sqrt{\left(x + \frac{5}{2}\right)^2} = \pm\sqrt{\frac{27}{4}}$$
$$x + \frac{5}{2} = \pm\frac{3\sqrt{3}}{2}$$
$$x = -\frac{5}{2} \pm \frac{3\sqrt{3}}{2} = \frac{-5 \pm 3\sqrt{3}}{2}$$
$$x = \frac{-5 + 3\sqrt{3}}{2} \text{ or } x = \frac{-5 - 3\sqrt{3}}{2}$$
The solution set is $\left\{\dfrac{-5 + 3\sqrt{3}}{2}, \dfrac{-5 - 3\sqrt{3}}{2}\right\}$.

69.
$$2x(x+3)-10=0$$
$$2x^2+6x=10$$
$$x^2+3x=5$$
$$x^2+3x+\left|\frac{3}{2}\right|^2=5+\left|\frac{3}{2}\right|^2$$
$$\left|x+\frac{3}{2}\right|^2=\frac{29}{4}$$
$$\sqrt{\left|x+\frac{3}{2}\right|^2}=\pm\sqrt{\frac{29}{4}}$$
$$x+\frac{3}{2}=\pm\frac{\sqrt{29}}{2}$$
$$x=\left|\frac{3}{2}\pm\frac{\sqrt{29}}{2}=\frac{|3\pm\sqrt{29}}{2}\right.$$
$$x=\frac{|3+\sqrt{29}}{2}\text{ or }x=\frac{|3|\sqrt{29}}{2}$$

The solution set is $\left\{\frac{|3+\sqrt{29}}{2},\frac{|3|\sqrt{29}}{2}\right\}$.

71.
$$h=16t^2;\quad h=64$$
$$16t^2=64$$
$$t^2=4$$
$$t=\pm2$$

The object hits the ground in 2 sec.

73.
$$100(1+r)^2=121$$
$$(1+r)^2=\frac{121}{100}$$
$$\sqrt{(1+r)^2}=\pm\sqrt{\frac{121}{100}}$$
$$1+r=\pm\frac{11}{10}$$
$$r=\frac{11}{10}-1=\frac{1}{10}=0.10$$

The rate of interest is 10%.

75.
$$\frac{\cdot b\pm\sqrt{b^2\cdot 4ac}}{2a};\quad a=1,\ b=\cdot 9,\ c=0$$
$$\frac{\cdot(\cdot 9)\pm\sqrt{(\cdot 9)^2\cdot 4\cdot 1\cdot 0}}{2\cdot 1}=\frac{9\pm\sqrt{81\cdot 0}}{2}$$
$$=\frac{9\pm 9}{2}=9\text{ or }0$$

Values are 9 or 0.

77.
$$\frac{\cdot b\pm\sqrt{b^2\cdot 4ac}}{2a};\quad a=1,\ b=\cdot 2,\ c=\cdot 2$$
$$\frac{\cdot(\cdot 2)\pm\sqrt{(\cdot 2)^2\cdot 4\cdot 1\cdot(\cdot 2)}}{2\cdot 1}=\frac{2\pm\sqrt{4+8}}{2}$$
$$=\frac{2\pm\sqrt{12}}{2}=\frac{2\pm 2\sqrt{3}}{2}=1\pm\sqrt{3}$$

Values are $1+\sqrt{3}$ or $1-\sqrt{3}$.

79.
$$\frac{\cdot b\pm\sqrt{b^2\cdot 4ac}}{2a};\quad a=8,\ b=7,\ c=\cdot 1$$
$$\frac{\cdot 7\pm\sqrt{7^2\cdot 4\cdot 8\cdot(\cdot 1)}}{2\cdot 8}=\frac{\cdot 7\pm\sqrt{49+32}}{16}$$
$$=\frac{\cdot 7\pm\sqrt{81}}{16}=\frac{\cdot 7\pm 9}{16}=\frac{1}{8}\text{ or }\cdot 1$$

Values are $\frac{1}{8}$ or -1.

81.

$$\frac{-b \pm \sqrt{b^2 - 4ac}}{2a}; \quad a = 3, \ b = -8, \ c = 7$$

$$\frac{-(-8) \pm \sqrt{(-8)^2 - 4 \cdot 3 \cdot 7}}{2 \cdot 3} = \frac{8 \pm \sqrt{64 - 84}}{6}$$

$$= \frac{8 \pm \sqrt{-20}}{6} = \frac{8 \pm 2i\sqrt{5}}{6} = \frac{4 \pm i\sqrt{5}}{3}$$

$$= \frac{4}{3} \pm \frac{\sqrt{5}}{3}i$$

Values are $\dfrac{4}{3} + \dfrac{\sqrt{5}}{3}i$ or $\dfrac{4}{3} - \dfrac{\sqrt{5}}{3}i$.

83.

$$\frac{-b \pm \sqrt{b^2 - 4ac}}{2a}; \quad a = 1, \ b = 2, \ c = 6$$

$$\frac{-2 \pm \sqrt{2^2 - 4 \cdot 1 \cdot 6}}{2 \cdot 1} = \frac{-2 \pm \sqrt{4 - 24}}{2}$$

$$= \frac{-2 \pm \sqrt{-20}}{2} = \frac{-2 \pm 2i\sqrt{5}}{2} = -1 \pm i\sqrt{5}$$

Values are $-1 + i\sqrt{5}$ or $-1 - i\sqrt{5}$.

85. a.

$$\overline{C} = x^2 - 4x + 6$$

$$= \left(x^2 - 4x + 4\right) + 6 - 4$$

$$= (x - 2)^2 + 2$$

The minimum cost occurs when 2000 items are produced.

b. The minimum average cost per unit is $2.

87.

$$B = 20t^2 - 120t + 200$$

$$= 20\left(t^2 - 6t\right) + 200$$

$$= 20\left(t^2 - 6t + 9\right) + 200 - 9 \cdot 20$$

$$= 20(t - 3)^2 + 20$$

The number of bacteria is lowest 3 days after the treatment.

89. Sample answer: The solution process requires factoring $x^2 - 49 = (x + 7)(x - 7) = 0$, setting each factor equal to zero, and solving which produces the two solutions 7 and –7. This method does not work when solving $x^2 - 2 = 0$ because the left side cannot be factored.

91.

$$5x^2 - 5x = 1$$

$$x^2 - x = \frac{1}{5}$$

$$x^2 - x + \left(-\frac{1}{2}\right)^2 = \frac{1}{5} + \left(-\frac{1}{2}\right)^2$$

$$\left(x - \frac{1}{2}\right)^2 = \frac{9}{20}$$

$$\sqrt{\left(x - \frac{1}{2}\right)^2} = \pm\sqrt{\frac{9}{20}}$$

$$x - \frac{1}{2} = \pm\sqrt{\frac{9 \cdot 5}{20 \cdot 5}} = \pm\frac{3\sqrt{5}}{10}$$

$$x = \frac{1}{2} \pm \frac{3\sqrt{5}}{10} = \frac{5 \pm 3\sqrt{5}}{10}$$

The solution set is $\left\{ \dfrac{5 + 3\sqrt{5}}{10}, \dfrac{5 - 3\sqrt{5}}{10} \right\}$.

93.

$$x^2 + 12x = 8$$

$$x^2 + 12x + 6^2 = 8 + 6^2$$

$$(x + 6)^2 = 44$$

$$\sqrt{(x + 6)^2} = \pm\sqrt{44} = \pm\sqrt{4 \cdot 11}$$

$$x + 6 = \pm 2\sqrt{11}$$

$$x = -6 \pm 2\sqrt{11}$$

The solution set is $\left\{ -6 + 2\sqrt{11}, -6 - 2\sqrt{11} \right\}$.

95.
$$2x^2 + 12x + 18 = 27$$
$$x^2 + 6x + 9 = \frac{27}{2}$$
$$(x+3)^2 = \frac{27}{2}$$
$$\sqrt{(x+3)^2} = \pm\sqrt{\frac{27}{2}} = \pm\sqrt{\frac{9 \cdot 3 \cdot 2}{2 \cdot 2}}$$
$$x + 3 = \pm\frac{3\sqrt{6}}{2}$$
$$x = -3 \pm \frac{3\sqrt{6}}{2} = \frac{-6 \pm 3\sqrt{6}}{2}$$
The solution set is $\left\{ \dfrac{6 + 3\sqrt{6}}{2}, \dfrac{6 \cdot 3\sqrt{6}}{2} \right\}$.

97.
$$5(x \cdot 3)^2 + 36 = 0$$
$$5(x \cdot 3)^2 = \cdot 36$$
$$(x \cdot 3)^2 = \cdot \frac{36}{5}$$
$$\sqrt{(x \cdot 3)^2} = \pm\sqrt{\cdot \frac{36}{5}} = \pm\sqrt{\cdot \frac{36 \cdot 5}{5 \cdot 5}}$$
$$x \cdot 3 = \pm\frac{6i\sqrt{5}}{5}$$
$$x = 3 \pm \frac{6\sqrt{5}}{5}i$$
The solution set is $\left\{ 3 + \dfrac{6\sqrt{5}}{5}i, \; 3 \cdot \dfrac{6\sqrt{5}}{5}i \right\}$.

99.
$$(x-2)^2 = 24$$
$$\sqrt{(x-2)^2} = \pm\sqrt{24}$$
$$x - 2 = \pm 2\sqrt{6}$$
$$x = 2 \pm 2\sqrt{6}$$
The solution set is $\left\{ 2 + 2\sqrt{6}, \; 2 - 2\sqrt{6} \right\}$.

101.
$$3x^2 \cdot 16 = 0$$
$$3x^2 = 16$$
$$x^2 = \frac{16}{3}$$
$$x = \pm\sqrt{\frac{16}{3}} = \pm\sqrt{\frac{16 \cdot 3}{3 \cdot 3}} = \pm\frac{4\sqrt{3}}{3}$$
The solution set is $\left\{ \dfrac{4\sqrt{3}}{3}, \; \cdot \dfrac{4\sqrt{3}}{3} \right\}$.

103.
$$5x^2 + 60 = 0$$
$$5x^2 = \cdot 60$$
$$x^2 = \cdot 12$$
$$x = \pm\sqrt{\cdot 12} = \pm\sqrt{\cdot 1 \cdot 4 \cdot 3} = \pm 2i\sqrt{3}$$
The solution set is $\left\{ 2i\sqrt{3}, \; -2i\sqrt{3} \right\}$.

8.2   The Quadratic Formula: Applications

Problems 8.2

1. $3x^2 + 2x - 5 = 0$; $a = 3, b = 2, c = -5$

$$x = \frac{-2 \pm \sqrt{2^2 - 4\cdot3\cdot(-5)}}{2\cdot3} = \frac{-2 \pm \sqrt{4+60}}{6}$$

$$= \frac{-2 \pm \sqrt{64}}{6} = \frac{-2 \pm 8}{6} = 1 \text{ or } -\frac{5}{3}$$

The solution set is $\left\{-\frac{5}{3}, 1\right\}$.

2. $x^2 = 4x + 4$

$x^2 - 4x - 4 = 0$; $a = 1, b = -4, c = -4$

$$x = \frac{-(-4) \pm \sqrt{(-4)^2 - 4\cdot1\cdot(-4)}}{2\cdot1}$$

$$= \frac{4 \pm \sqrt{16+16}}{2} = \frac{4 \pm \sqrt{32}}{2} = \frac{4 \pm 4\sqrt{2}}{2}$$

$$= 2 \pm 2\sqrt{2}$$

The solution set is $\left\{2+2\sqrt{2},\ 2-2\sqrt{2}\right\}$.

3. $x^2 = 15x$

$x^2 - 15x + 0 = 0$; $a = 1, b = -15, c = 0$

$$x = \frac{-(-15) \pm \sqrt{(-15)^2 - 4\cdot1\cdot0}}{2\cdot1}$$

$$= \frac{15 \pm \sqrt{225 - 0}}{2} = \frac{15 \pm 15}{2} = 15 \text{ or } 0$$

The solution set is $\{15, 0\}$.

4. $\dfrac{x^2}{4} - \dfrac{3}{8}x = \dfrac{1}{4}$

$2x^2 - 3x = 2$

$2x^2 - 3x - 2 = 0$; $a = 2, b = -3, c = -2$

$$x = \frac{-(-3) \pm \sqrt{(-3)^2 - 4\cdot2\cdot(-2)}}{2\cdot2} = \frac{3 \pm \sqrt{9+16}}{4}$$

$$= \frac{3 \pm \sqrt{25}}{4} = \frac{3 \pm 5}{4} = 2 \text{ or } -\frac{1}{2}$$

The solution set is $\left\{2, -\frac{1}{2}\right\}$.

5. $3x^2 + 2x = -1$

$3x^2 + 2x + 1 = 0$; $a = 3, b = 2, c = 1$

$$x = \frac{-2 \pm \sqrt{2^2 - 4\cdot3\cdot1}}{2\cdot3} = \frac{-2 \pm \sqrt{4-12}}{6}$$

$$= \frac{-2 \pm \sqrt{-8}}{6} = \frac{-2 \pm 2i\sqrt{2}}{6} = -\frac{1}{3} \pm \frac{\sqrt{2}}{3}i$$

The solution set is $\left\{-\frac{1}{3}+\frac{\sqrt{2}}{3}i,\ -\frac{1}{3}-\frac{\sqrt{2}}{3}i\right\}$.

6. $64x^3 + 1 = 0$

$(4x+1)(16x^2 - 4x + 1) = 0$

$4x + 1 = 0$ or $16x^2 - 4x + 1 = 0$

$x = -\dfrac{1}{4}$ or $16x^2 - 4x + 1 = 0$

$16x^2 - 4x + 1 = 0$; $a = 16, b = -4, c = 1$

$$x = \frac{-(-4) \pm \sqrt{(-4)^2 - 4\cdot16\cdot1}}{2\cdot16} = \frac{4 \pm \sqrt{16-64}}{32}$$

$$= \frac{4 \pm \sqrt{-48}}{32} = \frac{4 \pm 4i\sqrt{3}}{32} = \frac{1}{8} \pm \frac{\sqrt{3}}{8}i$$

The solution set is $\left\{-\frac{1}{4},\ \frac{1}{8}+\frac{\sqrt{3}}{8}i,\ \frac{1}{8}-\frac{\sqrt{3}}{8}i\right\}$.

7.
$$100p - 50 = \frac{50}{p}$$
$$p(100p - 50) = p \cdot \frac{50}{p}$$
$$100p^2 - 50p = 50$$
$$2p^2 - p = 1$$
$$2p^2 - p - 1 = 0; \quad a = 2, b = -1, c = -1$$
$$p = \frac{-(-1) \pm \sqrt{(-1)^2 - 4 \cdot 2 \cdot (-1)}}{2 \cdot 2} = \frac{1 \pm \sqrt{1+8}}{4}$$
$$= \frac{1 \pm \sqrt{9}}{4} = \frac{1 \pm 3}{4} = 1 \text{ or } -\frac{1}{2}$$
The price at the equilibrium point is \$1.

8.
$$p^2 - 8 = -p + 4$$
$$p^2 + p - 12 = 0; \quad a = 1, b = 1, c = -12$$
$$p = \frac{-1 \pm \sqrt{1^2 - 4 \cdot 1 \cdot (-12)}}{2 \cdot 1} = \frac{-1 \pm \sqrt{1+48}}{2}$$
$$= \frac{-1 \pm \sqrt{49}}{2} = \frac{-1 \pm 7}{2} = 3 \text{ or } -4$$
The price at the equilibrium point is \$3.

Exercises 8.2

1.
$$x^2 + x - 2 = 0; \quad a = 1, b = 1, c = -2$$
$$x = \frac{-1 \pm \sqrt{1^2 - 4 \cdot 1 \cdot (-2)}}{2 \cdot 1} = \frac{-1 \pm \sqrt{1+8}}{2}$$
$$= \frac{-1 \pm \sqrt{9}}{2} = \frac{-1 \pm 3}{2} = 1 \text{ or } -2$$
The solution set is $\{1, -2\}$.

3.
$$x^2 + 4x = -1$$
$$x^2 + 4x + 1 = 0; \quad a = 1, b = 4, c = 1$$
$$x = \frac{-4 \pm \sqrt{4^2 - 4 \cdot 1 \cdot 1}}{2 \cdot 1} = \frac{-4 \pm \sqrt{16 - 4}}{2}$$
$$= \frac{-4 \pm \sqrt{12}}{2} = \frac{-4 \pm 2\sqrt{3}}{2} = -2 \pm \sqrt{3}$$
The solution set is $\left\{-2 + \sqrt{3}, -2 - \sqrt{3}\right\}$.

5.
$$x^2 - 3x = 2$$
$$x^2 - 3x - 2 = 0; \quad a = 1, b = -3, c = -2$$
$$x = \frac{-(-3) \pm \sqrt{(-3)^2 - 4 \cdot 1 \cdot (-2)}}{2 \cdot 1}$$
$$= \frac{3 \pm \sqrt{9 + 8}}{2} = \frac{3 \pm \sqrt{17}}{2}$$
The solution set is $\left\{\dfrac{3 + \sqrt{17}}{2}, \dfrac{3 - \sqrt{17}}{2}\right\}$.

7.
$$7y^2 = 12y - 5$$
$$7y^2 - 12y + 5 = 0; \quad a = 7, b = -12, c = 5$$
$$y = \frac{-(-12) \pm \sqrt{(-12)^2 - 4 \cdot 7 \cdot 5}}{2 \cdot 7}$$
$$= \frac{12 \pm \sqrt{144 - 140}}{14} = \frac{12 \pm \sqrt{4}}{14} = \frac{12 \pm 2}{14}$$
$$= 1 \text{ or } \frac{5}{7}$$
The solution set is $\left\{1, \dfrac{5}{7}\right\}$.

9.
$$5y^2 + 8y = -5$$
$$5y^2 + 8y + 5 = 0; \quad a = 5, b = 8, c = 5$$
$$y = \frac{-8 \pm \sqrt{8^2 - 4 \cdot 5 \cdot 5}}{2 \cdot 5} = \frac{-8 \pm \sqrt{64 - 100}}{10}$$
$$= \frac{-8 \pm \sqrt{-36}}{10} = \frac{-8 \pm 6i}{10} = -\frac{4}{5} \pm \frac{3}{5}i$$
The solution set is $\left\{ -\frac{4}{5} + \frac{3}{5}i, \ -\frac{4}{5} - \frac{3}{5}i \right\}$.

11.
$$7y + 6 = -2y^2$$
$$2y^2 + 7y + 6 = 0; \quad a = 2, b = 7, c = 6$$
$$y = \frac{-7 \pm \sqrt{7^2 - 4 \cdot 2 \cdot 6}}{2 \cdot 2} = \frac{-7 \pm \sqrt{49 - 48}}{4}$$
$$= \frac{-7 \pm \sqrt{1}}{4} = \frac{-7 \pm 1}{4} = -\frac{3}{2} \text{ or } -2$$
The solution set is $\left\{ -\frac{3}{2}, -2 \right\}$.

13.
$$\frac{x^2}{5} - \frac{x}{2} = -\frac{3}{10}$$
$$2x^2 - 5x = -3$$
$$2x^2 - 5x + 3 = 0; \quad a = 2, b = -5, c = 3$$
$$x = \frac{-(-5) \pm \sqrt{(-5)^2 - 4 \cdot 2 \cdot 3}}{2 \cdot 2} = \frac{5 \pm \sqrt{25 - 24}}{4}$$
$$= \frac{5 \pm \sqrt{1}}{4} = \frac{5 \pm 1}{4} = \frac{3}{2} \text{ or } 1$$
The solution set is $\left\{ \frac{3}{2}, 1 \right\}$.

15.
$$\frac{x^2}{7} + \frac{x}{2} = -\frac{3}{14}$$
$$2x^2 + 7x = -3$$
$$2x^2 + 7x + 3 = 0; \quad a = 2, b = 7, c = 3$$
$$x = \frac{-7 \pm \sqrt{7^2 - 4 \cdot 2 \cdot 3}}{2 \cdot 2} = \frac{-7 \pm \sqrt{49 - 24}}{4}$$
$$= \frac{-7 \pm \sqrt{25}}{4} = \frac{-7 \pm 5}{4} = -\frac{1}{2} \text{ or } -3$$
The solution set is $\left\{ -\frac{1}{2}, -3 \right\}$.

17.
$$\frac{x^2}{2} - \frac{3x}{4} = -\frac{1}{8}$$
$$4x^2 - 6x = -1$$
$$4x^2 - 6x + 1 = 0; \quad a = 4, b = -6, c = 1$$
$$x = \frac{-(-6) \pm \sqrt{(-6)^2 - 4 \cdot 4 \cdot 1}}{2 \cdot 4} = \frac{6 \pm \sqrt{36 - 16}}{8}$$
$$= \frac{6 \pm \sqrt{20}}{8} = \frac{6 \pm 2\sqrt{5}}{8} = \frac{3 \pm \sqrt{5}}{4}$$
The solution set is $\left\{ \frac{3 + \sqrt{5}}{4}, \ \frac{3 - \sqrt{5}}{4} \right\}$.

19.
$$\frac{x^2}{8} = -\frac{x}{4} - \frac{1}{8}$$
$$x^2 = -2x - 1$$
$$x^2 + 2x + 1 = 0; \quad a = 1, b = 2, c = 1$$
$$x = \frac{-2 \pm \sqrt{2^2 - 4 \cdot 1 \cdot 1}}{2 \cdot 1} = \frac{-2 \pm \sqrt{4 - 4}}{2}$$
$$= \frac{-2 \pm \sqrt{0}}{2} = \frac{-2}{2} = -1$$
The solution set is $\{-1\}$.

21.
$$6x = 4x^2 + 1$$
$$4x^2 - 6x + 1 = 0; \quad a = 4, b = -6, c = 1$$
$$x = \frac{-(-6) \pm \sqrt{(-6)^2 - 4 \cdot 4 \cdot 1}}{2 \cdot 4} = \frac{6 \pm \sqrt{36 - 16}}{8}$$
$$= \frac{6 \pm \sqrt{20}}{8} = \frac{6 \pm 2\sqrt{5}}{8} = \frac{3 \pm \sqrt{5}}{4}$$
The solution set is $\left\{ \frac{3 + \sqrt{5}}{4}, \ \frac{3 - \sqrt{5}}{4} \right\}$.

23.
$$3x = 1 - 3x^2$$
$$3x^2 + 3x - 1 = 0; \quad a = 3, b = 3, c = -1$$
$$x = \frac{-3 \pm \sqrt{3^2 - 4 \cdot 3 \cdot (-1)}}{2 \cdot 3} = \frac{-3 \pm \sqrt{9 + 12}}{6}$$
$$= \frac{-3 \pm \sqrt{21}}{6}$$
The solution set is $\left\{ \frac{-3 + \sqrt{21}}{6}, \ \frac{-3 - \sqrt{21}}{6} \right\}$.

25. $x(x+2)=2x(x+1)-4$

$x^2+2x=2x^2+2x-4$

$x^2-4=0$; $a=1, b=0, c=-4$

$x=\dfrac{-0\pm\sqrt{0^2-4\cdot1\cdot(-4)}}{2\cdot1}=\dfrac{0\pm\sqrt{16}}{2}$

$=\dfrac{\pm4}{2}=2 \text{ or } -2$

The solution set is $\{2, -2\}$.

27. $6x(x+5)=(x+15)^2$

$6x^2+30x=x^2+30x+225$

$5x^2-225=0$

$x^2-45=0$; $a=1, b=0, c=-45$

$x=\dfrac{-0\pm\sqrt{0^2-4\cdot1\cdot(-45)}}{2\cdot1}=\dfrac{0\pm\sqrt{180}}{2}$

$=\dfrac{\pm6\sqrt5}{2}=\pm3\sqrt5$

The solution set is $\left\{3\sqrt5, -3\sqrt5\right\}$.

29. $(x-2)^2=4x(x-1)$

$x^2-4x+4=4x^2-4x$

$3x^2-4=0$; $a=3, b=0, c=-4$

$x=\dfrac{-0\pm\sqrt{0^2-4\cdot3\cdot(-4)}}{2\cdot3}=\dfrac{0\pm\sqrt{48}}{6}$

$=\dfrac{\pm4\sqrt3}{6}=\pm\dfrac{2\sqrt3}{3}$

The solution set is $\left\{\dfrac{2\sqrt3}{3}, -\dfrac{2\sqrt3}{3}\right\}$.

31. $x^3-8=0$

$(x-2)(x^2+2x+4)=0$

$x-2=0 \text{ or } x^2+2x+4=0$

$x=2 \text{ or } x^2+2x+4=0$

$x^2+2x+4=0$; $a=1, b=2, c=4$

$x=\dfrac{-2\pm\sqrt{2^2-4\cdot1\cdot4}}{2\cdot1}=\dfrac{-2\pm\sqrt{4-16}}{2}$

$=\dfrac{-2\pm\sqrt{-12}}{2}=\dfrac{-2\pm2i\sqrt3}{2}=-1\pm i\sqrt3$

The solution set is $\{2, -1+i\sqrt3, -1-i\sqrt3\}$.

33. $8x^3-1=0$

$(2x-1)(4x^2+2x+1)=0$

$2x-1=0 \text{ or } 4x^2+2x+1=0$

$x=\dfrac{1}{2} \text{ or } 4x^2+2x+1=0$

$4x^2+2x+1=0$; $a=4, b=2, c=1$

$x=\dfrac{-2\pm\sqrt{2^2-4\cdot4\cdot1}}{2\cdot4}=\dfrac{-2\pm\sqrt{4-16}}{8}$

$=\dfrac{-2\pm\sqrt{-12}}{8}=\dfrac{-2\pm2i\sqrt3}{8}=-\dfrac{1}{4}\pm\dfrac{\sqrt3}{4}i$

The solution set is

$\left\{\dfrac{1}{2}, -\dfrac{1}{4}+\dfrac{\sqrt3}{4}i, -\dfrac{1}{4}-\dfrac{\sqrt3}{4}i\right\}$.

35. $x^3+27=0$

$(x+3)(x^2-3x+9)=0$

$x+3=0 \text{ or } x^2-3x+9=0$

$x=-3 \text{ or } x^2-3x+9=0$

$x^2-3x+9=0$; $a=1, b=-3, c=9$

$x=\dfrac{-(-3)\pm\sqrt{(-3)^2-4\cdot1\cdot9}}{2\cdot1}=\dfrac{3\pm\sqrt{9-36}}{2}$

$=\dfrac{3\pm\sqrt{-27}}{2}=\dfrac{3\pm3i\sqrt3}{2}=\dfrac{3}{2}\pm\dfrac{3\sqrt3}{2}i$

The solution set is

$\left\{-3, \dfrac{3}{2}+\dfrac{3\sqrt3}{2}i, \dfrac{3}{2}-\dfrac{3\sqrt3}{2}i\right\}$.

37. **a.** Yes, the consumption can be the same.

b. $0.013t^2 \approx 0.96t + 25.4 = 0.4t + 6.03$

$0.013t^2 \approx 1.36t + 19.37 = 0$

$a = 0.013, \; b = \approx 1.36, \; c = 19.37$

$t = \dfrac{\approx(\approx 1.36) \pm \sqrt{(\approx 1.36)^2 \approx 4 \approx 0.013 \approx 19.37}}{2 \approx 0.013}$

$= \dfrac{1.36 \pm \sqrt{1.8496 \approx 1.00724}}{0.026}$

$= \dfrac{1.36 \pm \sqrt{0.84236}}{0.026} = \dfrac{1.36 \pm 0.9178}{0.026}$

≈ 87.6 or 17

Using $t = 17$ the consumption would have been the same in 1987.

39.

$30p \cdot 50 = \dfrac{10}{p}$

$p(30p \cdot 50) = p \cdot \dfrac{10}{p}$

$30p^2 \cdot 50p = 10$

$3p^2 \cdot 5p = 1$

$3p^2 \cdot 5p \cdot 1 = 0; \quad a = 3, b = \cdot 5, c = \cdot 1$

$p = \dfrac{\cdot(\cdot 5) \pm \sqrt{(\cdot 5)^2 \cdot 4 \cdot 3 \cdot (\cdot 1)}}{2 \cdot 3}$

$= \dfrac{5 \pm \sqrt{25 + 12}}{6} = \dfrac{5 \pm \sqrt{37}}{6} = \dfrac{5 \pm 6.08}{6}$

$= 1.85$ or $\cdot 0.18$

The price at the equilibrium point is \$1.85.

41. $40 = 20x \cdot x^2$

$x^2 \cdot 20x + 40 = 0; \quad a = 1, b = \cdot 20, c = 40$

$p = \dfrac{\cdot(\cdot 20) \pm \sqrt{(\cdot 20)^2 \cdot 4 \cdot 1 \cdot 40}}{2 \cdot 1}$

$= \dfrac{20 \pm \sqrt{400 \cdot 160}}{2} = \dfrac{20 \pm \sqrt{240}}{2}$

$= \dfrac{20 \pm 4\sqrt{15}}{2} = 10 \pm 2\sqrt{15}$

The bending moment is $10 + 2\sqrt{15}$ or $10 - 2\sqrt{15}$.

43. $aL^2 + bL + c = d;$

$a = 400, b = 200, c = 200, d = 800$

$400L^2 + 200L + 200 = 800$

$400L^2 + 200L \cdot 600 = 0$

$2L^2 + L \cdot 3 = 0; \quad a = 2, b = 1, c = \cdot 3$

$L = \dfrac{\cdot 1 \pm \sqrt{1^2 \cdot 4 \cdot 2 \cdot (\cdot 3)}}{2 \cdot 2} = \dfrac{\cdot 1 \pm \sqrt{1 + 24}}{4}$

$= \dfrac{\cdot 1 \pm \sqrt{25}}{4} = \dfrac{\cdot 1 \pm 5}{4} = 1$ or $\cdot \dfrac{3}{2}$

The maximum safe length is 1 unit.

45. $\sqrt{b^2 \cdot 4ac}; \quad a = 3, b = \cdot 2, c = \cdot 1$

$\sqrt{(\cdot 2)^2 \cdot 4 \cdot 3 \cdot (\cdot 1)} = \sqrt{4 + 12} = \sqrt{16} = 4$

47. $\sqrt{b^2 \cdot 4ac}; \quad a = 3, b = \cdot 5, c = 4$

$\sqrt{(\cdot 5)^2 \cdot 4 \cdot 3 \cdot 4} = \sqrt{25 \cdot 48} = \sqrt{\cdot 23} = i\sqrt{23}$

49. $(2x+1)(3x-4) = 6x^2 - 8x + 3x - 4$
$= 6x^2 - 5x - 4$

51. $(3x-7)(4x+3) = 12x^2 + 9x - 28x - 21$
$= 12x^2 - 19x - 21$

53. $4a^2x^2 + 4abx + 4ac = 0$ 　　Multiply both sides by $4a$.

55. $4a^2x^2 + 4abx + b^2 = b^2 - 4ac$ 　　Add b^2 to both sides.

57. $2ax + b = \pm\sqrt{b^2 - 4ac}$ 　　Take the square root of both sides.

59. $x = \dfrac{-b \pm \sqrt{b^2 - 4ac}}{2a}$ 　　Divide both sides by $2a$.

61. Sample answer: A linear equation is degree 1; a quadratic equation is degree 2; and a cubic equation is degree 3.

63.
$$27x^3 - 8 = 0$$
$$(3x - 2)(9x^2 + 6x + 4) = 0$$
$$3x - 2 = 0 \quad \text{or} \quad 9x^2 + 6x + 4 = 0$$
$$x = \frac{2}{3} \text{ or } 9x^2 + 6x + 4 = 0$$
$$9x^2 + 6x + 4 = 0; \quad a = 9, b = 6, c = 4$$
$$x = \frac{-6 \pm \sqrt{6^2 - 4 \cdot 9 \cdot 4}}{2 \cdot 9} = \frac{-6 \pm \sqrt{36 - 144}}{18}$$
$$= \frac{-6 \pm \sqrt{-108}}{18} = \frac{-6 \pm 6i\sqrt{3}}{18} = -\frac{1}{3} \pm \frac{\sqrt{3}}{3}i$$
The solution set is
$$\left\{ \frac{2}{3}, -\frac{1}{3} + \frac{\sqrt{3}}{3}i, -\frac{1}{3} - \frac{\sqrt{3}}{3}i \right\}.$$

65.
$$3x^2 + 2x = -1$$
$$3x^2 + 2x + 1 = 0; \quad a = 3, b = 2, c = 1$$
$$x = \frac{-2 \pm \sqrt{2^2 - 4 \cdot 3 \cdot 1}}{2 \cdot 3} = \frac{-2 \pm \sqrt{4 - 12}}{6}$$
$$= \frac{-2 \pm \sqrt{-8}}{6} = \frac{-2 \pm 2i\sqrt{2}}{6} = -\frac{1}{3} \pm \frac{\sqrt{2}}{3}i$$
The solution set is $\left\{ -\frac{1}{3} + \frac{\sqrt{2}}{3}i, -\frac{1}{3} - \frac{\sqrt{2}}{3}i \right\}.$

67.
$$\frac{x^2}{4} - \frac{3}{8}x = \frac{1}{4}$$
$$2x^2 - 3x = 2$$
$$2x^2 - 3x - 2 = 0; \quad a = 2, b = -3, c = -2$$
$$x = \frac{-(-3) \pm \sqrt{(-3)^2 - 4 \cdot 2 \cdot (-2)}}{2 \cdot 2} = \frac{3 \pm \sqrt{9 + 16}}{4}$$
$$= \frac{3 \pm \sqrt{25}}{4} = \frac{3 \pm 5}{4} = 2 \text{ or } -\frac{1}{2}$$
The solution set is $\left\{ 2, -\frac{1}{2} \right\}.$

69.
$$6x = x^2$$
$$x^2 - 6x + 0 = 0; \quad a = 1, b = -6, c = 0$$
$$x = \frac{-(-6) \pm \sqrt{(-6)^2 - 4 \cdot 1 \cdot 0}}{2 \cdot 1}$$
$$= \frac{6 \pm \sqrt{36 - 0}}{2} = \frac{6 \pm 6}{2} = 6 \text{ or } 0$$
The solution set is $\{6, 0\}.$

71.
$$x^2 = 4x + 4$$
$$x^2 - 4x - 4 = 0; \quad a = 1, b = -4, c = -4$$
$$x = \frac{-(-4) \pm \sqrt{(-4)^2 - 4 \cdot 1 \cdot (-4)}}{2 \cdot 1}$$
$$= \frac{4 \pm \sqrt{16 + 16}}{2} = \frac{4 \pm \sqrt{32}}{2} = \frac{4 \pm 4\sqrt{2}}{2}$$
$$= 2 \pm 2\sqrt{2}$$
The solution set is $\left\{ 2 + 2\sqrt{2}, 2 - 2\sqrt{2} \right\}.$

73.
$$3x^2 + 2x - 5 = 0; \quad a = 3, b = 2, c = -5$$
$$x = \frac{-2 \pm \sqrt{2^2 - 4 \cdot 3 \cdot (-5)}}{2 \cdot 3} = \frac{-2 \pm \sqrt{4 + 60}}{6}$$
$$= \frac{-2 \pm \sqrt{64}}{6} = \frac{-2 \pm 8}{6} = 1 \text{ or } -\frac{5}{3}$$
The solution set is $\left\{ -\frac{5}{3}, 1 \right\}.$

75.

$$100p \cdot 50 = \frac{50}{p}$$

$$p(100p \cdot 50) = 50$$

$$100p^2 \cdot 50p = 50$$

$$2p^2 \cdot p = 1$$

$$2p^2 \cdot p \cdot 1 = 0; \quad a = 2, b = \cdot 1, c = \cdot 1$$

$$p = \frac{\cdot(\cdot 1) \pm \sqrt{(\cdot 1)^2 \cdot 4 \cdot 2 \cdot (\cdot 1)}}{2 \cdot 2} = \frac{1 \pm \sqrt{1+8}}{4}$$

$$= \frac{1 \pm \sqrt{9}}{4} = \frac{1 \pm 3}{4} = 1 \text{ or } \cdot \frac{1}{2}$$

The price at the equilibrium point is \$1.

8.3   The Discriminant and Its Applications

Problems 8.3

1. a. $2x^2 = 6x \cdot 7$

$2x^2 \cdot 6x + 7 = 0; \quad a = 2, b = \cdot 6, c = 7$

$b^2 \cdot 4ac = (\cdot 6)^2 \cdot 4 \cdot 2 \cdot 7 = 36 \cdot 56 = \cdot 20$

Since the discriminant is negative, there are two non-real complex solutions.

b. $x^2 + kx = \cdot 9$

$x^2 + kx + 9 = 0; \quad a = 1, b = k, c = 9$

$b^2 \cdot 4ac = k^2 \cdot 4 \cdot 1 \cdot 9 = k^2 \cdot 36$

$k^2 \cdot 36 = 0$

$k^2 = 36$

$k = \pm 6$

There is one real rational solution when k is 6 or –6.

2. $20x^2 + 10x \cdot 30; \quad a = 20, b = 10, c = \cdot 30$

$b^2 \cdot 4ac = 10^2 \cdot 4 \cdot 20 \cdot (\cdot 30) = 100 + 2400$

$$= 2500 = (50)^2$$

Since the discriminant is a positive perfect square, the expression is factorable.

3. $21x^2 + x \cdot 10; \quad a = 21, b = 1, c = \cdot 10$

$b^2 \cdot 4ac = 1^2 \cdot 4 \cdot 21 \cdot (\cdot 10) = 1 + 840$

$$= 841 = 29^2$$

$$x = \frac{\cdot 1 \pm \sqrt{841}}{2 \cdot 21} = \frac{\cdot 1 \pm 29}{42}$$

$$x = \cdot \frac{5}{7} \text{ or } \quad x = \frac{2}{3}$$

$$7x = \cdot 5 \text{ or } \quad 3x = 2$$

$$7x + 5 = 0 \quad \text{or } 3x \cdot 2 = 0$$

The factorization is $(7x + 5)(3x \cdot 2)$

4.

$$x = -2 \text{ or } \quad x = \frac{3}{4}$$
$$x + 2 = 0 \quad \text{or} \quad 4x = 3$$
$$x + 2 = 0 \quad \text{or } 4x - 3 = 0$$
$$(x+2)(4x-3) = 0$$
$$4x^2 - 3x + 8x - 6 = 0$$
$$4x^2 + 5x - 6 = 0$$

b. $\quad 4x^2 - 12x + 5 = 0; \quad a = 4, b = -12, c = 5$

$$-\frac{b}{a} = -\frac{-12}{4} = 3 \qquad r_1 + r_2 = \frac{1}{2} + \frac{5}{2} = \frac{6}{2} = 3$$
$$\frac{c}{a} = \frac{5}{4} \qquad\qquad r_1 \cdot r_2 = \frac{1}{2} \cdot \frac{5}{2} = \frac{5}{4}$$

5. **a.** $\quad 4x^2 - 12x + 5 = 0; \quad a = 4, b = -12, c = 5$

$$-\frac{b}{a} = -\frac{-12}{4} = 3$$
$$r_1 + r_2 = \frac{1}{2} + \left(-\frac{5}{2}\right) = -\frac{4}{2} = -2$$

Since the sum of the solutions (-2) does not equal $-\frac{b}{a} = 3$, $\frac{1}{2}$ and $\frac{5}{2}$ cannot be the solutions.

Since the sum of the solutions does equal $-\dfrac{b}{a}$ and the product of the solutions does equal $\dfrac{c}{a}$, $\dfrac{1}{2}$ and $\dfrac{5}{2}$ are the solutions.

Exercises 8.3

1. $\quad 3x^2 + 5x - 2 = 0; \quad a = 3, b = 5, c = -2$

$D = 5^2 - 4\cdot3\cdot(-2) = 25 + 24 = 49 = 7^2$

Since the discriminant is a positive perfect square number, there are two rational number solutions.

3. $\quad 4x^2 = 4x - 1$

$4x^2 - 4x + 1 = 0; \quad a = 4, b = -4, c = 1$

$D = (-4)^2 - 4\cdot4\cdot1 = 16 - 16 = 0$

Since the discriminant is zero, there is one rational number solution.

5. $\quad 2x^2 = 2x + 5$

$2x^2 - 2x - 5 = 0; \quad a = 2, b = -2, c = -5$

$D = (-2)^2 - 4\cdot2\cdot(-5) = 4 + 40 = 44$

Since the discriminant is positive and not a perfect square number, there are two irrational number solutions.

7. $\quad 4x^2 - 5x + 3 = 0; \quad a = 4, b = -5, c = 3$

$D = (-5)^2 - 4\cdot4\cdot3 = 25 - 48 = -23$

Since the discriminant is negative, there are two non-real complex number solutions.

9.

$$x^2 - 2 = \frac{5}{2}x$$

$$x^2 - \frac{5}{2}x - 2 = 0; \quad a = 1, b = -\frac{5}{2}, c = -2$$

$$D = \left(-\frac{5}{2}\right)^2 - 4\cdot1\cdot(-2) = \frac{25}{4} + 8 = \frac{57}{4}$$

Since the discriminant is positive and not a perfect square number, there are two irrational number solutions.

11. $\quad x^2 - 4kx + 64 = 0; \quad a = 1, b = -4k, c = 64$

$b^2 - 4ac = (-4k)^2 - 4\cdot1\cdot64 = 16k^2 - 256$

$$16k^2 - 256 = 0$$
$$16k^2 = 256$$
$$k^2 = 16$$
$$k = \pm4$$

There is one real rational solution when k is 4 or –4.

13. $kx^2 - 10x = 5$

$kx^2 - 10x - 5 = 0;\quad a = k, b = -10, c = -5$

$b^2 - 4ac = (-10)^2 - 4\cdot k\cdot(-5) = 100 + 20k$

$100 + 20k = 0$

$\quad 20k = -100$

$\qquad k = -5$

There is one real rational solution when k is −5.

15. $2x^2 = kx - 8$

$2x^2 - kx + 8 = 0;\quad a = 2, b = -k, c = 8$

$b^2 - 4ac = (-k)^2 - 4\cdot2\cdot8 = k^2 - 64$

$k^2 - 64 = 0$

$\quad k^2 = 64$

$\quad k = \pm 8$

There is one real rational solution when k is 8 or −8.

17. $25x^2 - kx = -4$

$25x^2 - kx + 4 = 0;\quad a = 25, b = -k, c = 4$

$b^2 - 4ac = (-k)^2 - 4\cdot25\cdot4 = k^2 - 400$

$k^2 - 400 = 0$

$\quad k^2 = 400$

$\quad k = \pm 20$

There is one real rational solution when k is 20 or −20.

19. $x^2 + 8x = k$

$x^2 + 8x - k = 0;\quad a = 1, b = 8, c = -k$

$b^2 - 4ac = 8^2 - 4\cdot1\cdot(-k) = 64 + 4k$

$64 + 4k = 0$

$\quad 4k = -64$

$\quad k = -16$

There is one real rational solution when k is −16.

21. $10x^2 - 7x + 8;\quad a = 10, b = -7, c = 8$

$b^2 - 4ac = (-7)^2 - 4\cdot10\cdot8 = 49 - 320$

$\qquad\qquad = -271$

Since the discriminant is not a positive perfect square, the expression is not factorable.

23. $12x^2 - 17x + 6;\quad a = 12, b = -17, c = 6$

$b^2 - 4ac = (-17)^2 - 4\cdot12\cdot6 = 289 - 288 = 1$

$x = \dfrac{-(-17)\pm\sqrt{1}}{2\cdot12} = \dfrac{17\pm1}{24}$

$x = \dfrac{3}{4}$ or $\quad x = \dfrac{2}{3}$

$4x = 3$ or $\quad 3x = 2$

$4x - 3 = 0$ or $3x - 2 = 0$

The factorization is $(4x - 3)(3x - 2)$.

25. $12x^2 - 17x + 2;\quad a = 12, b = -17, c = 2$

$b^2 - 4ac = (-17)^2 - 4\cdot12\cdot2 = 289 - 96 = 193$

Since the discriminant is not a positive perfect square, the expression is not factorable.

27. $15x^2 + 52x - 84;\quad a = 15, b = 52, c = -84$

$b^2 - 4ac = 52^2 - 4\cdot15\cdot(-84) = 2704 + 5040$

$\qquad\qquad = 7744 = 88^2$

$x = \dfrac{-52\pm\sqrt{7744}}{2\cdot15} = \dfrac{-52\pm88}{30}$

$x = \dfrac{6}{5}$ or $\quad x = -\dfrac{14}{3}$

$5x = 6$ or $\quad 3x = -14$

$5x - 6 = 0$ or $3x + 14 = 0$

The factorization is $(5x - 6)(3x + 14)$.

29. $12x^2 \cdot 61x + 60;\quad a = 12,\ b = \cdot\,61,\ c = 60$

$b^2 \cdot 4ac = (\cdot\,61)^2 \cdot\ 4 \cdot 12 \cdot 60 = 3721 \cdot 2880$

$$= 841 = 29^2$$

$$x = \frac{\cdot\,(\cdot\,61) \pm \sqrt{841}}{2 \cdot 12} = \frac{61 \pm 29}{24}$$

$$x = \frac{15}{4} \quad \text{or} \quad x = \frac{4}{3}$$

$$4x = 15 \quad \text{or} \quad 3x = 4$$

$$4x \cdot 15 = 0 \quad \text{or } 3x \cdot 4 = 0$$

The factorization is $(4x \cdot 15)(3x \cdot 4)$.

31. $x = 3 \quad \text{or} \quad x = 4$

$$x - 3 = 0 \ \text{or}\ x - 4 = 0$$

$$(x - 3)(x - 4) = 0$$

$$x^2 - 4x - 3x + 12 = 0$$

$$x^2 - 7x + 12 = 0$$

33. $x = -5 \quad \text{or} \quad x = -7$

$$x + 5 = 0 \quad \text{or } x + 7 = 0$$

$$(x + 5)(x + 7) = 0$$

$$x^2 + 7x + 5x + 35 = 0$$

$$x^2 + 12x + 35 = 0$$

35. $x = 3 \text{ or} \quad x = -\dfrac{2}{3}$

$$x - 3 = 0 \ \text{or} \quad 3x = -2$$

$$x - 3 = 0 \ \text{or } 3x + 2 = 0$$

$$(x - 3)(3x + 2) = 0$$

$$3x^2 + 2x - 9x - 6 = 0$$

$$3x^2 - 7x - 6 = 0$$

37. $x = \dfrac{1}{2} \quad \text{or} \quad x = -\dfrac{1}{2}$

$$2x = 1 \quad \text{or} \quad 2x = -1$$

$$2x - 1 = 0 \ \text{or } 2x + 1 = 0$$

$$(2x - 1)(2x + 1) = 0$$

$$4x^2 + 2x - 2x - 1 = 0$$

$$4x^2 - 1 = 0$$

39. $x = 0 \ \text{or} \quad x = -\dfrac{1}{5}$

$$x = 0 \ \text{or} \quad 5x = -1$$

$$x = 0 \ \text{or } 5x + 1 = 0$$

$$x(5x + 1) = 0$$

$$5x^2 + x = 0$$

41. $4x^2 - 6x + 5 = 0$

a. $-\dfrac{b}{a} = -\dfrac{-6}{4} = \dfrac{3}{2}$

b. $\dfrac{c}{a} = \dfrac{5}{4}$

c. $\dfrac{1}{2} + \dfrac{5}{2} = \dfrac{6}{2} = 3 \neq \dfrac{3}{2}$

The given values are not solutions.

43. $5x^2 + 13x - 6 = 0$

a. $-\dfrac{b}{a} = -\dfrac{13}{5}$

b. $\dfrac{c}{a} = \dfrac{-6}{5} = -\dfrac{6}{5}$

c. $\dfrac{2}{5} + (\cdot\,3) = \dfrac{2}{5} \cdot \dfrac{15}{5} = \cdot\,\dfrac{13}{5} = \cdot\,\dfrac{13}{5}$

$\dfrac{2}{5} \cdot (\cdot\,3) = \cdot\,\dfrac{6}{5} = \cdot\,\dfrac{6}{5}$

The given values are the solutions.

45.　　$2x^2 + 5x + 2 = 0$

　a.　$-\dfrac{b}{a} = -\dfrac{5}{2}$

　b.　$\dfrac{c}{a} = \dfrac{2}{2} = 1$

　c.　$\dfrac{1}{2} + 2 = \dfrac{5}{2} - -\dfrac{5}{2}$

　　　The given values are not solutions.

47.　　$3x^2 + kx = 40$

　　　$3x^2 + kx - 40 = 0$

　　　$\cdot\, 5 \cdot r = \dfrac{\cdot\, 40}{3}$

　　　$r = \dfrac{\cdot\, 40}{3} \cdot \dfrac{1}{\cdot\, 5} = \dfrac{8}{3}$

49.　　$10x^2 + (k-2)x = 3$

　　　$10x^2 + (k-2)x - 3 = 0$

　　　$-\dfrac{k-2}{10} = -\dfrac{13}{10}$

　　　$k - 2 = 13$

　　　　$k = 15$

51.　　$x + 2\sqrt{x} - 3 = 0$

　　　　$x - 3 = -2\sqrt{x}$

　　　　$(x-3)^2 = \left(-2\sqrt{x}\right)^2$

　　　　$x^2 - 6x + 9 = 4x$

　　　　$x^2 - 10x + 9 = 0$

　　　　$(x-9)(x-1) = 0$

　　　　　$x - 9 = 0$ or $x - 1 = 0$

　　　　　$x = 9$ or　　$x = 1$

　　Check: $9 + 2\sqrt{9} \neq 3 = 9 + 6 \neq 3 = 12 \neq 0$

　　　Discard: $x = 9$

　　　$1 + 2\sqrt{1} \neq 3 = 1 + 2 \neq 3 = 0$

　　The solution is $x = 1$.

53.　　$x^2 + 6x + 5 = 0$

　　　$(x+5)(x+1) = 0$

　　　　$x + 5 = 0$ or $x + 1 = 0$

　　　　　$x = -5$ or　　$x = -1$

55.　　$27.5 = \cdot\, t^2 + 2t + 27$

　　　$t^2 \cdot 2t + 0.5 = 0;\quad a = 1, b = \cdot\, 2, c = 0.5$

　　　$D = (\cdot\, 2)^2 \cdot 4 \cdot 1 \cdot 0.5 = 4 \cdot 2 = 2$

　　Since the discriminant is a positive number, there are two solutions. The diver will be at a height of 27.5 m above the water twice.

57.　$29 = \cdot\, t^2 + 2t + 27$

　$t^2 \cdot 2t + 2 = 0;\quad a = 1, b = \cdot\, 2, c = 2$

　$D = (\cdot\, 2)^2 \cdot 4 \cdot 1 \cdot 2 = 4 \cdot 8 = \cdot\, 4$

　Since the discriminant is a negative number, there are no solutions. The diver will never be at a height of 29 m above the water.

59. Sample answer: It is impossible to have exactly one complex solution because of the ± in the quadratic formula. If there is to be a complex solution, the discriminant must be negative and two solutions result from the ± sign.

61. Sample answer: It is impossible to have exactly one rational and one irrational solution because the discriminant cannot be a perfect square and a non-perfect square at the same time.

63. $4x^2 - 12x + 5 = 0$

$$-\frac{b}{a} = -\frac{-12}{4} = 3$$

$$\frac{c}{a} = \frac{5}{4}$$

$$\frac{1}{2} + \frac{5}{2} = \frac{6}{2} = 3$$

$$\frac{1}{2} \cdot \frac{5}{2} = \frac{5}{4}$$

The given values are the solutions.

65. $12x^2 + x \cdot 35; \quad a = 12, b = 1, c = \cdot 35$

$$b^2 \cdot 4ac = 1^2 \cdot 4 \cdot 12 \cdot (\cdot 35) = 1 + 1680$$

$$= 1681 = 41^2$$

$$x = \frac{\cdot 1 \pm \sqrt{1681}}{2 \cdot 12} = \frac{\cdot 1 \pm 41}{24}$$

$$x = \frac{5}{3} \text{ or } \quad x = \cdot \frac{7}{4}$$

$$3x = 5 \text{ or } \quad 4x = \cdot 7$$

$$3x \cdot 5 = 0 \text{ or } 4x + 7 = 0$$

The factorization is $(3x \cdot 5)(4x + 7)$.

67.

$$x = -1 \text{ or } \quad x = \frac{2}{3}$$

$$x + 1 = 0 \quad \text{or} \quad 3x = 2$$

$$x + 1 = 0 \quad \text{or } 3x - 2 = 0$$

$$(x + 1)(3x - 2) = 0$$

$$3x^2 - 2x + 3x - 2 = 0$$

$$3x^2 + x - 2 = 0$$

8.4   Solving Equations in Quadratic Form

Problems 8.4

1.   a.

$$\frac{10}{x^2 \cdot 25} \cdot \frac{1}{x \cdot 5} = 1$$

$$\frac{10}{(x+5)(x \cdot 5)} \cdot \frac{1}{x \cdot 5} = 1$$

$$(x+5)(x \cdot 5) \cdot \frac{10}{(x+5)(x \cdot 5)} \cdot \frac{1}{x \cdot 5} = 1 \cdot (x+5)(x \cdot 5)$$

$$10 \cdot 1(x+5) = (x+5)(x \cdot 5)$$

$$10 \cdot x \cdot 5 = x^2 \cdot 25$$

$$0 = x^2 + x \cdot 30$$

$$(x+6)(x \cdot 5) = 0$$

$$x + 6 = 0 \quad \text{or } x \cdot 5 = 0$$

$$x = \cdot 6 \text{ or } \quad x = 5$$

Discard $x = 5$ because it makes the denominator zero. The solution is –6.

b.

$$\frac{10}{x^2-16}+\frac{1}{x-4}=1$$

$$\frac{10}{(x+4)(x-4)}+\frac{1}{x-4}=1$$

$$(x+4)(x-4)\cdot\frac{10}{(x+4)(x-4)}+\frac{1}{x-4}\cdot=1\cdot(x+4)(x-4)$$

$$10+1(x+4)=(x+4)(x-4)$$

$$10+x+4=x^2-16$$

$$0=x^2-x-30$$

$$(x-6)(x+5)=0$$

$$x-6=0 \text{ or } x+5=0$$

$$x=6 \text{ or } \quad x=-5$$

The solutions are –5 and 6.

2. $x^4-17x^2+16=0$

Substitute $u=x^2$

$$u^2-17u+16=0$$

$$(u-16)(u-1)=0$$

$$u-16=0 \quad \text{or } u-1=0$$

$$u=16 \text{ or } \quad u=1$$

$$x^2=16 \text{ or } \quad x^2=1$$

$$x=\pm4 \text{ or } \quad x=\pm1$$

The solutions are 4, –4, 1, and –1.

3. $\left(x^2+x\right)^2-11\left(x^2+x\right)-12=0$

Substitute $u=x^2+x$

$$u^2-11u-12=0$$

$$(u-12)(u+1)=0$$

$$u-12=0 \quad \text{or} \quad u+1=0$$

$$x^2+x-12=0 \text{ or } x^2+x+1=0$$

$$(x+4)(x-3)=0 \text{ or } x=\frac{-1\pm\sqrt{1^2-4\cdot1\cdot1}}{2\cdot1}$$

$$x+4=0 \text{ or } x-3=0 \text{ or } x=\frac{-1\pm\sqrt{-3}}{2}$$

$$x=-4 \text{ or } \quad x=3 \text{ or } x=-\frac{1}{2}\pm\frac{\sqrt{3}}{2}i$$

The solutions are:

$$-4,\ 3,\ -\frac{1}{2}+\frac{\sqrt{3}}{2}i,\ \text{ and } -\frac{1}{2}-\frac{\sqrt{3}}{2}i\,.$$

4. $x^{1/2}-7x^{1/4}+10=0$

Substitute $u=x^{1/4}$

$$u^2-7u+10=0$$

$$(u-5)(u-2)=0$$

$$u-5=0 \quad \text{or } u-2=0$$

$$u=5 \quad \text{or} \quad u=2$$

$$x^{1/4}=5 \quad \text{or} \quad x^{1/4}=2$$

$$x=5^4 \quad \text{or} \quad x=2^4$$

$$x=625 \text{ or} \quad x=16$$

The solutions are 625 and 16.

5.
$$x - 3\sqrt{x} - 10 = 0$$

Substitute $u = \sqrt{x}$

$$u^2 - 3u - 10 = 0$$
$$(u - 5)(u + 2) = 0$$
$$u - 5 = 0 \quad \text{or} \quad u + 2 = 0$$
$$u = 5 \quad \text{or} \quad u = -2$$
$$\sqrt{x} = 5 \quad \text{or} \quad \sqrt{x} = -2$$
$$x = 25 \quad \text{or} \quad x = 4$$

Check: $25 \neq 3\sqrt{25} \neq 10 = 25 \neq 15 \neq 10 = 0$

$\qquad 4 \neq 3\sqrt{4} \neq 10 = 4 \neq 6 \neq 10 = \neq 12 \neq 0$

Discard: $x = 4$

The solution is 25.

6.
$$2x^{-2} \cdot 5x^{-1} \cdot 3 = 0$$
$$\frac{2}{x^2} \cdot \frac{5}{x} \cdot 3 = 0$$
$$x^2 \cdot \frac{2}{x^2} \cdot \frac{5}{x} \cdot 3 = x^2 \cdot 0$$
$$2 \cdot 5x \cdot 3x^2 = 0$$
$$3x^2 + 5x \cdot 2 = 0$$
$$(3x \cdot 1)(x + 2) = 0$$
$$3x - 1 = 0 \quad \text{or} \quad x + 2 = 0$$
$$3x = 1 \quad \text{or} \quad x = -2$$
$$x = \frac{1}{3} \quad \text{or} \quad x = -2$$

The solutions are -2 and $\frac{1}{3}$.

Exercises 8.4

1.
$$\frac{x}{x+4} + \frac{x}{x+1} = 0$$
$$(x+4)(x+1)\frac{x}{x+4} + \frac{x}{x+1} = 0 \cdot (x+4)(x+1)$$
$$x(x+1) + x(x+4) = 0$$
$$x^2 + x + x^2 + 4x = 0$$
$$2x^2 + 5x = 0$$
$$x(2x+5) = 0$$
$$x = 0 \quad \text{or} \quad 2x + 5 = 0$$
$$x = 0 \quad \text{or} \quad 2x = -5$$
$$x = 0 \quad \text{or} \quad x = -\frac{5}{2}$$

The solutions are 0 and $-\frac{5}{2}$.

3.

$$\frac{x-1}{x+11}-\frac{2}{x-1}=0$$

$$(x+11)(x-1)\cdot\left[\frac{x-1}{x+11}-\frac{2}{x-1}\right]=0\cdot(x+11)(x-1)$$

$$(x-1)(x-1)-2(x+11)=0$$

$$x^2-2x+1-2x-22=0$$

$$x^2-4x-21=0$$

$$(x+3)(x-7)=0$$

$$x+3=0 \quad\text{or}\quad x-7=0$$

$$x=-3 \quad\text{or}\quad x=7$$

The solutions are -3 and 7.

5.

$$\frac{x}{x-1}-\frac{x}{x+1}=0$$

$$(x-1)(x+1)\cdot\left[\frac{x}{x-1}-\frac{x}{x+1}\right]=0\cdot(x-1)(x+1)$$

$$x(x+1)-x(x-1)=0$$

$$x^2+x-x^2+x=0$$

$$2x=0$$

$$x=0$$

The solution is 0.

7.

$$\frac{x}{x+2}-\frac{x}{x+1}=-\frac{1}{6}$$

$$6(x+2)(x+1)\cdot\left[\frac{x}{x+2}-\frac{x}{x+1}\right]=-\frac{1}{6}\cdot6(x+2)(x+1)$$

$$6x(x+1)-6x(x+2)=-\left(x^2+3x+2\right)$$

$$6x^2+6x-6x^2-12x=-x^2-3x-2$$

$$-6x=-x^2-3x-2$$

$$x^2-3x+2=0$$

$$(x-1)(x-2)=0$$

$$x-1=0 \quad\text{or}\quad x-2=0$$

$$x=1 \quad\text{or}\quad x=2$$

The solutions are 1 and 2.

9.

$$\frac{x}{x+4}+\frac{x}{x+2}=\frac{4}{3}$$

$$3(x+4)(x+2)\cdot\frac{x}{x+4}+\frac{x}{x+2}=\frac{4}{3}\cdot 3(x+4)(x+2)$$

$$3x(x+2)+3x(x+4)=4(x^2+6x+8)$$

$$3x^2+6x+3x^2+12x=4x^2+24x+32$$

$$6x^2+18x=4x^2+24x+32$$

$$10x^2+42x+32=0$$

$$5x^2+21x+16=0$$

$$(5x+16)(x+1)=0$$

$$5x+16=0 \text{ or } x+1=0$$

$$x=-\frac{16}{5} \text{ or } \quad x=-1$$

The solutions are $-\dfrac{16}{5}$ and -1.

11. $x^4-13x^2+36=0$

Substitute $u=x^2$

$u^2-13u+36=0$

$(u-9)(u-4)=0$

$u-9=0 \quad \text{or } u-4=0$

$u=9 \quad \text{or} \quad u=4$

$x^2=9 \quad \text{or} \quad x^2=4$

$x=\pm 3 \text{ or} \quad x=\pm 2$

The solutions are 3, –3, 2, and –2.

13. $4x^4+35x^2=9$

$4x^4+35x^2-9=0$

Substitute $u=x^2$

$4u^2+35u-9=0$

$(4u-1)(u+9)=0$

$4u-1=0 \quad \text{or } u+9=0$

$u=\dfrac{1}{4} \quad \text{or} \quad u=-9$

$x^2=\dfrac{1}{4} \quad \text{or} \quad x^2=-9$

$x=\pm\dfrac{1}{2} \text{ or} \quad x=\pm 3i$

The solutions are $\frac{1}{2}, -\frac{1}{2}, 3i,$ and $-3i$.

15.

$$3y^4=5y^2+2$$

$3y^4-5y^2-2=0$

Substitute $u=y^2$

$3u^2-5u-2=0$

$(3u+1)(u-2)=0$

$3u+1=0 \quad \text{or } u-2=0$

$u=-\dfrac{1}{3} \quad \text{or} \quad u=2$

$y^2=-\dfrac{1}{3} \quad \text{or} \quad y^2=2$

$y=\pm\dfrac{\sqrt{3}}{3}i \text{ or} \quad y=\pm\sqrt{2}$

The solutions are

$\dfrac{\sqrt{3}}{3}i, -\dfrac{\sqrt{3}}{3}i, \sqrt{2},$ and $-\sqrt{2}$.

17.

$$x^6 + 7x^3 \cdot 8 = 0$$

Substitute $u = x^3$

$$u^2 + 7u \cdot 8 = 0$$

$$(u \cdot 1)(u + 8) = 0$$

$u \cdot 1 = 0$ or $u + 8 = 0$

$x^3 \cdot 1 = 0$ or $x^3 + 8 = 0$

$(x \cdot 1)(x^2 + x + 1) = 0$ or $(x + 2)(x^2 \cdot 2x + 4) = 0$

$x \cdot 1 = 0$ or $x^2 + x + 1 = 0$ or $x + 2 = 0$ or $x^2 \cdot 2x + 4 = 0$

$$x = 1 \quad \text{or} \quad x = \frac{\cdot 1 \pm \sqrt{1 \cdot 4 \cdot 1 \cdot 1}}{2 \cdot 1} = \frac{\cdot 1 \pm \sqrt{\cdot 3}}{2} = \frac{\cdot 1 \pm i\sqrt{3}}{2} = \cdot \frac{1}{2} \pm \frac{\sqrt{3}}{2} i$$

$$x = \cdot 2 \text{ or } x = \frac{2 \pm \sqrt{4 \cdot 4 \cdot 1 \cdot 4}}{2 \cdot 1} = \frac{2 \pm \sqrt{\cdot 12}}{2} = \frac{2 \pm 2i\sqrt{3}}{2} = 1 \pm i\sqrt{3}$$

The solutions are $1, -2, 1 + i\sqrt{3}, 1 - i\sqrt{3}, -\frac{1}{2} + \frac{\sqrt{3}}{2} i$, and $-\frac{1}{2} - \frac{\sqrt{3}}{2} i$.

19.

$$(x+1)^2 - 3(x+1) = 40$$

$$(x+1)^2 - 3(x+1) - 40 = 0$$

Substitute $u = x + 1$

$$u^2 - 3u - 40 = 0$$

$$(u - 8)(u + 5) = 0$$

$u - 8 = 0$ or $u + 5 = 0$

$u = 8$ or $u = -5$

$x + 1 = 8$ or $x + 1 = -5$

$x = 7$ or $x = -6$

The solutions are 7, and –6.

21.

$$(y^2 - y)^2 - 8(y^2 - y) = 9$$

$$(y^2 - y)^2 - 8(y^2 - y) - 9 = 0$$

Substitute $u = y^2 - y$

$$u^2 - 8u - 9 = 0$$

$$(u - 9)(u + 1) = 0$$

$u - 9 = 0$ or $u + 1 = 0$

$u = 9$ or $u = -1$

$y^2 - y = 9$ or $y^2 - y = -1$

$y^2 - y - 9 = 0$ or $y^2 - y + 1 = 0$

$$y = \frac{1 \pm \sqrt{1 \cdot 4 \cdot 1 \cdot (\cdot 9)}}{2 \cdot 1} = \frac{1 \pm \sqrt{37}}{2}$$

$$y = \frac{1 \pm \sqrt{1 \cdot 4 \cdot 1 \cdot 1}}{2 \cdot 1} = \frac{1 \pm \sqrt{\cdot 3}}{2} = \frac{1}{2} \pm \frac{\sqrt{3}}{2} i$$

The solutions are

$\frac{1 + \sqrt{37}}{2}, \frac{1 - \sqrt{37}}{2}, \frac{1}{2} + \frac{\sqrt{3}}{2} i$, and $\frac{1}{2} - \frac{\sqrt{3}}{2} i$.

23.
$$x^{1/2} + 3x^{1/4} - 10 = 0$$
Substitute $u = x^{1/4}$
$$u^2 + 3u - 10 = 0$$
$$(u+5)(u-2) = 0$$
$$u + 5 = 0 \quad \text{or } u - 2 = 0$$
$$u = -5 \quad \text{or} \quad u = 2$$
$$x^{1/4} = -5 \quad \text{or} \quad x^{1/4} = 2$$
$$x = (-5)^4 \text{ or} \quad x = 2^4$$
$$x = 625 \quad \text{or} \quad x = 16$$
Check: $625^{1/2} + 3 \cdot 625^{1/4} \neq 10 = 25 + 15 \neq 10$
$$= 30 \neq 0$$
Discard: $x = 625$
$$16^{1/2} + 3 \cdot 16^{1/4} \neq 10 = 4 + 6 \neq 10 = 0$$
The solution is 16.

25.
$$y^{2/3} - 5y^{1/3} = -6$$
$$y^{2/3} - 5y^{1/3} + 6 = 0$$
Substitute $u = y^{1/3}$
$$u^2 - 5u + 6 = 0$$
$$(u-3)(u-2) = 0$$
$$u - 3 = 0 \quad \text{or } u - 2 = 0$$
$$u = 3 \quad \text{or} \quad u = 2$$
$$y^{1/3} = 3 \quad \text{or} \quad y^{1/3} = 2$$
$$y = 3^3 \text{ or} \quad y = 2^3$$
$$y = 27 \text{ or} \quad y = 8$$
The solutions are 27 and 8.
The check is left to the student.

27.
$$x + \sqrt{x} - 6 = 0$$
Substitute $u = \sqrt{x}$
$$u^2 + u - 6 = 0$$
$$(u-2)(u+3) = 0$$
$$u - 2 = 0 \text{ or } u + 3 = 0$$
$$u = 2 \text{ or} \quad u = -3$$
$$\sqrt{x} = 2 \text{ or} \quad \sqrt{x} = -3$$
$$x = 4 \text{ or} \quad x = 9$$
Check: $4 + \sqrt{4} \neq 6 = 4 + 2 \neq 6 = 0$
$$9 + \sqrt{9} \neq 6 = 9 + 3 \neq 6 = 6 \neq 0$$
Discard: $x = 9$
The solution is 4.

29.
$$(x^2 - 4x) - 8\sqrt{x^2 - 4x} + 15 = 0$$
Substitute $u = \sqrt{x^2 - 4x}$
$$u^2 - 8u + 15 = 0$$
$$(u-5)(u-3) = 0$$
$$u - 5 = 0 \quad \text{or} \quad u - 3 = 0$$
$$u = 5 \quad \text{or} \quad u = 3$$
$$\sqrt{x^2 - 4x} = 5 \quad \text{or} \quad \sqrt{x^2 - 4x} = 3$$
$$x^2 - 4x = 25 \text{ or} \quad x^2 - 4x = 9$$
$$x^2 - 4x - 25 = 0 \quad \text{or } x^2 - 4x - 9 = 0$$
$$x = \frac{4 \pm \sqrt{16 \cdot 4 \cdot 1 \cdot (\cdot 25)}}{2 \cdot 1} = \frac{4 \pm \sqrt{116}}{2}$$
$$= \frac{4 \pm 2\sqrt{29}}{2} = 2 \pm \sqrt{29}$$
$$x = \frac{4 \pm \sqrt{16 \cdot 4 \cdot 1 \cdot (\cdot 9)}}{2 \cdot 1} = \frac{4 \pm \sqrt{52}}{2}$$
$$= \frac{4 \pm 2\sqrt{13}}{2} = 2 \pm \sqrt{13}$$
The solutions are
$$2 + \sqrt{29}, \, 2 - \sqrt{29}, \, 2 + \sqrt{13}, \text{ and } 2 - \sqrt{13}.$$
The check is left to the student.

31. $z + 3 - \sqrt{z+3} - 6 = 0$

Substitute $u = \sqrt{z+3}$

$u^2 - u - 6 = 0$

$(u-3)(u+2) = 0$

$u - 3 = 0$ or $u + 2 = 0$

$u = 3$ or $u = -2$

$\sqrt{z+3} = 3$ or $\sqrt{z+3} = -2$

$z + 3 = 9$ or   No solution

$z = 6$

The solution is 6.
The check is left to the student.

33. $3\sqrt{x} - 5\sqrt[4]{x} + 2 = 0$

Substitute $u = \sqrt[4]{x}$

$3u^2 - 5u + 2 = 0$

$(3u - 2)(u - 1) = 0$

$3u - 2 = 0$ or $u - 1 = 0$

$u = \dfrac{2}{3}$ or $u = 1$

$\sqrt[4]{x} = \dfrac{2}{3}$ or $\sqrt[4]{x} = 1$

$x = \dfrac{16}{81}$ or $x = 1$

The solutions are $\frac{16}{81}$ and 1.
The check is left to the student.

35. $x^{-2} + 2x^{-1} - 8 = 0$

Substitute $u = x^{-1}$

$u^2 + 2u - 8 = 0$

$(u+4)(u-2) = 0$

$u + 4 = 0$ or $u - 2 = 0$

$u = -4$ or $u = 2$

$x^{-1} = -4$ or $x^{-1} = 2$

$x = -\dfrac{1}{4}$ or $x = \dfrac{1}{2}$

The solutions are $-\dfrac{1}{4}$ and $\dfrac{1}{2}$.

37. $3x^{-4} - 5x^{-2} - 2 = 0$

Substitute $u = x^{-2}$

$3u^2 - 5u - 2 = 0$

$(3u + 1)(u - 2) = 0$

$3u + 1 = 0$ or $u - 2 = 0$

$u = -\dfrac{1}{3}$ or $u = 2$

$x^{-2} = -\dfrac{1}{3}$ or $x^{-2} = 2$

$x^2 = -3$ or $x^2 = \dfrac{1}{2}$

$x = \pm\sqrt{-3}$ or $x = \pm\sqrt{\dfrac{1}{2}}$

$x = \pm i\sqrt{3}$ or $x = \pm\dfrac{\sqrt{2}}{2}$

The solutions are $i\sqrt{3}$, $-i\sqrt{3}$, $\dfrac{\sqrt{2}}{2}$, and $-\dfrac{\sqrt{2}}{2}$.

39. $6x^{-4} + 5x^{-2} - 4 = 0$

Substitute $u = x^{-2}$

$6u^2 + 5u - 4 = 0$

$(3u+4)(2u-1) = 0$

$3u+4 = 0 \qquad$ or $\quad 2u-1 = 0$

$u = -\dfrac{4}{3} \qquad$ or $\qquad u = \dfrac{1}{2}$

$x^{-2} = -\dfrac{4}{3} \qquad$ or $\qquad x^{-2} = \dfrac{1}{2}$

$x^2 = -\dfrac{3}{4} \qquad$ or $\qquad x^2 = 2$

$x = \pm\sqrt{-\dfrac{3}{4}} \quad$ or $\qquad x = \pm\sqrt{2}$

$x = \pm\dfrac{\sqrt{3}}{2}i \quad$ or $\qquad x = \pm\sqrt{2}$

The solutions are

$$\sqrt{2},\ -\sqrt{2},\ \frac{\sqrt{3}}{2}i,\ \text{and}\ -\frac{\sqrt{3}}{2}i\,.$$

41.   Let t = the time for Jill

$t+5$ = the time for Jack

$$\frac{1}{t} + \frac{1}{t+5} = \frac{1}{6}$$

$$6t(t+5)\left[\frac{1}{t} + \frac{1}{t+5}\right] = 6t(t+5)\frac{1}{6}$$

$6(t+5) + 6t = t(t+5)$

$6t + 30 + 6t = t^2 + 5t$

$0 = t^2 - 7t - 30$

$(t-10)(t+3) = 0$

$t-10 = 0$ or $t+3 = 0$

$t = 10$ hr or $t = -3$

$t+5 = 15$ hr

Jill takes 10 hr and Jack takes 15 hr when working alone.

43. $x^4 - 10x^2 + 25 = 0$

Substitute $u = x^2$

$u^2 - 10u + 25 = 0$

$(u-5)(u-5) = 0$

$u-5 = 0 \qquad$ or $u-5 = 0$

$u = 5 \qquad$ or $\quad u = 5$

$x^2 = 5 \qquad$ or $\quad x^2 = 5$

$x = \pm\sqrt{5}$ or $\quad x = \pm\sqrt{5}$

The x-intercepts are $\left(\sqrt{5},\ 0\right)$ and $\left(-\sqrt{5},\ 0\right)$.

45.

$$\frac{7x+2}{6} \geq \frac{3x-2}{4}$$

$$12\left|\frac{7x+2}{6}\right| \geq 12\left|\frac{3x-2}{4}\right|$$

$2(7x+2) \geq 3(3x-2)$

$14x+4 \geq 9x-6$

$5x \geq -10$

$\dfrac{5x}{5} \geq \dfrac{-10}{5}$

$x \geq -2$

47.   Let n = the number of students in the group

$n - 2$ = the number of student who paid

$\dfrac{1600}{n}$ = original cost per student

$\dfrac{1600}{n-2}$ = actual cost for those who paid

$$\frac{1600}{n} + 40 = \frac{1600}{n-2}$$

$$n(n-2)\left[\frac{1600}{n} + 40\right] = n(n-2)\left[\frac{1600}{n-2}\right]$$

$1600(n-2) + 40n(n-2) = 1600n$

$1600n - 3200 + 40n^2 - 80n = 1600n$

$40n^2 - 80n - 3200 = 0$

$n^2 - 2n - 80 = 0$

$(n-10)(n+8) = 0$

$n-10 = 0 \quad$ or $n+8 = 0$

$n = 10$ or $\qquad n = -8$

There are 10 students in the group.

49. $x - 1 = 0$ has one solution.
$x^2 - 1 = 0$ has two solutions.
$x^3 - 1 = 0$ has three solutions.
$x^4 - 1 = 0$ has four solutions.

51. Answers will vary.

53.
$$x - 5\sqrt{x} + 4 = 0$$
Substitute $u = \sqrt{x}$
$$u^2 - 5u + 4 = 0$$
$$(u - 1)(u - 4) = 0$$
$$u - 1 = 0 \text{ or } u - 4 = 0$$
$$u = 1 \text{ or } \quad u = 4$$
$$\sqrt{x} = 1 \text{ or } \quad \sqrt{x} = 4$$
$$x = 1 \text{ or } \quad x = 16$$
The solutions are 1 and 16.
The check is left to the student.

55.
$$x^{1/2} - 5x^{1/4} + 6 = 0$$
Substitute $u = x^{1/4}$
$$u^2 - 5u + 6 = 0$$
$$(u - 3)(u - 2) = 0$$
$$u - 3 = 0 \text{ or } u - 2 = 0$$
$$u = 3 \text{ or } \quad u = 2$$
$$x^{1/4} = 3 \text{ or } x^{1/4} = 2$$
$$x = 3^4 \text{ or } \quad x = 2^4$$
$$x = 81 \text{ or } \quad x = 16$$
The solutions are 81 and 16.
The check is left to the student.

57.
$$x^{-4} - 9x^{-2} + 14 = 0$$
Substitute $u = x^{-2}$
$$u^2 - 9u + 14 = 0$$
$$(u - 7)(u - 2) = 0$$
$$u - 7 = 0 \quad \text{or } u - 2 = 0$$
$$u = 7 \quad \text{or} \quad u = 2$$
$$x^{-2} = 7 \quad \text{or} \quad x^{-2} = 2$$
$$x^2 = \frac{1}{7} \quad \text{or} \quad x^2 = \frac{1}{2}$$
$$x = \pm\sqrt{\frac{1}{7}} \text{ or } \quad x = \pm\sqrt{\frac{1}{2}}$$
$$x = \pm\frac{\sqrt{7}}{7} \text{ or } \quad x = \pm\frac{\sqrt{2}}{2}$$
The solutions are
$$\frac{\sqrt{7}}{7}, -\frac{\sqrt{7}}{7}, \frac{\sqrt{2}}{2}, \text{ and } -\frac{\sqrt{2}}{2}.$$

59.
$$\left(x^2 - x\right)^2 - \left(x^2 - x\right) - 2 = 0$$
Substitute $u = x^2 - x$
$$u^2 - u - 2 = 0$$
$$(u - 2)(u + 1) = 0$$
$$u - 2 = 0 \text{ or } \quad u + 1 = 0$$
$$u = 2 \text{ or } \quad u = -1$$
$$x^2 - x = 2 \text{ or } x^2 - x = -1$$
$$x^2 - x - 2 = 0 \text{ or } x^2 - x + 1 = 0$$
$$(x - 2)(x + 1) = 0 \text{ or } x^2 - x + 1 = 0$$
$$x - 2 = 0 \text{ or } x + 1 = 0 \text{ or } x^2 - x + 1 = 0$$
$$x = 2 \text{ or } x = -1 \quad \text{or } x^2 - x + 1 = 0$$
$$x = \frac{1 \pm \sqrt{1 \cdot 4 \cdot 1 \cdot 1}}{2 \cdot 1} = \frac{1 \pm \sqrt{3}}{2} = \frac{1}{2} \pm \frac{\sqrt{3}}{2}i$$
The solutions are
$$2, -1, \frac{1}{2} + \frac{\sqrt{3}}{2}i, \text{ and } \frac{1}{2} - \frac{\sqrt{3}}{2}i.$$

61.

$$\frac{12}{x^2-36}\cdot\frac{1}{x-6}=1$$

$$\frac{12}{(x+6)(x-6)}\cdot\frac{1}{x-6}=1$$

$$(x+6)(x-6)\cdot\frac{12}{(x+6)(x-6)}\cdot\frac{1}{x-6}=1\cdot(x+6)(x-6)$$

$$12\cdot 1(x+6)=(x+6)(x-6)$$

$$12\cdot x-6=x^2-36$$

$$0=x^2+x-42$$

$$(x+7)(x-6)=0$$

$$x+7=0\quad\text{or}\ x-6=0$$

$$x=-7\ \text{or}\qquad x=6$$

Discard $x=6$ because it makes the denominator zero. The solution is –7.

63. $x^4-5x^2+4=0$

Substitute $u=x^2$.

$u^2-5u+4=0$

$(u-4)(u-1)=0$

$u-4=0\ \ \text{or}\ u-1=0$

$u=4\ \ \text{or}\quad u=1$

$x^2=4\ \ \text{or}\quad x^2=1$

$x=\pm2\ \text{or}\quad x=\pm1$

The solutions are 2, –2, 1, and –1.

8.5   Nonlinear Inequalities

Problems 8.5

1. $(x-4)(x+2)\ge 0$

$(x-4)(x+2)=0$ is the related equation. $x=4$, $x=-2$ are the critical values.

Separate into regions using –2 and 4 as boundaries.

Select test points: -3, 0, and 5.

When $x=-3$, $(-3-4)(-3+2)=-7(-1)=7\ge0$ False.

When $x=0$, $(0-4)(0+2)=-4\cdot2=-8\ge 0$ True.

When $x=5$, $(5-4)(5+2)=1\cdot7=7\ge0$ False.

The solution set is: $\{x\mid -2\le x\le 4\}$ or $[-2,4]$.

2. $x^2 - 5 > -2$

$x^2 - 3 > 0$

$\left(x - \sqrt{3}\right)\left(x + \sqrt{3}\right) > 0$

$\left(x - \sqrt{3}\right)\left(x + \sqrt{3}\right) = 0$ is the related equation. $x = \sqrt{3},\ x = -\sqrt{3}$ are the critical values.

Separate into regions using $-\sqrt{3}$ and $\sqrt{3}$ as boundaries.

Select test points: $-2,\ 0,$ and 2.

When $x = -2$, $\left(-2\right)^2 - 5 = 4 - 5 = -1 > -2$ True.

When $x = 0$, $0^2 - 5 = 0 - 5 = -5 \not> -2$ False.

When $x = 2$, $2^2 - 5 = 4 - 5 = -1 > -2$ True.

The solution set is: $\left\{x \mid x < -\sqrt{3} \cup x > \sqrt{3}\right\}$ or $\left(-\infty,\ -\sqrt{3}\right) \cup \left(\sqrt{3},\ +\infty\right)$.

3. $(x + 3)(x - 2)(x + 1) \le 0$

$(x + 3)(x - 2)(x + 1) = 0$ is the related equation. $x = -3,\ x = 2,\ x = -1$ are the critical values.

Separate into regions using $-3, -1,$ and 2 as boundaries.

Select test points: $-4,\ -2,\ 0,$ and 3.

When $x = -4$, $(-4 + 3)(-4 - 2)(-4 + 1) = -1(-6)(-3) = -18 \le 0$ True.

When $x = -2$, $(-2 + 3)(-2 - 2)(-2 + 1) = 1(-4)(-1) = 4 \not\le 0$ False.

When $x = 0$, $(0 + 3)(0 - 2)(0 + 1) = 3(-2)(1) = -6 \le 0$ True.

When $x = 3$, $(3 + 3)(3 - 2)(3 + 1) = 6(1)(4) = 24 \not\le 0$ False.

The solution set is: $\left\{x \mid x \le -3 \cup -1 \le x \le 2\right\}$ or $\left(-\infty,\ -3\right] \cup \left[-1,\ 2\right]$.

Chapter 8   Quadratic Equations and Inequalities

4.　**a.**　$\dfrac{x+3}{2x-1} \leq 0$

$x = -3,\ x = \frac{1}{2}$ are the critical values.

Separate into regions using -3 and $\frac{1}{2}$ as boundaries.

Select test points: $-4,\ 0,$ and 1.

When $x = -4$,　$\dfrac{-4+3}{2(-4)-1} = \dfrac{-1}{-9} = \dfrac{1}{9} \not\leq 0$　False.

When $x = 0$,　$\dfrac{0+3}{2(0)-1} = \dfrac{3}{-1} = -3 \leq 0$　True.

When $x = 1$,　$\dfrac{1+3}{2(1)-1} = \dfrac{4}{1} = 4 \not\leq 0$　　False.

The solution set is:　$\left\{x \ \middle|\ -3 \leq x < \dfrac{1}{2}\right\}$　or　$\left[-3, \dfrac{1}{2}\right)$.

b.

$\dfrac{x}{x-1} > 1$

$\dfrac{x}{x-1} - 1 > 0$

$\dfrac{x}{x-1} - 1 \cdot \dfrac{x-1}{x-1} > 0$

$\dfrac{x - x + 1}{x-1} > 0$

$\dfrac{1}{x-1} > 0$

$x = 1$ is the critical value.

Separate into regions using 1 as a boundary.

Select test points: 0 and 2.

When $x = 0$,　$\dfrac{0}{0-1} = 0 \not> 1$　　　　False.

When $x = 2$,　$\dfrac{2}{2-1} = \dfrac{2}{1} = 2 > 1$　　True.

The solution set is: $\{x \mid x > 1\}$ or $(1, +\infty)$.

5.　　$30 - x^2 > 5 \quad \rightarrow \quad 25 - x^2 > 0$

$(5+x)(5-x) > 0$

$(5+x)(5-x) = 0$ is the related equation. $x = -5,\ x = 5$ are the critical values.

Separate into regions using -5 and 5 as boundaries.

Select test points: $-6,\ 0,$ and 6.

When $x = -6$,　$30 - (-6)^2 = 30 - 36 = -6 \not> 5$　　　False.

When $x = 0$,　$30 - 0^2 = 30 - 0 = 30 > 5$　　　　True.

When $x = 6$,　$30 - 6^2 = 30 - 36 = -6 \not> 5$　　　False.

The solution set is: $\{x \mid -5 < x < 5\}$ or $(-5, 5)$.

Since the distance must be positive, the solution is $0 < x < 5$.

Exercises 8.5

1. $(x+1)(x-3)>0$

$(x+1)(x-3)=0$ is the related equation. $x=-1$, $x=3$ are the critical values.

Separate into regions using -1 and 3 as boundaries.

Select test points: -2, 0, and 4.

When $x=-2$, $(-2+1)(-2-3)=-1(-5)=5>0$ True.

When $x=0$, $(0+1)(0-3)=1(-3)=-3\not>0$ False.

When $x=4$, $(4+1)(4-3)=5\cdot 1=5>0$ True.

The solution set is: $\{x|x<-1\cup x>3\}$ or $(-\infty,\ -1)\cup(3,\ +\infty)$.

3. $x(x+4)\cdot 0$

$x(x+4)=0$ is the related equation. $x=0$, $x=-4$ are the critical values.

Separate into regions using -4 and 0 as boundaries.

Select test points: -5, -1, and 1.

When $x=-5$, $-5(-5+4)=-5(-1)=5\not> 0$ False.

When $x=-1$, $-1(-1+4)=-1(3)=-3\cdot 0$ True.

When $x=1$, $1(1+4)=1\cdot 5=5\not> 0$ False.

The solution set is: $\{x|-4\cdot x\cdot 0\}$ or $[-4,\ 0]$.

5. $x^2\leq x\leq 2\leq 0$

$(x\leq 2)(x+1)\leq 0$

$(x\leq 2)(x+1)=0$ is the related equation. $x=2$, $x=\leq 1$ are the critical values.

Separate into regions using -1 and 2 as boundaries.

Select test points: ≤ 2, 0, and 3.

When $x=\leq 2$, $(\leq 2)^2\leq(\leq 2)\leq 2=4+2\leq 2=4\not\leq 0$ False.

When $x=0$, $0^2\leq 0\leq 2=\leq 2\leq 0$ True.

When $x=3$, $3^2\leq 3\leq 2=9\leq 3\leq 2=4\not\leq 0$ False.

The solution set is: $\{x|\leq 1\leq x\leq 2\}$ or $[\leq 1,\ 2]$.

7. $x^2 - 3x \geq 0$

$x(x - 3) \geq 0$

$x(x - 3) = 0$ is the related equation. $x = 0$, $x = 3$ are the critical values.

Separate into regions using 0 and 3 as boundaries.

Select test points: -1, 1, and 4.

When $x = -1$, $(-1)^2 - 3(-1) = 1 + 3 = 4 \geq 0$ True.

When $x = 1$, $(1)^2 - 3(1) = 1 - 3 = -2 \not\geq 0$ False.

When $x = 4$, $(4)^2 - 3(4) = 16 - 12 = 4 \geq 0$ True.

The solution set is: $\{x \mid x \leq 0 \cup x \geq 3\}$ or $(-\infty,\ 0] \cup [3,\ +\infty)$.

9. $x^2 - 3x + 2 < 0$

$(x - 1)(x - 2) < 0$

$(x - 1)(x - 2) = 0$ is the related equation. $x = 1$, $x = 2$ are the critical values.

Separate into regions using 2 and 1 as boundaries.

Select test points: 0, 1.5, and 3.

When $x = 0$, $0^2 - 3(0) + 2 = 0 - 0 + 2 = 2 \not< 0$ False.

When $x = 1.5$, $1.5^2 - 3(1.5) + 2 = 2.25 - 4.5 + 2 = -0.25 < 0$ True.

When $x = 3$, $3^2 - 3(3) + 2 = 9 - 9 + 2 = 2 \not< 0$ False.

The solution set is: $\{x \mid 1 < x < 2\}$ or $(1,\ 2)$.

11. $x^2 + 2x - 3 < 0$

$(x - 1)(x + 3) < 0$

$(x - 1)(x + 3) = 0$ is the related equation. $x = 1$, $x = -3$ are the critical values.

Separate into regions using -3 and 1 as boundaries.

Select test points: -4, 0, and 2.

When $x = -4$, $(-4)^2 + 2(-4) - 3 = 16 - 8 - 3 = 5 \not< 0$ False.

When $x = 0$, $0^2 + 2(0) - 3 = 0 + 0 - 3 = -3 < 0$ True.

When $x = 2$, $2^2 + 2(2) - 3 = 4 + 4 - 3 = 5 \not< 0$ False.

The solution set is: $\{x \mid -3 < x < 1\}$ or $(-3,\ 1)$.

13. $x^2 + 10x \geq -25$

$x^2 + 10x + 25 \geq 0$

$(x+5)^2 \geq 0$

$(x+5)^2 = 0$ is the related equation. $x = -5$ is the critical value.

Separate into regions using -5 as a boundary.

Select test points: -6, and 0.

When $x = -6$, $(-6)^2 + 10(-6) = 36 - 60 = -24 \not\geq -25$ False.

When $x = 0$, $0^2 + 10(0) = 0 + 0 = 0 \not\geq -25$ False.

The solution set is: $\{x | x = -5\}$.

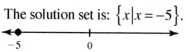

15. $x^2 - 8x \leq -16$

$x^2 - 8x + 16 \leq 0$

$(x - 4)^2 \leq 0$

$(x - 4)^2 = 0$ is the related equation. $x = 4$ is the critical value.

Separate into regions using 4 as a boundary.

Select test points: 0, and 5.

When $x = 0$, $0^2 - 8(0) = 0 - 0 = 0 \leq -16$ True.

When $x = 5$, $5^2 - 8(5) = 25 - 40 = -15 \leq -16$ True.

The solution set is: $\{x | x \text{ is a real number}\}$ or $(-\infty, +\infty)$.

17. $x^2 - x < 1$

$x^2 - x - 1 < 0$

$x^2 - x - 1 = 0$

$x = \dfrac{-(-1) \pm \sqrt{(-1)^2 - 4(1)(-1)}}{2(1)} = \dfrac{1 \pm \sqrt{1+4}}{2}$

$= \dfrac{1 \pm \sqrt{5}}{2} \approx \dfrac{1 \pm 2.2}{2} \approx 1.6 \text{ or } \approx 0.6$

19. $x^2 - x < 4$

$x^2 - x - 4 < 0$

$x^2 - x - 4 = 0$

$x = \dfrac{-(-1) \pm \sqrt{(-1)^2 - 4(1)(-4)}}{2(1)} = \dfrac{1 \pm \sqrt{1+16}}{2}$

$= \dfrac{1 \pm \sqrt{17}}{2} \approx \dfrac{1 \pm 4.1}{2} \approx 2.55 \text{ or } \approx 1.55$

21. $(x+1)(x-2)(x+3) \geq 0$

$(x+1)(x-2)(x+3) = 0$ is the related equation. $x = -1$, $x = 2$, $x = -3$ are the critical values.

Separate into regions using −3, −1, and 2 as boundaries.

Select test points: −4, −2, 0, and 3.

When $x = -4$, $(-4+1)(-4-2)(-4+3) = -3(-6)(-1) = -18 \not\geq 0$ False.

When $x = -2$, $(-2+1)(-2-2)(-2+3) = -1(-4)(1) = 4 \geq 0$ True.

When $x = 0$, $(0+1)(0-2)(0+3) = 1(-2)(3) = -6 \not\geq 0$ False.

When $x = 3$, $(3+1)(3-2)(3+3) = 4(1)(6) = 24 \geq 0$ True.

The solution set is: $\{x \mid -3 \leq x \leq -1 \cup x \geq 2\}$ or $[-3, -1] \cup [2, +\infty)$.

23. $(x-1)(x-2)(x-3) \leq 0$

$(x-1)(x-2)(x-3) = 0$ is the related equation. $x = 1$, $x = 2$, $x = 3$ are the critical values.

Separate into regions using 1, 2, and 3 as boundaries.

Select test points: 0, 1.5, 2.5, and 4.

When $x = 0$, $(0-1)(0-2)(0-3) = -1(-2)(-3) = -6 \leq 0$ True.

When $x = 1.5$, $(1.5-1)(1.5-2)(1.5-3) = 0.5(-0.5)(-1.5) = 0.375 \not\leq 0$ False.

When $x = 2.5$, $(2.5-1)(2.5-2)(2.5-3) = 1.5(0.5)(-0.5) = -0.375 \leq 0$ True.

When $x = 4$, $(4-1)(4-2)(4-3) = 3(2)(1) = 6 \not\leq 0$ False.

The solution set is: $\{x \mid x \leq 1 \cup 2 \leq x \leq 3\}$ or $(-\infty, 1] \cup [2, 3]$.

25. $\dfrac{2}{x-2} \geq 0$

$x = 2$ is the critical value.

Separate into regions using 2 as a boundary.

Select test points: 0 and 3.

When $x = 0$, $\dfrac{2}{0-2} = \dfrac{2}{-2} = -1 \not\geq 0$ False.

When $x = 3$, $\dfrac{2}{3-2} = \dfrac{2}{1} = 2 \geq 0$ True.

The solution set is: $\{x \mid x > 2\}$ or $(2, +\infty)$.

27. $\dfrac{x+5}{x-1} > 0$

$x = -5$, $x = 1$ are the critical values.

Separate into regions using −5 and 1 as boundaries.

Select test points: −6, 0, and 2.

When $x = -6$, $\dfrac{-6+5}{-6-1} = \dfrac{-1}{-7} = \dfrac{1}{7} > 0$ True.

When $x = 0$, $\dfrac{0+5}{0-1} = \dfrac{5}{-1} = -5 \not> 0$ False.

When $x = 2$, $\dfrac{2+5}{2-1} = \dfrac{7}{1} = 7 > 0$ True.

The solution set is: $\{x \mid x < -5 \cup x > 1\}$

or $(-\infty, -5) \cup (1, +\infty)$.

29. $\dfrac{3x-4}{2x-1} < 0$

$x = \frac{4}{3}$, $x = \frac{1}{2}$ are the critical values.

Separate into regions using $\frac{1}{2}$ and $\frac{4}{3}$ as
boundaries.

Select test points: 0, 1, and 2.

When $x = 0$, $\dfrac{3-0-4}{2-0-1} = \dfrac{-4}{-1} = 4 \not< 0$ False.

When $x = 1$, $\dfrac{3-1-4}{2-1-1} = \dfrac{-1}{1} = -1 < 0$ True.

When $x = 2$, $\dfrac{3-2-4}{2-2-1} = \dfrac{2}{3} \not< 0$ False.

The solution set is: $\left\{x \big| \frac{1}{2} < x < \frac{4}{3}\right\}$ or $\left(\frac{1}{2}, \frac{4}{3}\right)$.

31. $\dfrac{x+5}{x\cdot 1} > 2$

$\dfrac{x+5}{x\cdot 1} \cdot 2 > 0$

$\dfrac{x+5}{x\cdot 1} \cdot 2\cdot\dfrac{x\cdot 1}{x\cdot 1} > 0$

$\dfrac{x+5\cdot 2x+2}{x\cdot 1} > 0$

$\dfrac{\cdot x + 7}{x\cdot 1} > 0$

$x = 7$ and $x = 1$ are the critical values.

Separate into regions using 1 and 7 as boundaries.

Select test points: 0, 2, and 8.

When $x = 0$, $\dfrac{0+5}{0-1} = \dfrac{5}{-1} = -5 \not> 2$ False.

When $x = 2$, $\dfrac{2+5}{2-1} = \dfrac{7}{1} = 7 > 2$ True.

When $x = 8$, $\dfrac{8+5}{8-1} = \dfrac{13}{7} \not> 2$ False.

The solution set is: $\left\{x \big| 1 < x < 7\right\}$ or $(1, 7)$.

33. $\dfrac{1}{x\cdot 1} < \dfrac{1}{x\cdot 2}$

$\dfrac{1}{x\cdot 1} \cdot \dfrac{1}{x\cdot 2} < 0$

$\dfrac{1}{x\cdot 1}\cdot\dfrac{x\cdot 2}{x\cdot 2} \cdot \dfrac{1}{x\cdot 2}\cdot\dfrac{x\cdot 1}{x\cdot 1} < 0$

$\dfrac{x\cdot 2 \cdot x + 1}{(x\cdot 1)(x\cdot 2)} < 0$

$\dfrac{\cdot 1}{(x\cdot 1)(x\cdot 2)} < 0$

$x = 1$ and $x = 2$ are the critical values.

Separate into regions using 1 and 2 as boundaries.

Select test points: 0, 1.5, and 3.

When $x = 0$, $\dfrac{1}{0\infty1} = \dfrac{1}{\infty1} = \infty1 < \dfrac{1}{0\infty2} = \infty\dfrac{1}{2}$ True.

When $x = 1.5$, $\dfrac{1}{1.5\infty1} = \dfrac{1}{0.5} = 2 \not< \dfrac{1}{1.5\infty2} = \dfrac{1}{\infty0.5} = \infty2$ False.

When $x = 3$, $\dfrac{1}{3\infty1} = \dfrac{1}{2} < \dfrac{1}{3\infty2} = \dfrac{1}{1} = 1$ True.

The solution set is: $\left\{x \big| x < 1 \cup x > 2\right\}$ or $(\infty, 1) \cup (2, +\infty)$.

339

35. $x^2 - 9 \geq 0$

$(x+3)(x-3) \geq 0$

$(x+3)(x-3) = 0$ is the related equation. $x = -3$, $x = 3$ are the critical values.

Separate into regions using -3 and 3 as boundaries.

Select test points: -4, 0, and 4.

When $x = -4$, $(-4)^2 - 9 = 16 - 9 = 7 \geq 0$ True.

When $x = 0$, $0^2 - 9 = -9 \not\geq 0$ False.

When $x = 4$, $4^2 - 9 = 16 - 9 = 7 \geq 0$ True.

The solution set is: $\{x \mid x \leq -3 \cup x \geq 3\}$ or $(-\infty, -3] \cup [3, +\infty)$.

37. $x^2 - 6x + 5 \geq 0$

$(x-1)(x-5) \geq 0$

$(x-1)(x-5) = 0$ is the related equation. $x = 1$, $x = 5$ are the critical values.

Separate into regions using 1 and 5 as boundaries.

Select test points: 0, 2, and 6.

When $x = 0$, $0^2 - 6(0) + 5 = 0 - 0 + 5 = 5 \geq 0$ True.

When $x = 2$, $2^2 - 6(2) + 5 = 4 - 12 + 5 = -3 \not\geq 0$ False.

When $x = 6$, $6^2 - 6(6) + 5 = 36 - 36 + 5 = 5 \geq 0$ True.

The solution set is: $\{x \mid x \leq 1 \cup x \geq 5\}$ or $(-\infty, 1] \cup [5, +\infty)$.

39. $R^2 - 3R + 1 > 5$

$R^2 - 3R - 4 > 0$

$(R+1)(R-4) > 0$

$(R+1)(R-4) = 0$ is the related equation. $R = -1$, $R = 4$ are the critical values.

Separate into regions using -1 and 4 as boundaries.

Select test points: -2, 0, and 5.

When $R = -2$, $(-2)^2 - 3(-2) + 1 = 4 + 6 + 1 = 11 > 5$ True.

When $R = 0$, $0^2 - 3(0) + 1 = 0 - 0 + 1 = 1 \not> 5$ False.

When $R = 5$, $5^2 - 3(5) + 1 = 25 - 15 + 1 = 11 > 5$ True.

The solution set is: $\{R \mid R < -1 \cup R > 4\}$ or $(-\infty, -1) \cup (4, +\infty)$.

Since the resistance must be positive, the solution is $R > 4$.

41. $110T - T^2 > 1000$

$T^2 - 110T + 1000 < 0$

$(T - 100)(T - 10) < 0$

$(T - 100)(T - 10) = 0$ is the related equation. $T = 100$, $T = 10$ are the critical values.

Separate into regions using 10 and 100 as boundaries.

Select test points: 0, 11, and 101.

When $T = 0$, $110(0) - 0^2 = 0 - 0 = 0 \not> 1000$ False.

When $T = 11$, $110(11) - 11^2 = 1210 - 121 = 1089 > 1000$ True.

When $T = 101$, $110(101) - 101^2 = 11110 - 10201 = 909 \not> 1000$ False.

The solution set is: $\{T \mid 10 < T < 100\}$ or $(10, 100)$.

The solution is $10° < T < 100°$.

43. $48t - 16t^2 > 32$

$16t^2 - 48t + 32 < 0$

$t^2 - 3t + 2 < 0$

$(t - 1)(t - 2) < 0$

$(t - 1)(t - 2) = 0$ is the related equation. $t = 1$, $t = 2$ are the critical values.

Separate into regions using 1 and 2 as boundaries.

Select test points: 0, 1.5, and 3.

When $t = 0$, $48(0) - 16(0)^2 = 0 - 0 = 0 \not> 32$ False.

When $t = 1.5$, $48(1.5) - 16(1.5)^2 = 72 - 36 = 36 > 32$ True.

When $t = 3$, $48(3) - 16(3)^2 = 144 - 144 = 0 \not> 32$ False.

The solution set is: $\{t \mid 1 < t < 2\}$ or $(1, 2)$.

The projectile is above 32 ft when $1 < t < 2$.

45. $y = 3x + 6; \quad x = 2$

$y = 3(2) + 6 = 6 + 6 = 12$

47.
$y = -\dfrac{2}{3}x + 4; \quad x = 3$

$y = -\dfrac{2}{3}(3) + 4 = -2 + 4 = 2$

+
49. $y = 3x + 6; \quad y = 0$

     $0 = 3x + 6$

   $-6 = 3x$

   $-2 = x$

51. $50 \geq d \geq 60 \geq \quad 50 \geq 0.05v^2 + v \geq 60 \geq \quad 50 \geq 0.05v^2 + v$ and $0.05v^2 + v \geq 60$

Solve the equations $0.05v^2 + v \geq 50 = 0$ and $0.05v^2 + v \geq 60 = 0$

$$v = \frac{\geq 1 \pm \sqrt{1 \geq 4 \geq 0.05 \geq (\geq 50)}}{2 \geq 0.05} = \frac{\geq 1 \pm \sqrt{1 + 10}}{0.10} = \frac{\geq 1 \pm \sqrt{11}}{0.10} \geq \frac{\geq 1 \pm 3.32}{0.10} \geq 23.2 \text{ or } \geq 43.2$$

Select test points: 0 and 24

When $x = 0$, $0.05(0)^2 + 0 = 0$, so $50 \not\geq 0$ False.

When $x = 24$, $0.05(24)^2 + 24 = 28.8 + 24 = 52.8$, so $50 \geq 52.8$ True. $(v \geq 23.2)$

$$v = \frac{\geq 1 \pm \sqrt{1 \geq 4 \geq 0.05 \geq (\geq 60)}}{2 \geq 0.05} = \frac{\geq 1 \pm \sqrt{1 + 12}}{0.10} = \frac{\geq 1 \pm \sqrt{13}}{0.10} \geq \frac{\geq 1 \pm 3.61}{0.10} \geq 26.1 \text{ or } \geq 46.1$$

Select test points: 0 and 27· When $x = 0$, $0.05(0)^2 + 0 = 0 \geq 60$ True. $(v \geq 26.1)$

When $x = 27$, $0.05(27)^2 + 27 = 36.45 + 27 = 63.45 \not\geq 60$ False.

Since the velocity must be positive, 23.2 mi/hr $\geq v \geq 26.1$ mi/hr.

53. $0.05v^2 + v = 950$

Solve the equation $0.05v^2 + v \approx 950 = 0$.

$$v = \frac{\approx 1 \pm \sqrt{1 \approx 4 \approx 0.05 \approx (\approx 950)}}{2 \approx 0.05} = \frac{\approx 1 \pm \sqrt{1 + 190}}{0.10} = \frac{\approx 1 \pm \sqrt{191}}{0.10} \approx \frac{\approx 1 \pm 13.82}{0.10} \approx 128.2 \text{ or } \approx 148.2$$

Since the velocity must be positive, $v = 128.2$ mi/hr.

55. Sample answer: Since the square of any real number is positive or zero, the answer can never be negative. So the only solution will occur when $(x-1)^2 = 0$ or when $x = 1$.

57. Sample answer; The denominator of the fraction cannot be zero. Thus the solution of the denominator cannot be included in the solution set and thus only one endpoint is included in the solution.

59. $\dfrac{x+2}{x} > 0$

$x = -2$ and $x = 0$ are the critical values.

Separate into regions using -2 and 0 as
            boundaries.

Select test points: -3, -1, and 1.

When $x = -3$, $\dfrac{-3+2}{-3} = \dfrac{-1}{-3} = 3 > 0$ True.

When $x = -1$, $\dfrac{-1+2}{-1} = \dfrac{1}{-1} = -1 \not> 0$ False.

When $x = 1$, $\dfrac{1+2}{1} = \dfrac{3}{1} = 3 > 0$ True.

The solution set is: $\{x \mid x < \infty 2 \cup x > 0\}$

or $(\infty\infty, \infty 2) \cup (0, +\infty)$.

342

61. $(x-4)(x+3)(x-1) \le 0$

$(x-4)(x+3)(x-1)=0$ is the related equation. $x=4$, $x=-3$, $x=1$ are the critical values.

Separate into regions using -3, 1, and 4 as boundaries.

Select test points: -4, 0, 2, and 5.

When $x=-4$, $(-4-4)(-4+3)(-4-1)=-8(-1)(-5)=-40 \le 0$ True.

When $x=0$, $(0-4)(0+3)(0-1)=-4(3)(-1)=12 \not\le 0$ False.

When $x=2$, $(2-4)(2+3)(2-1)=-2(5)(1)=-10 \le 0$ True.

When $x=5$, $(5-4)(5+3)(5-1)=1(8)(4)=32 \not\le 0$ False.

The solution set is: $\{x|x \le -3 \cup 1 \le x \le 4\}$ or $(-\infty,\ -3]\cup[1,\ 4]$.

63. $x^2+x \le 6$

$x^2+x-6 \le 0$

$(x+3)(x-2) \le 0$

$(x+3)(x-2)=0$ is the related equation. $x=-3$, $x=2$ are the critical values.

Separate into regions using -3 and 2 as boundaries.

Select test points: -4, 0, and 3.

When $x=-4$, $(-4)^2+(-4)=16-4=12 \not\le 6$ False.

When $x=0$, $(0)^2+0=0 \le 6$ True.

When $x=3$, $(3)^2+3=9+3=12 \not\le 6$ False.

The solution set is: $\{x|-3 \le x \le 2\}$ or $[-3,\ 2]$.

65. $(x+1)(x-3) \ge 0$

$(x+1)(x-3)=0$ is the related equation. $x=-1$, $x=3$ are the critical values.

Separate into regions using -1 and 3 as boundaries.

Select test points: -2, 0, and 4.

When $x=-2$, $(-2+1)(-2-3)=-1(-5)=5 \ge 0$ True.

When $x=0$, $(0+1)(0-3)=1(-3)=-3 \not\ge 0$ False.

When $x=4$, $(4+1)(4-3)=5 \cdot 1=5 \ge 0$ True.

The solution set is: $\{x|x \le -1 \cup x \ge 3\}$ or $(-\infty,\ -1]\cup[3,\ +\infty)$.

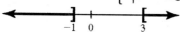

Review Exercises

1. a. $16x^2 - 49 = 0$

$$16x^2 = 49$$

$$x^2 = \frac{49}{16}$$

$$x = \pm\sqrt{\frac{49}{16}} = \pm\frac{7}{4}$$

The solution set is $\left\{\frac{7}{4}, -\frac{7}{4}\right\}$.

b. $25x^2 - 16 = 0$

$$25x^2 = 16$$

$$x^2 = \frac{16}{25}$$

$$x = \pm\sqrt{\frac{16}{25}} = \pm\frac{4}{5}$$

The solution set is $\left\{\frac{4}{5}, -\frac{4}{5}\right\}$.

2. a. $5x^2 + 30 = 0$

$$5x^2 = -30$$

$$x^2 = -6$$

$$x = \pm\sqrt{-6} = \pm i\sqrt{6}$$

The solution set is $\left\{i\sqrt{6}, -i\sqrt{6}\right\}$.

b. $6x^2 + 42 = 0$

$$6x^2 = -42$$

$$x^2 = -7$$

$$x = \pm\sqrt{-7} = \pm i\sqrt{7}$$

The solution set is $\left\{i\sqrt{7}, -i\sqrt{7}\right\}$.

3. a. $(x-3)^2 = 32$

$$x - 3 = \pm\sqrt{32}$$

$$x - 3 = \pm 4\sqrt{2}$$

$$x = 3 \pm 4\sqrt{2}$$

The solution set is $\left\{3+4\sqrt{2},\, 3-4\sqrt{2}\right\}$.

b. $(x-5)^2 = 50$

$$x - 5 = \pm\sqrt{50}$$

$$x - 5 = \pm 5\sqrt{2}$$

$$x = 5 \pm 5\sqrt{2}$$

The solution set is $\left\{5+5\sqrt{2},\, 5-5\sqrt{2}\right\}$.

4. a. $2(x \cdot 2)^2 + 25 = 0$

$$2(x \cdot 2)^2 = \cdot\, 25$$

$$(x \cdot 2)^2 = \cdot\, \frac{25}{2}$$

$$x \cdot 2 = \pm\sqrt{\cdot\, \frac{25}{2}} = \pm\sqrt{\cdot\, \frac{25 \cdot 2}{2 \cdot 2}}$$

$$x \cdot 2 = \pm\frac{5i\sqrt{2}}{2}$$

$$x = 2 \pm \frac{5\sqrt{2}}{2}i$$

The solution set is $\left\{2 + \frac{5\sqrt{2}}{2}i,\, 2 \mathrel{\rfloor} \frac{5\sqrt{2}}{2}i\right\}$.

b. $3(x \cdot 3)^2 + 64 = 0$

$$3(x \cdot 3)^2 = \cdot\, 64$$

$$(x \cdot 3)^2 = \cdot\, \frac{64}{3}$$

$$x \cdot 3 = \pm\sqrt{\cdot\, \frac{64}{3}} = \pm\sqrt{\cdot\, \frac{64 \cdot 3}{3 \cdot 3}}$$

$$x \cdot 3 = \pm\frac{8i\sqrt{3}}{3}$$

$$x = 3 \pm \frac{8\sqrt{3}}{3}i$$

The solution set is $\left\{3 + \frac{8\sqrt{3}}{3}i,\, 3 \mathrel{\rfloor} \frac{8\sqrt{3}}{3}i\right\}$.

5. **a.** $5x^2 - 10x + 5 = 12$

$5(x^2 - 2x + 1) = 12$

$5(x-1)^2 = 12$

$(x-1)^2 = \dfrac{12}{5}$

$x - 1 = \pm\sqrt{\dfrac{12}{5}} = \pm\sqrt{\dfrac{4\cdot3\cdot5}{5\cdot5}}$

$x - 1 = \pm\dfrac{2\sqrt{15}}{5}$

$x = 1 \pm \dfrac{2\sqrt{15}}{5} = \dfrac{5 \pm 2\sqrt{15}}{5}$

The solution set is $\left\{\dfrac{5+2\sqrt{15}}{5},\ \dfrac{5-2\sqrt{15}}{5}\right\}$.

b. $12x^2 + 12x + 3 = 16$

$3(4x^2 + 4x + 1) = 16$

$3(2x+1)^2 = 16$

$(2x+1)^2 = \dfrac{16}{3}$

$2x+1 = \pm\sqrt{\dfrac{16}{3}} = \pm\sqrt{\dfrac{16\cdot3}{3\cdot3}}$

$2x+1 = \pm\dfrac{4\sqrt{3}}{3}$

$2x = -1 \pm \dfrac{4\sqrt{3}}{3} = \dfrac{-3\pm4\sqrt{3}}{3}$

$x = \dfrac{-3\pm4\sqrt{3}}{6}$

The solution set is $\left\{\dfrac{-3+4\sqrt{3}}{6},\ \dfrac{-3-4\sqrt{3}}{6}\right\}$.

6. **a.** $x^2 - 8x - 9 = 0$

$x^2 - 8x + 4^2 = 9 + 4^2$

$(x-4)^2 = 25$

$x - 4 = \pm\sqrt{25} = \pm5$

$x = 4 \pm 5 = 9$ or -1

The solution set is $\{9, -1\}$.

b. $x^2 + 12x + 32 = 0$

$x^2 + 12x + 6^2 = -32 + 6^2$

$(x+6)^2 = 4$

$x + 6 = \pm\sqrt{4} = \pm2$

$x = -6 \pm 2 = -4$ or -8

The solution set is $\{-4, -8\}$.

7. **a.** $4x^2 + 4x - 3 = 0$

$4x^2 + 4x = 3$

$x^2 + x = \dfrac{3}{4}$

$x^2 + x + \left|\dfrac{1}{2}\right|^2 = \dfrac{3}{4} + \left|\dfrac{1}{2}\right|^2$

$\left|x + \dfrac{1}{2}\right|^2 = 1$

$x + \dfrac{1}{2} = \pm\sqrt{1} = \pm1$

$x = -\dfrac{1}{2} \pm 1 = \dfrac{1}{2}$ or $-\dfrac{3}{2}$

The solution set is $\left\{\dfrac{1}{2},\ -\dfrac{3}{2}\right\}$.

b. $16x^2 - 24x + 7 = 0$

$16x^2 - 24x = -7$

$x^2 - \dfrac{3}{2}x = -\dfrac{7}{16}$

$x^2 - \dfrac{3}{2}x + \left|\dfrac{3}{4}\right|^2 = -\dfrac{7}{16} + \left|\dfrac{3}{4}\right|^2$

$\left|x - \dfrac{3}{4}\right|^2 = \dfrac{1}{8}$

$x - \dfrac{3}{4} = \pm\sqrt{\dfrac{1}{8}} = \pm\sqrt{\dfrac{2}{16}}$

$x = \dfrac{3}{4} \pm \dfrac{\sqrt{2}}{4} = \dfrac{3\pm\sqrt{2}}{4}$

The solution set is $\left\{\dfrac{3+\sqrt{2}}{4},\ \dfrac{3-\sqrt{2}}{4}\right\}$.

8. **a.** $3x^2 + 5x - 2 = 0;$ $a = 3, b = 5, c = -2$

$$x = \frac{-5 \pm \sqrt{5^2 - 4 \cdot 3 \cdot (-2)}}{2 \cdot 3} = \frac{-5 \pm \sqrt{25 + 24}}{6}$$

$$= \frac{-5 \pm \sqrt{49}}{6} = \frac{-5 \pm 7}{6} = \frac{1}{3} \text{ or } -2$$

The solution set is $\left\{ \frac{1}{3}, -2 \right\}$.

b. $5x^2 - 9x - 2 = 0;$ $a = 5, b = -9, c = -2$

$$x = \frac{-(-9) \pm \sqrt{(-9)^2 - 4 \cdot 5 \cdot (-2)}}{2 \cdot 5}$$

$$= \frac{9 \pm \sqrt{81 + 40}}{10} = \frac{9 \pm \sqrt{121}}{10} = \frac{9 \pm 11}{10}$$

$$= 2 \text{ or } -\frac{1}{5}$$

The solution set is $\left\{ 2, -\frac{1}{5} \right\}$.

9. **a.** $3x^2 = 2x + 4$

$3x^2 - 2x - 4 = 0;$ $a = 3, b = -2, c = -4$

$$x = \frac{-(-2) \pm \sqrt{(-2)^2 - 4 \cdot 3 \cdot (-4)}}{2 \cdot 3}$$

$$= \frac{2 \pm \sqrt{4 + 48}}{6} = \frac{2 \pm \sqrt{52}}{6} = \frac{2 \pm 2\sqrt{13}}{6}$$

$$= \frac{1 \pm \sqrt{13}}{3}$$

The solution set is $\left\{ \frac{1 + \sqrt{13}}{3}, \frac{1 - \sqrt{13}}{3} \right\}$.

b. $4x^2 = 6x + 3$

$4x^2 - 6x - 3 = 0;$ $a = 4, b = -6, c = -3$

$$x = \frac{-(-6) \pm \sqrt{(-6)^2 - 4 \cdot 4 \cdot (-3)}}{2 \cdot 4}$$

$$= \frac{6 \pm \sqrt{36 + 48}}{8} = \frac{6 \pm \sqrt{84}}{8} = \frac{6 \pm 2\sqrt{21}}{8}$$

$$= \frac{3 \pm \sqrt{21}}{4}$$

The solution set is $\left\{ \frac{3 + \sqrt{21}}{4}, \frac{3 - \sqrt{21}}{4} \right\}$.

10. **a.** $16x = x^2$

$x^2 - 16x = 0;$ $a = 1, b = -16, c = 0$

$$x = \frac{-(-16) \pm \sqrt{(-16)^2 - 4 \cdot 1 \cdot 0}}{2 \cdot 1}$$

$$= \frac{16 \pm \sqrt{256}}{2} = \frac{16 \pm 16}{2} = 16 \text{ or } 0$$

The solution set is $\{16, 0\}$.

b. $12x = x^2$

$x^2 - 12x = 0;$ $a = 1, b = -12, c = 0$

$$x = \frac{-(-12) \pm \sqrt{(-12)^2 - 4 \cdot 1 \cdot 0}}{2 \cdot 1}$$

$$= \frac{12 \pm \sqrt{144}}{2} = \frac{12 \pm 12}{2} = 12 \text{ or } 0$$

The solution set is $\{12, 0\}$.

11. **a.** $x^2 + \dfrac{x}{15} = \dfrac{1}{3}$

$15x^2 + x = 5$

$15x^2 + x - 5 = 0;$ $a = 15, b = 1, c = -5$

$$x = \frac{-1 \pm \sqrt{1^2 - 4 \cdot 15 \cdot (-5)}}{2 \cdot 15}$$

$$= \frac{-1 \pm \sqrt{1 + 300}}{30} = \frac{-1 \pm \sqrt{301}}{30}$$

The solutions: $\left\{ \frac{-1 + \sqrt{301}}{30}, \frac{-1 - \sqrt{301}}{30} \right\}$.

b. $\dfrac{x^2}{2} + \dfrac{9x}{10} = \dfrac{1}{5}$

$5x^2 + 9x = 2$

$5x^2 + 9x - 2 = 0;$ $a = 5, b = 9, c = -2$

$$x = \frac{-9 \pm \sqrt{9^2 - 4 \cdot 5 \cdot (-2)}}{2 \cdot 5} = \frac{-9 \pm \sqrt{81 + 40}}{10}$$

$$= \frac{-9 \pm \sqrt{121}}{10} = \frac{-9 \pm 11}{10} = \frac{1}{5} \text{ or } -2$$

The solution set is $\left\{ \frac{1}{5}, -2 \right\}$.

12. a.

$$3x^2 - 2x = -1$$
$$3x^2 - 2x + 1 = 0; \quad a = 3, b = -2, c = 1$$
$$x = \frac{-(-2)\pm\sqrt{(-2)^2 - 4\cdot3\cdot1}}{2\cdot3} = \frac{2\pm\sqrt{4-12}}{6}$$
$$= \frac{2\pm\sqrt{-8}}{6} = \frac{2\pm2i\sqrt{2}}{6} = \frac{1}{3}\pm\frac{\sqrt{2}}{3}i$$

The solution set is $\left\{\dfrac{1}{3}+\dfrac{\sqrt{2}}{3}i,\ \dfrac{1}{3}-\dfrac{\sqrt{2}}{3}i\right\}$.

b.

$$5x^2 - 2x = -4$$
$$5x^2 - 2x + 4 = 0; \quad a = 5, b = -2, c = 4$$
$$x = \frac{-(-2)\pm\sqrt{(-2)^2 - 4\cdot5\cdot4}}{2\cdot5} = \frac{2\pm\sqrt{4-80}}{10}$$
$$= \frac{2\pm\sqrt{-76}}{10} = \frac{2\pm2i\sqrt{19}}{10} = \frac{1}{5}\pm\frac{\sqrt{19}}{5}i$$

The solution set is $\left\{\dfrac{1}{5}+\dfrac{\sqrt{19}}{5}i,\ \dfrac{1}{5}-\dfrac{\sqrt{19}}{5}i\right\}$.

13. a.

$$8x^3 - 125 = 0$$
$$(2x - 5)(4x^2 + 10x + 25) = 0$$
$$2x - 5 = 0 \text{ or } 4x^2 + 10x + 25 = 0$$
$$x = \frac{5}{2} \text{ or } 4x^2 + 10x + 25 = 0$$
$$4x^2 + 10x + 25 = 0; \quad a = 4, b = 10, c = 25$$
$$x = \frac{-10\pm\sqrt{10^2 - 4\cdot4\cdot25}}{2\cdot4}$$
$$= \frac{-10\pm\sqrt{100-400}}{8} = \frac{-10\pm\sqrt{-300}}{8}$$
$$= \frac{-10\pm10i\sqrt{3}}{8} = -\frac{5}{4}\pm\frac{5\sqrt{3}}{4}i$$

The solution set is

$$\left\{\frac{5}{2},\ -\frac{5}{4}+\frac{5\sqrt{3}}{4}i,\ -\frac{5}{4}-\frac{5\sqrt{3}}{4}i\right\}.$$

b.

$$125x^3 - 8 = 0$$
$$(5x - 2)(25x^2 + 10x + 4) = 0$$
$$5x - 2 = 0 \text{ or } 25x^2 + 10x + 4 = 0$$
$$x = \frac{2}{5} \text{ or } 25x^2 + 10x + 4 = 0$$
$$25x^2 + 10x + 4 = 0; \quad a = 25, b = 10, c = 4$$
$$x = \frac{-10\pm\sqrt{10^2 - 4\cdot25\cdot4}}{2\cdot25}$$
$$= \frac{-10\pm\sqrt{100-400}}{50} = \frac{-10\pm\sqrt{-300}}{50}$$
$$= \frac{-10\pm10i\sqrt{3}}{50} = -\frac{1}{5}\pm\frac{\sqrt{3}}{5}i$$

The solution set is

$$\left\{\frac{2}{5},\ -\frac{1}{5}+\frac{\sqrt{3}}{5}i,\ -\frac{1}{5}-\frac{\sqrt{3}}{5}i\right\}.$$

14. a.

$$100p - 150 = \frac{450}{p}$$
$$p(100p - 150) = p\cdot\frac{450}{p}$$
$$100p^2 - 150p = 450$$
$$2p^2 - 3p = 9$$
$$2p^2 - 3p - 9 = 0; \quad a = 2, b = -3, c = -9$$
$$p = \frac{-(-3)\pm\sqrt{(-3)^2 - 4\cdot2\cdot(-9)}}{2\cdot2}$$
$$= \frac{3\pm\sqrt{9+72}}{4} = \frac{3\pm\sqrt{81}}{4} = \frac{3\pm9}{4}$$
$$= 3 \text{ or } -1.5$$

The price at the equilibrium point is $3.00.

b.

$$150p - 100 = \frac{50}{p}$$
$$p(150p - 100) = p\cdot\frac{50}{p}$$
$$150p^2 - 100p = 50$$
$$3p^2 - 2p = 1$$
$$3p^2 - 2p - 1 = 0; \quad a = 3, b = -2, c = -1$$
$$p = \frac{-(-2)\pm\sqrt{(-2)^2 - 4\cdot3\cdot(-1)}}{2\cdot3}$$
$$= \frac{2\pm\sqrt{4+12}}{6} = \frac{2\pm\sqrt{16}}{6} = \frac{2\pm4}{6}$$
$$= 1 \text{ or } -0.33$$

The price at the equilibrium point is $1.00.

15. **a.** $16x^2 - 3x = -1$

$16x^2 - 3x + 1 = 0$; $a = 16, b = -3, c = 1$

$b^2 - 4ac = (-3)^2 - 4 \cdot 16 \cdot 1 = 9 - 64 = -55$

Since the discriminant is negative, there are two non-real complex number solutions.

b. $8x^2 - kx + 2 = 0$; $a = 8, b = -k, c = 2$

$b^2 - 4ac = (-k)^2 - 4 \cdot 8 \cdot 2 = k^2 - 64$

$k^2 - 64 = 0$

$\qquad k^2 = 64$

$\qquad k = \pm 8$

There is one real rational solution when k is 8 or −8.

16. **a.** $3x^2 - 11x - 6$; $a = 3, b = -11, c = -6$

$b^2 - 4ac = (-11)^2 - 4 \cdot 3 \cdot (-6) = 121 + 72$

$\qquad\qquad = 193$

Since the discriminant is not a positive perfect square, the expression is not factorable.

b. $18x^2 + 13x + 2$; $a = 18, b = 13, c = 2$

$b^2 - 4ac = 13^2 - 4 \cdot 18 \cdot 2 = 169 - 144 = 25$

$x = \dfrac{-13 \pm \sqrt{25}}{2 \cdot 18} = \dfrac{-13 \pm 5}{36}$

$x = -\dfrac{2}{9}$ or $\quad x = -\dfrac{1}{2}$

$9x = -2$ or $\quad 2x = -1$

$9x + 2 = 0$ or $2x + 1 = 0$

The factorization is $(9x + 2)(2x + 1)$.

17. **a.** $18x^2 - 9x - 5$; $a = 18, b = -9, c = -5$

$b^2 - 4ac = (-9)^2 - 4 \cdot 18 \cdot (-5) = 81 + 360$

$\qquad\qquad = 441 = 21^2$

$x = \dfrac{-(-9) \pm \sqrt{441}}{2 \cdot 18} = \dfrac{9 \pm 21}{36}$

$x = \dfrac{5}{6}$ or $\quad x = -\dfrac{1}{3}$

$6x = 5$ or $\quad 3x = -1$

$6x - 5 = 0$ or $3x + 1 = 0$

The factorization is $(6x - 5)(3x + 1)$.

b. $18x^2 + 13x + 1$; $a = 18, b = 13, c = 1$

$b^2 - 4ac = 13^2 - 4 \cdot 18 \cdot 1 = 169 - 72 = 97$

Since the discriminant is not a positive perfect square, the expression is not factorable.

18. **a.** $x = 3$ or $\quad x = -2$

$x - 3 = 0$ or $x + 2 = 0$

$(x - 3)(x + 2) = 0$

$x^2 + 2x - 3x - 6 = 0$

$x^2 - x - 6 = 0$

b. $x = \dfrac{1}{4}$ or $\quad x = -\dfrac{2}{3}$

$4x = 1$ or $\quad 3x = -2$

$4x - 1 = 0$ or $3x + 2 = 0$

$(4x - 1)(3x + 2) = 0$

$12x^2 + 8x - 3x - 2 = 0$

$12x^2 + 5x - 2 = 0$

19. a. $15x^2 + 4x - 3 = 0$

$\text{Sum} = -\dfrac{b}{a} = -\dfrac{4}{15}$

$\text{Product} = \dfrac{c}{a} = \dfrac{-3}{15} = -\dfrac{1}{5}$

b. $9x^2 - 12x - 5 = 0$

$\text{Sum} = -\dfrac{b}{a} = -\dfrac{-12}{9} = \dfrac{4}{3}$

$\text{Product} = \dfrac{c}{a} = \dfrac{-5}{9} = -\dfrac{5}{9}$

20. a. $15x^2 + 4x - 3 = 0$

$\text{Sum} = -\dfrac{b}{a} = -\dfrac{4}{15}$

$\text{Product} = \dfrac{c}{a} = \dfrac{-3}{15} = -\dfrac{1}{5}$

$-\dfrac{1}{3} + \dfrac{3}{5} = -\dfrac{5}{15} \cdot \dfrac{9}{15} = -\dfrac{4}{15} = -\dfrac{4}{15}$

$-\dfrac{1}{3} \cdot \dfrac{3}{5} = -\dfrac{1}{5} = -\dfrac{1}{5}$

The given values are the solutions.

b. $9x^2 - 12x - 5 = 0$

$\text{Sum} = -\dfrac{b}{a} = -\dfrac{-12}{9} = \dfrac{4}{3}$

$\text{Product} = \dfrac{c}{a} = \dfrac{-5}{9} = -\dfrac{5}{9}$

$-\dfrac{1}{3} + \dfrac{5}{3} = \dfrac{4}{3} \neq \dfrac{4}{3}$

The given values are not solutions.

21. a.

$$\dfrac{8}{x^2 - 16} \cdot \dfrac{1}{x - 4} = 1$$

$$\dfrac{8}{(x+4)(x-4)} \cdot \dfrac{1}{x-4} = 1$$

$$(x+4)(x-4) \cdot \dfrac{8}{(x+4)(x-4)} \cdot \dfrac{1}{x-4} = 1 \cdot (x+4)(x-4)$$

$$8 \cdot 1(x+4) = (x+4)(x-4)$$

$$8 \cdot x - 4 = x^2 - 16$$

$$0 = x^2 + x - 20$$

$$(x+5)(x-4) = 0$$

$$x + 5 = 0 \quad \text{or} \quad x - 4 = 0$$

$$x = -5 \text{ or} \qquad x = 4$$

Discard $x = 4$ because it makes the denominator zero. The solution is -5.

b.

$$\frac{-24}{x^2-36}+\frac{2}{x+6}=1$$

$$\frac{-24}{(x+6)(x-6)}+\frac{2}{x+6}=1$$

$$(x+6)(x-6)\cdot\frac{-24}{(x+6)(x-6)}+\frac{2}{x+6}\cdot=1\cdot(x+6)(x-6)$$

$$-24+2(x-6)=(x+6)(x-6)$$

$$-24+2x-12=x^2-36$$

$$0=x^2-2x$$

$$x(x-2)=0$$

$$x=0 \text{ or } x-2=0$$

$$x=0 \text{ or } \quad x=2$$

The solution are 0 and 2.

22. a. $(x^2+x)^2+2(x^2+x)-8=0$

Substitute $u=x^2+x$

$$u^2+2u-8=0$$

$$(u-2)(u+4)=0$$

$$u-2=0 \text{ or } u+4=0$$

$$u=2 \text{ or } \quad u=-4$$

$$x^2+x=2 \text{ or } x^2+x=-4$$

$$x^2+x-2=0 \text{ or } x^2+x+4=0$$

$$(x+2)(x-1)=0 \text{ or } x^2+x+4=0$$

$$x+2=0 \quad \text{or } x-1=0 \text{ or } x^2+x+4=0$$

$$x=-2 \text{ or } x=-1 \quad \text{or } x^2+x+4=0$$

$$x=\frac{-1\pm\sqrt{1-4\cdot1\cdot4}}{2\cdot1}=\frac{-1\pm\sqrt{-15}}{2}$$

$$=-\frac{1}{2}\pm\frac{\sqrt{15}}{2}i$$

The solutions are

$$-2, 1, -\frac{1}{2}+\frac{\sqrt{15}}{2}i, \text{ and } -\frac{1}{2}-\frac{\sqrt{15}}{2}i.$$

b. $(x^2-3x)^2-4(x^2-3x)-12=0$

Substitute $u=x^2-3x$

$$u^2-4u-12=0$$

$$(u-6)(u+2)=0$$

$$u-6=0 \text{ or } \quad u+2=0$$

$$u=6 \text{ or } \quad u=-2$$

$$x^2-3x=6 \text{ or } x^2-3x=-2$$

$$x^2-3x-6=0 \text{ or } \quad x^2-3x+2=0$$

$$x^2-3x-6=0 \text{ or } (x-2)(x-1)=0$$

$$x^2-3x-6=0 \text{ or } x-2=0 \text{ or } x-1=0$$

$$x^2-3x-6=0 \text{ or } \quad x=2 \text{ or } \quad x=1$$

$$x=\frac{3\pm\sqrt{9-4\cdot1\cdot(-6)}}{2\cdot1}=\frac{3\pm\sqrt{33}}{2}$$

The solutions are

$$2, 1, \frac{3+\sqrt{33}}{2}, \text{ and } \frac{3-\sqrt{33}}{2}.$$

23. a.
$$x^{1/2} - 4x^{1/4} = -3$$
$$x^{1/2} - 4x^{1/4} + 3 = 0$$
Substitute $u = x^{1/4}$
$$u^2 - 4u + 3 = 0$$
$$(u - 3)(u - 1) = 0$$
$$u - 3 = 0 \quad \text{or} \quad u - 1 = 0$$
$$u = 3 \quad \text{or} \quad u = 1$$
$$x^{1/4} = 3 \quad \text{or} \quad x^{1/4} = 1$$
$$x = 3^4 \quad \text{or} \quad x = 1^4$$
$$x = 81 \quad \text{or} \quad x = 1$$
The solutions are 81 and 1.
The check is left for the student.

b.
$$x^{1/2} + x^{1/4} = 6$$
$$x^{1/2} + x^{1/4} - 6 = 0$$
Substitute $u = x^{1/4}$
$$u^2 + u - 6 = 0$$
$$(u - 2)(u + 3) = 0$$
$$u - 2 = 0 \quad \text{or} \quad u + 3 = 0$$
$$u = 2 \quad \text{or} \quad u = -3$$
$$x^{1/4} = 2 \quad \text{or} \quad x^{1/4} = -3$$
$$x = 2^4 \quad \text{or} \quad x = (-3)^4$$
$$x = 16 \quad \text{or} \quad x = 81$$
Check: $(81)^{1/2} + (81)^{1/4} = 9 + 3 = 12 \neq 6$
Discard: $x = 81$
$$16^{1/2} + 16^{1/4} = 4 + 2 = 6 = 6$$
The solution is 16.

24. a.
$$x^{2/3} + x^{1/3} = 12$$
$$x^{2/3} + x^{1/3} - 12 = 0$$
Substitute $u = x^{1/3}$
$$u^2 + u - 12 = 0$$
$$(u - 3)(u + 4) = 0$$
$$u - 3 = 0 \quad \text{or} \quad u + 4 = 0$$
$$u = 3 \quad \text{or} \quad u = -4$$
$$x^{1/3} = 3 \quad \text{or} \quad x^{1/3} = -4$$
$$x = 3^3 \quad \text{or} \quad x = (-4)^3$$
$$x = 27 \quad \text{or} \quad x = -64$$
The solutions are 27 and –64.

b.
$$x^{2/3} - 5x^{1/3} = 6$$
$$x^{2/3} - 5x^{1/3} - 6 = 0$$
Substitute $u = x^{1/3}$
$$u^2 - 5u - 6 = 0$$
$$(u - 6)(u + 1) = 0$$
$$u - 6 = 0 \quad \text{or} \quad u + 1 = 0$$
$$u = 6 \quad \text{or} \quad u = -1$$
$$x^{1/3} = 6 \quad \text{or} \quad x^{1/3} = -1$$
$$x = 6^3 \quad \text{or} \quad x = (-1)^3$$
$$x = 216 \quad \text{or} \quad x = -1$$
The solutions are 216 and –1.

25. a.
$$x - 2\sqrt{x} = 3$$
$$x - 2\sqrt{x} - 3 = 0$$
Substitute $u = \sqrt{x}$
$$u^2 - 2u - 3 = 0$$
$$(u - 3)(u + 1) = 0$$
$$u - 3 = 0 \quad \text{or} \quad u + 1 = 0$$
$$u = 3 \quad \text{or} \quad u = -1$$
$$\sqrt{x} = 3 \quad \text{or} \quad \sqrt{x} = -1$$
$$x = 9 \quad \text{or} \quad x = 1$$

Check: $9 \neq 2\sqrt{9} = 9 \neq 6 = 3 = 3$
$$1 \neq 2\sqrt{1} = 1 \neq 2 = \neq 1 \neq 3$$
Discard: $x = 1$
The solution is 9.

b.
$$x - 4\sqrt{x} = 5$$
$$x - 4\sqrt{x} - 5 = 0$$
Substitute $u = \sqrt{x}$
$$u^2 - 4u - 5 = 0$$
$$(u - 5)(u + 1) = 0$$

$u - 5 = 0$ or $u + 1 = 0$

$u = 5$ or $u = -1$

$\sqrt{x} = 5$ or $\sqrt{x} = -1$

$x = 25$ or $x = 1$

Check: $25 \neq 4\sqrt{25} = 25 \neq 20 = 5 = 5$

$1 \neq 4\sqrt{1} = 1 \neq 4 = \neq 3 \neq 5$

Discard: $x = 1$

The solution is 25.

26. a. $3x^{-4} - 4x^{-2} + 1 = 0$

Substitute $u = x^{-2}$

$3u^2 - 4u + 1 = 0$

$(3u - 1)(u - 1) = 0$

$3u - 1 = 0$ or $u - 1 = 0$

$u = \dfrac{1}{3}$ or $u = 1$

$x^{-2} = \dfrac{1}{3}$ or $x^{-2} = 1$

$x^2 = 3$ or $x^2 = 1$

$x = \pm\sqrt{3}$ or $x = \pm 1$

The solutions are $\sqrt{3}, -\sqrt{3}, 1,$ and -1.

b. $3x^{-4} - 2x^{-2} - 1 = 0$

Substitute $u = x^{-2}$

$3u^2 - 2u - 1 = 0$

$(3u + 1)(u - 1) = 0$

$3u + 1 = 0$ or $u - 1 = 0$

$u = -\dfrac{1}{3}$ or $u = 1$

$x^{-2} = -\dfrac{1}{3}$ or $x^{-2} = 1$

$x^2 = -3$ or $x^2 = 1$

$x = \pm\sqrt{-3}$ or $x = \pm\sqrt{1}$

$x = \pm i\sqrt{3}$ or $x = \pm 1$

The solutions are $i\sqrt{3}, -i\sqrt{3}, 1,$ and -1.

27. a. $(x \cdot 2)(x + 3) < 0$

$(x \cdot 2)(x + 3) = 0$ is the related equation. $x = 2$, $x = \cdot\ 3$ are the critical values.

Separate into regions using –3 and 2 as boundaries.

Select test points: $\cdot$ 4, 0, and 3.

When $x = \cdot\ 4$, $(\cdot\ 4 \cdot 2)(\cdot\ 4 + 3) = \cdot\ 6(\cdot\ 1) = 6 \not< 0$ False.

When $x = 0$, $(0 \cdot 2)(0 + 3) = \cdot\ 2(3) = \cdot\ 6 < 0$ True.

When $x = 3$, $(3 \cdot 2)(3 + 3) = 1 \cdot 6 = 6 \not< 0$ False.

The solution set is: $\{x|\cdot\ 3 < x < 2\}$ or $(\cdot\ 3, 2)$.

b. $(x+2)(x-3)<0$

$(x+2)(x-3)=0$ is the related equation. $x=-2$, $x=3$ are the critical values.

Separate into regions using -2 and 3 as boundaries.

Select test points: -4, 0, and 4.

When $x=-4$, $(-4+2)(-4-3)=-2(-7)=14 \not< 0$ False.

When $x=0$, $(0+2)(0-3)=2(-3)=-6<0$ True.

When $x=4$, $(4+2)(4-3)=6\cdot 1=6 \not< 0$ False.

The solution set is: $\{x| -2<x<3\}$ or $(-2, 3)$.

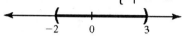

28. a. $x^2+4x \geq 0$

$x(x+4) \geq 0$

$x(x+4)=0$ is the related equation. $x=0$, $x=-4$ are the critical values.

Separate into regions using -4 and 0 as boundaries.

Select test points: -5, -1, and 1.

When $x=-5$, $(-5)^2+4(-5)=25-20=5 \geq 0$ True.

When $x=-1$, $(-1)^2+4(-1)=1-4=-3 \not\geq 0$ False.

When $x=1$, $1^2+4(1)=1+4=5 \geq 0$ True.

The solution set is: $\{x|x \leq -4 \cup x \geq 0\}$ or $(-\infty, -4]\cup[0, +\infty)$.

b. $x^2 - 3x \geq 0$

$x(x-3) \geq 0$

$x(x-3)=0$ is the related equation. $x=0$, $x=3$ are the critical numbers.

Separate into regions using 0 and 3 as boundaries.

Select test points: -1, 1, and 4.

When $x=-1$, $(-1)^2-3(-1)=1+3=4 \geq 0$ True.

When $x=1$, $1^2-3(1)=1-3=-2 \not\geq 0$ False.

When $x=4$, $4^2-3(4)=16-12=4 \geq 0$ True.

The solution set is: $\{x|x \leq 0 \cup x \geq 3\}$ or $(-\infty, 0]\cup[3, +\infty)$.

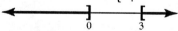

29. a. $x^2 + 4x < 45$

$x^2 + 4x - 45 < 0$

$(x-5)(x+9) < 0$

$(x-5)(x+9) = 0$ is the related equation. $x = 5$, $x = -9$ are the critical values.

Separate into regions using -9 and 5 as boundaries.

Select test points: $-10, 0,$ and 6.

When $x = -10$, $(-10)^2 + 4(-10) = 100 - 40 = 60 \not< 45$ False.

When $x = 0$, $0^2 + 4(0) = 0 + 0 = 0 < 45$ True.

When $x = 6$, $(6)^2 + 4(6) = 36 + 24 = 60 \not< 45$ False.

The solution set is: $\{x | -9 < x < 5\}$ or $(-9, 5)$.

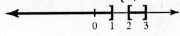

b. $x^2 - 2x < 2$

$x^2 - 2x + 1 < 2 + 1$

$(x-1)^2 < 3$

$x - 1 = \pm\sqrt{3}$ is the related equation. $x = 1 + \sqrt{3}$, $x = 1 - \sqrt{3}$ are the critical values.

Separate into regions using $1 + \sqrt{3}$ and $1 - \sqrt{3}$ as boundaries.

Select test points: $-1, 0,$ and 3.

When $x = -1$, $(-1)^2 - 2(-1) = 1 + 2 = 3 \not< 2$ False.

When $x = 0$, $0^2 - 2(0) = 0 - 0 = 0 < 2$ True.

When $x = 3$, $3^2 - 2(3) = 9 - 6 = 3 \not< 2$ False.

The solution set is: $\{x | 1 - \sqrt{3} < x < 1 + \sqrt{3}\}$ or $(1 - \sqrt{3}, 1 + \sqrt{3})$.

30. a. $(x-1)(x-2)(x-3) \le 0$

$(x-1)(x-2)(x-3) = 0$ is the related equation. $x = 1$, $x = 2$, $x = 3$ are the critical values.

Separate into regions using $1, 2,$ and 3 as boundaries.

Select test points: $0, 1.5, 2.5,$ and 4.

When $x = 0$, $(0-1)(0-2)(0-3) = -1(-2)(-3) = -6 \le 0$ True.

When $x = 1.5$, $(1.5-1)(1.5-2)(1.5-3) = 0.5(-0.5)(-1.5) = 0.375 \not\le 0$ False.

When $x = 2.5$, $(2.5-1)(2.5-2)(2.5-3) = 1.5(0.5)(-0.5) = -0.375 \le 0$ True.

When $x = 4$, $(4-1)(4-2)(4-3) = 3(2)(1) = 6 \not\le 0$ False.

The solution set is: $\{x | x \le 1 \cup 2 \le x \le 3\}$ or $(-\infty, 1] \cup [2, 3]$.

b. $(x+1)(x+2)(x-3) \le 0$

$(x+1)(x+2)(x-3) = 0$ is the related equation. $x = -1$, $x = -2$, $x = 3$ are the critical values.

Separate into regions using -1, -2, and 3 as boundaries.

Select test points: -3, -1.5, 0, and 4.

When $x = -3$, $\quad (-3+1)(-3+2)(-3-3) = -2(-1)(-6) = -12 \le 0$ — True.

When $x = -1.5$, $\quad (-1.5+1)(-1.5+2)(-1.5-3) = -0.5(0.5)(-4.5) = 1.125 \not\le 0$ — False.

When $x = 0$, $\quad (0+1)(0+2)(0-3) = 1(2)(-3) = -6 \le 0$ — True.

When $x = 4$, $\quad (4+1)(4+2)(4-3) = 5(6)(1) = 30 \not\le 0$ — False.

The solution set is: $\{x \mid x \le -2 \cup -1 \le x \le 3\}$ or $(-\infty, -2] \cup [-1, 3]$.

31. a. $(x+1)(x+2)(x-3) > 0$

$(x+1)(x+2)(x-3) = 0$ is the related equation. $x = -1$, $x = -2$, $x = 3$ are the critical values.

Separate into regions using -1, -2, and 3 as boundaries.

Select test points: -3, -1.5, 0, and 4.

When $x = -3$, $\quad (-3+1)(-3+2)(-3-3) = -2(-1)(-6) = -12 \not> 0$ — False.

When $x = -1.5$, $\quad (-1.5+1)(-1.5+2)(-1.5-3) = -0.5(0.5)(-4.5) = 1.125 > 0$ — True.

When $x = 0$, $\quad (0+1)(0+2)(0-3) = 1(2)(-3) = -6 \not> 0$ — False.

When $x = 4$, $\quad (4+1)(4+2)(4-3) = 5(6)(1) = 30 > 0$ — True.

The solution set is: $\{x \mid -2 < x < -1 \cup x > 3\}$ or $(-2, -1) \cup (3, +\infty)$.

b. $(x-1)(x+2)(x+3) > 0$

$(x-1)(x+2)(x+3) = 0$ is the related equation. $x = 1$, $x = -2$, $x = -3$ are the critical values.

Separate into regions using -3, -2, and 1 as boundaries.

Select test points: -4, -2.5, 0, and 4.

When $x = -4$, $\quad (-4-1)(-4+2)(-4+3) = -5(-2)(-1) = -10 \not> 0$ — False.

When $x = -2.5$, $\quad (-2.5-1)(-2.5+2)(-2.5+3) = -3.5(-0.5)(0.5) = 0.875 > 0$ — True.

When $x = 0$, $\quad (0-1)(0+2)(0+3) = -1(2)(3) = -6 \not> 0$ — False.

When $x = 4$, $\quad (4-1)(4+2)(4+3) = 3(6)(7) = 126 > 0$ — True.

The solution set is: $\{x \mid -3 < x < -2 \cup x > 1\}$ or $(-3, -2) \cup (1, +\infty)$.

32. a. $\dfrac{x+2}{x\le 2}\le 0$

$x=\le 2,\ x=2$ are the critical values.

Separate into regions using -2 and 2 as boundaries.

Select test points: $\le 3,\ 0,$ and 3.

When $x=\le 3$, $\dfrac{\le 3+2}{\le 3\le 2}=\dfrac{\le 1}{\le 5}=\dfrac{1}{5}\not\le 0$ False.

When $x=0$, $\dfrac{0+2}{0\le 2}=\dfrac{2}{\le 2}=\le 1\le 0$ True.

When $x=3$, $\dfrac{3+2}{3\le 2}=\dfrac{5}{1}=5\not\le 0$ False.

The solution set is: $\left\{x\,|\,\le 2\le x<2\right\}$ or $\left[\le 2,\,2\right)$.

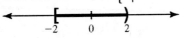

b.

$\dfrac{\cdot\,12}{x\cdot 6}<3$

$\dfrac{\cdot\,12}{x\cdot 6}\cdot 3<0$

$\dfrac{\cdot\,12}{x\cdot 6}\cdot 3\cdot\dfrac{x\cdot 6}{x\cdot 6}<0$

$\dfrac{\cdot\,12\cdot 3x+18}{x\cdot 6}<0$

$\dfrac{\cdot\,3x+6}{x\cdot 6}<0$

$x=2$ and $x=6$ are the critical values.

Separate into regions using 2 and 6 as boundaries.

Select test points: $0,\ 3,$ and 7.

When $x=0$, $\dfrac{\infty l2}{0\,\infty6}=\dfrac{\infty l2}{\infty6}=2<3$ True.

When $x=3$, $\dfrac{\infty l2}{3\,\infty6}=\dfrac{\infty l2}{\infty3}=4\not<3$ False.

When $x=7$, $\dfrac{\infty l2}{7\,\infty6}=\dfrac{\infty l2}{1}=\infty l2<3$ True.

The solution set is: $\left\{x\,|\,x<2\cup x>6\right\}$ or $\left(\infty\infty,\,2\right)\cup\left(6,\,+\infty\right)$.

Cumulative Review Chapters 1 – 8

1. $\{2,\,4,\,6,\,8,\,10,\,12\}$

2. $\dfrac{11}{15}=0.7\overline{3}$

3. $|-11|=11$

4. $\lfloor\!\lfloor 3(7+6)\rfloor+7=\lfloor\!\lfloor 3(13)\rfloor+7=\lfloor 39+7=\lfloor 32$

5.
$$x + 4 = 2(5x - 3)$$
$$x + 4 = 10x - 6$$
$$-9x = -10$$
$$x = \frac{10}{9}$$

6. $\{x \mid x > -1 \text{ and } x < 1\}$

7. $|3x + 1| > 2$
$$3x + 1 > 2 \text{ or } 3x + 1 < -2$$
$$3x > 1 \text{ or } \quad 3x < -3$$
$$x > \frac{1}{3} \text{ or } \quad x < -1$$

8.

Investment	Principal	Rate	Interest
Bonds	x	0.07	$0.07x$
CD's	18,000–x	0.09	$0.09(18,000-x)$

$$0.07x + 0.09(18,000 - x) = 1420$$
$$0.07x + 1620 - 0.09x = 1420$$
$$1620 - 0.02x = 1420$$
$$-0.02x = -200$$
$$x = \$10,000 \text{ in bonds}$$
$$18,000 - x = \$8,000 \text{ in CD's}$$
$10,000 is invested at 7% and $8,000 at 9%.

9. $2x - y = 2$

10.
$$m = \frac{-6 - 5}{-4 - (-7)} = \frac{-11}{3} = -\frac{11}{3}$$

Since the slopes are negative reciprocals of each other, the lines are perpendicular.

11.
$$y - 2 = -1(x - (-3))$$
$$y - 2 = -1(x + 3)$$
$$y - 2 = -x - 3$$
$$x + y = -1$$

12. $8x - y = 3 \rightarrow 8x - 3 = y$

The slope of the line is 8, so the parallel line has slope 8.
$$y - 4 = 8(x - (-6))$$
$$y - 4 = 8(x + 6)$$
$$y - 4 = 8x + 48$$
$$-8x + y = 52$$
$$8x - y = -52$$

13. $\{(x, y) \mid y = 5 + x\}$

Domain: $\{\text{all real numbers}\}$

Range: $\{\text{all real numbers}\}$

14. $x + y - -1$ and $y - 3x - 5$

Graph the line $x + y = -1$. The boundary is a solid line.

Test point $(0,0)$: $0 + 0 - -1$, $0 - -1$.

False, so shade the oppposite side of the line.

Graph the line $y = 3x - 5$. The boundary is a solid line.

Test point $(0,0)$: $0 - 3(0) - 5$, $0 - -5$. False, so shade

the opposite side of the line.

The solution is the intersection of the graphs.

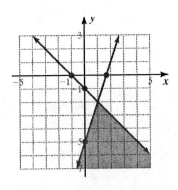

15. To eliminate y, multiply the second equation by -4, add the result to the first equation, and solve:

$$3x + 4y = -1 \rightarrow \quad 3x + 4y = -1$$
$$x + \ y = -1 \rightarrow \ -4x - 4y = \ 4$$
$$\hphantom{x + y = -1 \rightarrow} -x = 3$$
$$\hphantom{x + y = -1 \rightarrow} x = -3$$
$$\hphantom{x + y} -3 + y = -1 \rightarrow y = 2$$

Solution: $(-3, 2)$.

16.
$$D = \begin{vmatrix} 1 & 2 \\ 4 & 2 \end{vmatrix} = 2 - 8 = -6$$

$$D_x = \begin{vmatrix} 6 & 2 \\ 7 & 2 \end{vmatrix} = 12 - 14 = -2$$

$$x = \frac{D_x}{D} = \frac{-2}{-6} = \frac{1}{3}$$

17. Let n = the number of nickels

Let d = the number of dimes

$$5n + 10d = 320$$

$$n = d + 10$$

Substitute the second equation into the first equation and solve:

$$5(d + 10) + 10d = 320$$
$$5d + 50 + 10d = 320$$
$$15d = 270$$
$$d = 18$$
$$n = d + 10 = 18 + 10 = 28$$

Jose has 28 nickels and 18 dimes.

18. $(4x^3 + 3x^2 + 9) - (6x^3 - 6x^2 + 4x - 9)$

$$= 4x^3 + 3x^2 + 9 - 6x^3 + 6x^2 - 4x + 9$$
$$= -2x^3 + 9x^2 - 4x + 18$$

19. $36x^2 + 60xy + 25y^2 = (6x + 5y)^2$

20. $2x^4 + 8x^2 = 2x^2(x^2 + 4)$

21.
$$x^3 + x^2 - 9x - 9 = 0$$
$$x^2(x + 1) - 9(x + 1) = 0$$
$$(x + 1)(x^2 - 9) = 0$$
$$(x + 1)(x + 3)(x - 3) = 0$$
$$x + 1 = 0 \ \text{ or } x + 3 = 0 \ \text{ or } x - 3 = 0$$
$$x = -1 \text{ or } \quad x = -3 \text{ or } \quad x = 3$$

The solutions are -3, -1, and 3.

22.
$$\frac{-12x^7 + 12x^3 + 6x^2}{6x^5} = \frac{-12x^7}{6x^5} + \frac{12x^3}{6x^5} + \frac{6x^2}{6x^5}$$

$$= -2x^2 + \frac{2}{x^2} + \frac{1}{x^3}$$

23.
$$\frac{x}{x^2-81}-\frac{9}{x^2-81}=\frac{x-9}{x^2-81}=\frac{\cancel{x-9}}{\cancel{(x-9)}(x+9)}=\frac{1}{x+9}$$

24. Multiply the numerator and denominator by the LCD which is x^3:

$$\frac{x\cdot\frac{1}{x^2}}{x\cdot\frac{1}{x^3}}=\frac{x^3\cdot x\cdot\frac{1}{x^2}}{x^3\cdot x\cdot\frac{1}{x^3}}=\frac{x^3\cdot x\cdot\cancel{x}\cdot x\cdot\frac{1}{\cancel{x^2}}}{x^3\cdot x\cdot\cancel{x}\cdot\frac{1}{\cancel{x^3}}}=\frac{x^4\cdot x}{x^4\cdot 1}=\frac{x(x^3\cdot 1)}{(x^2+1)(x^2\cdot 1)}=\frac{x\cancel{(x-1)}(x^2+x+1)}{(x^2+1)\cancel{(x-1)}(x+1)}$$

$$=\frac{x(x^2+x+1)}{(x^2+1)(x+1)}$$

25. Multiply both sides by the LCD $(x+1)(x-1)$:

$$1\;\Big|\;\frac{1}{x+1}=\frac{2}{x^2\,\big|\,1}$$

$$\frac{(x+1)(x\,|\,1)}{1}\Big|\Big|1\,\Big|\,\frac{1}{x+1}\Big|=\frac{(x+1)(x\,|\,1)}{1}\Big|\frac{2}{(x+1)(x\,|\,1)}$$

$$(x+1)(x\,|\,1)\,|\,1(x\,|\,1)=2$$

$$x^2\,|\,1\,|\,x+1=2$$

$$x^2\,|\,x\,|\,2=0$$

$$(x\,|\,2)(x+1)=0$$

$$x\,|\,2=0\ \text{ or }\ x+1=0$$

$$x=2\ \text{ or }\quad x=|1$$

-1 makes the denominator $x+1=0$ and thus is extraneous. The solution is 2.

26. Let $x=$ the first even integer
Let $x+2=$ the consecutive even integer

$$\frac{1}{x}+\frac{1}{x+2}=\frac{9}{40}$$

$$40x(x+2)\frac{1}{x}+\frac{1}{x+2}=40x(x+2)\frac{9}{40}$$

$$40(x+2)+40x=9x(x+2)$$

$$40x+80+40x=9x^2+18x$$

$$0=9x^2-62x-80$$

$$(9x+10)(x-8)=0$$

$$9x+10=0\qquad\text{or}\ x-8=0$$

$$9x=-10\ \text{ or }\quad x=8$$

$$x=-\frac{10}{9}\ \text{ or }\quad x=8$$

The integers are 8 and 10.

27. $(-16)^{1/2}=\sqrt{-16}$

This is not a real number.

28. $\sqrt[3]{192a^7b^9}=\sqrt[3]{64a^6b^9\cdot 3a}=4a^2b^3\sqrt[3]{3a}$

29. $\sqrt{\dfrac{5}{6k}}=\sqrt{\dfrac{5\cdot 6k}{6k\cdot 6k}}=\dfrac{\sqrt{30k}}{6k}$

30. $\sqrt[3]{3x}\left(\sqrt[3]{9x^2}+\sqrt[3]{16x}\right)=\sqrt[3]{27x^3}+\sqrt[3]{48x^2}$

$=3x+\sqrt[3]{8\cdot 6x^2}=3x+2\sqrt[3]{6x^2}$

31. $\sqrt{x+8}-x=8$

$\sqrt{x+8}=x+8$

$\left(\sqrt{x+8}\right)^2=(x+8)^2$

$x+8=x^2+16x+64$

$0=x^2+15x+56$

$0=(x+8)(x+7)$

$x+8=0 \quad \text{or } x+7=0$

$x=-8 \quad \text{or} \qquad x=-7$

Check: $\sqrt{-8+8}-(-8)=\sqrt{0}+8=8=8$

$\sqrt{-7+8}-(-7)=\sqrt{1}+7=8=8$

Solutions are $x=-8$ and $x=-7$.

32. $(1-4i)-(4+i)=(1-4)+(-4-1)i$

$=-3-5i$

33. $(-1-8i)(-4-3i)=4+3i+32i+24i^2$

$=4+35i-24=-20+35i$

34. $16x^2-9=0$

$16x^2=9$

$x^2=\dfrac{9}{16}$

$x=\pm\dfrac{3}{4}$

The solutions are $\frac{3}{4}$ and $-\frac{3}{4}$.

35. $x^2+2x-3=0$

$x^2+2x=3$

$x^2+2x+1=3+1$

$(x+1)^2=4$

$x+1=\pm 2$

$x=-1\pm 2=1\text{ or }-3$

The solutions are 1 and -3.

36. $x^2=-4x+2$

$x^2+4x-2=0; \quad a=1, b=4, c=-2$

$x=\dfrac{-4\pm\sqrt{4^2-4\cdot 1\cdot(-2)}}{2\cdot 1}=\dfrac{-4\pm\sqrt{16+8}}{2}$

$=\dfrac{-4\pm\sqrt{24}}{2}=\dfrac{-4\pm 2\sqrt{6}}{2}=-2\pm\sqrt{6}$

The solution set is $\left\{-2+\sqrt{6},-2-\sqrt{6}\right\}$.

37. $200p-680=\dfrac{480}{p}$

$p(200p-680)=480$

$200p^2-680p=480$

$5p^2-17p=12$

$5p^2-17p-12=0; \quad a=5, b=-17, c=-12$

$p=\dfrac{-(-17)\pm\sqrt{(-17)^2-4\cdot 5\cdot(-12)}}{2\cdot 5}$

$=\dfrac{17\pm\sqrt{289+240}}{10}=\dfrac{17\pm\sqrt{529}}{10}=\dfrac{17\pm 23}{10}$

$=4\text{ or }-\dfrac{3}{5}$

The price at the equilibrium point is $4.

38. $12x^2 + 7x - 10;\quad a = 12,\ b = 7,\ c = -10$

$b^2 - 4ac = 7^2 - 4 \cdot 12 \cdot (-10) = 49 + 480$

$$= 529 = 23^2$$

Since the discriminant is a positive perfect square, the expression is factorable.

39. $x^4 - 8x^2 + 7 = 0$

Substitute $u = x^2$

$u^2 - 8u + 7 = 0$

$(u-1)(u-7) = 0$

$u - 1 = 0 \ \text{or}\ u - 7 = 0$

$u = 1 \quad \text{or} \quad u = 7$

$x^2 = 1 \quad \text{or} \quad x^2 = 7$

$x = \pm 1 \ \text{or} \quad x = \pm\sqrt{7}$

The solutions are $1,\ -1,\ \sqrt{7},$ and $-\sqrt{7}$.

40. $(x+6)(x-4) < 0$

$(x+6)(x-4) = 0$ is the related equation. $x = -6$, $x = 4$ are the critical values.

Separate into regions using -6 and 4 as boundaries.

Select test points: -7, 0, and 5.

When $x = -7$, $(-7+6)(-7-4) = -1(-11) = 11 \not< 0$ False.

When $x = 0$, $(0+6)(0-4) = 6(-4) = -24 < 0$ True.

When $x = 5$, $(5+6)(5-4) = 11 \cdot 1 = 11 \not< 0$ False.

The solution set is: $\{x \mid -6 < x < 4\}$ or $(-6, 4)$.

Chapter 9   Quadratic Functions and the Conic Sections

9.1   Quadratic Functions and Their Graphs

Problems 9.1

1. $y = -2x^2$

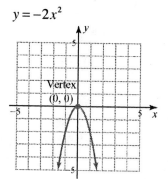

2.   **a.** $f(x) = -4x^2$

     b. $g(x) = 4x^2$

     c. $h(x) = \dfrac{1}{4}x^2$

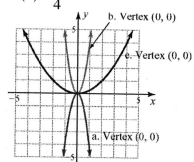

3. $y = -x^2 - 1$

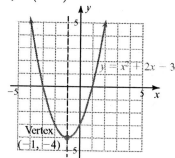

4. $y = (x-2)^2 - 1$

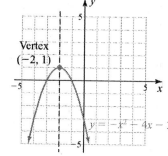

5. $y = x^2 + 2x - 3$

$y = (x^2 + 2x + 1) - 3 - 1$

$y = (x+1)^2 - 4$

6. $y = -x^2 - 4x - 3$

$y = -(x^2 + 4x + 4 - 4) - 3$

$y = -(x+2)^2 + 4 - 3$

$y = -(x+2)^2 + 1$

363

7. $x = (y+1)^2 - 2$

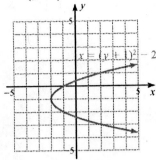

Vertex: $(-2, -1)$

8. $x = y^2 + 2y - 3$

$x = (y^2 + 2y + 1) - 3 - 1$

$x = (y+1)^2 - 4$

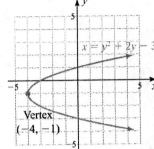

Vertex: $(-4, -1)$

9. $2l + 2w = 200$

$l + w = 100$

$l = 100 - w$

$A = (100 - w)w$

$\quad = 100w - w^2$

$\quad = -(w^2 - 100w)$

$\quad = -(w^2 - 100w + 2500 - 2500)$

$\quad = -(w - 50)^2 + 2500$

Vertex: $(50, 2500)$

The farmer will have maximum area when the width is 50 ft and the length is 50 ft.

Exercises 9.1

1.   **a.** $y = 2x^2$ vertex: $(0, 0)$

  b. $y = 2x^2 + 2$ vertex: $(0, 2)$

  c. $y = 2x^2 - 2$ vertex: $(0, -2)$

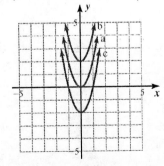

3.   **a.** $y = -2x^2$ vertex: $(0, 0)$

  b. $y = -2x^2 + 1$ vertex: $(0, 1)$

  c. $y = -2x^2 - 1$ vertex: $(0, -1)$

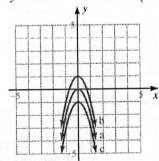

5. **a.** $y = \dfrac{1}{4}x^2$ vertex: $(0, 0)$

b. $y = -\dfrac{1}{4}x^2$ vertex: $(0, 0)$

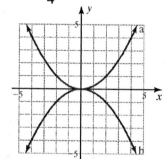

7. **a.** $y = \dfrac{1}{3}x^2 + 1$ vertex: $(0, 1)$

b. $y = -\dfrac{1}{3}x^2 + 1$ vertex: $(0, 1)$

9. **a.** $y = (x+2)^2 + 3$ vertex: $(-2, 3)$

b. $y = (x+2)^2$ vertex: $(-2, 0)$

c. $y = (x+2)^2 - 2$ vertex: $(-2, -2)$

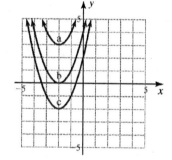

11. **a.** $y = -(x+2)^2 - 2$ vertex: $(-2, -2)$

b. $y = -(x+2)^2$ vertex: $(-2, 0)$

c. $y = -(x+2)^2 - 4$ vertex: $(-2, -4)$

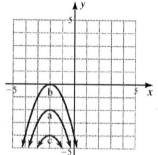

13. **a.** $y = -2(x+2)^2 - 2$ vertex: $(-2, -2)$

b. $y = -2(x+2)^2$ vertex: $(-2, 0)$

c. $y = -2(x+2)^2 - 4$ vertex: $(-2, -4)$

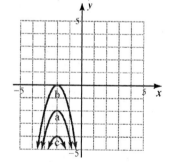

15. **a.** $y = 2(x+1)^2 + \dfrac{1}{2}$ vertex: $\left| 1, \dfrac{1}{2} \right|$

b. $y = 2(x+1)^2$ vertex: $(\, |1, 0)$

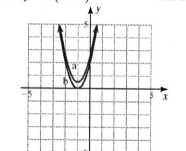

17. $y = x^2 + 2x + 1$

$y = (x+1)^2$ vertex: $(-1, 0)$

y-intercept: $(0, 1)$

axis: $x = -1$ another point: $(-2, 1)$

Solve: $0 = x^2 + 2x + 1$

$0 = (x+1)^2$

$x + 1 = 0$

$x = -1$

x-intercept: $(-1, 0)$

$a > 0$ parabola opens upward

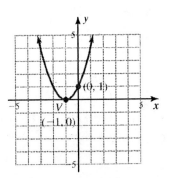

19. $y = -x^2 + 2x + 1$

$y = -(x^2 - 2x) + 1$

$y = -(x^2 - 2x + 1 - 1) + 1$

$y = -(x-1)^2 + 1 + 1$

$y = -(x-1)^2 + 2$ vertex: $(1, 2)$

y-intercept: $(0, 1)$

axis: $x = 1$ another point: $(2, 1)$

Solve: $0 = -(x-1)^2 + 2$

$(x-1)^2 = 2$

$x - 1 = \pm\sqrt{2}$

$x = 1 \pm \sqrt{2}$

x-intercept: $(1+\sqrt{2}, 0) \approx (2.4, 0)$

and $(1 - \sqrt{2}, 0) \approx (-0.4, 0)$

$a < 0$ parabola opens downward

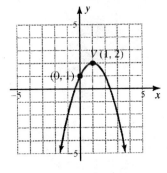

21. $y = -x^2 + 4x - 5$

$y = -(x^2 - 4x) - 5$

$y = -(x^2 - 4x + 4 - 4) - 5$

$y = -(x-2)^2 + 4 - 5$

$y = -(x-2)^2 - 1$ vertex: $(2, -1)$

y-intercept: $(0, -5)$

axis: $x = 2$ another point: $(4, -5)$

Solve: $0 = -(x-2)^2 - 1$

$(x-2)^2 = -1$

No solution since a square

cannot be negative.

No x-intercepts

$a < 0$ parabola opens downward

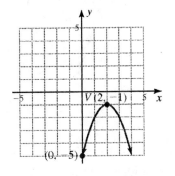

23. $y = 2x^2 - 5x + 3$

$y = 2\left(x^2 - \frac{5}{2}x\right) + 3$

$y = 2\left(x^2 - \frac{5}{2}x + \frac{25}{16} - \frac{25}{16}\right) + 3$

$y = 2\left(x - \frac{5}{4}\right)^2 - \frac{25}{8} + 3$

$y = 2\left(x - \frac{5}{4}\right)^2 - \frac{1}{8}$ vertex: $\left(\frac{5}{4}, -\frac{1}{8}\right)$

y-intercept: $(0, 3)$

axis: $x = \frac{5}{4}$ another point: $\left(\frac{5}{2}, 3\right)$

Solve: $0 = 2\left(x - \frac{5}{4}\right)^2 - \frac{1}{8}$

$2\left(x - \frac{5}{4}\right)^2 = \frac{1}{8}$

$\left(x - \frac{5}{4}\right)^2 = \frac{1}{16}$

$x - \frac{5}{4} = \pm\frac{1}{4}$

$x = \frac{5}{4} \pm \frac{1}{4}$

x-intercept: $\left(\frac{3}{2}, 0\right)$ and $(1, 0)$

$a > 0$ parabola opens upward

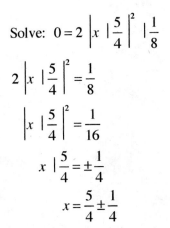

25. $y = -2x^2 - 4x + 5$

$y = -2(x^2 + 2x) + 5$

$y = -2(x^2 + 2x + 1 - 1) + 5$

$y = -2(x+1)^2 + 2 + 5$

$y = -2(x+1)^2 + 7$ vertex: $(-1, 7)$

y-intercept: $(0, 5)$

axis: $x = -1$ another point: $(-2, 5)$

Solve: $0 = -2(x+1)^2 + 7$

$2(x+1)^2 = 7$

$(x+1)^2 = \dfrac{7}{2}$

$x + 1 = \pm\sqrt{\dfrac{7}{2}}$

$x = -1 \pm \sqrt{\dfrac{7}{2}}$

x-intercept: $(0.9, 0)$

and $(-2.9, 0)$

$a < 0$ parabola opens downward

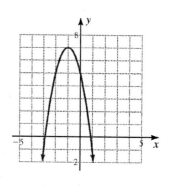

27. $y = 3x^2 + 3x + 2$

$y = 3(x^2 + x) + 2$

$y = 3\left(x^2 + x + \dfrac{1}{4} - \dfrac{1}{4}\right) + 2$

$y = 3\left(x + \dfrac{1}{2}\right)^2 + \dfrac{3}{4} + 2$

$y = 3\left(x + \dfrac{1}{2}\right)^2 + \dfrac{11}{4}$ vertex: $\left(-\dfrac{1}{2}, \dfrac{11}{4}\right)$

y-intercept: $(0, 2)$

axis: $x = -\dfrac{1}{2}$ another point: $(1, 2)$

Solve: $0 = 3\left(x + \dfrac{1}{2}\right)^2 + \dfrac{11}{4}$

$3\left(x + \dfrac{1}{2}\right)^2 = \dfrac{11}{4}$

$\left(x + \dfrac{1}{2}\right)^2 = \dfrac{11}{12}$

$x + \dfrac{1}{2} = \pm\sqrt{\dfrac{11}{12}}$

$x = -\dfrac{1}{2} \pm \sqrt{\dfrac{11}{12}}$

x-intercept: $(1.5, 0)$

and $(-0.5, 0)$

$a < 0$ parabola opens downward

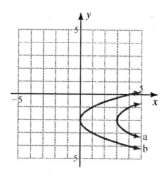

29. a. $x = (y+2)^2 + 3$ vertex: $(3, -2)$

 b. $x = (y+2)^2$ vertex: $(0, -2)$

31. a. $x = -(y+2)^2 - 2$ vertex: $(-2, -2)$

b. $x = -(y+2)^2$ vertex: $(0, -2)$

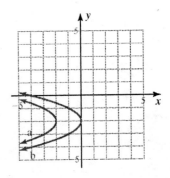

33. a. $x = -y^2 + 2y + 1$

$x = -(y^2 - 2y) + 1$

$x = -(y^2 - 2y + 1 - 1) + 1$

$x = -(y-1)^2 + 1 + 1$

$x = -(y-1)^2 + 2$

vertex: $(2, 1)$

b. $x = -y^2 + 2y + 4$

$x = -(y^2 - 2y) + 4$

$x = -(y^2 - 2y + 1 - 1) + 4$

$x = -(y-1)^2 + 1 + 4$

$x = -(y-1)^2 + 5$

vertex: $(5, 1)$

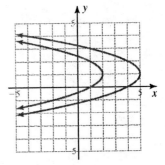

35. Find the vertex to determine the number of items for maximum profit.

$P = -5000 + 8x - 0.001x^2$

$P = -0.001(x^2 - 8000x) - 5000$

$P = -0.001(x^2 - 8000x + 16,000,000 - 16,000,000) - 5000$

$P = -0.001(x - 4000)^2 + 16,000 - 5000$

$P = -0.001(x - 4000)^2 + 11,000$

Vertex: $(4000, 11000)$

The company must produce 4000 items for a profit of $11,000.

37. Find the vertex to determine the amount to be spent on advertising to obtain maximum sales.

$N = 50x - x^2$

$N = -(x^2 - 50x)$

$N = -(x^2 - 50x + 625 - 625)$

$N = -(x - 25)^2 + 625$

Vertex: $(25, 625)$

$25,000 should be spent on advertising to obtain maximum sales.

39. Find the vertex to find the maximum height the ball reaches.

$h = -16t^2 + 160t$

$h = -16(t^2 - 10t)$

$h = -16(t^2 - 10t + 25 - 25)$

$h = -16(t - 5)^2 + 400$

Vertex: $(5, 400)$

The maximum height is 400 ft.

41. Let P = money received from sale of potatoes

Let W = number of weeks elapsed for the sale

$600 + 100W$ = number of bushels of potatoes

$1 - 0.10W$ = the price per bushel of potatoes

Find the vertex to find maximum income

$P = (600 + 100W)(1 - 0.10W)$

$P = -10W^2 + 40W + 600$

$P = -10(W^2 - 4W) + 600$

$P = -10(W^2 - 4W + 4 - 4) + 600$

$P = -10(W - 2)^2 + 40 + 600$

$P = -10(W - 2)^2 + 640$

Vertex: $(2, 640)$

The maximum income occurs at the end of 2 weeks.

45. **a.**
$d = -\dfrac{1}{400}x^2 + x$

$d = -\dfrac{1}{400}(x^2 - 400x)$

$d = -\dfrac{1}{400}(x^2 - 400x + 40000 - 40000)$

$d = -\dfrac{1}{400}(x - 200)^2 + 100$

Vertex: $(200, 100)$

b. The maximum height is 100 ft.

43. **a.**
$R = -\dfrac{1}{98}x^2 + \dfrac{6}{7}x$

$R = -\dfrac{1}{98}(x^2 - 84x)$

$R = -\dfrac{1}{98}(x^2 - 84x + 1764 - 1764)$

$R = -\dfrac{1}{98}(x - 42)^2 + 18$

Vertex: $(42, 18)$

b. The maximum height is 18 in.

c. Since the graph passes through (0, 0) and the axis is $x = 42$, the symmetric point is (84, 0). Horizontal length is 84 in.

d.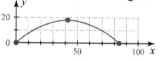

c. Since the graph passes through (0, 0) and the axis is $x = 200$, the symmetric point is (400, 0). Horizontal distance is 400 ft.

d.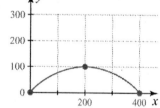

47.
$d = \sqrt{(x_2 - x_1)^2 + (y_2 - y_1)^2}$

$d = \sqrt{(6 - 3)^2 + (8 - 4)^2} = \sqrt{3^2 + 4^2}$

$\quad = \sqrt{9 + 16} = \sqrt{25} = 5$

49.
$d = \sqrt{(x_2 \approx x_1)^2 + (y_2 \approx y_1)^2}$

$d = \sqrt{(4 \approx 2)^2 + (2 \approx (\approx 3))^2} = \sqrt{2^2 + 5^2}$

$\quad = \sqrt{4 + 25} = \sqrt{29} \approx 5.4$

51.
$d = \sqrt{(x - 0)^2 + (y - p)^2} = \sqrt{x^2 + (y - p)^2}$

53.
$\sqrt{x^2 + (y - p)^2} = y + p$

$x^2 + (y - p)^2 = (y + p)^2$

$x^2 + y^2 - 2py + p^2 = y^2 + 2py + p^2$

$x^2 = 4py$

55. The parabola passes through the point (5, 2).

$$5^2 = 4p(2)$$
$$25 = 8p$$
$$3.125 = p$$
$$x^2 = 4(3.125)y = 12.5y$$

Focus: $(0, 3.125)$

57. Sample answer: Given: $f(x) = ax^2 + bx + c$. When the coefficient of the square term is positive, the graph opens upward. When the coefficient of the square term is negative, the graph opens downward.

59. Sample answer: The value of k in the function $f(x) = ax^2 + k$ moves the graph up k units if k is positive and moves the graph down k units if k is negative.

61. Sample answer: The graph of a function has only one y value paired with each x value, so there can be only one y-intercept.

63. $x = y^2 + 2y - 3$
$$x = (y^2 + 2y) - 3$$
$$x = (y^2 + 2y + 1 - 1) - 3$$
$$x = (y+1)^2 - 1 - 3$$
$$x = (y+1)^2 - 4$$

vertex: $(-4, -1)$

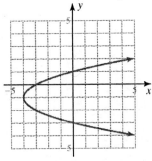

65. $x = 2(y-1)^2 + 3$

vertex: $(3, 1)$

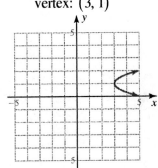

67. $y = -2x^2 - 4x - 3$
$$y = -2(x^2 + 2x) - 3$$
$$y = -2(x^2 + 2x + 1 - 1) - 3$$
$$y = -2(x+1)^2 + 2 - 3$$
$$y = -2(x+1)^2 - 1$$

vertex: $(-1, -1)$

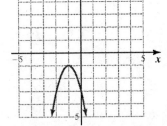

69. $y = (x-2)^2 + 3$

vertex: $(2, 3)$

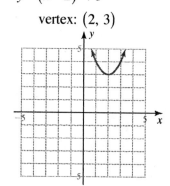

71. $f(x) = -x^2 + 4$

vertex: $(0, 4)$

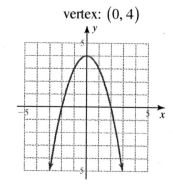

73. $f(x) = 2x^2$

vertex: $(0, 0)$

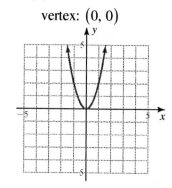

75. $h(x) = -2x^2$

vertex: $(0, 0)$

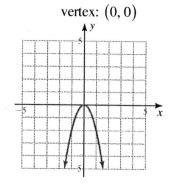

77. Find the vertex to determine the maximum revenue.

$R = 300p - 15p^2$

$R = -15(p^2 - 20p)$

$R = -15(p^2 - 20p + 100 - 100)$

$R = -15(p - 10)^2 + 1500$

Vertex: $(10, 1500)$

The price should be $10 for a maximum revenue of $1500.

9.2   Circles and Ellipses

Problems 9.2

1.   **a.** $d = \sqrt{(8 \cdot 2)^2 + (10 \cdot 2)^2} = \sqrt{6^2 + 8^2} = \sqrt{36 + 64} = \sqrt{100} = 10$

   b. $d = \sqrt{(5 \cdot (\cdot 3))^2 + (4 \cdot 2)^2} = \sqrt{8^2 + 2^2} = \sqrt{64 + 4} = \sqrt{68} = \sqrt{4 \cdot 17} = 2\sqrt{17}$

   c. $d = \sqrt{(\cdot 4 \cdot (\cdot 4))^2 + (7 \cdot 3)^2} = \sqrt{0^2 + 4^2} = \sqrt{0 + 16} = \sqrt{16} = 4$

2. $h = -1,\ k = 2,\ r = 3$

$\left(x - (-1)\right)^2 + (y - 2)^2 = 3^2$

$(x + 1)^2 + (y - 2)^2 = 9$

3. $h = 0,\ k = 0,\ r = \sqrt{3}$

$(x - 0)^2 + (y - 0)^2 = \left(\sqrt{3}\right)^2$

$x^2 + y^2 = 3$

4. $(x - 3)^2 + (y + 5)^2 = 1$

$(x - 3)^2 + \left(y - (-5)\right)^2 = 1^2$

$h = 3,\ k = -5,\ r = 1$

Center: $(3, -5);\quad r = 1$

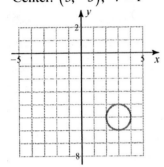

5. $x^2 + y^2 = 7$

$(x \approx 0)^2 + (y \approx 0)^2 = \left(\sqrt{7}\right)^2$

$h = 0,\ k = 0,\ r = \sqrt{7}$

Center: $(0, 0);\quad r = \sqrt{7} \approx 2.6$

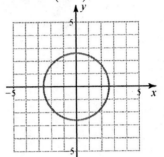

6. $x^2 - 4x + y^2 - 6y + 9 = 0$

$\left(x^2 - 4x\right) + \left(y^2 - 6y\right) = -9$

$\left(x^2 - 4x + 4\right) + \left(y^2 - 6y + 9\right) = -9 + 4 + 9$

$(x - 2)^2 + (y - 3)^2 = 4$

$(x - 2)^2 + (y - 3)^2 = 2^2$

$h = 2,\ k = 3,\ r = 2$

Center: $(2, 3);\quad r = 2$

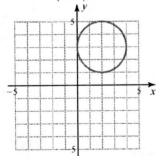

7. $9x^2 + 4y^2 = 36$

$\dfrac{9x^2}{36} + \dfrac{4y^2}{36} = \dfrac{36}{36}$

$\dfrac{x^2}{4} + \dfrac{y^2}{9} = 1$

$h = 0,\ k = 0,\ a = 3,\ b = 2$

Center: $(0, 0);$ x-intercepts: $(2, 0), (-2, 0)$

y-intercepts: $(0, 3), (0, -3)$

vertices: $(0, 3), (0, -3)$

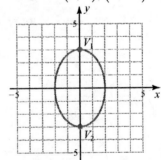

8.

$$\frac{(x+2)^2}{4}+\frac{(y-1)^2}{9}=1$$

$h=-2,\ k=1,\ a=3,\ b=2$

Center: $(-2, 1)$

vertices: $(-2, 4), (-2, -2)$

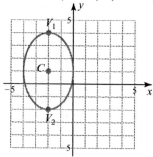

9. $h=0,\ k=0,\ a=2,\ b=1$

$$\frac{x^2}{2^2}+\frac{y^2}{1^2}=1$$

$$\frac{x^2}{4}+\frac{y^2}{1}=1$$

Exercises 9.2

1. $d=\sqrt{(-1-2)^2+(0-4)^2}=\sqrt{(-3)^2+(-4)^2}=\sqrt{9+16}=\sqrt{25}=5$

3. $d=\sqrt{(-1-(-4))^2+(3-(-5))^2}=\sqrt{3^2+8^2}=\sqrt{9+64}=\sqrt{73}$

5. $d=\sqrt{(1-4)^2+(-1-8)^2}=\sqrt{(-3)^2+(-9)^2}=\sqrt{9+81}=\sqrt{90}=3\sqrt{10}$

7. $d=\sqrt{(-2-3)^2+(-1-(-1))^2}=\sqrt{(-5)^2+0^2}=\sqrt{25+0}=\sqrt{25}=5$

9. $d=\sqrt{(-1-(-1))^2+(-4-2)^2}=\sqrt{0^2+(-6)^2}=\sqrt{0+36}=\sqrt{36}=6$

11. $h=3,\ k=8,\ r=2$

$(x-3)^2+(y-8)^2=2^2$

$(x-3)^2+(y-8)^2=4$

13. $h=-3,\ k=4,\ r=5$

$(x-(-3))^2+(y-4)^2=5^2$

$(x+3)^2+(y-4)^2=25$

15. $h=-3,\ k=-2,\ r=4$

$(x-(-3))^2+(y-(-2))^2=4^2$

$(x+3)^2+(y+2)^2=16$

17. $h=2,\ k=-4,\ r=\sqrt{5}$

$(x-2)^2+(y-(-4))^2=(\sqrt{5})^2$

$(x-2)^2+(y+4)^2=5$

19. $h=0,\ k=0,\ r=3$

$(x-0)^2+(y-0)^2=3^2$

$x^2+y^2=9$

21. $(x-1)^2+(y-2)^2=9$

$(x-1)^2+(y-2)^2=3^2$

$h=1,\ k=2,\ r=3$

Center: $(1, 2);\ r=3$

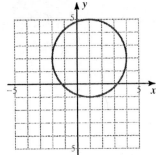

23. $(x+1)^2+(y-2)^2=4$

$(x-(-1))^2+(y-2)^2=2^2$

$h=-1,\ k=2,\ r=2$

Center: $(-1, 2);\ r=2$

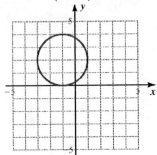

25. $(x-1)^2+(y+2)^2=1$

$(x-1)^2+(y-(-2))^2=1^2$

$h=1,\ k=-2,\ r=1$

Center: $(1, -2);\ r=1$

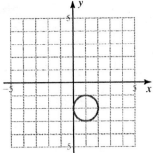

27. $(x+2)^2+(y+1)^2=9$

$(x-(-2))^2+(y-(-1))^2=3^2$

$h=-2,\ k=-1,\ r=3$

Center: $(-2, -1);\ r=3$

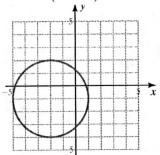

29. $(x\approx 1)^2+(y\approx 1)^2=7$

$(x\approx 1)^2+(y\approx 1)^2=(\sqrt{7})^2$

$h=1,\ k=1,\ r=\sqrt{7}$

Center: $(1, 1);\ r=\sqrt{7}\approx 2.6$

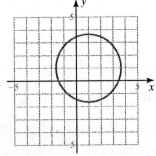

31. $x^2-6x+y^2-4y+9=0$

$(x^2-6x+9)+(y^2-4y+4)=-9+9+4$

$(x-3)^2+(y-2)^2=2^2$

$h=3,\ k=2,\ r=2$

Center: $(3, 2);\ r=2$

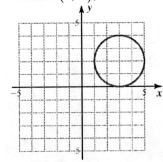

33.
$$x^2 + y^2 - 4x + 2y - 4 = 0$$
$$(x^2 - 4x + 4) + (y^2 + 2y + 1) = 4 + 4 + 1$$
$$(x-2)^2 + (y+1)^2 = 9$$
$$(x-2)^2 + (y-(-1))^2 = 3^2$$
$h = 2, \ k = -1, \ r = 3$
Center: $(2, -1); \ r = 3$

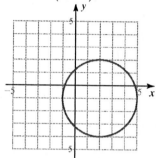

35.
$$x^2 + y^2 - 25 = 0$$
$$x^2 + y^2 = 25$$
$$(x-0)^2 + (y-0)^2 = 5^2$$
$h = 0, \ k = 0, \ r = 5$
Center: $(0, 0); \ r = 5$

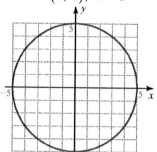

37.
$$x^2 + y^2 \approx 7 = 0$$
$$x^2 + y^2 = 7$$
$$(x \approx 0)^2 + (y \approx 0)^2 = (\sqrt{7})^2$$
$h = 0, \ k = 0, \ r = \sqrt{7}$
Center: $(0, 0); \ r = \sqrt{7} \approx 2.6$

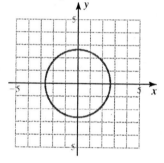

39.
$$x^2 + y^2 + 6x - 2y = -6$$
$$(x^2 + 6x + 9) + (y^2 - 2y + 1) = -6 + 9 + 1$$
$$(x+3)^2 + (y-1)^2 = 4$$
$$(x-(-3))^2 + (y-1)^2 = 2^2$$
$h = -3, \ k = 1, \ r = 2$
Center: $(-3, 1); \ r = 2$

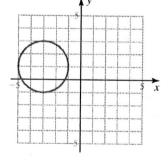

41.
$$x^2 + y^2 - 6x - 2y + 6 = 0$$
$$(x^2 - 6x + 9) + (y^2 - 2y + 1) = -6 + 9 + 1$$
$$(x-3)^2 + (y-1)^2 = 4$$
$$(x-3)^2 + (y-1)^2 = 2^2$$
$h = 3, \ k = 1, \ r = 2$
Center: $(3, 1); \ r = 2$

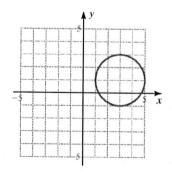

43. $25x^2 + 4y^2 = 100$

$$\frac{25x^2}{100} + \frac{4y^2}{100} = \frac{100}{100}$$

$$\frac{x^2}{4} + \frac{y^2}{25} = 1$$

$h = 0, \ k = 0, \ a = 5, \ b = 2$

Center: $(0, 0)$

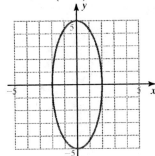

45. $x^2 + 4y^2 = 4$

$$\frac{x^2}{4} + \frac{4y^2}{4} = \frac{4}{4}$$

$$\frac{x^2}{4} + \frac{y^2}{1} = 1$$

$h = 0, \ k = 0, \ a = 2, \ b = 1$

Center: $(0, 0)$

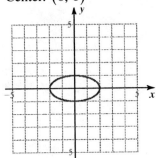

47. $x^2 + 4y^2 = 16$

$$\frac{x^2}{16} + \frac{4y^2}{16} = \frac{16}{16}$$

$$\frac{x^2}{16} + \frac{y^2}{4} = 1$$

$h = 0, \ k = 0, \ a = 4, \ b = 2$

Center: $(0, 0)$

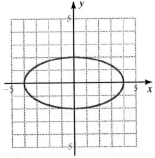

49. $\dfrac{x^2}{9} + \dfrac{y^2}{16} = 1$

$h = 0, \ k = 0, \ a = 4, \ b = 3$

Center: $(0, 0)$

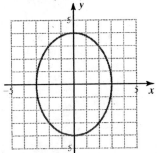

51. $\dfrac{(x-1)^2}{4} + \dfrac{(y-2)^2}{9} = 1$

$h = 1, \ k = 2, \ a = 3, \ b = 2$

Center: $(1, 2)$

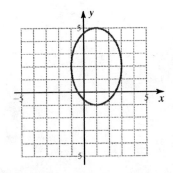

53.
$$\frac{(x-2)^2}{9}+\frac{(y+3)^2}{4}=1$$

$$\frac{(x-2)^2}{9}+\frac{(y-(-3))^2}{4}=1$$

$h=2,\ k=-3,\ a=3,\ b=2$

Center: $(2,-3)$

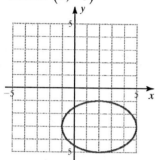

55.
$$\frac{(x-1)^2}{16}+\frac{(y-1)^2}{9}=1$$

$h=1,\ k=1,\ a=4,\ b=3$

Center: $(1,1)$

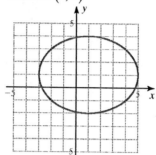

57. Find the distance from the center to the given point to determine the radius:

$$r=\sqrt{(4-0)^2+(3-0)^2}=\sqrt{4^2+3^2}$$
$$=\sqrt{16+9}=\sqrt{25}=5$$
$h=0,\ k=0,\ r=5$
$$(x-0)^2+(y-0)^2=5^2$$
$$x^2+y^2=25$$

59. Find the distance from the center to the given point to determine the radius:

$$r=\sqrt{((-5)-0)^2+((-12)-0)^2}$$
$$=\sqrt{(-5)^2+(-12)^2}=\sqrt{25+144}=\sqrt{169}=13$$
$h=0,\ k=0,\ r=13$
$$(x-0)^2+(y-0)^2=13^2$$
$$x^2+y^2=169$$

61. Find the distance from the center to the given point to determine the radius:

$$r=\sqrt{(0-0)^2+(3-0)^2}=\sqrt{0^2+3^2}$$
$$=\sqrt{0+9}=\sqrt{9}=3$$
$h=0,\ k=0,\ r=3$
$$(x-0)^2+(y-0)^2=3^2$$
$$x^2+y^2=9$$

63. Since a is determined by the y-intercepts, the equation is of the form $\dfrac{x^2}{b^2}+\dfrac{y^2}{a^2}=1$.

$h=0,\ k=0,\ a=6,\ b=2$

$$\frac{x^2}{2^2}+\frac{y^2}{6^2}=1$$

$$\frac{x^2}{4}+\frac{y^2}{36}=1$$

65. Since a is determined by the x-intercepts, the equation is of the form $\dfrac{x^2}{a^2}+\dfrac{y^2}{b^2}=1$.

$h=0,\ k=0,\ a=6,\ b=4$

$$\frac{x^2}{6^2}+\frac{y^2}{4^2}=1$$

$$\frac{x^2}{36}+\frac{y^2}{16}=1$$

67. Place a coordinate system on the drum so that the origin is at the center of the circular end. If r is the radius of the drum, then the equation of the circle is $x^2 + y^2 = r^2$ and $(10, r-5)$ is a point on the circle. Solving for r, we have:

$$x^2 + y^2 = r^2$$
$$10^2 + (r-5)^2 = r^2$$
$$100 + r^2 - 10r + 25 = r^2$$
$$-10r = -125$$
$$r = 12.5$$

The radius of the drum is 12.5 in.

69. The equation of the circle is $x^2 + y^2 = 15^2$. Substituting the pair $(x, 10)$ into the equation and solving, we have:

$$x^2 + y^2 = 15^2$$
$$x^2 + 10^2 = 15^2$$
$$x^2 + 100 = 225$$
$$x^2 = 125$$
$$x = \sqrt{125} = 5\sqrt{5}$$

Since the horizontal bar is $2x$ in length, the bar is $2(5\sqrt{5}) = 10\sqrt{5} \approx 22.4$ ft long.

71.

The equation is of the form $\dfrac{x^2}{a^2} + \dfrac{y^2}{b^2} = 1$.

$h = 0, \ k = 0, \ a = 4, \ b = 2.5$

$\dfrac{x^2}{4^2} + \dfrac{y^2}{2.5^2} = 1$ or $\dfrac{x^2}{16} + \dfrac{y^2}{6.25} = 1$

73.

The equation is of the form $\dfrac{x^2}{a^2} + \dfrac{y^2}{b^2} = 1$.

$h = 0, \ k = 0, \ a = 4, \ b = 3$

$\dfrac{x^2}{4^2} + \dfrac{y^2}{3^2} = 1$ or $\dfrac{x^2}{16} + \dfrac{y^2}{9} = 1$

75. a. $2a = 94.5 + 91.5$

$2a = 186$

$a = 93$ million miles

b. $c = 93 - 91.5 = 1.5$ million miles

c. $b^2 = a^2 - c^2$

$b = \sqrt{a^2 - c^2}$

$b = \sqrt{93^2 - 1.5^2} = \sqrt{8649 - 2.25}$

$= \sqrt{8646.75} = 92.99$ million miles

77. a. $h = 0, \ k = 0, \ a = 6, \ b = 4.5$

$\dfrac{x^2}{6^2} + \dfrac{y^2}{4.5^2} = 1$ or $\dfrac{x^2}{36} + \dfrac{y^2}{20.25} = 1$

b. Solve for y when $x = 4$ in.

$$\dfrac{4^2}{6^2} + \dfrac{y^2}{20.25} = 1$$

$$\dfrac{y^2}{20.25} = 1 \cdot \dfrac{16}{36} = \dfrac{20}{36} = \dfrac{5}{9}$$

$$y^2 = \dfrac{5}{9} \cdot 20.25 = 11.25 = \dfrac{45}{4}$$

$$y = \dfrac{3\sqrt{5}}{2}$$

$$2y = \dfrac{6\sqrt{5}}{2}$$

The width is $\dfrac{6\sqrt{5}}{2} \approx 6.71$ in.

79.

The equation of the bridge is $\dfrac{x^2}{625}+\dfrac{y^2}{400}=1$.

Let $y = 15$ and solve for x:

$$\frac{x^2}{625}+\frac{15^2}{400}=1$$

$$\frac{x^2}{625}=1\cdot\frac{225}{400}=\frac{175}{400}=\frac{7}{16}$$

$$x^2=\frac{7}{16}\cdot 625$$

$$x=\sqrt{\frac{625\cdot 7}{16}}=\frac{25\sqrt{7}}{4}$$

The distance from the right side of the river is 4 ft less than (because of the width of the boat) the difference of 25 and the x-coordinate, so the distance is: $21-\dfrac{25\sqrt{7}}{4}$ ft

or approximately 4.5 ft.

81. When the answer is positive, the x-coordinate is positive. Thus the graph would be the right half of the circle.

83. $d_1 = PF_1 = \sqrt{(x-c)^2+(y-0)^2}=\sqrt{(x-c)^2+y^2}$

85.

$$\sqrt{(x+c)^2+y^2}=2a-\sqrt{(x-c)^2+y^2}$$

$$\left(\sqrt{(x+c)^2+y^2}\right)^2=\left(2a-\sqrt{(x-c)^2+y^2}\right)^2$$

$$(x+c)^2+y^2=4a^2-4a\sqrt{(x-c)^2+y^2}+(x-c)^2+y^2$$

$$x^2+2cx+c^2+y^2=4a^2-4a\sqrt{(x-c)^2+y^2}+x^2-2cx+c^2+y^2$$

$$4cx-4a^2=-4a\sqrt{(x-c)^2+y^2}$$

$$a^2-cx=a\sqrt{(x-c)^2+y^2}$$

87. $x^2\left(a^2-c^2\right)+a^2y^2=a^2\left(a^2-c^2\right)$

Let $a^2-c^2=b^2$

$$b^2x^2+a^2y^2=a^2b^2$$

89. Sample answer: When $a > b$, the graph of the ellipse will be elongated in the horizontal direction.

91. Sample answer: When $a = b$, the graph will become a circle with radius a.

93. $\dfrac{(x+3)^2}{4}+\dfrac{(y+1)^2}{9}=1$

$h=-3,\ k=-1,\ a=3,\ b=2$

Center: $(-3,-1)$

vertices: $(-3,2),(-3,-4)$

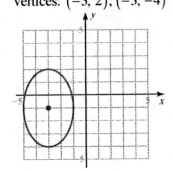

95. $4x^2+9y^2=36$

$\dfrac{4x^2}{36}+\dfrac{9y^2}{36}=\dfrac{36}{36}$

$\dfrac{x^2}{9}+\dfrac{y^2}{4}=1$

$h=0,\ k=0,\ a=3,\ b=2$

Center: $(0,0)$; x-intercepts: $(3,0),(-3,0)$

y-intercepts: $(0,2),(0,-2)$

vertices: $(3,0),(-3,0)$

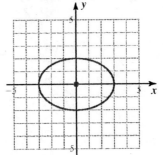

97. $x^2-6x+y^2-4y+9=0$

$\left(x^2-6x+9\right)+\left(y^2-4y+4\right)=-9+9+4$

$(x-3)^2+(y-2)^2=4$

$(x-3)^2+(y-2)^2=2^2$

$h=3,\ k=2,\ r=2$

Center: $(3,2)$; $r=2$

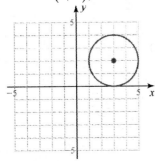

99. $x^2+y^2=4$

$(x-0)^2+(y-0)^2=2^2$

$h=0,\ k=0,\ r=2$

Center: $(0,0)$; $r=2$

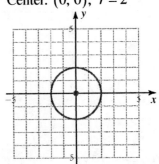

101. $(x-3)^2+(y-1)^2=4$

$(x-3)^2+(y-1)^2=2^2$

$h=3,\ k=1,\ r=2$

Center: $(3,1)$; $r=2$

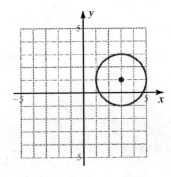

103. $h = 0, \; k = 0, \; r = 5$

$$(x-0)^2 + (y-0)^2 = 5^2$$
$$x^2 + y^2 = 25$$

105. $h = -3, \; k = 6, \; r = 3$

$$(x-(-3))^2 + (y-6)^2 = 3^2$$
$$(x+3)^2 + (y-6)^2 = 9$$

9.3   Hyperbolas and Identification of Conics

Problems 9.3

1.   **a.** $\dfrac{y^2}{16} - \dfrac{x^2}{9} = 1$

Center: $(0, 0)$

y^2 term is positive, so $a = 4$, $b = 3$

Vertices: $(0, 4), (0, -4)$

Auxiliary rectangle passes through

$y = \pm 4$ and $x = \pm 3$

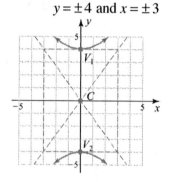

b. $\dfrac{x^2}{16} - \dfrac{y^2}{9} = 1$

Center: $(0, 0)$

x^2 term is positive, so $a = 4$, $b = 3$

Vertices: $(4, 0), (-4, 0)$

Auxiliary rectangle passes through

$x = \pm 4$ and $y = \pm 3$

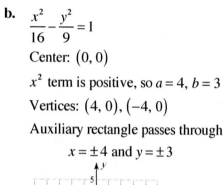

2. $\dfrac{(x+1)^2}{9} - \dfrac{(y+3)^2}{9} = 1$

Center: $(-1. \; -3)$

x^2 term is positive, so $a = 3$, $b = 3$

Vertices: $(-4, -3), (2, -3)$

The auxiliary rectangle has its sides on

the lines $y = 0$, $y = -6$, $x = 2$ and $x = -4$.

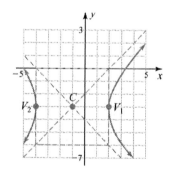

3. a.
$$4x^2 = 36 + 9y^2$$
$$4x^2 - 9y^2 = 36$$
Minus sign indicates a hyperbola.
$$\frac{4x^2}{36} - \frac{9y^2}{36} = 1$$
$$\frac{x^2}{9} - \frac{y^2}{4} = 1$$
Center: $(0, 0)$
x^2 term is positive, so $a = 3$, $b = 2$
Vertices: $(3, 0), (-3, 0)$

b. $y = x^2 + 3$
Only one variable is squared, thus the conic is a parabola.
Vertex: $(0, 3)$, opens upward

c.
$$y^2 = 9 - x^2$$
$$x^2 + y^2 = 9$$
Square terms have same positive
coefficient, thus the conic is a circle.
Center: $(0, 0)$; $r = 3$

d.
$$9x^2 = 36 - 4y^2$$
$$9x^2 + 4y^2 = 36$$
Square terms are different positive
numbers, thus the conic is an ellipse.
$$\frac{9x^2}{36} + \frac{4y^2}{36} = 1$$
$$\frac{x^2}{4} + \frac{y^2}{9} = 1$$
Center: $(0, 0)$; $a = 3$, $b = 2$
x-intercepts: $(2, 0), (-2, 0)$
y-intercepts: $(0, 3), (0, -3)$

Exercises 9.3

1.
$$\frac{x^2}{25} - \frac{y^2}{9} = 1$$
Center: $(0, 0)$
x^2 term is positive, so $a = 5$, $b = 3$
Vertices: $(5, 0), (-5, 0)$
Auxiliary rectangle passes through
$x = \pm 5$ and $y = \pm 3$

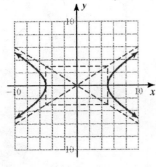

3.
$$\frac{y^2}{9} - \frac{x^2}{9} = 1$$
Center: $(0, 0)$
y^2 term is positive, so $a = 3$, $b = 3$
Vertices: $(0, 3), (0, -3)$
Auxiliary rectangle passes through
$x = \pm 3$ and $y = \pm 3$

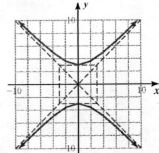

5. $\dfrac{x^2}{9} - \dfrac{y^2}{1} = 1$

Center: $(0, 0)$

x^2 term is positive, so $a = 3$, $b = 1$

Vertices: $(3, 0), (-3, 0)$

Auxiliary rectangle passes through

$$x = \pm 3 \text{ and } y = \pm 1$$

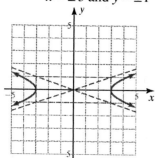

7. $\dfrac{x^2}{64} - \dfrac{y^2}{49} = 1$

Center: $(0, 0)$

x^2 term is positive, so $a = 8$, $b = 7$

Vertices: $(8, 0), (-8, 0)$

Auxiliary rectangle passes through

$$x = \pm 8 \text{ and } y = \pm 7$$

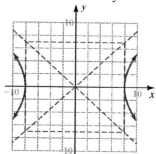

9. $\dfrac{y^2}{\frac{16}{9}} - \dfrac{x^2}{\frac{9}{16}} = 1$

Center: $(0, 0)$

y^2 term is positive, so $a = \frac{4}{3}$, $b = \frac{3}{4}$

Vertices: $\left(0, \frac{4}{3}\right), \left(0, -\frac{4}{3}\right)$

Auxiliary rectangle passes through

$$x = \pm \tfrac{3}{4} \text{ and } y = \pm \tfrac{4}{3}$$

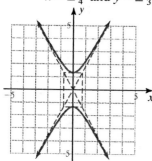

11. $y^2 - 9x^2 = 9$

$\dfrac{y^2}{9} - \dfrac{9x^2}{9} = \dfrac{9}{9}$

$\dfrac{y^2}{9} - \dfrac{x^2}{1} = 1$

Center: $(0, 0)$

y^2 term is positive, so $a = 3$, $b = 1$

Vertices: $(0, 3), (0, -3)$

Auxiliary rectangle passes through

$$x = \pm 1 \text{ and } y = \pm 3$$

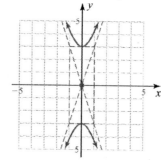

13. $\dfrac{(x-1)^2}{4} - \dfrac{(y+1)^2}{9} = 1$

Center: $(1, -1)$

x^2 term is positive, so $a = 2$, $b = 3$

Vertices: $(3, -1), (-1, -1)$

The auxiliary rectangle has its sides on the
lines $y = 2$, $y = -4$, $x = 3$ and $x = -1$.

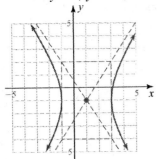

15. $\dfrac{(y-1)^2}{9} - \dfrac{(x-2)^2}{4} = 1$

Center: $(2, 1)$

y^2 term is positive, so $a = 3$, $b = 2$

Vertices: $(2, 4), (2, -2)$

The auxiliary rectangle has its sides on the
lines $y = 4$, $y = -2$, $x = 4$ and $x = 0$.

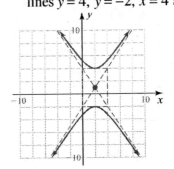

17. $x^2 + y^2 = 25$

Square terms have same positive
coefficient, thus the conic is a circle.

Center: $(0, 0)$; $r = 5$

Intercepts: $(5, 0), (-5, 0), (0, 5), (0, -5)$

19. $x^2 - y^2 = 36$

Minus sign indicates a hyperbola.

$\dfrac{x^2}{36} - \dfrac{y^2}{36} = 1$

Center: $(0, 0)$

x^2 term is positive, so $a = 6$, $b = 6$

Intercepts: $(6, 0), (-6, 0)$

21. $x^2 - y = 9$

$\qquad y = x^2 - 9$

Only one variable is squared, thus the conic
is a parabola.

Vertex: $(0, -9)$

23. $y^2 - x = 4$

$\qquad x = y^2 - 4$

Only one variable is squared, thus the conic
is a parabola.

Vertex: $(-4, 0)$

25. $\qquad 9x^2 = 36 - 9y^2$

$9x^2 + 9y^2 = 36$

$\qquad x^2 + y^2 = 4$

Square terms have same positive
coefficient, thus the conic is a circle.

Center: $(0, 0)$; $r = 2$

Intercepts: $(2, 0), (-2, 0), (0, 2), (0, -2)$

27. $9x^2 = 36 + 9y^2$

$9x^2 - 9y^2 = 36$

$x^2 - y^2 = 4$

Minus sign indicates a hyperbola.

$\dfrac{x^2}{4} - \dfrac{y^2}{4} = 1$

Center: $(0, 0)$

x^2 term is positive, so $a = 2$, $b = 2$

Intercepts: $(2, 0), (-2, 0)$

29.
$$x^2 = 9 - 9y^2$$
$$x^2 + 9y^2 = 9$$

Square terms are different positive
  numbers, thus the conic is an ellipse.
$$\frac{x^2}{9} + \frac{9y^2}{9} = 1$$
$$\frac{x^2}{9} + \frac{y^2}{1} = 1$$
Center: $(0, 0)$; $a = 3, b = 1$

Intercepts: $(3, 0), (-3, 0), (0, 1), (0, -1)$

31. a. Hyperbola (minus sign)

b. $\dfrac{D^2}{8} \approx \dfrac{d^2}{4} = 1$

Center: $(0, 0)$

D^2 term is positive, so $a = \sqrt{8} \approx 2.8, b = 2$

Intercepts: $(2.8, 0), (\approx 2.8, 0)$

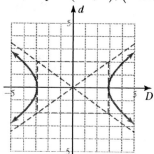

33. $4v^2 + 9-^2 = 144$

Square terms are different positive
  numbers, thus the conic is an ellipse.
$$\frac{4v^2}{144} + \frac{9-^2}{144} = 1$$
$$\frac{v^2}{36} + \frac{-^2}{16} = 1$$
Center: $(0, 0)$; $a = 6, b = 4$

Intercepts: $(6, 0), (-6, 0), (0, 4), (0, -4)$

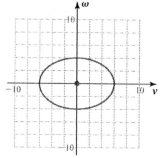

35. $y = x - 4$

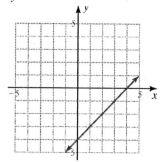

37. $y = x^2 + 1$

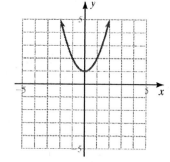

39.
$$4x^2 + 9y^2 = 36$$
$$\frac{4x^2}{36} + \frac{9y^2}{36} = \frac{36}{36}$$
$$\frac{x^2}{9} + \frac{y^2}{4} = 1$$
Center: $(0, 0)$; $a = 3, b = 2$

Intercepts: $(3, 0), (-3, 0), (0, 2), (0, -2)$

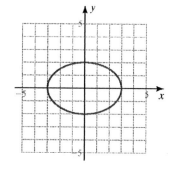

41.
$$d_1 = PF_1 = \sqrt{(x-c)^2 + (y-0)^2} = \sqrt{(x-c)^2 + y^2}$$

43.
$$\sqrt{(x-c)^2 + y^2} = 2a + \sqrt{(x+c)^2 + y^2}$$
$$\left(\sqrt{(x-c)^2 + y^2}\right)^2 = \left(2a + \sqrt{(x+c)^2 + y^2}\right)^2$$
$$(x-c)^2 + y^2 = 4a^2 + 4a\sqrt{(x+c)^2 + y^2} + (x+c)^2 + y^2$$
$$x^2 - 2cx + c^2 + y^2 = 4a^2 + 4a\sqrt{(x+c)^2 + y^2} + x^2 + 2cx + c^2 + y^2$$
$$-4cx - 4a^2 = 4a\sqrt{(x+c)^2 + y^2}$$
$$a^2 + cx = -a\sqrt{(x+c)^2 + y^2}$$

45.
$$x^2(c^2 - a^2) - a^2 y^2 = a^2(c^2 - a^2)$$
Let $c^2 - a^2 = b^2$
$$b^2 x^2 - a^2 y^2 = a^2 b^2$$

47.
$$\frac{x^2}{a^2} \mid \frac{y^2}{b^2} = 1$$
$$\left|\frac{x}{a} + \frac{y}{b}\right| \left|\frac{x}{a} \mid \frac{y}{b}\right| = 1$$

49. The denominator becomes very large. As a result the fraction approaches a value of zero.

51.
$$\frac{-x}{-a} + \frac{y-}{b-} = 0$$
$$\frac{y}{b} = -\frac{x}{a}$$
$$y = -\frac{b}{a} x$$

53. Sample answer: If the equation is a difference of two squares, the graph of the equation is a hyperbola. If the equation is a sum of two squares with different coefficients, the graph of the equation is an ellipse.

55. **a.** Sample answer: The graph of the equation is a circle, if $A = B$.
    b. Sample answer; The graph of the equation is an ellipse, if $AB > 0$.
    c. Sample answer: The graph of the equation is a hyperbola, if $AB < 0$.

57.
$$y^2 = 9 - x^2$$
$$x^2 + y^2 = 9$$
Square terms have same positive coefficient, thus the conic is a circle.

59.
$$y = x^2 + 3$$
Only one variable is squared, thus the conic is a parabola.

61.

$$\frac{y^2}{16} - \frac{x^2}{9} = 1$$

Center: $(0, 0)$

y^2 term is positive, so $a = 4$, $b = 3$

Vertices: $(0, 4), (0, -4)$

Auxiliary rectangle passes through

$$y = \pm 4 \text{ and } x = \pm 3$$

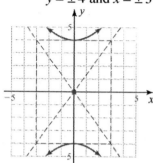

63.

$$9x^2 - 25y^2 = 225$$

$$\frac{9x^2}{225} - \frac{25y^2}{225} = \frac{225}{225}$$

$$\frac{x^2}{25} - \frac{y^2}{9} = 1$$

Center: $(0, 0)$

x^2 term is positive, so $a = 5$, $b = 3$

Vertices: $(5, 0), (-5, 0)$

Auxiliary rectangle passes through

$$x = \pm 5 \text{ and } y = \pm 3$$

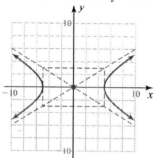

9.4   Nonlinear Systems of Equations

Problems 9.4

1. Solve the first equation for y, substitute into the second equation and solve:

$$x + y = 4 \quad - \quad y = 4 - x$$

$$x^2 + y^2 = 16$$

$$x^2 + (4 - x)^2 = 16$$

$$x^2 + 16 - 8x + x^2 = 16$$

$$2x^2 - 8x = 0$$

$$2x(x - 4) = 0$$

$$2x = 0 \qquad \text{or } x - 4 = 0$$

$$x = 0 \qquad \text{or} \qquad x = 4$$

$$y = 4 - 0 = 4 \text{ or } y = 4 - 4 = 0$$

The solutions are $(0, 4)$ and $(4, 0)$.

2. Solve the first equation for y, substitute into the second equation and solve:

$$x + y = 3 \quad - \quad y = 3 - x$$

$$x^2 + y^2 = 3$$

$$x^2 + (3 - x)^2 = 3$$

$$x^2 + 9 - 6x + x^2 = 3$$

$$2x^2 - 6x + 6 = 0$$

$$x^2 - 3x + 3 = 0$$

$$x = \frac{|(|3) \pm \sqrt{(|3)^2 |4 |\, 3}}{2 \,|} = \frac{3 \pm \sqrt{9 \,|\, 12}}{2}$$

$$= \frac{3 \pm \sqrt{|3}}{2} = \frac{3}{2} \pm \frac{\sqrt{3}}{2} i$$

$$x = \frac{3}{2} + \frac{\sqrt{3}}{2} i \qquad \text{or } x = \frac{3}{2} |\, \frac{\sqrt{3}}{2} i$$

$$y = 3 \,|\, \left| \frac{3}{2} + \frac{\sqrt{3}}{2} i \right| \text{ or } y = 3 \,|\, \left| \frac{3}{2} |\, \frac{\sqrt{3}}{2} i \right|$$

$$y = \frac{3}{2} |\, \frac{\sqrt{3}}{2} i \qquad \text{or } y = \frac{3}{2} + \frac{\sqrt{3}}{2} i$$

The solutions are $\left| \frac{3}{2} + \frac{\sqrt{3}}{2} i, \frac{3}{2} |\, \frac{\sqrt{3}}{2} i \right|$

and $\left| \frac{3}{2} |\, \frac{\sqrt{3}}{2} i, \frac{3}{2} + \frac{\sqrt{3}}{2} i \right|$.

3.
$$x^2 + y^2 = 9$$
$$x^2 - 9y^2 = 9$$

To eliminate y^2, multiply the first equation by 9 and add the result to the second equation:

$$9x^2 + 9y^2 = 81$$
$$\underline{x^2 - 9y^2 = 9}$$
$$10x^2 = 90$$
$$x^2 = 9$$
$$x = \pm 3$$

Substitute into the original first equation and solve for y:

$$(\pm 3)^2 + y^2 = 9$$
$$9 + y^2 = 9$$
$$y^2 = 0$$
$$y = 0$$

The solutions are $(3, 0)$ and $(-3, 0)$.

4. Find x when $C = R$:

$$40x - 500 = 135x - x^2$$

$$x^2 - 95x - 500 = 0$$

$$(x - 100)(x + 5) = 0$$

$$x - 100 = 0 \quad \text{or } x + 5 = 0$$

$$x = 100 \text{ or } \qquad x = -5$$

100 units must be manufactured and sold to break even.

5. Let $L =$ the length of the picture frame.
Let $W =$ the width of the picture frame.
The area is $A = L \cdot W = 88$.
The perimeter is $P = 2L + 2W = 38$.
Solve the area equation for W and substitute into the perimeter equation:

$$W = \frac{88}{L}$$

$$2L + 2 \frac{-88-}{-L-} = 38$$

$$2L^2 + 176 = 38L$$

$$L^2 - 19L + 88 = 0$$

$$(L - 11)(L - 8) = 0$$

$$L = 11 \qquad \text{or } L = 8$$

$$W = \frac{88}{11} = 8 \text{ or } W = \frac{88}{8} = 11$$

The dimensions of the picture frame are 11 in by 8 in.

Exercises 9.4

1. Solve the second equation for y, substitute into the first equation and solve:
$$x^2 + y^2 = 16$$
$$x + y = 4 \; - \quad y = 4 - x$$
$$x^2 + (4 - x)^2 = 16$$
$$x^2 + 16 - 8x + x^2 = 16$$
$$2x^2 - 8x = 0$$
$$2x(x - 4) = 0$$
$$x = 0 \qquad \text{or } x = 4$$
$$y = 4 - 0 = 4 \text{ or } y = 4 - 4 = 0$$
The solutions are $(0, 4)$ and $(4, 0)$.

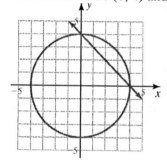

3. Solve the second equation for y, substitute into the first equation and solve:
$$x^2 + y^2 = 25$$
$$y - x = 5 \;\rightarrow\; y = x + 5$$
$$x^2 + (x + 5)^2 = 25$$
$$x^2 + x^2 + 10x + 25 = 25$$
$$2x^2 + 10x = 0$$
$$2x(x + 5) = 0$$
$$x = 0 \qquad \text{or } x = -5$$
$$y = 0 + 5 = 5 \text{ or } y = -5 + 5 = 0$$
The solutions are $(0, 5)$ and $(-5, 0)$.

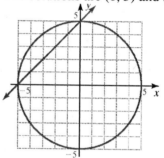

5. Solve the second equation for y, substitute into the first equation and solve:
$$x^2 + y^2 = 25$$
$$y - x = 1 \;\rightarrow\; y = x + 1$$
$$x^2 + (x + 1)^2 = 25$$
$$x^2 + x^2 + 2x + 1 = 25$$
$$2x^2 + 2x - 24 = 0$$
$$x^2 + x - 12 = 0$$
$$(x - 3)(x + 4) = 0$$
$$x = 3 \qquad \text{or } x = -4$$
$$y = 3 + 1 = 4 \text{ or } y = -4 + 1 = -3$$
The solutions are $(3, 4)$ and $(-4, -3)$.

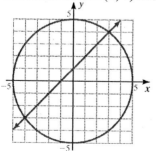

7. Solve the second equation for y, substitute into the first equation and solve:
$$y = x^2 - 5x + 4$$
$$x - y = 1 \;\rightarrow\; y = x - 1$$
$$x - 1 = x^2 - 5x + 4$$
$$0 = x^2 - 6x + 5$$
$$0 = (x - 5)(x - 1)$$
$$x = 5 \qquad \text{or } x = 1$$
$$y = 5 - 1 = 4 \text{ or } y = 1 - 1 = 0$$
The solutions are $(5, 4)$ and $(1, 0)$.

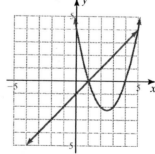

9. Substitute the first equation into the second equation and solve:

$$y = (x-1)^2$$
$$y - x = 1$$
$$(x-1)^2 - x = 1$$
$$x^2 - 2x + 1 - x = 1$$
$$x^2 - 3x = 0$$
$$x(x-3) = 0$$
$$x = 0 \text{ or } x = 3$$
$$y = (0-1)^2 = 1 \text{ or } y = (3-1)^2 = 4$$

The solutions are (0, 1) and (3, 4).

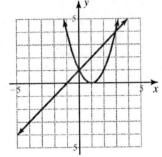

11. Solve the second equation for y, substitute into the first equation and solve:

$$4x^2 + 9y^2 = 36$$
$$3y \mid 2x = 6 \mid y = \frac{2}{3}x + 2$$
$$4x^2 + 9\left|\frac{2}{3}x + 2\right|^2 = 36$$
$$4x^2 + 9\left|\frac{4}{9}x^2 + \frac{8}{3}x + 4\right| = 36$$
$$4x^2 + 4x^2 + 24x + 36 = 36$$
$$8x^2 + 24x = 0$$
$$8x(x+3) = 0$$
$$x = 0 \text{ or } x = \cdot\ 3$$
$$y = \frac{2}{3}\cdot 0 + 2 = 2 \text{ or } y = \frac{2}{3}(\cdot\ 3) + 2 = 0$$

The solutions are (0, 2) and (–3, 0).

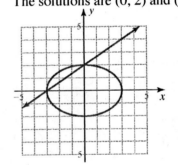

13. Solve the second equation for x, substitute into the first equation and solve:

$$x^2 - y^2 = 16$$
$$x + 4y = 4 \rightarrow x = 4 - 4y$$
$$(4 - 4y)^2 - y^2 = 16$$
$$16 - 32y + 16y^2 - y^2 = 16$$
$$15y^2 - 32y = 0$$
$$y(15y - 32) = 0$$
$$y = 0 \text{ or } y = \frac{32}{15}$$

If $y = 0$, $x = 4 \cdot\ 4 \cdot 0 = 4$

If $y = \frac{32}{15}$, $x = 4 \cdot\ 4 \cdot \frac{32}{15} = \frac{60}{15} \cdot \frac{128}{15} = \cdot\ \frac{68}{15}$

The solutions are $(4, 0)$ and $\left|\left|\frac{68}{15}, \frac{32}{15}\right|\right|$.

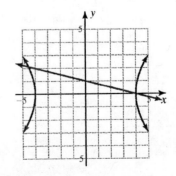

15. Solve the second equation for y, substitute into the first equation and solve:

$$x^2 + y^2 = 4$$

$$y - x = 5 \;\rightarrow\; y = x + 5$$

$$x^2 + (x+5)^2 = 4$$

$$x^2 + x^2 + 10x + 25 = 4$$

$$2x^2 + 10x + 21 = 0$$

$$x = \frac{-10 \pm \sqrt{10^2 - 4 \cdot 2 \cdot 21}}{2 \cdot 2} = \frac{-10 \pm \sqrt{100 - 168}}{4} = \frac{-10 \pm \sqrt{-68}}{4} = -\frac{5}{2} \pm \frac{\sqrt{17}}{2}i$$

If $x = -\dfrac{5}{2} + \dfrac{\sqrt{17}}{2}i, \;\; y = -\dfrac{5}{2} + \dfrac{\sqrt{17}}{2}i + 5 = \dfrac{5}{2} + \dfrac{\sqrt{17}}{2}i$

If $x = -\dfrac{5}{2} - \dfrac{\sqrt{17}}{2}i, \;\; y = -\dfrac{5}{2} - \dfrac{\sqrt{17}}{2}i + 5 = \dfrac{5}{2} - \dfrac{\sqrt{17}}{2}i$

The solutions are $\left(-\dfrac{5}{2} + \dfrac{\sqrt{17}}{2}i, \dfrac{5}{2} + \dfrac{\sqrt{17}}{2}i \right)$ and $\left(-\dfrac{5}{2} - \dfrac{\sqrt{17}}{2}i, \dfrac{5}{2} - \dfrac{\sqrt{17}}{2}i \right)$.

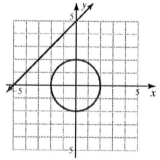

17. Substitute the second equation into the first equation and solve:

$$y = 4 - x^2$$

$$y = x^2 - 4$$

$$x^2 - 4 = 4 - x^2$$

$$2x^2 = 8$$

$$x^2 = 4$$

$$x = \pm 2$$

$$y = 4 - (\pm 2)^2 = 4 - 4 = 0$$

The solutions are $(2, 0)$ and $(-2, 0)$.

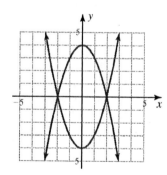

19. $x^2 + y^2 = 25$

$x^2 - y^2 = 7$

To eliminate y^2, add the first equation to the second equation and solve:

$$x^2 + y^2 = 25$$
$$\underline{x^2 - y^2 = 7}$$
$$2x^2 \quad\quad = 32$$
$$x^2 = 16$$
$$x = \pm 4$$

Substitute into the first equation and solve for y:

$$(\pm 4)^2 + y^2 = 25$$
$$16 + y^2 = 25$$
$$y^2 = 9$$
$$y = \pm 3$$

The solutions are (4, 3), (–4, 3), (4, –3), and (–4, –3).

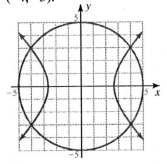

21. $x^2 + \quad y^2 = 16$

$x^2 + 16y^2 = 16$

To eliminate x^2, multiply the first equation by –1 and add to the second equation and solve:

$$-x^2 - \quad y^2 = -16$$
$$\underline{x^2 + 16y^2 = \quad 16}$$
$$15y^2 = \quad 0$$
$$y^2 = 0$$
$$y = 0$$

Substitute into the original first equation and solve for x:

$$x^2 + 0^2 = 16$$
$$x^2 = 16$$
$$x = \pm 4$$

The solutions are (4, 0) and (–4, 0).

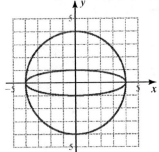

23. $3x^2 - y^2 = 2$

$x^2 + 2y^2 = 3$

To eliminate y^2, multiply the first equation by 2 and add to the second equation and solve:

$$6x^2 - 2y^2 = 4$$
$$\underline{x^2 + 2y^2 = 3}$$
$$7x^2 \quad\quad = 7$$
$$x^2 = 1$$
$$x = \pm 1$$

Substitute into the original first equation and solve for y:

$$3(\pm 1)^2 - y^2 = 2$$
$$1 = y^2$$
$$\pm 1 = y$$

The solutions are (1, 1), (1, –1), (–1, 1) and (–1, –1).

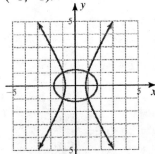

25.
$$x^2 + 2y^2 = 11$$
$$2x^2 + y^2 = 19$$

To eliminate x^2, multiply the first equation by -2, add to the second equation and solve:

$$-2x^2 - 4y^2 = -22$$
$$\underline{2x^2 + y^2 = 19}$$
$$-3y^2 = -3$$
$$y^2 = 1$$
$$y = \pm 1$$

Substitute into the original first equation and solve for x:

$$x^2 + 2(\pm 1)^2 = 11$$
$$x^2 = 9$$
$$x = \pm 3$$

The solutions are $(3, 1)$, $(3, -1)$, $(-3, 1)$, and $(-3, -1)$.

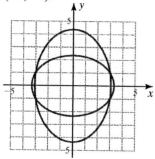

27.
$$x^2 + y^2 = 4$$
$$x^2 - y^2 = 9$$

To eliminate y^2, add the first equation to the second equation and solve:

$$x^2 + y^2 = 4$$
$$\underline{x^2 - y^2 = 9}$$
$$2x^2 = 13$$
$$x^2 = \frac{13}{2} \rightarrow x = \pm \frac{\sqrt{26}}{2}$$

Substitute into the first equation, solve for y:

$$\left(\pm \frac{\sqrt{26}}{2}\right)^2 + y^2 = 4$$
$$\frac{13}{2} + y^2 = 4$$
$$y^2 = -\frac{5}{2} \rightarrow y = \pm \frac{\sqrt{10}}{2}i$$

The solutions are:

$$\frac{-\sqrt{26}}{2}, \frac{\sqrt{10}}{2}i, \; -\frac{\sqrt{26}}{2}, -\frac{\sqrt{10}}{2}i,$$

$$-\frac{\sqrt{26}}{2}, \frac{\sqrt{10}}{2}i, \text{ and } -\frac{\sqrt{26}}{2}, -\frac{\sqrt{10}}{2}i$$

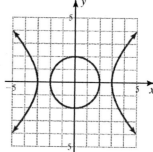

29. $x^2 + y^2 = 1$

$4x^2 + 9y^2 = 36$

To eliminate x^2, multiply the first equation by –4, add to the second equation and solve:

$$-4x^2 - 4y^2 = -4$$

$$\underline{4x^2 + 9y^2 = 36}$$

$$5y^2 = 32$$

$$y^2 = \frac{32}{5}$$

$$y = \pm\frac{4\sqrt{10}}{5}$$

Substitute into the original first equation and solve for x:

$$x^2 + \left(\pm\frac{4\sqrt{10}}{5}\right)^2 = 1$$

$$x^2 + \frac{32}{5} = 1$$

$$x^2 = -\frac{27}{5}$$

$$x = \pm\frac{3\sqrt{15}}{5}i$$

The solutions are:

$$-\frac{3\sqrt{15}}{5}i, \frac{4\sqrt{10}}{5}, -\frac{3\sqrt{15}}{5}i, -\frac{4\sqrt{10}}{5},$$

$$-\frac{3\sqrt{15}}{5}i, \frac{4\sqrt{10}}{5}, \text{ and } -\frac{3\sqrt{15}}{5}i, -\frac{4\sqrt{10}}{5}$$

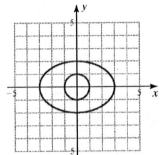

31.   Find x when $C = R$:

$$x + 4 = 6x - x^2$$

$$x^2 - 5x + 4 = 0$$

$$(x - 4)(x - 1) = 0$$

$$x - 4 = 0 \text{ or } x - 1 = 0$$

$$x = 4 \text{ or } \quad x = 1$$

4000 or 1000 units must be manufactured and sold to break even.

33.   Let x = the first number
Let y = the second number

$$x + y = 15$$

$$x^2 - y^2 = 15$$

Solve the first equation for y and substitute into the second equation:

$$x^2 - (15 - x)^2 = 15$$

$$x^2 - (225 - 30x + x^2) = 15$$

$$x^2 - 225 + 30x - x^2 = 15$$

$$30x = 240$$

$$x = 8$$

$$y = 15 - 8 = 7$$

The two numbers are 7 and 8.

35. Let x = the first number
Let y = the second number
$$xy = 176$$
$$x^2 + y^2 = 377$$
Solve the first equation for y and substitute into the second equation:
$$x^2 + \left(\frac{176}{x}\right)^2 = 377$$
$$x^2 + \frac{30976}{x^2} = 377$$
$$x^4 + 30976 = 377x^2$$
$$x^4 - 377x^2 + 30976 = 0$$
$$(x^2 - 121)(x^2 - 256) = 0$$
$$(x+11)(x-11)(x+16)(x-16) = 0$$
$$x = -11; \quad y = \frac{176}{-11} = -16$$
$$x = 11; \quad y = \frac{176}{11} = 16$$
$$x = 16; \quad y = \frac{176}{16} = 11$$
$$x = -16; \quad y = \frac{176}{-16} = -11$$
The numbers are 11 and 16 or -11 and -16.

37. Let L = the length of the check.
Let W = the width of the check.
The area is $A = L \cdot W = 2170$.
The perimeter is $P = 2L + 2W = 202$.
Solve the area equation for W and substitute into the perimeter equation:
$$W = \frac{2170}{L}$$
$$2L + 2\left(\frac{2170}{L}\right) = 202$$
$$2L^2 + 4340 = 202L$$
$$L^2 - 101L + 2170 = 0$$
$$(L - 70)(L - 31) = 0$$
$$L = 70 \qquad \text{or} \quad L = 31$$
$$W = \frac{2170}{70} = 31 \text{ or } W = \frac{2170}{31} = 70$$
The dimensions of the check are 70 ft by 31 ft.

39. Let r = the rate of interest
Let P = the principal (amount loaned)
$$Pr = 340$$
$$P(r + 0.01) = 476$$
Solve the first equation for P and substitute into the second equation:
$$\frac{340}{r}(r + 0.01) = 476$$
$$340(r + 0.01) = 476r$$
$$340r + 3.40 = 476r$$
$$3.40 = 136r$$
$$r = 0.025$$
$$P = \frac{340}{0.025} = 13,600$$
The amount loaned is $13,600 at a rate of interest of 2.5%.

41. $x - y < 4$

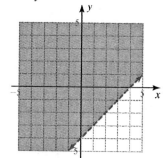

43. $2x \geq 3y \geq 6$

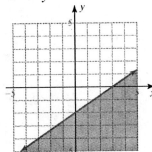

45. $y \geq 2x + 4$

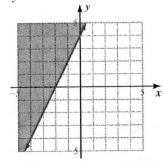

47.   Let p = the price of eggs

Let $\frac{25}{p}$ = number of eggs bought for 25 cents

$$\frac{25}{p} = 2 \cdot 8p$$

$$25 = 16p^2$$

$$p^2 = \frac{25}{16}$$

$$p = \pm\frac{5}{4}$$

$$\frac{25}{\frac{5}{4}} = 25 \cdot \frac{4}{5} = 20$$

20 eggs can be bought for 25 cents.

49. a.   Sample answer: If $b \geq d$, the lines either intersect or are parallel to each other. If they intersect, there is one solution. If they are parallel, there are no solutions. Thus, the maximum number of solutions is one. If $b = d$, the lines could be the same if $a = c$. Then the number of solutions is infinite.

b.   Sample answer: The system will have no solutions if $a = c$ and if $b \geq d$.

51. a.   Sample answer: The maximum number of solutions for a system with two hyperbolas is 4.

b.   Sample answer: The maximum number of solutions for a system with two circles is 2.

c.   Sample answer: The maximum number of solutions for a system with two ellipses is 4.

d.   Sample answer: The maximum number of solutions for a system with a hyperbola and a circle is 4.

53. Let L = the length of the rectangle.
Let W = the width of the rectangle.
The area is $A = L \cdot W = 120$.
The perimeter is $P = 2L + 2W = 44$.
Solve the area equation for W and substitute into the perimeter equation:

$$W = \frac{120}{L}$$

$$2L + 2 \cdot \frac{120}{L} = 44$$

$$2L^2 + 240 = 44L$$

$$L^2 - 22L + 120 = 0$$

$$(L - 12)(L - 10) = 0$$

$$L = 12 \quad \text{or} \quad L = 10$$

$$W = \frac{120}{12} = 10 \quad \text{or} \quad W = \frac{120}{10} = 12$$

The dimensions of the rectangle are 12 cm by 10 cm.

55.
$$4y^2 - x^2 = 4$$
$$x^2 + y^2 = 1$$

To eliminate x^2, add the first equation to the second equation and solve:

$$-x^2 + 4y^2 = 4$$
$$\underline{x^2 + y^2 = 1}$$
$$5y^2 = 5$$
$$y^2 = 1$$
$$y = \pm 1$$

Substitute into the original second equation and solve for x:

$$x^2 + (\pm 1)^2 = 1$$
$$x^2 + 1 = 1$$
$$x^2 = 0$$
$$x = 0$$

The solutions are $(0, 1)$ and $(0, -1)$.

57. Solve the second equation for y, substitute into the first equation and solve:

$$x^2 + y^2 = 4$$
$$x + y = 4 \quad - \quad y = 4 - x$$
$$x^2 + (4 - x)^2 = 4$$
$$x^2 + 16 - 8x + x^2 = 4$$
$$2x^2 - 8x + 12 = 0$$
$$x^2 - 4x + 6 = 0$$

$$x = \frac{-(-4) \pm \sqrt{(-4)^2 - 4 \cdot 1 \cdot 6}}{2 \cdot 1} = \frac{4 \pm \sqrt{16 - 24}}{2}$$

$$= \frac{4 \pm \sqrt{-8}}{2} = 2 \pm i\sqrt{2}$$

$$x = 2 + i\sqrt{2} \quad \text{or } x = 2 - i\sqrt{2}$$

$$y = 4 - \left(2 + i\sqrt{2}\right) \text{ or } y = 4 - \left(2 - i\sqrt{2}\right)$$

$$y = 2 - i\sqrt{2} \quad \text{or } y = 2 + i\sqrt{2}$$

The solutions are $\left(2 + i\sqrt{2},\, 2 - i\sqrt{2}\right)$
and $\left(2 - i\sqrt{2},\, 2 + i\sqrt{2}\right)$.

59. Solve the second equation for y, substitute into the first equation and solve:

$$x^2 + y^2 = 1$$
$$x + y = 1 \quad - \quad y = 1 - x$$
$$x^2 + (1 - x)^2 = 1$$
$$x^2 + 1 - 2x + x^2 = 1$$
$$2x^2 - 2x = 0$$
$$2x(x - 1) = 0$$
$$2x = 0 \quad \text{or } x - 1 = 0$$
$$x = 0 \quad \text{or} \quad x = 1$$
$$y = 1 - 0 = 1 \text{ or } y = 1 - 1 = 0$$

The solutions are $(0, 1)$ and $(1, 0)$.

9.5   Nonlinear Systems of Inequalities

Problems 9.5

1. $y - x^2 - 2$
The boundary of the region is the parabola
$y = x^2 - 2$ with vertex $(0, -2)$. The boundary
is included in the solution.
Select test point $(0, 0)$: $0 - 0^2 - 2 = -2$ is
true, so all points inside the parabola are
included. Shade this side.

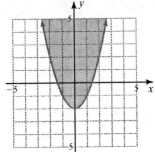

2. $x^2 + y^2 < 9$
The boundary of the region is the circle
$x^2 + y^2 = 9$ with radius 3 centered at the
origin. The boundary is not included in the
solution.
Select test point $(0, 0)$: $0^2 + 0^2 = 0 + 0 = 0 < 9$
is true, so all points inside the circle are
included. Shade this side.

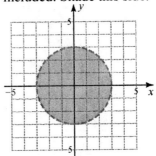

3. $4x^2 \geq 4y^2 \geq 16$
The boundary of the region is the hyperbola
$\frac{x^2}{4} - \frac{y^2}{4} = 1$ with center $(0, 0)$ and vertices at
$(2, 0)$ and $(-2, 0)$. The boundary is included
in the solution.
Select test point $(0, 0)$: $0 \geq 0 \geq 16$ is false, so
all points on the opposite side are included.
Shade this side.

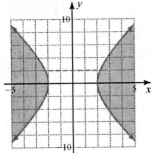

4. $x^2 + y^2 \geq 9$
$\qquad y \geq x^2$
The boundary for the first inequality is the
circle $x^2 + y^2 = 9$ with radius 3 centered at the
origin. The boundary for the second inequality
is the parabola $y = x^2$ with a vertex of $(0, 0)$.
The boundaries of both regions are included in
the solution.
Select test point $(0, 1)$: $0^2 + 1^2 = 0 + 1 = 1 \leq 9$
is true, so all points inside the circle are
included. Select the test point $(0, 1)$: $1 \geq 0^2$ is
true, so all points inside the parabola are
included. The final solution is the intersection
of these two areas.

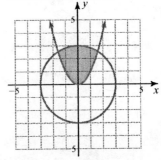

Exercises 9.5

1. $x^2 + y^2 > 16$

The boundary of the region is the circle $x^2 + y^2 = 16$ with radius 4 centered at the origin. The boundary is not included in the solution.

Select test point (0, 0): $0^2 + 0^2 = 0 > 16$ is false, so all points outside the circle are included. Shade this side.

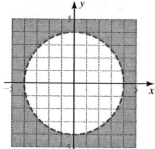

3. $x^2 + y^2 \le 1$

The boundary of the region is the circle $x^2 + y^2 = 1$ with radius 1 centered at the origin. The boundary is included in the solution.

Select test point (0, 0): $0^2 + 0^2 = 0 \le 1$ is true, so all points inside the circle are included. Shade this side.

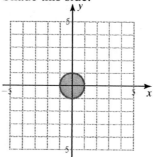

5. $y < x^2 - 2$

The boundary of the region is the parabola $y = x^2 - 2$ with vertex (0, –2). The boundary is not included in the solution.

Select test point (0, 0): $0 < 0^2 - 2 = -2$ is false, so all points outside the parabola are included. Shade this side.

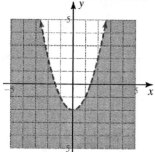

7. $y - -x^2 + 3$

The boundary of the region is the parabola $y = -x^2 + 3$ with vertex (0, 3). The boundary is included in the solution.

Select test point (0, 0): $0 - -(0)^2 + 3 = 3$ is true, so all points inside the parabola are included. Shade this side.

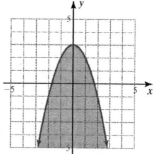

9. $4x^2 - 9y^2 > 36$

The boundary of the region is the hyperbola $\frac{x^2}{9} - \frac{y^2}{4} = 1$ with center (0, 0) and vertices at (3, 0) and (–3, 0). The boundary is not included in the solution.

Select test point (0, 0): $0 - 0 > 36$ is false, so all points on the opposite side are included. Shade this side.

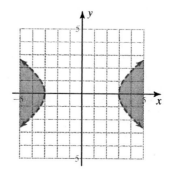

399

11. $x^2 \geq y^2 \geq 1$

The boundary of the region is the hyperbola $x^2 - y^2 = 1$ with center $(0, 0)$ and vertices at $(1, 0)$ and $(-1, 0)$. The boundary is included in the solution.
Select test point $(0, 0)$: $0 \geq 0 \geq 1$ is false, so all points on the opposite side are included. Shade this side.

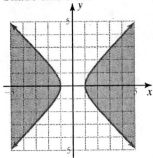

13. $x^2 + y^2 \geq 25$

$\qquad y \geq x^2$

The boundary for the first inequality is the circle $x^2 + y^2 = 25$ with radius 5 centered at the origin. The boundary for the second inequality is the parabola $y = x^2$ with a vertex of $(0, 0)$. The boundaries of both regions are included in the solution.
Select test point $(0, 1)$: $0^2 + 1^2 = 0 + 1 = 1 \leq 25$ is true, so all points inside the circle are included. Select the test point $(0, 1)$: $1 \geq 0^2$ is true, so all points inside the parabola are included. The final solution is the intersection of these two areas.

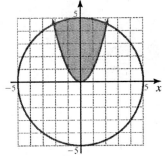

15. $x^2 + y^2 \leq 25$

$\qquad y \leq x^2$

The boundary for the first inequality is the circle $x^2 + y^2 = 25$ with radius 5 centered at the origin. The boundary for the second inequality is the parabola $y = x^2$ with a vertex of $(0, 0)$. The boundaries of both regions are included in the solution.
Select test point $(0, 1)$: $0^2 + 1^2 = 1 \geq 25$ is false, so all points outside the circle are included. Select the test point $(0, 1)$: $1 \leq 0^2$ is false, so all points outside the parabola are included. The final solution is the intersection of these two areas.

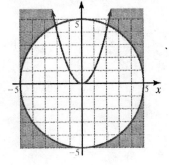

17. $y < x^2 + 2$

$\qquad y > x^2 - 2$

The boundary for the first inequality is the parabola $y = x^2 + 2$ with a vertex of $(0, 2)$. The boundary for the second inequality is the parabola $y = x^2 - 2$ with a vertex of $(0, -2)$. The boundaries of both regions are not included in the solution.
Select test point $(0, 0)$: $0 < 0^2 + 2 = 2$ is true, so all points below this parabola are included. Select the test point $(0, 0)$: $0 > 0^2 - 2 = -2$ is true, so all points inside this parabola are included. The final solution is the intersection of these two areas.

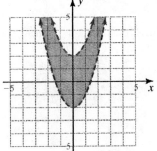

19. $y \geq x^2 + 2$

$y \leq x^2 - 2$

The boundary for the first inequality is the parabola $y = x^2 + 2$ with a vertex of (0, 2). The boundary for the second inequality is the parabola $y = x^2 - 2$ with a vertex of (0, –2). The boundaries of both regions are included in the solution.

Select test point (0, 0): $0 \geq 0^2 + 2 = 2$ is false, so all points inside this parabola are included. Select the test point (0, 0): $0 - 0^2 - 2 = -2$ is true, so all points inside this parabola are included. The final solution is the intersection of these two areas.

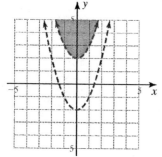

21. $\dfrac{x^2}{4} \leq \dfrac{y^2}{4} \leq 1$

$\dfrac{x^2}{25} + \dfrac{y^2}{4} \leq 1$

The boundary for the first inequality is the hyperbola $\frac{x^2}{4} - \frac{y^2}{4} = 1$ with a center of (0, 0) and vertices of (2, 0) and (–2, 0). The boundary for the second inequality is the ellipse $\frac{x^2}{25} + \frac{y^2}{4} = 1$ with a center of (0, 0) and vertices of (5, 0) and (–5, 0). The boundaries of both regions are included in the solution. Select test point (0, 0): $0 \geq 0 \geq 1$ is false, so all points on the opposite side of the hyperbola are included. Select the test point (0, 0): $0 + 0 \leq 1$ is true, so all points inside the ellipse are included. The final solution is the intersection of these two areas.

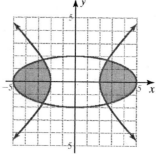

23. $\dfrac{x^2}{36} + \dfrac{y^2}{16} < 1$

$\dfrac{x^2}{16} + \dfrac{y^2}{36} < 1$

The boundary for the first inequality is the ellipse $\frac{x^2}{36} + \frac{y^2}{16} = 1$ with a center of (0, 0) and vertices of (6, 0) and (–6, 0). The boundary for the second inequality is the ellipse $\frac{x^2}{16} + \frac{y^2}{36} = 1$ with a center of (0, 0) and vertices of (0, 6) and (0, –6). The boundaries of both regions are not included in the solution. Select test point (0, 0): $0 + 0 < 1$ is true, so all points inside the first ellipse are included. Select the test point (0, 0): $0 + 0 < 1$ is true, so all points inside the second ellipse are included. The final solution is the intersection of these two areas.

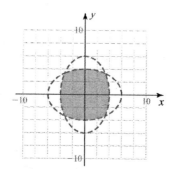

25. a. $\dfrac{x^2}{36}+\dfrac{y^2}{16}>1$ (outside the ellipse)

$\dfrac{x^2}{16}+\dfrac{y^2}{16}<1$ (inside the circle)

These have no points in common.

b. $\dfrac{x^2}{36}+\dfrac{y^2}{16}\le 1$ (outside the ellipse)

$\dfrac{x^2}{16}+\dfrac{y^2}{16}\le 1$ (inside the circle)

These just touch on the boundaries and have two solutions.

27. $y=3x+5$

$y=3(-2)+5=-6+5=-1$

29. $P(x)+Q(x)=\left(x^2-9\right)+(x+3)=x^2+x-6$

$P(x)-Q(x)=\left(x^2-9\right)-(x+3)$

$=x^2-9-x-3=x^2-x-12$

31. $C=72{,}000+60x$

$=72{,}000+60\left(6000-30p\right)$

$=72{,}000+360{,}000-1800p$

$=432{,}000-1800p$

33.

$$R=C$$
$$6000p-30p^2=432{,}000-1800p$$
$$-30p^2+7800p-432{,}000=0$$
$$p^2-260p+14400=0$$
$$(p-180)(p-80)=0$$
$$p=180 \text{ or } p=80$$

35. Sample answer: The boundary of a region is included if the inequality is $\ge$ or $\ge$.

37. a. Answers will vary.
 b. Answers will vary.

39. $x^2+4y^2\ge 4$

$y\ge x^2+1$

The boundary for the first inequality is the ellipse $\frac{x^2}{4}+y^2=1$ with center of (0, 0) and vertices of (2, 0) and (–2, 0). The boundary for the second inequality is the parabola $y=x^2+1$ with a vertex of (0, 1). The boundaries of both regions are included in the solution.
Select test point (0, 0): $0+0\ge 4$ is false, so all points outside the ellipse are included.
Select the test point (0, 0): $0\ge 0^2+1$ is false, so all points inside the parabola are included.
The final solution is the intersection of these two areas.

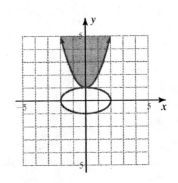

41. $x^2 \ge 9y^2 \ge 9$

The boundary of the region is the hyperbola $\frac{x^2}{9} - y^2 = 1$ with center (0, 0) and vertices at (3, 0) and (–3, 0). The boundary is included in the solution.

Select test point (0, 0): $0 \ge 0 \ge 9$ is false, so all points on the opposite side are included. Shade this side.

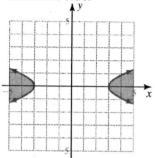

45. $y - -x^2 + 2$

The boundary of the region is the parabola $y = -x^2 + 2$ with vertex (0, 2). The boundary is included in the solution.

Select test point (0, 0): $0 - -0^2 + 2 = 2$ is false, so all points outside the parabola are included. Shade this side.

43. $x^2 > 4 - y^2$

The boundary of the region is the circle $x^2 + y^2 = 4$ with radius 2 centered at the origin. The boundary is not included in the solution.

Select test point (0, 0): $0^2 > 4 - 0^2$ is false, so all points outside the circle are included. Shade this side.

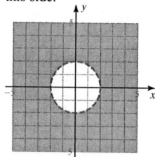

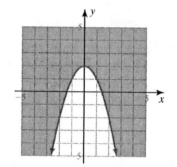

Review Exercises

1. **a.** $y = 9x^2$ Vertex: (0, 0)

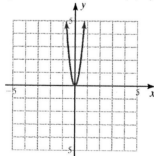

b. $y = -9x^2$ Vertex: (0, 0)

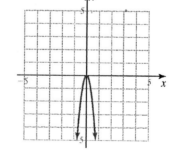

2.   a. $y = (x-1)^2 - 2$ Vertex: $(1, -2)$

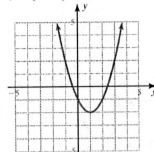

b. $y = -(x-1)^2 + 2$ Vertex: $(1, 2)$

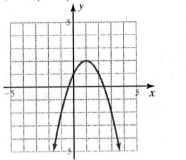

3.   a. $y = x^2 - 4x + 2$

$\qquad = (x^2 - 4x + 4) + 2 - 4$

$\qquad = (x-2)^2 - 2$ Vertex: $(2, -2)$

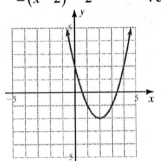

b. $y = -x^2 + 6x - 5$

$\qquad = -(x^2 - 6x + 9 - 9) - 5$

$\qquad = -(x-3)^2 + 9 - 5$

$\qquad = -(x-3)^2 + 4$ Vertex: $(3, 4)$

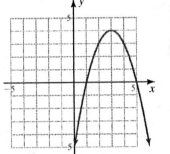

4.   a. $y = 2x^2 - 4x + 3$

$\qquad = 2(x^2 - 2x + 1 - 1) + 3$

$\qquad = 2(x-1)^2 - 2 + 3$

$\qquad = 2(x-1)^2 + 1$ Vertex: $(1, 1)$

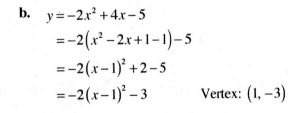

b. $y = -2x^2 + 4x - 5$

$\qquad = -2(x^2 - 2x + 1 - 1) - 5$

$\qquad = -2(x-1)^2 + 2 - 5$

$\qquad = -2(x-1)^2 - 3$ Vertex: $(1, -3)$

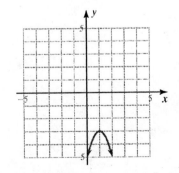

5. **a.** $x = 2(y-2)^2 - 2$ Vertex: $(-2, 2)$

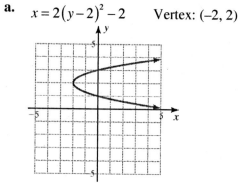

b. $x = -2(y-3)^2 + 1$ Vertex: $(1, 3)$

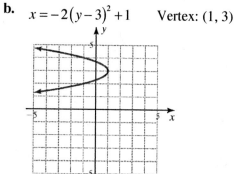

6. **a.** $x = y^2 - 4y + 1$

$= (y^2 - 4y + 4) + 1 - 4$

$= (y-2)^2 - 3$ Vertex: $(-3, 2)$

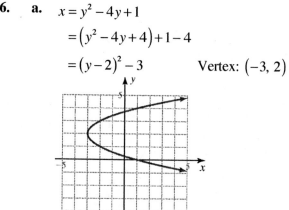

b. $x = y^2 - 2y + 3$

$= (y^2 - 2y + 1) + 3 - 1$

$= (y-1)^2 + 2$ Vertex: $(2, 1)$

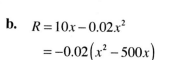

7. **a.** $R = 20x - 0.01x^2$

$= -0.01(x^2 - 2000x)$

$= -0.01(x^2 - 2000x + 1000000 - 1000000)$

$= -0.01(x - 1000)^2 + 10000$

Vertex: $(1000, 10000)$

$x = 1000$ for maximum revenue

b. $R = 10x - 0.02x^2$

$= -0.02(x^2 - 500x)$

$= -0.02(x^2 - 500x + 62500 - 62500)$

$= -0.02(x - 250)^2 + 1250$

Vertex: $(250, 1250)$

$x = 250$ for maximum revenue

8. $h = -16t^2 + 24t$

$= -16\left(t^2 - \frac{3}{2}t\right)$

$= -16\left(t^2 - \frac{3}{2}t + \frac{9}{16} - \frac{9}{16}\right)$

$= -16\left(t - \frac{3}{4}\right)^2 + 9$

Vertex: $\left(\frac{3}{4}, 9\right)$

The ball reaches a maximum height of 9 ft in $\frac{3}{4}$ sec.

9.   **a.** $d = \sqrt{(2-5)^2 + (8-(-3))^2} = \sqrt{(-3)^2 + 11^2} = \sqrt{9+121} = \sqrt{130}$

b. $d = \sqrt{(2-10)^2 + (-4-10)^2} = \sqrt{(-8)^2 + (-14)^2} = \sqrt{64+196} = \sqrt{260} = 2\sqrt{65}$

c. $d = \sqrt{(-3-(-3))^2 + (8-3)^2} = \sqrt{0^2 + 5^2} = \sqrt{0+25} = \sqrt{25} = 5$

10.   **a.**   Center: $(-2, 2)$; $r = 3$

$\left(x-(-2)\right)^2 + (y-2)^2 = 3^2$

$(x+2)^2 + (y-2)^2 = 9$

b.   Center: $(3, -2)$; $r = 3$

$(x-3)^2 + \left(y-(-2)\right)^2 = 3^2$

$(x-3)^2 + (y+2)^2 = 9$

11.   **a.** $(x+2)^2 + (y-1)^2 = 4$

$\left(x-(-2)\right)^2 + (y-1)^2 = 2^2$

$h = -2,\ k = 1,\ r = 2$

Center: $(-2, 1)$; $r = 2$

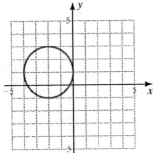

b. $(x-1)^2 + (y+2)^2 = 9$

$(x-1)^2 + \left(y-(-2)\right)^2 = 3^2$

$h = 1,\ k = -2,\ r = 3$

Center: $(1, -2)$; $r = 3$

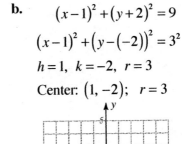

12.   **a.** $x^2 + y^2 = 4$

Center: $(0, 0)$; $r = 2$

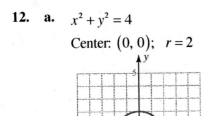

b. $x^2 + y^2 = 25$

Center: $(0, 0)$; $r = 5$

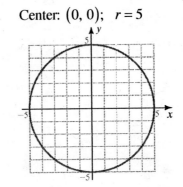

13. **a.**

$$x^2 + y^2 + 2x + 2y - 2 = 0$$
$$\left(x^2 + 2x\right) + \left(y^2 + 2y\right) = 2$$
$$\left(x^2 + 2x + 1\right) + \left(y^2 + 2y + 1\right) = 2 + 1 + 1$$
$$\left(x + 1\right)^2 + \left(y + 1\right)^2 = 4$$
$$\left(x - (-1)\right)^2 + \left(y - (-1)\right)^2 = 2^2$$

$h = -1,\ k = -1,\ r = 2$

Center: $(-1, -1)$; $r = 2$

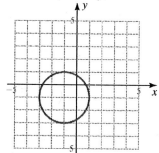

b.

$$x^2 + y^2 - 4x + 6y + 9 = 0$$
$$\left(x^2 - 4x\right) + \left(y^2 + 6y\right) = -9$$
$$\left(x^2 - 4x + 4\right) + \left(y^2 + 6y + 9\right) = -9 + 4 + 9$$
$$\left(x - 2\right)^2 + \left(y + 3\right)^2 = 4$$
$$\left(x - 2\right)^2 + \left(y - (-3)\right)^2 = 2^2$$

$h = 2,\ k = -3,\ r = 2$

Center: $(2, -3)$; $r = 2$

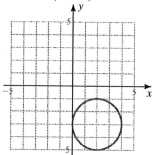

14. **a.**

$$4x^2 + 9y^2 = 36$$
$$\frac{4x^2}{36} + \frac{9y^2}{36} = \frac{36}{36}$$
$$\frac{x^2}{9} + \frac{y^2}{4} = 1$$

$h = 0,\ k = 0,\ a = 3,\ b = 2$

Center: $(0, 0)$; vertices: $(3, 0), (-3, 0)$

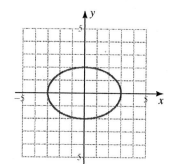

b.

$$9x^2 + y^2 = 9$$
$$\frac{9x^2}{9} + \frac{y^2}{9} = \frac{9}{9}$$
$$\frac{x^2}{1} + \frac{y^2}{9} = 1$$

$h = 0,\ k = 0,\ a = 3,\ b = 1$

Center: $(0, 0)$; vertices: $(0, 3), (0, -3)$

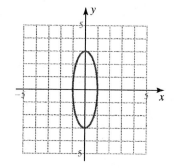

15. **a.**

$$\frac{\left(x - 1\right)^2}{4} + \frac{\left(y - 2\right)^2}{9} = 1$$

$h = 1,\ k = 2,\ a = 3,\ b = 2$

Center: $(1, 2)$

vertices: $(1, 5), (1, -1)$

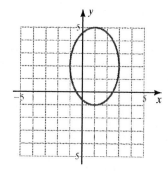

b. $\dfrac{(x+2)^2}{9} + \dfrac{(y-2)^2}{4} = 1$

$h = -2,\; k = 2,\; a = 3,\; b = 2$

Center: $(-2, 2)$

vertices: $(1, 2),\, (-5, 2)$

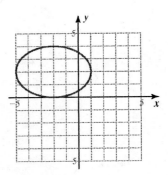

16. a. $\dfrac{x^2}{9} - \dfrac{y^2}{16} = 1$

Center: $(0, 0)$

x^2 term is positive, so $a = 3,\, b = 4$

Vertices: $(3, 0),\, (-3, 0)$

Auxiliary rectangle passes through

$x = \pm 3$ and $y = \pm 4$

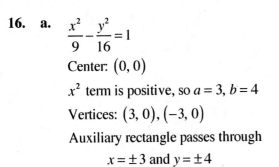

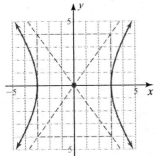

b. $\dfrac{x^2}{16} - \dfrac{y^2}{9} = 1$

Center: $(0, 0)$

x^2 term is positive, so $a = 4,\, b = 3$

Vertices: $(4, 0),\, (-4, 0)$

Auxiliary rectangle passes through

$x = \pm 4$ and $y = \pm 3$

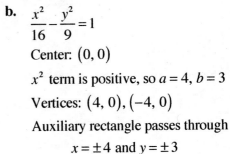

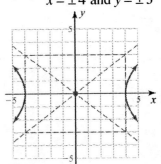

17. a. $\dfrac{y^2}{9} - \dfrac{x^2}{16} = 1$

Center: $(0, 0)$

y^2 term is positive, so $a = 3,\, b = 4$

Vertices: $(0, 3),\, (0, -3)$

Auxiliary rectangle passes through

$x = \pm 4$ and $y = \pm 3$

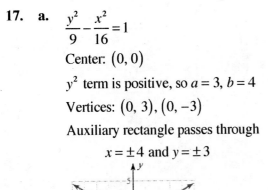

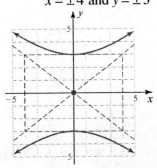

b. $\dfrac{y^2}{16} - \dfrac{x^2}{9} = 1$

Center: $(0, 0)$

y^2 term is positive, so $a = 4,\, b = 3$

Vertices: $(0, 4),\, (0, -4)$

Auxiliary rectangle passes through

$x = \pm 3$ and $y = \pm 4$

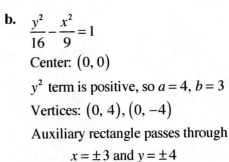

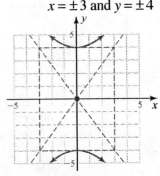

18. a. $x = 1 - y^2$

$x = -y^2 + 1$

One variable is squared; conic is a parabola.

b. $x^2 = 4y^2 - 4$

$4y^2 - x^2 = 4$

Minus sign indicates a hyperbola.

c. $y^2 = 9 - x^2$

$x^2 + y^2 = 9$

Square terms have same positive
coefficient, thus the conic is a circle.

d. $4y^2 = 36 - 9x^2$

$9x^2 + 4y^2 = 36$

Square terms are different positive
numbers, thus the conic is an ellipse.

19. a. Solve the second equation for y, substitute
into the first equation and solve:

$$x^2 + y^2 = 1$$
$$x + y = 1 \quad - \quad y = 1 - x$$
$$x^2 + (1 - x)^2 = 1$$
$$x^2 + 1 - 2x + x^2 = 1$$
$$2x^2 - 2x = 0$$
$$2x(x - 1) = 0$$
$$x = 0 \qquad \text{or } x = 1$$
$$y = 1 - 0 = 1 \text{ or } y = 1 - 1 = 0$$

The solutions are (0, 1) and (1, 0).

b. Solve the second equation for y, substitute
into the first equation and solve:

$$x^2 + y^2 = 10$$
$$x + y = 4 \quad - \quad y = 4 - x$$
$$x^2 + (4 - x)^2 = 10$$
$$x^2 + 16 - 8x + x^2 = 10$$
$$2x^2 - 8x + 6 = 0$$
$$x^2 - 4x + 3 = 0$$
$$(x - 3)(x - 1) = 0$$
$$x = 3 \qquad \text{or } x = 1$$
$$y = 4 - 3 = 1 \text{ or } y = 4 - 1 = 3$$

The solutions are (3, 1) and (1, 3).

20. a. Substitute the second equation into the first
equation and solve:

$$x^2 - y^2 = 16$$
$$2x = y$$
$$x^2 - (2x)^2 = 16$$
$$x^2 - 4x^2 = 16$$
$$-3x^2 = 16$$
$$x^2 = -\frac{16}{3}$$
$$x = \pm \frac{4\sqrt{3}}{3} i$$

If $x = \frac{4\sqrt{3}}{3} i$, then $y = \frac{8\sqrt{3}}{3} i$

If $x = -\frac{4\sqrt{3}}{3} i$, then $y = -\frac{8\sqrt{3}}{3} i$

The solutions are:

$$\frac{-4\sqrt{3}}{3} i, \frac{8\sqrt{3}}{3} i \text{ and } -\frac{4\sqrt{3}}{3} i, -\frac{8\sqrt{3}}{3} i.$$

b. Solve the second equation for y, substitute
into the first equation and solve:

$$x^2 + y^2 = 4$$
$$x + y = 3 \quad - \quad y = 3 - x$$
$$x^2 + (3 - x)^2 = 4$$
$$x^2 + 9 - 6x + x^2 = 4$$
$$2x^2 - 6x + 5 = 0$$
$$x = \frac{|(|6) \pm \sqrt{(|6) |4|2|5}}{2|2}$$
$$= \frac{6 \pm \sqrt{36 |40}}{4} = \frac{6 \pm \sqrt{|4}}{4} = \frac{3}{2} \pm \frac{1}{2} i$$

If $x = \frac{3}{2} + \frac{1}{2} i$, then $y = 3 | \left| \frac{3}{2} + \frac{1}{2} i \right| = \frac{3}{2} | \frac{1}{2} i$

If $x = \frac{3}{2} | \frac{1}{2} i$, then $y = 3 | \left| \frac{3}{2} | \frac{1}{2} i \right| = \frac{3}{2} + \frac{1}{2} i$

The solutions are:

$$\left| \frac{3}{2} + \frac{1}{2} i, \frac{3}{2} | \frac{1}{2} i \right| \text{ and } \left| \frac{3}{2} | \frac{1}{2} i, \frac{3}{2} + \frac{1}{2} i \right|.$$

21. a. $x^2 - y^2 = 5$

$x^2 + 2y^2 = 17$

To eliminate y^2, multiply the first equation by 2, add to the second equation and solve:

$$2x^2 - 2y^2 = 10$$
$$\underline{x^2 + 2y^2 = 17}$$
$$3x^2 \qquad = 27$$
$$x^2 = 9$$
$$x = \pm 3$$

Substitute into the original first equation and solve for y:

$$(\pm 3)^2 - y^2 = 5$$
$$9 - y^2 = 5$$
$$y^2 = 4$$
$$y = \pm 2$$

The solutions are (3, 2), (3, –2), (–3, 2) and (–3, –2).

b. $x^2 - y^2 = 3$

$2x^2 + y^2 = 9$

To eliminate y^2, add the first equation to the second equation and solve:

$$x^2 - y^2 = 3$$
$$\underline{2x^2 + y^2 = 9}$$
$$3x^2 \qquad = 12$$
$$x^2 = 4$$
$$x = \pm 2$$

Substitute into the first equation and solve for y:

$$(\pm 2)^2 - y^2 = 3$$
$$4 - y^2 = 3$$
$$y^2 = 1$$
$$y = \pm 1$$

The solutions are (2, 1), (2, –1), (–2, 1), and (–2, –1).

22. a.   Find x when $R = C$:

$$x^2 - 200 = 10x + 400$$
$$x^2 - 10x - 600 = 0$$
$$(x - 30)(x + 20) = 0$$
$$x - 30 = 0 \quad \text{or } x + 20 = 0$$
$$x = 30 \text{ or} \qquad x = -20$$

30 units must be manufactured and sold to break even.

b.   Find x when $R = C$:

$$x^2 - 200 = 6x + 80$$
$$x^2 - 6x - 280 = 0$$
$$(x - 20)(x + 14) = 0$$
$$x - 20 = 0 \quad \text{or } x + 14 = 0$$
$$x = 20 \text{ or} \qquad x = -14$$

20 units must be manufactured and sold to break even.

23. a. $y - 1 - x^2$

The boundary of the region is the parabola $y = 1 - x^2$ with vertex (0, 1). The boundary is included in the solution.

Select test point (0, 0): $0 - 1 - 0^2 = 1$ is true, so all points inside the parabola are included. Shade this side.

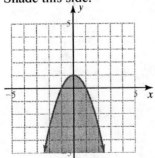

b. $x - 4 - y^2$

The boundary of the region is the parabola $x = 4 - y^2$ with vertex (4, 0). The boundary is included in the solution.

Select test point (0, 0): $0 - 4 - 0^2 = 4$ is true, so all points inside the parabola are included. Shade this side.

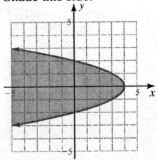

410

24. a. $x^2 + y^2 \leq 4$

The boundary of the region is the circle $x^2 + y^2 = 4$ with radius 2 centered at the origin. The boundary is included in the solution.

Select test point (0, 0): $0^2 + 0^2 = 0 \leq 4$ is true, so all points inside the circle are included. Shade this side.

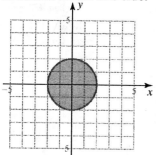

b. $x^2 + y^2 > 9$

The boundary of the region is the circle $x^2 + y^2 = 9$ with radius 3 centered at the origin. The boundary is not included in the solution.

Select test point (0, 0): $0^2 + 0^2 = 0 > 9$ is false, so all points outside the circle are included. Shade this side.

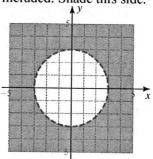

25. a. $4x^2 \leq y^2 \leq 4$

The boundary of the region is the hyperbola $\frac{x^2}{1} - \frac{y^2}{4} = 1$ with center (0, 0) and vertices at (1, 0) and (–1, 0). The boundary is included in the solution.

Select test point (0, 0): $0 \leq 0 \leq 4$ is true, so all points between the two halves of the hyperbola are included. Shade this side.

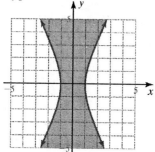

b. $x^2 \leq 4y^2 \leq 4$

The boundary of the region is the hyperbola $\frac{x^2}{4} - \frac{y^2}{1} = 1$ with center (0, 0) and vertices at (2, 0) and (–2, 0). The boundary is included in the solution.

Select test point (0, 0): $0 \leq 0 \leq 4$ is true, so all points between the two halves of the hyperbola are included. Shade this side.

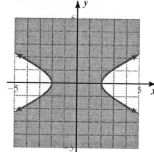

411

26. a. $x^2 + y^2 - 4$

$y - 2 - x^2$

The boundary for the first inequality is the circle $x^2 + y^2 = 4$ with radius 2 centered at the origin. The boundary for the second inequality is the parabola $y = 2 - x^2$ with a vertex of (0, 2). The boundaries of both regions are included in the solution. Select test point (0, 1): $0^2 + 1^2 = 1 \le 4$ is true, so all points inside the circle are included. Select the test point (0, 1): $1 - 2 - 0^2 = 2$ is true, so all points inside the parabola are included. The final solution is the intersection of these two areas.

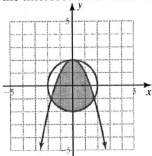

b. $x^2 + y^2 \ge 4$

$y \ge 4x^2$

The boundary for the first inequality is the circle $x^2 + y^2 = 4$ with radius 2 centered at the origin. The boundary for the second inequality is the parabola $y = 4x^2$ with a vertex of (0, 0). The boundaries of both regions are included in the solution. Select test point (0, 1): $0^2 + 1^2 = 1 \le 4$ is true, so all points inside the circle are included. Select the test point (0, 1): $1 \cdot 4 \cdot 0^2$ is true, so all points inside the parabola are included. The final solution is the intersection of these two areas.

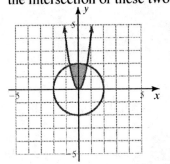

Cumulative Review Chapters 1 – 9

1. The additive inverse of $\frac{4}{3}$ is $-\frac{4}{3}$.

2. $\dfrac{45x^8}{15x^{-7}} = 3x^{8-(-7)} = 3x^{15}$

3. $-4^3 + \dfrac{12-8}{2} + 15 \div 3 = -4^3 + \dfrac{4}{2} + 15 \div 3$

$= -64 + \dfrac{4}{2} + 15 \div 3 = -64 + 2 + 5 = -57$

4. $\left|\dfrac{3}{2}x + 7\right| + 3 = 8$

$\left|\dfrac{3}{2}x + 7\right| = 5$

$\dfrac{3}{2}x + 7 = 5 \quad$ or $\quad \dfrac{3}{2}x + 7 = -5$

$\dfrac{3}{2}x = -2 \quad$ or $\quad \dfrac{3}{2}x = -12$

$3x = -4 \quad$ or $\quad 3x = -24$

$x = -\dfrac{4}{3} \quad$ or $\quad x = -8$

412

5. $x+1-4$ and $-2x < 8$

$x - 3$ and $x > -4$

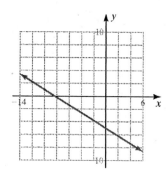

6.
$$B = \frac{7}{8}(A-10)$$
$$8B = 7(A-10)$$
$$8B = 7A - 70$$
$$8B + 70 = 7A$$
$$\frac{8B+70}{7} = A$$

7. Let $x =$ the woman's salary before increase
$$x + 0.10x = 31,900$$
$$1.10x = 31,900$$
$$x = 29,000$$
The woman's salary before the increase was $29,000.

8. For $x = 0$, $y = \cdot\, 2x \cdot 5$ becomes
$$y = \cdot\, 2 \cdot 0 \cdot 5 = \cdot\, 5.$$
The y-intercept is $(0, \cdot\, 5)$.

For $y = 0$, $y = \cdot\, 2x \cdot 5$ becomes
$$0 = \cdot\, 2x \cdot 5$$
$$5 = \cdot\, 2x$$
$$\cdot\,\frac{5}{2} = x$$
The x-intercept is $\left(\cdot\,\frac{5}{2}, 0\right)$.

9.

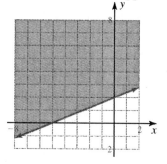

10. $2x \leq 5y \leq\, \leq 10$

When $x = 0$, $\leq 5y =\, \leq 10$ and $y = 2$

When $y = 0$, $2x =\, \leq 10$ and $x =\, \leq 5$

The boundary is a solid line.

Test point $(0,0)$: $2(0) \leq 2(0) \leq\, \leq 10$, $0 \leq\, \leq 10$

False, so shade the opposite side of the line.

11. $m = \dfrac{3-6}{7-3} = \dfrac{-3}{4} = -\dfrac{3}{4}$

12. $D = \{-2, 5, -3\}$

13. $f(x) = -2x + 2$
$$f(3) = -2(3) + 2 = -6 + 2 = -4$$

14. Multiply the first equation by -2 and add to the second equation:

$5x + 2y = 14 \quad - \qquad -10x - 4y = -28$

$10x + 4y = 4 \quad - \qquad \underline{10x + 4y = \quad 4}$

$\qquad\qquad\qquad\qquad\qquad\qquad 0 = -24$

This result is false, so there is no solution.

15.
$$x + y + z = 4$$
$$-x + 2y + z = 3$$
$$2x + y + z = 6$$

Multiply the first equation by -1, add to the second equation to eliminate z:
$$-x - y - z = -4$$
$$\underline{-x + 2y + z = 3}$$
$$-2x + y = -1$$

Multiply the first equation by -1, add to the third equation to eliminate z:
$$-x - y - z = -4$$
$$\underline{2x + y + z = 6}$$
$$x = 2$$

Substitute into the first result and solve:
$$-2(2) + y = -1 \;\to\; y = 3$$

Substitute into the first equation and solve:
$$2 + 3 + z = 4 \;-\; z = -1$$
The solution is $(2, 3, -1)$.

16.
$$\begin{vmatrix} 5 & 2 & -4 \\ 5 & 5 & -1 \\ 4 & 4 & 5 \end{vmatrix}$$

$$= 5\begin{vmatrix} 5 & -1 \\ 4 & 5 \end{vmatrix} - 2\begin{vmatrix} 5 & -1 \\ 4 & 5 \end{vmatrix} + (-4)\begin{vmatrix} 5 & 5 \\ 4 & 4 \end{vmatrix}$$

$$= 5(25 + 4) - 2(25 + 4) - 4(20 - 20)$$

$$= 145 - 58 - 0 = 87$$

17. Let $x =$ the speed of the boat in still water
Let $y =$ the speed of the current
$x + y =$ the speed downstream

$x - y =$ the speed upstream

$$(x + y)\frac{1}{3} = 12 \;-\; x + y = 36$$

$$(x - y)\frac{1}{2} = 12 \;-\; x - y = 24$$

Add the two equations and solve:
$$2x = 60$$
$$x = 30$$
The speed of the boat in still water is 30 mi/hr.

18. $(3j + 1)(3j - 1) = 9j^2 - 3j + 3j - 1 = 9j^2 - 1$

19. $12x^6 - 3x^4 + 8x^3 - 2x$
$$= x(12x^5 - 3x^3 + 8x^2 - 2)$$
$$= x\left[3x^3(4x^2 - 1) + 2(4x^2 - 1)\right]$$
$$= x(4x^2 - 1)(3x^3 + 2)$$
$$= x(2x + 1)(2x - 1)(3x^3 + 2)$$

20. $32x^2y^2 - 48xy^3 + 18y^4$
$$= 2y^2(16x^2 - 24xy + 9y^2)$$
$$= 2y^2(4x - 3y)^2$$

21.
$$2x^2 + 9x = -9$$
$$2x^2 + 9x + 9 = 0$$
$$(2x+3)(x+3) = 0$$
$$2x+3 = 0 \quad \text{or } x+3 = 0$$
$$x = -\frac{3}{2} \text{ or} \quad x = -3$$

22.
$$\frac{x+1}{3x+y} \cdot \frac{9x^2 - y^2}{2x^2 - x - 3}$$
$$= \frac{\cancel{x+1}}{\cancel{3x+y}} \cdot \frac{\cancel{(3x+y)}\,(3x-y)}{(2x-3)\cancel{(x+1)}}$$
$$= \frac{3x-y}{2x-3}$$

23. Multiply both sides by the LCD $(x+7)(x-7)$:
$$\frac{x}{x+7} - \frac{x}{x-7} = \frac{x^2+49}{x^2-49}$$
$$(x+7)(x-7)\left[\frac{x}{x+7} - \frac{x}{x-7}\right] = (x+7)(x-7)\cdot\frac{x^2+49}{(x+7)(x-7)}$$
$$x(x-7) - x(x+7) = x^2+49$$
$$x^2 - 7x - x^2 - 7x = x^2+49$$
$$-14x = x^2+49$$
$$0 = x^2+14x+49$$
$$0 = (x+7)(x+7)$$
$$x+7 = 0 \quad \text{or } x+7 = 0$$
$$x = -7 \text{ or} \quad x = -7$$

-7 makes the denominator $x+7 = 0$ and thus is extraneous. There is no solution.

24. Let t = the time when working together
$$\frac{1}{5} + \frac{1}{4} = \frac{1}{t}$$
$$20t\left[\frac{1}{5} + \frac{1}{4}\right] = 20t\cdot\frac{1}{t}$$
$$4t + 5t = 20$$
$$9t = 20$$
$$t = \frac{20}{9} = 2\frac{2}{9} \text{ hr}$$

Working together they do the job in $2\frac{2}{9}$ hr.

25.
$$P = kT$$
$$6 = k\cdot 360$$
$$k = \frac{6}{360} = \frac{1}{60}$$

26. $x^{2/3} \cdot x^{5/2} = x^{2/3 \,+\, 5/2} = x^{19/6}$

27. $\sqrt[6]{16c^4 d^4} = \left(2^4 c^4 d^4\right)^{1/6} = 2^{2/3} c^{2/3} d^{2/3} = \sqrt[3]{4c^2 d^2}$

28. $\left(\sqrt{125} + \sqrt{18}\right)\left(\sqrt{20} + \sqrt{50}\right) = \left(5\sqrt{5} + 3\sqrt{2}\right)\left(2\sqrt{5} + 5\sqrt{2}\right) = 10\sqrt{25} + 25\sqrt{10} + 6\sqrt{10} + 15\sqrt{4}$
$$= 10\cdot 5 + 31\sqrt{10} + 15\cdot 2 = 50 + 31\sqrt{10} + 30 = 80 + 31\sqrt{10}$$

29.
$$\sqrt{x+3} = -2$$
$$\left(\sqrt{x+3}\right)^2 = \left(-2\right)^2$$
$$x+3 = 4$$
$$x = 1$$
Check: $\sqrt{1+3} = \sqrt{4} = 2 \neq -2$
There are no real number solutions.

30.
$$\frac{1+2i}{9+4i} = \frac{(1+2i)(9-4i)}{(9+4i)(9-4i)}$$
$$= \frac{9-4i+18i-8i^2}{81-36i+36i-16i^2}$$
$$= \frac{17+14i}{97} = \frac{17}{97} + \frac{14}{97}i$$

31.
$$7x^2 + 42x + 63 = 11$$
$$x^2 + 6x + 9 = \frac{11}{7}$$
$$\left(x+3\right)^2 = \frac{11}{7}$$
$$\sqrt{\left(x+3\right)^2} = \pm\sqrt{\frac{11}{7}} = \pm\sqrt{\frac{77}{49}}$$
$$x+3 = \pm\frac{\sqrt{77}}{7}$$
$$x = -3 \pm \frac{\sqrt{77}}{7}$$
$$x = \frac{-21+\sqrt{77}}{7} \text{ or } x = \frac{-21-\sqrt{77}}{7}$$

32.
$$27x^3 \cdot 64 = 0$$
$$\left(3x \cdot 4\right)\left(9x^2 + 12x + 16\right) = 0$$
$$3x \cdot 4 = 0 \text{ or } 9x^2 + 12x + 16 = 0$$
$$x = \frac{4}{3} \text{ or } 9x^2 + 12x + 16 = 0$$
$$9x^2 + 12x + 16 = 0; \quad a = 9,\ b = 12,\ c = 16$$
$$x = \frac{\cdot 12 \pm \sqrt{12^2 \cdot 4 \cdot 9 \cdot 16}}{2 \cdot 9} = \frac{\cdot 12 \pm \sqrt{144 \cdot 576}}{18}$$
$$= \frac{\cdot 12 \pm \sqrt{\cdot 432}}{18} = \frac{\cdot 12 \pm 12i\sqrt{3}}{18} = \cdot\frac{2}{3} \pm \frac{2\sqrt{3}}{3}i$$
The solutions are:
$$\frac{4}{3}, -\frac{2}{3} + \frac{2\sqrt{3}}{3}i, \text{ and } -\frac{2}{3} - \frac{2\sqrt{3}}{3}i.$$

33.
$$4x^2 - 5x - 4 = 0$$
$$\text{Sum} = -\frac{b}{a} = -\frac{-5}{4} = \frac{5}{4}$$
$$\text{Product} = \frac{c}{a} = \frac{-4}{4} = -1$$

34.
$$x^2 + x \infty 56$$
$$x^2 + x \infty 56 \infty 0$$
$$\left(x \infty 7\right)\left(x+8\right) \infty 0$$

$\left(x\infty 7\right)\left(x+8\right) = 0$ is the related equation. $x = 7$, $x = \infty 8$ are the critical values.

Separate into regions using -8 and 7 as boundaries.

Select test points: $\infty 9$, 0, and 8.

When $x = \infty 9$, $\left(\infty 9\right)^2 + \left(\infty 9\right) = 81 \infty 9 = 72 \infty 56$ True.

When $x = 0$, $0^2 + 0 = 0 \not\infty 56$ False.

When $x = 8$, $8^2 + 8 = 64 + 8 = 72 \infty 56$ True.

The solution set is: $\left\{x \mid x \infty\infty 8 \cup x \infty 7\right\}$ or $\left(\infty\infty, \infty 8\right] \cup [7, +\infty)$.

35.
$$y = 3x^2 + 4x + 3$$
$$y = 3\left(x^2 + \frac{4}{3}x\right) + 3$$
$$y = 3\left(x^2 + \frac{4}{3}x + \frac{4}{9} - \frac{4}{9}\right) + 3$$
$$y = 3\left(x + \frac{2}{3}\right)^2 + \frac{4}{3} + 3$$
$$y = 3\left(x + \frac{2}{3}\right)^2 + \frac{13}{3} \quad \text{vertex: } \left(-\frac{2}{3}, \frac{13}{3}\right)$$

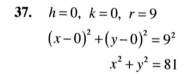

Vertex at $\left(\frac{-2}{3}, 4\frac{1}{3}\right)$

36.
$$R = 20x - 0.02x^2$$
$$= -0.02\left(x^2 - 1000x\right)$$
$$= -0.02\left(x^2 - 1000x + 250000 - 250000\right)$$
$$= -0.02\left(x - 500\right)^2 + 5000$$
Vertex: $\left(500, 5000\right)$

$x = 500$ for maximum revenue

37.
$$h = 0, \ k = 0, \ r = 9$$
$$\left(x - 0\right)^2 + \left(y - 0\right)^2 = 9^2$$
$$x^2 + y^2 = 81$$

38.
$$\frac{\left(x - 2\right)^2}{25} + \frac{\left(y + 4\right)^2}{16} = 1$$
$$h = 2, \ k = -4, \ a = 5, \ b = 4$$
Center: $\left(2, -4\right)$

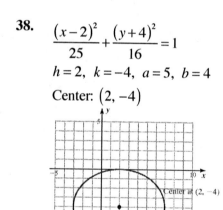

Center at $(2, -4)$

39. Solve the second equation for y, substitute into the first equation and solve:

$$x^2 + y^2 = 25$$
$$x + y = 1 \ - \quad y = 1 - x$$
$$x^2 + (1-x)^2 = 25$$
$$x^2 + 1 - 2x + x^2 = 25$$
$$2x^2 - 2x - 24 = 0$$
$$x^2 - x - 12 = 0$$
$$(x-4)(x+3) = 0$$
$$x = 4 \qquad \text{or } x = -3$$
$$y = 1 - 4 = -3 \text{ or } y = 1 - (-3) = 4$$

The solutions are $(4, -3)$ and $(-3, 4)$.

40. $x^2 + y^2 < 4$

The boundary of the region is the circle $x^2 + y^2 = 4$ with radius 2 centered at the origin. The boundary is not included in the solution.

Select test point $(0, 0)$: $0^2 + 0^2 = 0 < 4$ is true, so all points inside the circle are included. Shade this side.

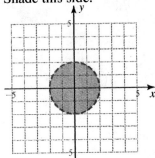

Chapter 10   Functions — Inverse, Exponential, and Logarithmic

10.1 The Algebra of Functions

Problems 10.1

1. a.
$$f(x)=x^2-9;\ \ g(x)=x-3$$
$$(f+g)(x)=f(x)+g(x)$$
$$=(x^2-9)+(x-3)$$
$$=x^2-9+x-3=x^2+x-12$$

b.
$$f(x)=x^2-9;\ \ g(x)=x-3$$
$$(f-g)(x)=f(x)-g(x)$$
$$=(x^2-9)-(x-3)$$
$$=x^2-9-x+3=x^2-x-6$$

c.
$$f(x)=x^2\cdot 9;\ \ g(x)=x\cdot 3$$
$$(fg)(x)=f(x)\cdot g(x)=(x^2\cdot 9)(x\cdot 3)$$
$$=x^3\cdot 3x^2\cdot 9x+27$$

d.
$$f(x)=x^2\neq 9;\ \ g(x)=x\neq 3$$
$$\neq\frac{f}{g}\neq(x)=\frac{f(x)}{g(x)}=\frac{x^2\neq 9}{x\neq 3}=\frac{(x+3)\,\cancel{(x\neq 3)}}{\cancel{x\neq 3}}$$
$$=x+3,\ x\neq 3$$

2.
$$f(x)=2x\neq 3$$
$$\frac{f(x)\neq f(a)}{x\neq a}=\frac{2x\neq 3\neq(2a\neq 3)}{x\neq a}$$
$$=\frac{2x\neq 3\neq 2a+3}{x\neq a}=\frac{2x\neq 2a}{x\neq a}$$
$$=\frac{2(x\neq a)}{x\neq a}=2,\ x\neq a$$

3. a.
$$f(x)=x^2-1;\ \ g(x)=3x+1$$
$$(f\circ g)(-1)=f(g(-1))=f(3(-1)+1)$$
$$=f(-2)=(-2)^2-1=4-1=3$$

b.
$$f(x)=x^2-1;\ \ g(x)=3x+1$$
$$(f\circ g)(x)=f(g(x))=f(3x+1)$$
$$=(3x+1)^2-1=9x^2+6x+1-1$$
$$=9x^2+6x$$

c.
$$f(x)=x^2-1;\ \ g(x)=3x+1$$
$$(g\circ f)(x)=g(f(x))=g(x^2-1)$$
$$=3(x^2-1)+1=3x^2-3+1$$
$$=3x^2-2$$

4.
$$f(x)=\frac{1}{2x};\ \ g(x)=\frac{3x\neq 1}{x+4}$$
Domain f: $\{x\,|\,x \text{ is a real number and } x\neq 0\}$
Domain g: $\{x\,|\,x \text{ is a real number and } x\neq\neq 4\}$
Domain $f+g=$ domain $f\neq g=$ domain fg
$=\{x\,|\,x \text{ is a real number and } x\neq 0 \text{ and } x\neq\neq 4\}$

5.
$$f(x)=\frac{5}{x+6};\ \ g(x)=\frac{10}{x\neq 1}$$
Domain f: $\{x\,|\,x \text{ is a real number and } x\neq\neq 6\}$
Domain g: $\{x\,|\,x \text{ is a real number and } x\neq 1\}$
Domain $\dfrac{f}{g}$
$=\{x\,|\,x \text{ is a real number and } x\neq\neq 6 \text{ and } x\neq 1\}$

6.

$$f(x) = \frac{1}{x+7}; \quad g(x) = \frac{x+4}{x \neq 1}$$

Domain f: $\{x \mid x$ is a real number and $x \neq \neq 7\}$

Domain g: $\{x \mid x$ is a real number and $x \neq 1\}$

Also exclude the values of x where $g(x) = 0$. So, $x \neq \neq 4$

Domain $\dfrac{f}{g} = \{x \mid x$ is a real number and $x \neq \neq 7$ and $x \neq 1$ and $x \neq \neq 4\}$

7.   a. $P(x) = R(x) \mid C(x)$

$$= \left|200x \mid \frac{x^2}{30}\right| \mid (50,000 + 50x)$$

$$= \left|\frac{x^2}{30} + 150x \mid 50,000\right.$$

b. Find the vertex:

$$x = \cdot \frac{b}{2a} = \cdot \frac{150}{2\left(\cdot \frac{1}{30}\right)} = \cdot \frac{150}{\cdot \frac{1}{15}} = 150 \cdot 15$$

$$= 2250 \text{ units}$$

$$P(2250) = -\frac{2250^2}{30} + 150(2250) - 50,000$$

$$= \$118,750$$

Exercises 10.1

1. $(f+g)(x) = f(x) + g(x)$

$$= (x+4) + (x^2 - 5x + 4)$$

$$= x + 4 + x^2 - 5x + 4 = x^2 - 4x + 8$$

3. $(hf)(x) = h(x) \cdot f(x) = (x^2 + 16)(x+4)$

$$= x^3 + 4x^2 + 16x + 64$$

5. $\left|\dfrac{f}{h}\right|(x) = \dfrac{f(x)}{h(x)} = \dfrac{x+4}{x^2+16}$

7.

$$f(x) = 3x - 2$$

$$\frac{f(x) - f(a)}{x-a} = \frac{3x - 2 - (3a - 2)}{x - a}$$

$$= \frac{3x - 2 - 3a + 2}{x - a} = \frac{3x - 3a}{x - a}$$

$$= \frac{3(x-a)}{x-a} = 3$$

9.

$$f(x) = x^2$$

$$\frac{f(x) - f(a)}{x-a} = \frac{x^2 - a^2}{x-a} = \frac{(x+a)(x-a)}{x-a}$$

$$= x + a$$

11.

$$f(x) = x^2 + 3x$$

$$\frac{f(x) - f(a)}{x-a} = \frac{x^2 + 3x - (a^2 + 3a)}{x - a}$$

$$= \frac{x^2 + 3x - a^2 - 3a}{x - a} = \frac{x^2 - a^2 + 3x - 3a}{x - a}$$

$$= \frac{(x+a)(x-a) + 3(x-a)}{x - a}$$

$$= \frac{(x-a)(x+a+3)}{x - a} = x + a + 3$$

13.
$$f(x)=x^2; \quad g(x)=\sqrt{x}; \quad x>0$$

a. $(f \circ g)(1)=f(g(1))=f(\sqrt{1})=f(1)$

$$=1^2=1$$

b. $(f \circ g)(x)=f(g(x))=f(\sqrt{x})$

$$=(\sqrt{x})^2=x$$

c. $(g \circ f)(x)=g(f(x))=g(x^2)=\sqrt{x^2}=x$

15.
$$f(x)=3x \cdot 2; \quad g(x)=x+1$$

a. $(f \circ g)(1)=f(g(1))=f(1+1)=f(2)$

$$=3 \cdot 2 \cdot 2=6 \cdot 2=4$$

b. $(f \circ g)(x)=f(g(x))=f(x+1)$

$$=3(x+1)-2=3x+3-2$$

$$=3x+1$$

c. $(g \circ f)(x)=g(f(x))=g(3x-2)$

$$=3x-2+1=3x-1$$

17.
$$f(x)=\sqrt{x+1}, \quad x>-1; \quad g(x)=x^2-1$$

a. $(f \circ g)(1)=f(g(1))=f(1^2-1)=f(0)$

$$=\sqrt{0+1}=\sqrt{1}=1$$

b. $(f \circ g)(x)=f(g(x))=f(x^2-1)$

$$=\sqrt{x^2-1+1}=\sqrt{x^2}=|x|$$

c. $(g \circ f)(x)=g(f(x))=g(\sqrt{x+1})$

$$=(\sqrt{x+1})^2-1=x+1-1=x$$

19.
$$f(x)=3; \quad g(x)=-1$$

a. $(f \circ g)(1)=f(g(1))=f(-1)=3$

b. $(f \circ g)(x)=f(g(x))=f(-1)=3$

c. $(g \circ f)(x)=g(f(x))=g(3)=-1$

21.
$$f(x)=\sqrt{x}; \quad x>0 \quad g(x)=x^2-1$$

a. $(f+g)(4)=f(4)+g(4)=\sqrt{4}+4^2-1$

$$=2+16-1=17$$

b. $(f-g)(4)=f(4)-g(4)=\sqrt{4}-(4^2-1)$

$$=2-(16-1)=-13$$

23.
$$f(x)=|x|; \quad g(x)=3$$

a. $\left|\dfrac{f}{g}\right|(3)=\dfrac{f(3)}{g(3)}=\dfrac{|3|}{3}=\dfrac{3}{3}=1$

b. $\left|\dfrac{f}{g}\right|(0)=\dfrac{f(0)}{g(0)}=\dfrac{|0|}{3}=\dfrac{0}{3}=0$

25.
$$f(x)=x \neq 3; \quad g(x)=(x+3)(x \neq 3)$$

a. $\dfrac{\neq f \neq}{\not\equiv g \not\equiv}(3)=\dfrac{f(3)}{g(3)}=\dfrac{3 \neq 3}{(3+3)(3 \neq 3)}=\dfrac{0}{0}$

Undefined: $x \neq 3$

b. $\dfrac{\neq g \neq}{\not\equiv f \not\equiv}(3)=\dfrac{g(3)}{f(3)}=\dfrac{(3+3)(3 \neq 3)}{3 \neq 3}=\dfrac{0}{0}$

Undefined: $x \neq 3$

27.
$$f(x)=\sqrt{x}, \quad x>0; \quad g(x)=x^2+1$$

a. $(f \circ g)(1)=f(g(1))=f(1^2+1)=f(2)=\sqrt{2}$

b. $(g \circ f)(1)=g(f(1))=g(\sqrt{1})=g(1)=1^2+1$

$$=2$$

c. $(f \circ g)(x)=f(g(x))=f(x^2+1)=\sqrt{x^2+1}$

d. $(g \circ f)(x)=g(f(x))=g(\sqrt{x})=(\sqrt{x})^2+1$

$$=x+1$$

29.

$$f(x)=\frac{1}{x^2-2};\quad g(x)=\sqrt{x},\ x>0$$

a. $(f\circ g)(2)=f(g(2))=f(\sqrt{2})=\dfrac{1}{(\sqrt{2})^2-2}$

$$=\frac{1}{0}\ \text{Undefined: } x-2$$

b. $(g\circ f)(2)=g(f(2))=g\left|\dfrac{1}{2^2-2}\right|=g\left|\dfrac{1}{2}\right|$

$$=\sqrt{\frac{1}{2}}=\frac{\sqrt{2}}{2}$$

c. $(f\circ g)(x)=f(g(x))=f(\sqrt{x})=\dfrac{1}{(\sqrt{x})^2-2}$

$$=\frac{1}{x-2},\ x>0$$

d. $(g\circ f)(x)=g(f(x))=g\left|\dfrac{1}{x^2-2}\right|$

$$=\sqrt{\frac{1}{x^2-2}}=\frac{\sqrt{x^2-2}}{x^2-2}$$

31. $f(x)=x^2;\quad g(x)=x\neq1$

Domain f: $\{x\,|\,x \text{ is a real number}\}$

Domain g: $\{x\,|\,x \text{ is a real number}\}$

Domain $f+g=$ domain $f\neq g=$ domain $fg=\{x\,|\,x \text{ is a real number}\}$

Exclude the values of x where $g(x)=0$. So, $x\neq1$

Domain $\dfrac{f}{g}=\{x\,|\,x \text{ is a real number and } x\neq1\}$

33. $f(x)=2x+1;\quad g(x)=\neq2x+2$

Domain f: $\{x\,|\,x \text{ is a real number}\}$

Domain g: $\{x\,|\,x \text{ is a real number}\}$

Domain $f+g=$ domain $f\neq g=$ domain $fg=\{x\,|\,x \text{ is a real number}\}$

Exclude the values of x where $g(x)=0$. So, $x\neq1$

Domain $\dfrac{f}{g}=\{x\,|\,x \text{ is a real number and } x\neq1\}$

35. $f(x)=\dfrac{1}{x\neq1};\quad g(x)=\dfrac{\neq3}{x+2}$

Domain f: $\{x\,|\,x \text{ is a real number and } x\neq1\}$

Domain g: $\{x\,|\,x \text{ is a real number and } x\neq\neq2\}$

Domain $f+g=$ domain $f\neq g=$ domain $fg=\{x\,|\,x \text{ is a real number and } x\neq1 \text{ and } x\neq\neq2\}$

Exclude the values of x where $g(x)=0$. There are no such values.

Domain $\dfrac{f}{g}=\{x\,|\,x \text{ is a real number and } x\neq1 \text{ and } x\neq\neq2\}$

37.
$$f(x)=\frac{2x}{x+4};\quad g(x)=\frac{x}{x\neq 1}$$

Domain f: $\{x\,|\,x \text{ is a real number and } x\neq \neq 4\}$

Domain g: $\{x\,|\,x \text{ is a real number and } x\neq 1\}$

Domain $f+g=$ domain $f\neq g=$ domain $fg=\{x\,|\,x \text{ is a real number and } x\neq \neq 4 \text{ and } x\neq 1\}$

Exclude the values of x where $g(x)=0$. So, $x\neq 0$.

Domain $\dfrac{f}{g}=\{x\,|\,x \text{ is a real number and } x\neq \neq 4 \text{ and } x\neq 1 \text{ and } x\neq 0\}$

39.
$$f(x)=\frac{x\neq 2}{x+3};\quad g(x)=\frac{x\neq 1}{x\neq 2}$$

Domain f: $\{x\,|\,x \text{ is a real number and } x\neq \neq 3\}$

Domain g: $\{x\,|\,x \text{ is a real number and } x\neq 2\}$

Domain $f+g=$ domain $f\neq g=$ domain $fg=\{x\,|\,x \text{ is a real number and } x\neq \neq 3 \text{ and } x\neq 2\}$

Exclude the values of x where $g(x)=0$. So, $x\neq 1$.

Domain $\dfrac{f}{g}=\{x\,|\,x \text{ is a real number and } x\neq \neq 3 \text{ and } x\neq 2 \text{ and } x\neq 1\}$

41. $P(x)=R(x)-C(x)$

$$=\left(40x-0.0005x^2\right)-\left(120{,}000+6x\right)$$
$$=40x-0.0005x^2-120{,}000-6x$$
$$=-0.0005x^2+34x-120{,}000$$

43.

a. $L(x)+R(x)=\left(-0.28x^2+2.8x+15\right)+\left(0.2x+7\right)=-0.28x^2+3x+22$

b. $L(0)+R(0)=-0.28(0)^2+3(0)+22=22$ million pounds

c. $L(10)+R(10)=-0.28(10)^2+3(10)+22=-28+30+22=24$ million pounds

d. $L(10)-R(10)=-0.28(10)^2+2.8(10)+15-\left(0.2(10)+7\right)=-28+28+15-9=6$ million lbs

45.

a. $\dfrac{C(t)}{P(t)}=\dfrac{2.5t^2+8.5t+111}{-0.46t^2+1.14t+31.08}$ (in thousands)

b. $\dfrac{C(0)}{P(0)}=\dfrac{2.5(0)^2+8.5(0)+111}{-0.46(0)^2+1.14(0)+31.08}=\dfrac{111}{31.08}=3.571$ thousand $=\$3571$

c. $\dfrac{C(5)}{P(5)}=\dfrac{2.5(5)^2+8.5(5)+111}{-0.46(5)^2+1.14(5)+31.08}=\dfrac{216}{25.28}=8.544$ thousand $=\$8544$

47. $C(F) = \frac{5}{9}(F - 32),\ \ K(C) = C + 273$

  a. $(K \circ C)(F) = K(C(F)) = K\left(\frac{5}{9}(F - 32)\right) = \frac{5}{9}(F - 32) + 273$

  b. $C(41) = \frac{5}{9}(41 - 32) = \frac{5}{9}(9) = 5°C$

  c. $(K \circ C)(212) = \frac{5}{9}(212 - 32) + 273 = \frac{5}{9}(180) + 273 = 100 + 273 = 373°\,K$

49. $F(x) = x + 32,\ \ E(F) = F - 30$

  a. $F(6) = 6 + 32 = 38$

  b. $(E \circ F)(x) = E(F(x)) = E(x + 32) = x + 32 - 30 = x + 2$

  c. $(E \circ F)(8) = 8 + 2 = 10$

51.   Reading the table:

  a. $f(40) = 0$

  b. $f(42) = 8$

  c. $f(44) = 16$

53. $F(x) = \frac{9}{5}x + 32,\ \ f(F) = 4(F - 40)$

  a. $(f \circ F)(x) = f(F(x)) = f\left(\frac{9}{5}x + 32\right) = 4\left(\frac{9}{5}x + 32 - 40\right) = 4\left(\frac{9}{5}x - 8\right)$

  b. $(f \circ F)(10) = 4\left(\frac{9}{5}(10) - 8\right) = 4(18 - 8) = 4(10) = 40$ chirps

55. $x + y = 3$

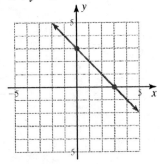

57. $2x + \frac{1}{2}y = 2$

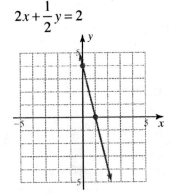

59. $y = -2x + 4$

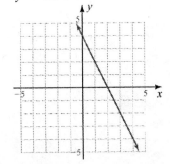

61. Domain of f: $\{x \mid 0 \le x \le 5\}$

63. Domain of $f + g$, $f \le g$, and fg
$$= \{x \mid 2 \le x \le 5\}$$

65. Sample answer. Addition of functions is commutative. Subtraction of functions is not commutative.

67. Sample answer. In general the order of composition of functions will produce different functions because it makes a difference which function is applied first. For example, if $f(x)=x^2$ and $g(x)=x+1$, the result will be different when 1 is added first or when the squaring is done first in the composition.

69. $P(x)=R(x)-C(x)$

$$= \left[300x-\frac{x^2}{30}\right]-(72{,}000+60x)$$

$$= 300x-\frac{x^2}{30}-72{,}000-60x$$

$$= -\frac{x^2}{30}+240x-72{,}000$$

71. $f(x)=x^3 \quad g(x)=x+1$

$(g\circ f)(x)=g(f(x))=g(x^3)=x^3+1$

73. $f(x)=x^3 \quad g(x)=x+1$

$(g\circ f)(-3)=g(f(-3))=g\left((-3)^3\right)$

$\qquad = g(-27)=-27+1=-26$

75. $f(x)=x^2+1$

$$\frac{f(x)-f(a)}{x-a}=\frac{x^2+1-\left(a^2+1\right)}{x-a}=\frac{x^2-a^2}{x-a}$$

$$=\frac{(x+a)(x-a)}{x-a}=x+a,\ x\neq a$$

77. $f(x)=x^2+4;\ g(x)=x-2$

$(f-g)(x)=f(x)-g(x)$

$\qquad =(x^2+4)-(x-2)$

$\qquad =x^2+4-x+2=x^2-x+6$

79. $f(x)=x^2+4;\ g(x)=x-2$

$$\left(\frac{f}{g}\right)(-2)=\frac{f(-2)}{g(-2)}=\frac{(-2)^2+4}{-2-2}=\frac{8}{-4}=-2$$

81. $f(x)=x^2+4;\ g(x)=x-2$

$$\left(\frac{g}{f}\right)(-2)=\frac{g(-2)}{f(-2)}=\frac{-2-2}{(-2)^2+4}=\frac{-4}{8}=-\frac{1}{2}$$

83. $f(x)=\dfrac{1}{x};\ g(x)=\dfrac{3}{x+1}$

Domain f: $\{x\,|\,x$ is a real number and $x\neq 0\}$

Domain g: $\{x\,|\,x$ is a real number and $x\neq -1\}$

Domain $f+g=$ domain $f-g=$ domain $fg=\{x\,|\,x$ is a real number and $x\neq 0$ and $x\neq -1\}$

Exclude the values of x where $g(x)=0$. There are no such values.

Domain $\dfrac{f}{g}=\{x\,|\,x$ is a real number and $x\neq 0$ and $x\neq -1\}$

10.2   Inverse Functions

Problems 10.2

1. **a.** Domain: $\{2, 3, 4\}$

Range: $\{1, 2, 3\}$

b. $S^{-1} = \{(3, 4), (2, 3), (1, 2)\}$

c. Domain: $\{1, 2, 3\}$

Range: $\{2, 3, 4\}$

d.

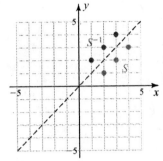

2. **a.** $f(x) = y = 2x - 3$

Interchange x and y and solve for y:

$$x = 2y - 3$$

$$x + 3 = 2y$$

$$\frac{x + 3}{2} = y$$

$$f^{-1}(x) = \frac{x + 3}{2} = \frac{1}{2}x + \frac{3}{2}$$

b.

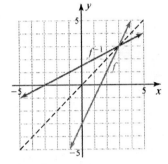

3. $f(x) = y = x^3$

Interchange x and y and solve for y:

$$x = y^3 \quad \rightarrow \quad y = \sqrt[3]{x}$$

The inverse is a function.

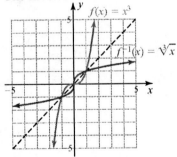

4. The inverse is a function.

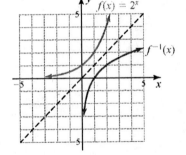

5. **a.** $n(C) = S = \dfrac{1}{6}(C - 4)$

Interchange C and S and solve for S:

$$C = \frac{1}{6}(S - 4)$$

$$6C = S - 4$$

$$6C + 4 = S$$

$$n^{-1}(C) = 6C + 4$$

b. $S = \dfrac{1}{6}(C - 4)$

$$6S = C - 4$$

$$6S + 4 = C$$

$$f(S) = 6S + 4$$

$$f(2) = 6(2) + 4 = 12 + 4 = 16°C$$

Exercises 10.2

1. $f = \{(1, 3), (2, 4), (3, 5)\}$

$f^{-1} = \{(3, 1), (4, 2), (5, 3)\}$

The inverse is a function.

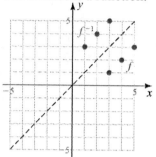

3. $f = \{(-1, 5), (-3, 4), (-4, 4)\}$

$f^{-1} = \{(5, -1), (4, -3), (4, -4)\}$

The inverse is not a function.

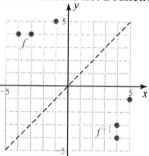

5. $f(x) = y = 3x + 3$

Interchange x and y and solve for y:

$x = 3y + 3$

$x - 3 = 3y$

$\dfrac{x - 3}{3} = y$

$f^{-1}(x) = \dfrac{x - 3}{3} = \dfrac{1}{3}x - 1$

The inverse is a function.

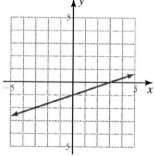

7. $f(x) = y = 2x - 4$

Interchange x and y and solve for y:

$x = 2y - 4$

$x + 4 = 2y$

$\dfrac{x + 4}{2} = y$

$f^{-1}(x) = \dfrac{x + 4}{2} = \dfrac{1}{2}x + 2$

The inverse is a function.

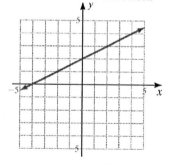

9. $f(x) = y = 2x^2$

Interchange x and y and solve for y:

$x = 2y^2$

$\dfrac{x}{2} = y^2$

$y = \pm\sqrt{\dfrac{x}{2}} = \pm\dfrac{\sqrt{2x}}{2}$

The inverse is not a function.

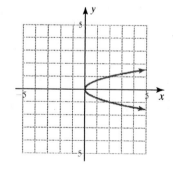

11. $f(x) = y = x^2 - 1$

Interchange x and y and solve for y:

$$x = y^2 - 1$$
$$x + 1 = y^2$$
$$y = \pm\sqrt{x+1}$$

The inverse is not a function.

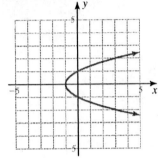

13. $f(x) = y = -x^3$

Interchange x and y and solve for y:

$$x = -y^3$$
$$-x = y^3$$
$$y = \sqrt[3]{-x}$$
$$f^{-1}(x) = \sqrt[3]{-x}$$

The inverse is a function.

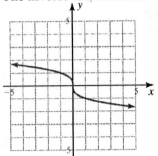

15. $y = f(x) = 2^{x+1}$

Interchange x and y:

$$x = 2^{y+1}$$

The inverse is a function.

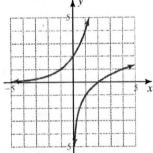

17. $y = f(x) = \left|\dfrac{1}{3}\right|^{x}$

Interchange x and y:

$$x = \left|\dfrac{1}{3}\right|^{y}$$

The inverse is a function.

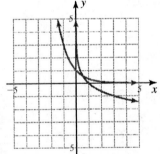

19. $y = f(x) = 2^{-x}$

Interchange x and y:

$$x = 2^{-y}$$

The inverse is a function.

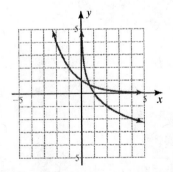

21.

$$f(x) = 4x + 4, \quad f^{-1}(x) = \frac{x-4}{4}$$

a.

$$f(f^{-1}(3)) = f\left(\frac{3-4}{4}\right) = f\left(-\frac{1}{4}\right)$$

$$= 4\left(-\frac{1}{4}\right) + 4 = -1 + 4 = 3$$

b.

$$f^{-1}(f(-1)) = f^{-1}(4(-1) + 4) = f^{-1}(0)$$

$$= \frac{0-4}{4} = -1$$

23.

$$y = f(x) = \frac{1}{x}$$

Interchange x and y and solve for y:

$$x = \frac{1}{y}$$

$$xy = 1$$

$$y = \frac{1}{x} \qquad f^{-1}(x) = \frac{1}{x}$$

25. $f(x) = y = 2x$

Interchange x and y and solve for y:

$$x = 2y$$

$$\frac{x}{2} = y$$

$$f^{-1}(x) = \frac{x}{2}$$

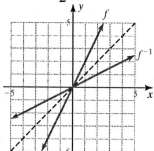

27. $f(x) = y = -x^2 + 2$

Interchange x and y and solve for y:

$$x = -y^2 + 2$$

$$y^2 = 2 - x$$

$$y = \pm\sqrt{2-x}$$

$$f^{-1}(x) = \pm\sqrt{2-x}$$

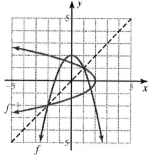

29. $f(x) = y = \sqrt{4 - x^2}$

Interchange x and y:

$$f^{-1}: x = \sqrt{4 - y^2}$$

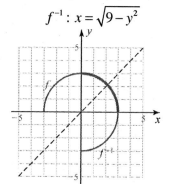

31. $f(x) = y = \sqrt{9 - x^2}$

Interchange x and y:

$$f^{-1}: x = \sqrt{9 - y^2}$$

33. $f(x) = y = x^3$

Interchange x and y and solve for y:

$$x = y^3$$
$$\sqrt[3]{x} = y$$
$$f^{-1}(x) = \sqrt[3]{x}$$

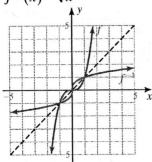

35. $f(x) = y = |x|$

Interchange x and y:

$$x = |y|$$
$$f^{-1}: \ x = |y|$$

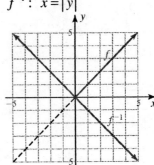

37. a. $S = f(L) = 3L - 21$

Interchange S and L and solve for L:

$$L = 3S - 21$$
$$L + 21 = 3S$$
$$\frac{L + 21}{3} = S$$
$$f^{-1}(L) = \frac{L + 21}{3}$$

b. $7 = 3L - 21$
$$28 = 3L$$
$$9\tfrac{1}{3} \text{ in} = L$$

39. a. $d = 32 - 16 = 16$

b. $d = f(w) = w - 16$

Interchange w and d and solve for d:

$$w = d - 16$$
$$w + 16 = d$$
$$f^{-1}(w) = w + 16$$

c. $12 = w - 16$
$$28 = w$$

41. a. $w = -0.12(1988) + 280 = 41.44 \text{ sec}$

b. $w = f(x) = -0.12x + 280$

Interchange x and w and solve for w:

$$x = -0.12w + 280$$
$$0.12w = 280 - x$$
$$w = \frac{280 - x}{0.12}$$
$$f^{-1}(x) = \frac{280 - x}{0.12}$$

c. $40 = -0.12x + 280$
$$0.12x = 240$$
$$x = 2000$$

43. $3^2 = 9$

45. $\left(\dfrac{1}{3}\right)^{-3} = 3^3 = 27$

47. $\left(\dfrac{1}{2}\right)^{x/3}$ when $x = -6$

$$\left(\dfrac{1}{2}\right)^{-6/3} = \left(\dfrac{1}{2}\right)^{-2} = 2^2 = 4$$

49. $f(x)=3x-2$

Multiply by 3 $\xrightarrow{undo}$ Divide by 3 $f^{-1}(x)=\dfrac{x+2}{3}$

Subtract 2 $\xrightarrow{undo}$ Add 2 $x+2$

                                                x

51. $f(x)=\dfrac{x+1}{2}$

Add 1 $-\overset{undo}{=}-$ Subtract 1 $f^{-1}(x)=2x-1$

Divide by 2 $-\overset{undo}{=}-$ Multiply by 2 $2x$

                                                x

53. $f(x)=x^{3}+1$

Cube $-\overset{undo}{=}-$ Cube root $f^{-1}(x)=\sqrt[3]{x-1}$

Add 1 $-\overset{undo}{=}-$ Subtract 1 $x-1$

                                            x

55. $f(x)=\sqrt{x}$

Square root $\geq\overset{undo}{2}\geq$ Square $f^{-1}(x)=x^{2},\ x\geq 0$

                                            x

57.   Sample answer: A line is a function and its reflection over the line $y=x$ will also be a function. If the original line is a horizontal line, then its reflection is a vertical line, which is not a function. Thus the restriction $a\neq 0$ is placed on the problem.

59.   Sample answer: The composition of a function and its inverse will always be x. Applying the inverse of a function to a function undoes what the original function did.

61.   Sample answer: To verify that two functions are inverses of each other, perform the composition of the functions. If the result is x, the functions are inverses of each other.

63. $f(x)=x^{3}$

Interchange x and y and solve for y:

$\qquad x=y^{3}$

$\qquad y=\sqrt[3]{x}$

$f^{-1}(x)=\sqrt[3]{x}$

The inverse is a function.

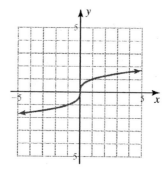

65.   Domain: $\{2,\ 3,\ 4\}$

Range: $\{1,\ 2,\ 3\}$

67.   Domain: $\{1,\ 2,\ 3\}$

Range: $\{2,\ 3,\ 4\}$

10.3   Exponential Functions

Problems 10.3

1. $f(x) = 3^x$, $g(x) = \left(\frac{1}{3}\right)^x$

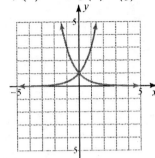

2. $f(x) = -e^x$, $g(x) = -e^{-x}$
Find the opposite of each y-coordinate in the tables.

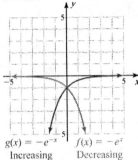

Increasing $g(x) = -e^{-x}$ Decreasing $f(x) = -e^x$

3.   a. $A = Pe^{rt}$ when $P = 100$, $r = 0.06$, $t = 2.5$
$A = 100e^{(0.06)(2.5)} = 100e^{0.15}$
$= 100(1.1618) = 116.18$
The compounded amount is \$116.18.

b.
$A = P\left[1 + \dfrac{r}{n}\right]^{nt}$ when $P = 100$, $r = 0.06$,
$t = 2.5$, $n = 4$

$A = 100\left[1 + \dfrac{0.06}{4}\right]^{4(2.5)} = 100(1 + 0.015)^{10}$
$= 100(1.015)^{10} = 100(1.1605) = 116.05$
The compounded amount is \$116.05.

4. $G = 1000e^{-1.2t}$ when $t = 1.5$
$G = 1000e^{-1.2(1.5)} = 1000e^{-1.8}$
$= 1000(0.165) = 165$
There are about 165 grams remaining after 18 months.

5. $P = P_0 e^{kt}$ when $P_0 = 226,546,000$, $t = 40$
$k = 0.016 - 0.0086 = 0.0074$
$P = 226,546,000e^{0.0074(40)}$
$= 226,546,000e^{0.296} = 304,584,336$
The predicted population in 2020 is 304,584,000.

Exercises 10.3

1.   a. $5^{-1} = \dfrac{1}{5}$

b. $5^0 = 1$

c. $5^1 = 5$

3.   a. $3^{-2} = \dfrac{1}{3^2} = \dfrac{1}{9}$

b. $3^0 = 1$

c. $3^2 = 9$

5.
   a. $10^{-2/2} = 10^{-1} = \dfrac{1}{10}$

   b. $10^{0/2} = 10^{0} = 1$

   c. $10^{2/2} = 10^{1} = 10$

7.
   a. $f(x) = 5^{x}$ Increasing

   b. $g(x) = 5^{-x}$ Decreasing

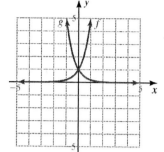

9.
   a. $f(x) = 10^{x}$ Increasing

   b. $g(x) = 10^{-x}$ Decreasing

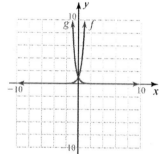

11.
   a. $f(x) = e^{2x}$ Increasing

   b. $g(x) = e^{-2x}$ Decreasing

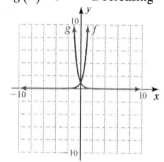

13.
   a. $f(x) = 3^{x} + 1$ Increasing

   b. $g(x) = 3^{-x} + 1$ Decreasing

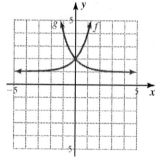

15.
   a. $f(x) = 2^{x+1}$ Increasing

   b. $g(x) = 2^{-x+1}$ Decreasing

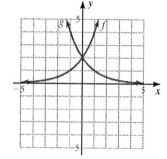

17.
   a. $A = Pe^{rt}$ when $P = 1000$, $r = 0.09$, $t = 10$

$$A = 1000e^{(0.09)(10)} = 1000e^{0.90}$$

$$= 1000(2.4596) = 2459.60$$

The compounded amount is \$2459.60.

   b. $A = P\left|1 + \dfrac{r}{n}\right|^{nt}$ when $P = 1000$, $r = 0.09$,

$$t = 10,\ n = 4$$

$$A = 1000\left|1 + \dfrac{0.09}{4}\right|^{4(10)} = 1000(1 + 0.0225)^{40}$$

$$= 1000(1.0225)^{40} = 2435.19$$

The compounded amount is \$2435.19.

19. a. $A = Pe^{rt}$ when $P = 1000$, $r = 0.06$, $t = 10$

$A = 1000e^{(0.06)(10)} = 1000e^{0.60}$

$= 1000(1.82212) = 1822.12$

The compounded amount is \$1822.12.

b. $A = P\left|1 + \dfrac{r}{n}\right|^{nt}$ when $P = 1000$, $r = 0.06$,

$t = 10$, $n = 4$

$A = 1000\left|1 + \dfrac{0.06}{4}\right|^{4(10)} = 1000(1 + 0.015)^{40}$

$= 1000(1.015)^{40} = 1814.02$

The compounded amount is \$1814.02.

21. a. $P = 2000\left(2^{0.2t}\right)$, $t = 0$

$P = 2000\left(2^{0.2(0)}\right) = 2000\left(2^0\right) = 2000$

b. $P = 2000\left(2^{0.2t}\right)$, $t = 5$

$P = 2000\left(2^{0.2(5)}\right) = 2000\left(2^1\right) = 4000$

c. $P = 2000\left(2^{0.2t}\right)$, $t = 10$

$P = 2000\left(2^{0.2(10)}\right) = 2000\left(2^2\right) = 8000$

23. a. $G = 2000e^{-1.05t}$, $t = 0$

$G = 2000e^{-1.05(0)} = 2000e^0 = 2000$ g

b. $G = 2000e^{-1.05t}$, $t = 1$

$G = 2000e^{-1.05(1)} = 2000e^{-1.05} = 699.9$ g

c. $G = 2000e^{-1.05t}$, $t = 2$

$G = 2000e^{-1.05(2)} = 2000e^{-2.10} = 244.9$ g

25. $P = P_0e^{kt}$ when $P_0 = 14{,}609{,}000$, $t = 20$

$k = 0.0232 - 0.004 = 0.0192$

$P = 14{,}609{,}000e^{0.0192(20)}$

$= 14{,}609{,}000e^{0.384} = 21{,}448{,}137$

The predicted population in 2000 is
21,448,000.

27. a. $S(t) = 32(10)^{0.19t}$ when $t = 5$

$S(5) = 32(10)^{0.19(5)} = 32(10)^{0.95}$

$= 32(8.9125) = 285.2$ million CDs

b. $S(t) = 32(10)^{0.19t}$ when $t = 25$

$S(25) = 32(10)^{0.19(25)} = 32(10)^{4.75}$

$= 32(56234) = 1{,}799{,}492$ million CDs

29. $N(t) = 500{,}000\left(\frac{2}{3}\right)^t$

a. $N(1) = 500{,}000\left(\frac{2}{3}\right)^1 = 500{,}000\left(\frac{2}{3}\right)$

$= 333{,}333$ cans in use after 1 yr

b. $N(2) = 500{,}000\left(\frac{2}{3}\right)^2 = 500{,}000\left(\frac{4}{9}\right)$

$= 222{,}222$ cans in use after 2 yr

c. $N(10) = 500{,}000\left(\frac{2}{3}\right)^{10} = 500{,}000\left(\frac{1024}{59049}\right)$

$= 8671$ cans in use after 10 yr

31. $A(a) = 14.7(10)^{-0.000018a}$

a. $A(29{,}000) = 14.7(10)^{-0.000018(29{,}000)}$

$= 14.7(10)^{-0.522} = 4.42$ lb/in^2

b. $A(20{,}000) = 14.7(10)^{-0.000018(20{,}000)}$

$= 14.7(10)^{-0.36} = 6.42$ lb/in^2

33. $\left(10^x\right)\left(10^y\right) = 10^{x+y}$ Product property

35. $\left(10^x\right)^3 = 10^{3x}$ Power property

37. $A = Pe^{rt}$ when $P = 1000$, $r = 0.06$, $t = 10$

$A = 1000e^{(0.06)(10)} = 1000e^{0.60}$

$\quad = 1000(1.82212) = 1822.12$

The continuously compounded amount is $1822.12.

$A = P\left|1 + \dfrac{r}{n}\right|^{nt}$ when $P = 1000$, $r = 0.06$,

$t = 10$, $n = 12$

$A = 1000\left|1 + \dfrac{0.06}{12}\right|^{12(10)} = 1000(1 + 0.005)^{120}$

$\quad = 1000(1.005)^{120} = 1819.40$

The compounded monthly amount is $1819.40.
Continuous is $2.72 more.

39. a. Sample answer: When $b = 1$, the graph of $f(x)$ will be a horizontal line.

b. Sample answer: $f(x) = b^x$ is a function when $b = 1$, because a horizontal line is a function.

c. Sample answer: $f(x) = b^x$ does not have an inverse function when $b = 1$, because the graph of the inverse is a vertical line which is not a function.

41. Sample answer: The graphs of $f(x) = b^x$ and $g(x) = b^{-x}$ are symmetric to each about the y-axis.

43. $G = 1000e^{-1.2t}$, $t = 1.5$

$G = 1000e^{-1.2(1.5)} = 1000e^{-1.8} = 165$ g

The radioactive substance decays to 165 grams after 18 months.

45. $f(x) = e^{x/2}$

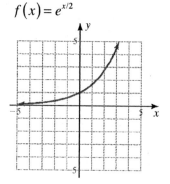

47. $f(x) = 6^x$

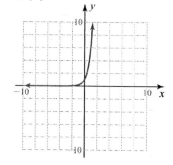

10.4   Logarithmic Functions and Their Properties

Problems 10.4

1. $y = f(x) = \log_3 x$ is equivalent to $3^y = x$

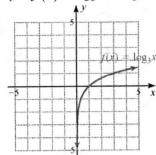

2. $2^x = 1024$

logarithmic form: $\log_2 1024 = x$

3. $\log_4 32 = \dfrac{5}{2}$

exponential form: $4^{5/2} = 32$

or $\left(\sqrt{4}\right)^5 = 2^5 = 32 = 32$ (checks)

4.   **a.** $\log_3 x = -5$ is equivalent to $3^{-5} = x$

$x = \dfrac{1}{3^5} = \dfrac{1}{243}$

b. $\log_x 81 = 2$ is equivalent to $x^2 = 81$

$x = \sqrt{81} = \pm 9$

The solution is $x = 9$ because the definition states that the base must be positive.

5.   **a.** $\log_2 64 = x$

$2^x = 64$

$2^x = 2^6$

$x = 6$

b. $\log_2 \dfrac{1}{8} = x$

$2^x = \dfrac{1}{8}$

$2^x = 2^{-3}$

$x = -3$

c. $\log_{10} 100 = x$

$10^x = 100$

$10^x = 10^2$

$x = 2$

d. $\log_{10} 0.1 = x$

$10^x = \dfrac{1}{10}$

$10^x = 10^{-1}$

$x = -1$

6. $\log_b \dfrac{a^2}{c^3} = 2\log_b a - 3\log_b c$

$\qquad = \log_b a^2 - \log_b c^3$

$\qquad = \log_b \dfrac{a^2}{c^3}$

7. $\log \dfrac{\sqrt{xy}}{z^3} = \dfrac{1}{2}\log x + \dfrac{1}{2}\log y - 3\log z$

$\qquad = \log x^{1/2} + \log y^{1/2} - \log z^3$

$\qquad = \log(xy)^{1/2} - \log z^3$

$\qquad = \log \dfrac{\sqrt{xy}}{z^3}$

8.
$$R = \log_{10} \frac{I}{I_0} = \log_{10} 10^{7.2} = 7.2$$

9.
$$\log_{10} 24(1.06)^{375} = \log_{10} 24 + \log_{10} 1.06^{375}$$
$$= \log_{10} 24 + 375\left(\log_{10} 1.06\right)$$
$$= 1.3802 + 375(0.0253)$$
$$= 1.3802 + 9.4875$$
$$= 10.8677$$

Exercises 10.4

1. $y = \log_2 x$ is equivalent to $2^y = x$

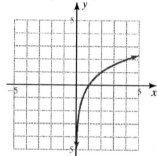

3. $y = \log_5 x$ is equivalent to $5^y = x$

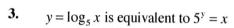
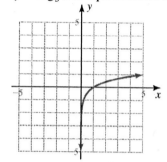

5. $f(x) = \log_{1/2} x$ is equivalent to $\left(\frac{1}{2}\right)^y = x$

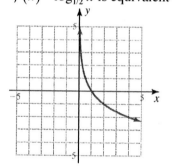

7. $y = \log_{1.5} x$ is equivalent to $1.5^y = x$

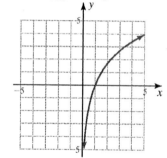

9. $2^x = 128$

logarithmic form: $\log_2 128 = x$

11. $10^t = 1000$

logarithmic form: $\log_{10} 1000 = t$

13. $81^{1/2} = 9$

logarithmic form: $\log_{81} 9 = \frac{1}{2}$

15. $216^{1/3} = 6$

logarithmic form: $\log_{216} 6 = \frac{1}{3}$

17. $e^3 = t$

logarithmic form: $\log_e t = 3$

19. $\log_9 729 = 3$

exponential form: $9^3 = 729$

Check: $9^3 = 729 = 729$

21.

$$\log_2 \frac{1}{256} = -8$$

exponential form: $2^{-8} = \frac{1}{256}$

Check: $2^{-8} = \frac{1}{2^8} = \frac{1}{256} = \frac{1}{256}$

23.

$$\log_{81} 27 = \frac{3}{4}$$

exponential form: $81^{3/4} = 27$

Check: $81^{3/4} = \left(\sqrt[4]{81}\right)^3 = 3^3 = 27 = 27$

25. $x = \log_4 16$

exponential form: $4^x = 16$

27. $-2 = \log_{10} 0.01$

exponential form: $10^{-2} = 0.01$

Check: $10^{-2} = \frac{1}{10^2} = \frac{1}{100} = 0.01$

29. $\log_e 30 = 3.4012$

exponential form: $e^{3.4012} = 30$

Check: $e^{3.4012} = 30 = 30$

31. $\log_e 0.3166 = -1.15$

exponential form: $e^{-1.15} = 0.3166$

Check: $e^{-1.15} = 0.3166 = 0.3166$

33. $\log_3 x = 2$

$$x = 3^2 = 9$$

35. $\log_3 x = -3$

$$x = 3^{-3} = \frac{1}{3^3} = \frac{1}{27}$$

37.

$$\log_4 x = \frac{1}{2}$$

$$x = 4^{1/2} = \sqrt{4} = 2$$

39.

$$\log_8 x = \frac{1}{3}$$

$$x = 8^{1/3} = \sqrt[3]{8} = 2$$

41. $\log_x 16 = 4$

$$x^4 = 16$$

$$x = \sqrt[4]{16} = 2$$

43. $\log_x 27 = 3$

$$x^3 = 27$$

$$x = \sqrt[3]{27} = 3$$

45.

$$\log_2 \frac{1}{4} = x$$

$$2^x = \frac{1}{4}$$

$$2^x = 2^{-2}$$

$$x = -2$$

47.

$$\log_{16} \frac{1}{2} = x$$

$$16^x = \frac{1}{2}$$

$$\left(2^4\right)^x = 2^{-1}$$

$$2^{4x} = 2^{-1}$$

$$4x = -1$$

$$x = -\frac{1}{4}$$

49.
$$\log_3 \frac{1}{9} = x$$
$$3^x = \frac{1}{9}$$
$$3^x = 3^{-2}$$
$$x = -2$$

51.
$$\log_2 256 = x$$
$$2^x = 256$$
$$2^x = 2^8$$
$$x = 8$$

53.
$$\log_3 81 = x$$
$$3^x = 81$$
$$3^x = 3^4$$
$$x = 4$$

55.
$$\log_2 \frac{1}{8} = x$$
$$2^x = \frac{1}{8}$$
$$2^x = 2^{-3}$$
$$x = -3$$

57.
$$\log_{10} 1,000,000 = x$$
$$10^x = 1,000,000$$
$$10^x = 10^6$$
$$x = 6$$

59.
$$\log_3 1 = x$$
$$3^x = 1$$
$$3^x = 3^0$$
$$x = 0$$

61.
$$\log_{10} 10 = x$$
$$10^x = 10$$
$$10^x = 10^1$$
$$x = 1$$

63.
$$\log_e e = x$$
$$e^x = e$$
$$e^x = e^1$$
$$x = 1$$

65.
$$\log_5 \frac{1}{5} = x$$
$$5^x = \frac{1}{5}$$
$$5^x = 5^{-1}$$
$$x = -1$$

67.
$$\log_8 2 = x$$
$$8^x = 2$$
$$\left(2^3\right)^x = 2^1$$
$$2^{3x} = 2^1$$
$$3x = 1$$
$$x = \frac{1}{3}$$

69.
$$\log_e e^{-3} = x$$
$$e^x = e^{-3}$$
$$x = -3$$

71.
$$\log_{10} 10^t = x$$
$$10^x = 10^t$$
$$x = t$$

73.
$$\log_4 4^t = x$$
$$4^x = 4^t$$
$$x = t$$

75.
$$\log \frac{26}{7} \mid \log \frac{15}{63} + \log \frac{5}{26} = \log \left| \frac{26}{7} \div \frac{15}{63} \mid \frac{5}{26} \right|$$
$$= \log \left| \frac{\cancel{26}}{7} \mid \frac{\cancel{63}}{15} \mid \frac{5}{\cancel{26}} \right| = \log \left| \frac{3 \cancel{\cancel{26}} \cancel{\cancel{7}} \cancel{\cancel{26}}}{\cancel{7} \cancel{\cancel{26}} \cancel{\cancel{26}}} \right|$$
$$= \log 3$$

439

77.

$$\log b^3 + \log 2 \mid \log \sqrt{b} + \log \frac{\sqrt{b^3}}{2} = \log \left| b^3 \cdot 2 \div \sqrt{b} \cdot \frac{\sqrt{b^3}}{2} \right| = \log \left| b^3 \cdot \cancel{2} \cdot \frac{1}{\cancel{\sqrt{b}}} \cdot \frac{b \cancel{\sqrt{b}}}{\cancel{2}} \right|$$

$$= \log b^4 = 4 \log b$$

79.

$$\log k^{3/2} + \log r \mid \log k \mid \log r^{3/4} = \log \left(k^{3/2} \cdot r \div k \div r^{3/4} \right) = \log \left| \frac{k^{3/2} r}{k \cdot r^{3/4}} \right| = \log \left(k^{3/2-1} r^{1-3/4} \right)$$

$$= \log \left(k^{1/2} r^{1/4} \right) = \log \left(k^{2/4} r^{1/4} \right) = \log \left(k^2 r \right)^{1/4} = \frac{1}{4} \log \left(k^2 r \right)$$

81.

$$2 \log b + 6 \log a \mid 3 \log b = 6 \log a \mid \log b = \log a^6 \mid \log b = \log \left| \frac{a^6}{b} \right|$$

83.

$$\frac{1}{3} \log x + \frac{1}{3} \log y^2 \mid \log z = \log x^{1/3} + \log y^{2/3} \mid \log z = \log \left(x^{1/3} \cdot y^{2/3} \right) \mid \log z = \log \left| \frac{x^{1/3} y^{2/3}}{z} \right|$$

$$= \log \left| \frac{\sqrt[3]{xy^2}}{z} \right|$$

85.

$$R = \log_{10} \frac{I}{I_0} = \log_{10} 10^{8.9} = 8.9$$

87.

$$\log_{10} A = \log_{10} 5000 (1.005)^{216}$$
$$= \log_{10} 5000 + \log_{10} 1.005^{216}$$
$$= \log_{10} 5000 + 216 \left(\log_{10} 1.005 \right)$$
$$= 3.69897 + 216 (0.00217)$$
$$= 3.69897 + 0.46872$$
$$= 4.16769$$

89.
$$P(A) = 29 + 50 \log_{10} (A+1)$$

a.
$$P(2) = 29 + 50 \log_{10} (2+1)$$
$$= 29 + 50 \log_{10} 3 = 29 + 50 (0.4771)$$
$$= 29 + 23.855 \approx 53\%$$

b.
$$P(8) = 29 + 50 \log_{10} (8+1)$$
$$= 29 + 50 \log_{10} 9 = 29 + 50 (0.9542)$$
$$= 29 + 47.71 \approx 77\%$$

c.
$$P(16) = 29 + 50 \log_{10} (16+1)$$
$$= 29 + 50 \log_{10} 17 = 29 + 50 (1.2304)$$
$$= 29 + 61.52 \approx 91\%$$

d.

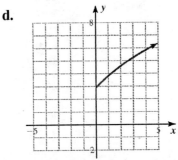

e.
$$P(12) = 29 + 50 \log_{10} (12+1)$$
$$= 29 + 50 \log_{10} 13 = 29 + 50 (1.1139)$$
$$= 29 + 55.695 \approx 85\%$$
$$0.85h = 60$$
$$h \approx 71 \text{ in}$$

91. $$32.68 = 3.268 \times 10^1$$

93. $$0.002387 = 2.387 - 10^{-3}$$

95. $0.0000569 = 5.69 - 10^{-5}$

97.

$$R = \frac{1000(1.02)^{4(10)}}{1000(1.085)^{10}} = \frac{1.02^{40}}{1.085^{10}}$$

$\log R = \log 1.02^{40} - \log 1.085^{10} = 40 \log 1.02 - 10 \log 1.085 = 40(0.00860) - 10(0.03543)$

$\qquad = 0.3440 - 0.3543 = -0.0103$

The 8.5% annually rate is greater.

99. Sample answer: The exponential function $f(x) = b^x$ and the logarithmic function $g(x) = \log_b x$ are inverse functions. The one undoes the other.

101. Sample answer: The number 1 cannot be the base of the logarithmic function because 1 to a power can never have a result other than 1.

103.

$$R = \log_{10} \frac{I}{I_0} = \log_{10} 10^{6.4} = 6.4$$

105.

$$\log \left[\frac{xy^2}{z^3} \right]^2 = 2 \log \left[\frac{xy^2}{z^3} \right] = 2 \left[\log (xy^2) \right] \log z^3 = 2 \left[\log x + \log y^2 \right] \log z^3 \right]$$

$$\qquad = 2 \left[\log x + 2 \log y \right] 3 \log z \right] = 2 \log x + 4 \log y \right] 6 \log z$$

107. $2^{10} = 1024$

logarithmic form: $\log_2 1024 = 10$

109.

$$\log_2 \frac{1}{8} = x$$

$$2^x = \frac{1}{8}$$

$$2^x = 2^{-3}$$

$$x = -3$$

111. $\log_{10} 0.1 = x$

$$10^x = 0.1$$

$$10^x = 10^{-1}$$

$$x = -1$$

113. $\log_3 x = -3$

$$x = 3^{-3} = \frac{1}{3^3} = \frac{1}{27}$$

115.

$$\log_3 \frac{1}{27} = x$$

$$3^x = \frac{1}{27}$$

$$3^x = 3^{-3}$$

$$x = -3$$

117. $y = \log_7 x$

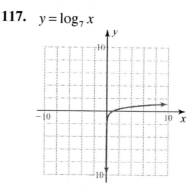

441

10.5   Common and Natural Logarithms

Problems 10.5

1. a. $\log 734,000 \approx 5.8657$

b. $\log 0.0000734 - -4.1343$

2. a. inv $\log 0.6243 \approx 4.2102$

b. inv $\log (\approx 3.1963) \approx 0.0006$

3. $\ln 4563 \approx 8.4257$

4. inv $\ln (\approx 4.3874) \approx 0.0124$

5. $\log_5 20 = \dfrac{\log_{10} 20}{\log_{10} 5} \approx \dfrac{1.3010}{0.6990} \approx 1.8612$

6. $\log_5 20 = \dfrac{\ln 20}{\ln 5} \approx \dfrac{2.9957}{1.6094} \approx 1.8614$

7. a. $f(x) = e^{(1/4)x}$

b. $f(x) = -e^{(1/4)x}$

c. $f(x) = e^{(1/4)x} + 2$

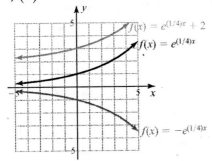

8. a. $f(x) = \ln(x+1)$

b. $f(x) = \ln x + 1$

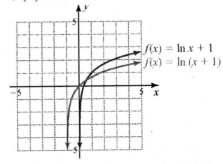

9. $\begin{aligned}
\text{pH} &= \approx \log\left(4.2 \approx 10^{\approx 7}\right) \\
&= \approx \left(\log 4.2 + \log 10^{\approx 7}\right) \\
&= \approx \left(\log 4.2 \approx 7\right) \\
&= 7 \approx \log 4.2 \\
&\approx 7 \approx 0.6232 \\
&\approx 6.3768
\end{aligned}$

10. a. $P = 90e^{\approx 0.5(0.35)} = 90e^{\approx 0.175}$
≈ 75.55

b. $P = 90e^{\approx 0.5(0.4)} = 90e^{\approx 0.2}$
≈ 73.69

Exercises 10.5

1. $\log 74.5 \approx 1.8722$

3. $\log 1840 \approx 3.2648$

5. $\log 0.0437 - -1.3595$

7. $\log 50.18 \approx 1.7005$

9. $\log 0.01238 - -1.9073$

11. $\log 0.008606 - -2.0652$

13. inv $\log 1.2672 \approx 18.50$

15. inv $\log (\approx 2.2328) \approx 0.01$

17. inv $\log 1.4630 \approx 29.04$

19. inv $\log(\approx 0.134) \approx 0.73$

21. $\ln 3 \approx 1.0986$

23. $\ln 52 \approx 3.9512$

25. $\ln 2356 \approx 7.7647$

27. $\ln 0.054 - -2.9188$

29. $\ln 0.00062 - -7.3858$

31. inv $\ln 1.2528 \approx 3.50$

33. inv $\ln 4.1744 \approx 65.00$

35. inv $\ln 0.0392 \approx 1.04$

37. inv $\ln(\approx 2.3025) \approx 0.10$

39. inv $\ln(\approx 4.6051) \approx 0.01$

41.
$$\log_3 20 = \frac{\log_{10} 20}{\log_{10} 3} \approx \frac{1.3010}{0.4771} \approx 2.7268$$
$$\log_3 20 = \frac{\ln 20}{\ln 3} \approx \frac{2.9957}{1.0986} \approx 2.7268$$

43.
$$\log_{100} 40 = \frac{\log_{10} 40}{\log_{10} 100} \approx \frac{1.6021}{2} \approx 0.8010$$
$$\log_{100} 40 = \frac{\ln 40}{\ln 100} \approx \frac{3.6889}{4.6052} \approx 0.8010$$

45.
$$\log_{0.2} 3 = \frac{\log_{10} 3}{\log_{10} 0.2} - \frac{0.4771}{-0.6990} - -0.6826$$
$$\log_{0.2} 3 = \frac{\ln 3}{\ln 0.2} - \frac{1.0986}{-1.6094} - -0.6826$$

47.
$$\log_4 0.4 = \frac{\log_{10} 0.4}{\log_{10} 4} - \frac{-0.3979}{0.6021} - -0.6610$$
$$\log_4 0.4 = \frac{\ln 0.4}{\ln 4} - \frac{-0.9163}{1.3863} - -0.6610$$

49.
$$\log_{\sqrt{2}} 0.8 = \frac{\log_{10} 0.8}{\log_{10} \sqrt{2}} - \frac{-0.0969}{0.1505} - -0.6439$$
$$\log_{\sqrt{2}} 0.8 = \frac{\ln 0.8}{\ln \sqrt{2}} - \frac{-0.2231}{0.3466} - -0.6439$$

51. $f(x) = e^{3x}$

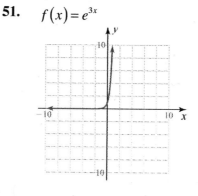

53. $f(x) = -e^{3x}$

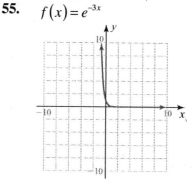

55. $f(x) = e^{-3x}$

443

57. $f(x) = e^{(1/2)x}$

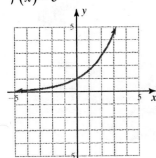

59. $f(x) = e^x + 1$

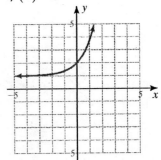

61. $f(x) = e^{0.5x} + 1$

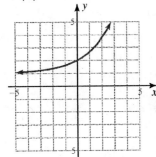

63. $f(x) = 2e^x$

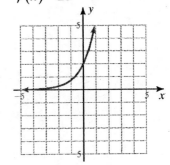

65. $f(x) = \ln(x + 1)$

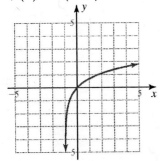

67. $f(x) = \ln x + 4$

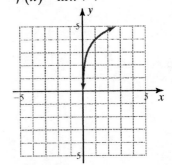

69. $f(x) = \ln(x - 1)$

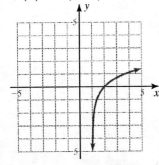

71.
$$\text{pH} = \approx \log\left(7 \approx 10^{\approx 7}\right)$$
$$= \approx \left(\log 7 + \log 10^{\approx 7}\right)$$
$$= \approx \left(\log 7 \approx 7\right)$$
$$= 7 \approx \log 7$$
$$\approx 7 \approx 0.8451$$
$$\approx 6.2$$

73.
$$\text{pH} = -\log\left(1.6 \approx 10^{-8}\right)$$
$$= -\left(\log 1.6 + \log 10^{-8}\right)$$
$$= -\left(\log 1.6 \approx 8\right)$$
$$= 8 \approx \log 1.6$$
$$\approx 8 \approx 0.2041$$
$$\approx 7.8$$

75.
$$\text{pH} = -\log\left(4 \approx 10^{-7}\right)$$
$$= -\left(\log 4 + \log 10^{-7}\right)$$
$$= -\left(\log 4 \approx 7\right)$$
$$= 7 \approx \log 4$$
$$\approx 7 \approx 0.6021$$
$$\approx 6.4$$

77. $H = 33{,}000{,}000 e^{0.03t}$

  a. $H = 33{,}000{,}000 e^{0.03(0)}$
$$= 33{,}000{,}000(1) = 33{,}000{,}000$$

  b. $H = 33{,}000{,}000 e^{0.03(10)}$
$$= 44{,}545{,}341$$

79. $B = 50{,}000 e^{0.2t}$

  a. $B = 50{,}000 e^{0.2(0)} = 50{,}000(1) = 50{,}000$

  b. $B = 50{,}000 e^{0.2(2)}$
$$= 74{,}591$$

  c. $B = 50{,}000 e^{0.2(6)}$
$$= 166{,}006$$

81. $S = 1000 e^{-0.1d}$

  a. $S = 1000 e^{-0.1(0)} = 1000(1) = 1000$

  b. $S = 1000 e^{-0.1(10)} = 368$

83. $C = 100\left(1 \approx e^{-0.5t}\right)$

  a. $C = 100\left(1 \approx e^{-0.5(0)}\right) = 100(1 \approx 1) = 0$

  b. $C = 100\left(1 \approx e^{-0.5(1)}\right)$
$$\approx 39.3$$

85.
$$M = -2.5 \log \approx \frac{B}{B_0} \approx$$

  a. $M = -2.5 \log \approx \dfrac{2.1 B_0}{B_0} \approx\ = -2.5 \log 2.1$
$$\approx\ -0.8055$$

  b. $M = -2.5 \log \approx \dfrac{36.2 B_0}{B_0} \approx\ = -2.5 \log 36.2$
$$\approx\ -3.8968$$

87. $\log 3 = k \log 2$
$$k = \frac{\log 3}{\log 2} \approx \frac{0.4771}{0.3010} \approx 1.5850$$

89. $25 = 10^k$
$$k = \log_{10} 25 \approx 1.3979$$

91. **a.** $f(x) = 2^x$

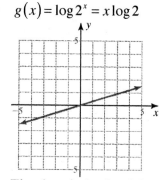

  b. $g(x) = \log 2^x = x \log 2$

  c. The slope is $\log 2 \approx 0.3$

93. The slope of $g(x) = \log$ of the base of $f(x)$.

95. Sample answer: The antilog_{10} of a number is the inverse logarithm of the number.

97. Sample answer: The graphs of $f(x) = 10^x$ and $g(x) = \log x$ are symmetric about the line $y = x$. They are inverses of each other.

99. Sample answer: The graph of $f(x) = e^{2x}$ will increase more quickly when $x > 0$ than the graph of $g(x) = e^{0.5x}$.

101. $\text{inv} \ln(-1.5960) = 0.2027$

103. 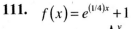 $\log_6 20 = \dfrac{\ln 20}{\ln 6} \approx \dfrac{2.9957}{1.7918} \approx 1.6720$

105. $\log 0.000415 - -3.3820$

107. $\text{inv} \log(\approx 2.4683) \approx 0.0034$

109. $f(x) = -e^{(1/4)x}$

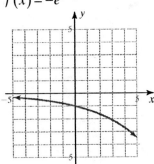

111. $f(x) = e^{(1/4)x} + 1$

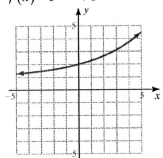

113. $f(x) = \ln x + 3$

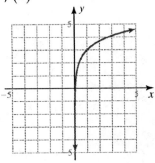

10.6   Exponential and Logarithmic Equations and Applications

Problems 10.6

1. a.
$$3^{3x-4} = 9$$
$$3^{3x-4} = 3^2$$
$$3x - 4 = 2$$
$$3x = 6$$
$$x = 2$$

b.
$$2^{x+3} = 8^{x-1}$$
$$2^{x+3} = \left(2^3\right)^{x-1}$$
$$2^{x+3} = 2^{3x-3}$$
$$x + 3 = 3x - 3$$
$$6 = 2x$$
$$3 = x$$

2.
$$5^x = 12$$
$$\log 5^x = \log 12$$
$$x \log 5 = \log 12$$
$$x = \frac{\log 12}{\log 5} \approx 1.5440$$

3.
$$20 = e^{0.15k}$$
$$\ln 20 = \ln e^{015k}$$
$$\ln 20 = 0.15k$$
$$k = \frac{\ln 20}{0.15} \approx 19.97$$

4.
$$50 = e^{21.5b}$$
$$\ln 50 = \ln e^{21.5b}$$
$$\ln 50 = 21.5b$$
$$b = \frac{\ln 50}{21.5} \approx 0.18\%$$

5.
$$\log_5(2x - 1) = 2$$
$$2x - 1 = 5^2$$
$$2x - 1 = 25$$
$$2x = 26$$
$$x = 13$$

Check:
$$\log_5(2 \cdot 13 - 1) = 2$$
$$\log_5 25 = 2$$
$$5^2 = 25$$

6. a.
$$\log(x - 3) + \log x = 1$$
$$\log(x - 3)x = 1$$
$$(x - 3)x = 10^1$$
$$x^2 - 3x = 10$$
$$x^2 - 3x - 10 = 0$$
$$(x - 5)(x + 2) = 0$$
$$x - 5 = 0 \text{ or } x + 2 = 0$$
$$x = 5 \text{ or } \quad x = -2$$

Check: $x = 5$
$$\log(5 \cdot 3) + \log 5 = 1$$
$$\log 2 + \log 5 = 1$$
$$\log(2 \cdot 5) = 1$$
$$\log 10 = 1$$
$$1 = 1$$

Discard: $x = -2$ since $\log(-2)$ is not defined.
The solution is 5.

b.
$$\log_3(x + 1) - \log_3(x - 3) = 1$$
$$\log_3 \left| \frac{x + 1}{x - 3} \right| = 1$$
$$\frac{x + 1}{x - 3} = 3^1$$
$$x + 1 = 3(x - 3)$$
$$x + 1 = 3x - 9$$
$$-2x = -10$$
$$x = 5$$

Check: $x = 5$
$$\log_3(5 + 1) - \log_3(5 - 3) = 1$$
$$\log_3 6 - \log_3 2 = 1$$
$$\log_3 \frac{6}{2} = 1$$
$$\log_3 3 = 1$$
$$1 = 1$$
The solution is 5.

7.
$$2P = Pe^{0.08t}$$
$$2 = e^{0.08t}$$
$$\ln 2 = \ln\left(e^{0.08t}\right)$$
$$\ln 2 = 0.08t$$
$$t = \frac{\ln 2}{0.08} \approx 8.7 \text{ yr}$$

8.
$$P = 6.1e^{0.012t}$$
$$P = 6.1e^{0.012(20)}$$
$$= 6.1e^{0.24}$$
$$\approx 7.8 \text{ billion}$$

9.
$$B = 1000e^{0.05t}$$
$$100{,}000 = 1000e^{0.05t}$$
$$100 = e^{0.05t}$$
$$\ln 100 = \ln e^{0.05t}$$
$$\ln 100 = 0.05t$$
$$t = \frac{\ln 100}{0.05} \approx 92.1 \text{ min}$$

10.
$$A(t) = A_0 e^{\approx 0.04t}$$
$$\frac{1}{2}A_0 = A_0 e^{\approx 0.04t}$$
$$\frac{1}{2} = e^{\approx 0.04t}$$
$$\ln \approx \frac{1}{2} \approx = \ln e^{\approx 0.04t}$$
$$\ln 0.5 = \approx 0.04t$$
$$t = \frac{\ln 0.5}{\approx 0.04} \approx 17.3 \text{ yr}$$

Exercises 10.6

1.
$$5^x = 25$$
$$5^x = 5^2$$
$$x = 2$$

3.
$$2^x = 32$$
$$2^x = 2^5$$
$$x = 5$$

5.
$$5^{3x} = 625$$
$$5^{3x} = 5^4$$
$$3x = 4$$
$$x = \frac{4}{3}$$

7.
$$7^{-3x} = 343$$
$$7^{-3x} = 7^3$$
$$-3x = 3$$
$$x = -1$$

9.
$$7^x = 512$$
$$\log 7^x = \log 512$$
$$x \log 7 = \log 512$$
$$x = \frac{\log 512}{\log 7} \approx 3.2059$$

11.
$$5^{x-2} = 625$$
$$5^{x-2} = 5^4$$
$$x - 2 = 4$$
$$x = 6$$

13.

$$3^x = 2$$

$$\log 3^x = \log 2$$

$$x \log 3 = \log 2$$

$$x = \frac{\log 2}{\log 3} \approx 0.6309$$

15.

$$2^{3x-2} = 32$$

$$2^{3x-2} = 2^5$$

$$3x - 2 = 5$$

$$3x = 7$$

$$x = \frac{7}{3}$$

17.

$$5^{3x} \cdot 5^{x^2} = 25$$

$$5^{x^2+3x} = 5^2$$

$$x^2 + 3x = 2$$

$$x^2 + 3x - 2 = 0$$

$$x = \frac{-3 \pm \sqrt{3^2 - 4 \cdot 1 \cdot (-2)}}{2 \cdot 1}$$

$$= \frac{-3 \pm \sqrt{9+8}}{2} = \frac{-3 \pm \sqrt{17}}{2}$$

19.

$$e^x = 10$$

$$\ln e^x = \ln 10$$

$$x = \ln 10 \approx 2.3026$$

21.

$$e^{\approx x} = 0.1$$

$$\ln e^{\approx x} = \ln 0.1$$

$$\approx x = \ln 0.1$$

$$x = \approx \ln 0.1 \approx 2.3026$$

23.

$$30 = e^{2k}$$

$$\ln 30 = \ln e^{2k}$$

$$\ln 30 = 2k$$

$$k = \frac{\ln 30}{2} \approx 1.7006$$

25.

$$10 = e^{\approx 2k}$$

$$\ln 10 = \ln e^{\approx 2k}$$

$$\ln 10 = \approx 2k$$

$$k = \frac{\ln 10}{\approx 2} \approx \approx 1.1513$$

27.

$$\log_2 x = 3$$

$$x = 2^3 = 8$$

29.

$$\log_2 x = -3$$

$$x = 2^{-3} = \frac{1}{2^3} = \frac{1}{8}$$

31.

$$\ln x = 1$$

$$x = e^1 = e \approx 2.7183$$

33.

$$\ln x = 3$$

$$x = e^3 \approx 20.0855$$

35.

$$\log_2 (3x - 5) = 1$$

$$3x - 5 = 2^1$$

$$3x = 7$$

$$x = \frac{7}{3}$$

37.
$$\log_4(3x-1)=2$$
$$3x-1=4^2$$
$$3x-1=16$$
$$3x=17$$
$$x=\frac{17}{3}$$

39.
$$\log_5(3x+1)=2$$
$$3x+1=5^2$$
$$3x+1=25$$
$$3x=24$$
$$x=8$$

41.
$$\log x+\log(x-3)=1$$
$$\log\left[x(x-3)\right]=1$$
$$x(x-3)=10^1$$
$$x^2-3x-10=0$$
$$(x-5)(x+2)=0$$
$$x-5=0 \text{ or } x+2=0$$
$$x=5 \text{ or } \quad x=-2$$

Check: $x=5$
$$\log 5+\log(5-3)=1$$
$$\log 5+\log 2=1$$
$$\log(5\cdot2)=1$$
$$\log 10=1$$
$$1=1$$

Discard: $x=-2$ since $\log(-2)$ is
not defined.
The solution is 5.

43.
$$\log_2(x+1)+\log_2(x+3)=3$$
$$\log_2\left[(x+1)(x+3)\right]=3$$
$$(x+1)(x+3)=2^3$$
$$x^2+4x+3=8$$
$$x^2+4x-5=0$$
$$(x+5)(x-1)=0$$
$$x+5=0 \quad \text{or } x-1=0$$
$$x=-5 \text{ or } \quad x=1$$

Discard: $x=-5$ since $\log(-5+1)$ is
not defined.

Check: $x=1$
$$\log_2(1+1)+\log_2(1+3)=3$$
$$\log_2 2+\log_2 4=3$$
$$1+2=3$$

The solution is 1.

45.
$$\log(x+1)-\log x=1$$
$$\log\left|\frac{x+1}{x}\right|=1$$
$$\frac{x+1}{x}=10^1$$
$$x+1=10x$$
$$1=9x$$
$$x=\frac{1}{9}$$

Check: $x=\frac{1}{9}$
$$\log\frac{-1}{-9}+1-\log\frac{1}{9}=1$$
$$\log\frac{10}{9}-\log\frac{1}{9}=1$$
$$\log\frac{-10}{-9}\div\frac{1}{9}=1$$
$$\log 10=1$$
$$1=1$$

The solution is $\frac{1}{9}$.

47. $\log_2(3+x) - \log_2(7-x) = 2$

$$\log_2\left|\frac{3+x}{7-x}\right| = 2$$

$$\frac{3+x}{7-x} = 2^2$$

$$3+x = 4(7-x)$$

$$3+x = 28-4x$$

$$5x = 25$$

$$x = 5$$

Check: $x = 5$

$$\log_2(3+5)-\log_2(7-5) = 2$$

$$\log_2 8 - \log_2 2 = 2$$

$$3-1 = 2$$

The solution is 5.

49. $\log_2(x^2+4x+7) = 2$

$$x^2+4x+7 = 2^2$$

$$x^2+4x+3 = 0$$

$$(x+3)(x+1) = 0$$

$$x+3 = 0 \quad \text{or} \ x+1 = 0$$

$$x = -3 \text{ or} \quad x = -1$$

Check: $x = -3$

$$\log_2\left((-3)^2+4(-3)+7\right) = 2$$

$$\log_2(9-12+7) = 2$$

$$\log_2 4 = 2$$

$$2 = 2$$

Check: $x = -1$

$$\log_2\left((-1)^2+4(-1)+7\right) = 2$$

$$\log_2(1-4+7) = 2$$

$$\log_2 4 = 2$$

$$2 = 2$$

The solutions are -3 and -1.

51. $2P = Pe^{0.05t}$

$$2 = e^{0.05t}$$

$$\ln 2 = 0.05t$$

$$t = \frac{\ln 2}{0.05} \approx 13.86 \text{ yr}$$

53. $2P = Pe^{0.065t}$

$$2 = e^{0.065t}$$

$$\ln 2 = 0.065t$$

$$t = \frac{\ln 2}{0.065} \approx 10.66 \text{ yr}$$

55. $P = 4.8e^{0.015t}$

$$P = 4.8e^{0.015(16)}$$

$$= 4.8e^{0.24}$$

$$\approx 6.1 \text{ billion}$$

57. $2000 = 1000e^{0.04t}$

$$2 = e^{0.04t}$$

$$\ln 2 = 0.04t$$

$$t = \frac{\ln 2}{0.04} \approx 17.3 \text{ min}$$

59. $25,000 = 1000e^{0.04t}$

$$25 = e^{0.04t}$$

$$\ln 25 = 0.04t$$

$$t = \frac{\ln 25}{0.04} \approx 80.5 \text{ min}$$

61. a. $B = 100,000e^{-0.2t}$

$B = 100,000e^{-0.2(0)}$

$B = 100,000$ bacteria

b. $B = 100,000e^{-0.2t}$

$B = 100,000e^{-0.2(2)}$

$B = 100,000e^{-0.4}$

$B = 67,032$ bacteria

c. $B = 100,000e^{-0.2t}$

$B = 100,000e^{-0.2(10)}$

$B = 100,000e^{-2}$

$B = 13,534$ bacteria

d. $B = 100,000e^{-0.2t}$

$B = 100,000e^{-0.2(20)}$

$B = 100,000e^{-4}$

$B = 1832$ bacteria

63. $A(t) = A_0 e^{\approx 0.00003t}$

$\dfrac{1}{2}A_0 = A_0 e^{\approx 0.00003t}$

$\dfrac{1}{2} = e^{\approx 0.00003t}$

$\ln\genfrac{}{}{0pt}{}{\approx}{\approx}\dfrac{1}{2}\genfrac{}{}{0pt}{}{\approx}{\approx} = \ln e^{\approx 0.00003t}$

$\ln 0.5 = \approx 0.00003t$

$t = \dfrac{\ln 0.5}{\approx 0.00003} \approx 23,105$ yr

65. $A(t) = A_0 e^{\approx 0.052t}$

$\dfrac{1}{2}A_0 = A_0 e^{\approx 0.052t}$

$\dfrac{1}{2} = e^{\approx 0.052t}$

$\ln\genfrac{}{}{0pt}{}{\approx}{\approx}\dfrac{1}{2}\genfrac{}{}{0pt}{}{\approx}{\approx} = \ln e^{\approx 0.052t}$

$\ln 0.5 = \approx 0.052t$

$t = \dfrac{\ln 0.5}{\approx 0.052} \approx 13.3$ yr

67. $P = 14.7e^{\approx 0.00005h}$

a. $P = 14.7e^{\approx 0.00005(0)} = 14.7$ lb/in^2

b. $P = 14.7e^{\approx 0.00005(5000)} = 14.7e^{\approx 0.25}$

≈ 11.4 lb/in^2

c. $P = 14.7e^{\approx 0.00005(10,000)} = 14.7e^{\approx 0.5}$

≈ 8.9 lb/in^2

69. $S = 540,000(1.09)^t$

a. $800,000 = 540,000(1.09)^t$

$\dfrac{80}{54} = 1.09^t$

$\ln\genfrac{}{}{0pt}{}{\approx}{\approx}\dfrac{80}{54}\genfrac{}{}{0pt}{}{\approx}{\approx} = t\ln 1.09$

$t = \dfrac{\ln\left(\frac{40}{27}\right)}{\ln 1.09} \approx 4.56 \approx 5\text{yr } (1997)$

b. $1,000,000 = 540,000(1.09)^t$

$\dfrac{100}{54} = 1.09^t$

$\ln\genfrac{}{}{0pt}{}{\approx}{\approx}\dfrac{100}{54}\genfrac{}{}{0pt}{}{\approx}{\approx} = t\ln 1.09$

$t = \dfrac{\ln\left(\frac{50}{27}\right)}{\ln 1.09} \approx 7.15 \approx 7\text{yr } (1999)$

71. $A = 5000e^{\approx 0.04t}$

a. $A = 5000e^{\approx 0.04(0)} = 5000$

b. $A = 5000e^{\approx 0.04(20)} = 5000e^{\approx 0.8}$

≈ 2247

c. $1000 = 5000e^{\approx 0.04t}$

$0.2 = e^{\approx 0.04t}$

$\ln 0.2 = \approx 0.04t$

$t = \dfrac{\ln 0.2}{\approx 0.04} \approx 40$ yr (2010)

73. $S = 54e^{0.15t}$

   a. $S = 54e^{0.15(0)} = \$54$ million

   b. $500 = 54e^{0.15t}$

$$\frac{500}{54} = e^{0.15t}$$

$$\ln \frac{500}{54} \approx = 0.15t$$

$$t = \frac{\ln\left(\frac{250}{27}\right)}{0.15} \approx 15 \text{ yr } (1995)$$

75. $H = h(A) = 11 + 19.44 \ln A$

   a. $60 = 11 + 19.44 \ln A$

$$49 = 19.44 \ln A$$

$$\ln A = \frac{49}{19.44} \approx 2.5206$$

$$A = e^{2.5206} \approx 12 \text{ yr old}$$

   b. $50 = 11 + 19.44 \ln A$

$$39 = 19.44 \ln A$$

$$\ln A = \frac{39}{19.44} \approx 2.0062$$

$$A = e^{2.0062} \approx 7 \text{ yr old}$$

77. $t = \approx 694.2 + 231.4 \log A$

$$t = \approx 694.2 + 231.4 \log(1{,}000{,}000)$$

$$= \approx 694.2 + 231.4(6)$$

$$= \approx 694.2 + 1388.4$$

$$\approx 694 \text{ months}$$

79. $f(n) = 2n + 1$

   a. $f(1) = 2 \cdot 1 + 1 = 2 + 1 = 3$

   b. $f(5) = 2 \cdot 5 + 1 = 10 + 1 = 11$

   c. $f(10) = 2 \cdot 10 + 1 = 20 + 1 = 21$

81. $11{,}000 = 580e^{5k}$

$$\frac{11{,}000}{580} = e^{5k}$$

$$\ln \frac{11{,}00}{58} \approx = 5k$$

$$k = \frac{\ln\left(\frac{1100}{58}\right)}{5} \approx 0.59$$

83. $N = 580e^{0.59t}$

$$N = 580e^{0.59(3)} = 580e^{1.77}$$

$$\approx 3405 \text{ thousand}$$

In 1990 approximately 3,405,000 fax machines were sold.

85. $N = 580e^{0.59t}$

$$N = 580e^{0.59(13)} = 580e^{7.67}$$

$$\approx 1{,}242{,}987 \text{ thousand}$$

In 2000 approximately 1,242,987,000 fax machines were sold.

87.   Sample answer: Set $t = 0$ in the exponential model and solve for S_0. This should result in the sales of cellular phones in 1985.

89.   Sample answer: The graph indicates that this is a logarithmic model. As the number of years increases, the number of cigarettes per capita is increasing very slightly.

91.   Sample answer: To find a it is necessary to know another set of data for the graph, i.e. the number of cigarettes per capita for another year after 1969.

93.
$$B = 1000e^{0.05t}$$
$$20,000 = 1000e^{0.05t}$$
$$20 = e^{0.05t}$$
$$\ln 20 = 0.05t$$
$$t = \frac{\ln 20}{0.05} \approx 59.9 \text{ min}$$

95.
$$\log(x+9) + \log x = 1$$
$$\log(x+9)x = 1$$
$$(x+9)x = 10^1$$
$$x^2 + 9x = 10$$
$$x^2 + 9x - 10 = 0$$
$$(x-1)(x+10) = 0$$
$$x - 1 = 0 \text{ or } x + 10 = 0$$
$$x = 1 \text{ or } \quad x = -10$$

Check: $x = 1$
$$\log(1+9) + \log 1 = 1$$
$$\log 10 + \log 1 = 1$$
$$1 + 0 = 1$$

Discard: $x = -10$ since $\log(-10)$ is
not defined.
The solution is 1.

97.
$$\log_6(4x-4) = 2$$
$$4x - 4 = 6^2$$
$$4x - 4 = 36$$
$$4x = 40$$
$$x = 10$$

Check: $x = 10$
$$\log_6(4\cdot 10 - 4) = 2$$
$$\log_6 36 = 2$$
$$2 = 2$$

The solution is 10.

99.
$$5^x = 10$$
$$\log 5^x = \log 10$$
$$x \log 5 = 1$$
$$x = \frac{1}{\log 5} \approx 1.4307$$

101.
$$2^{2x+1} = 8^{x-1}$$
$$2^{2x+1} = \left(2^3\right)^{x-1}$$
$$2^{2x+1} = 2^{3x-3}$$
$$2x + 1 = 3x - 3$$
$$4 = x$$

Review Exercises

1.
$$f(x)=2-x^2; \quad g(x)=2+x$$

a. $(f+g)(x)=f(x)+g(x)=(2-x^2)+(2+x)=-x^2+x+4$

b. $(f-g)(x)=f(x)-g(x)=(2-x^2)-(2+x)=2-x^2-2-x=-x^2-x$

c. $(fg)(x)=f(x)\cdot g(x)=(2-x^2)(2+x)=4+2x-2x^2-x^3$

d. $\left(\dfrac{f}{g}\right)(x)=\dfrac{f(x)}{g(x)}=\dfrac{2-x^2}{2+x}, \quad x\neq-2$

2.
$$f(x)=3-x^2; \quad g(x)=3+x$$

a. $(f+g)(x)=f(x)+g(x)=(3-x^2)+(3+x)=-x^2+x+6$

b. $(f-g)(x)=f(x)-g(x)=(3-x^2)-(3+x)=3-x^2-3-x=-x^2-x$

c. $(fg)(x)=f(x)\cdot g(x)=(3-x^2)(3+x)=9+3x-3x^2-x^3$

d. $\left(\dfrac{f}{g}\right)(x)=\dfrac{f(x)}{g(x)}=\dfrac{3-x^2}{3+x}, \quad x\neq-3$

3. a. $f(x)=6x+1$

$$\frac{f(x)-f(a)}{x-a}=\frac{6x+1-(6a+1)}{x-a}=\frac{6x+1-6a-1}{x-a}=\frac{6x-6a}{x-a}=\frac{6(x-a)}{x-a}=6$$

b. $f(x)=7x+1$

$$\frac{f(x)-f(a)}{x-a}=\frac{7x+1-(7a+1)}{x-a}=\frac{7x+1-7a-1}{x-a}=\frac{7x-7a}{x-a}=\frac{7(x-a)}{x-a}=7$$

4.
$$f(x)=x^3; \quad g(x)=2-x$$

a. $(g\circ f)(2)=g(f(2))=g(2^3)=g(8)=2-8=-6$

b. $(f\circ g)(x)=f(g(x))=f(2-x)=(2-x)^3$

c. $(g\circ f)(x)=g(f(x))=g(x^3)=2-x^3$

5.
$$f(x)=x^3; \quad g(x)=3-x$$

a. $(f\circ g)(x)=f(g(x))=f(3-x)=(3-x)^3$

b. $(g\circ f)(x)=g(f(x))=g(x^3)=3-x^3$

c. $(g\circ f)(2)=g(f(2))=g(2^3)=g(8)=3-8=-5$

6.
$$f(x)=\frac{9}{x-1}; \quad g(x)=\frac{x+3}{x-4}$$

Domain f: $\{x \mid x \text{ is a real number and } x\neq1\}$ Domain g: $\{x \mid x \text{ is a real number and } x\neq4\}$

Domain $f+g=$ domain $f-g=$ domain $fg=\{x \mid x \text{ is a real number and } x\neq1 \text{ and } x\neq4\}$

7.
$$f(x) = \frac{x+1}{x \neq 3}; \quad g(x) = \frac{x \neq 1}{x+4}$$

Domain f: $\{x \mid x$ is a real number and $x \neq 3\}$

Domain g: $\{x \mid x$ is a real number and $x \neq \neq 4\}$

Exclude the values of x where $g(x) = 0$. So, $x \neq 1$.

Domain $\dfrac{f}{g} = \{x \mid x$ is a real number and $x \neq \neq 4$ and $x \neq 1$ and $x \neq 3\}$

8.　**a.**
$$\begin{aligned} P(x) &= R(x) - C(x) \\ &= (100x - 0.02x^2) - (30{,}000 + 30x) \\ &= -0.02x^2 + 70x - 30{,}000 \end{aligned}$$

b.
$$\begin{aligned} P(x) &= R(x) - C(x) \\ &= (100x - 0.02x^2) - (40{,}000 + 40x) \\ &= -0.02x^2 + 60x - 40{,}000 \end{aligned}$$

9.　**a.** Domain: $\{4, 6, 8\}$　Range: $\{4, 6, 8\}$

b. $S^{-1} = \{(4, 4), (6, 6), (8, 8)\}$

c. Domain: $\{4, 6, 8\}$　Range: $\{4, 6, 8\}$

d.
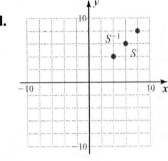

10.　**a.** Domain: $\{4, 6, 8\}$　Range: $\{5, 7, 9\}$

b. $S^{-1} = \{(5, 4), (7, 6), (9, 8)\}$

c. Domain: $\{5, 7, 9\}$　Range: $\{4, 6, 8\}$

d.
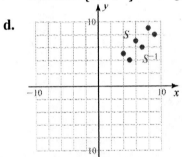

11.　**a.** $f(x) = y = 3x - 3$

Interchange x and y and solve for y:
$$\begin{aligned} x &= 3y - 3 \\ x + 3 &= 3y \\ \frac{x+3}{3} &= y \end{aligned}$$
$$f^{-1}(x) = \frac{x+3}{3}$$

b.
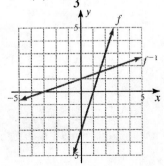

12.　**a.** $f(x) = y = 4x - 4$

Interchange x and y and solve for y:
$$\begin{aligned} x &= 4y - 4 \\ x + 4 &= 4y \\ \frac{x+4}{4} &= y \end{aligned}$$
$$f^{-1}(x) = \frac{x+4}{4}$$

b.
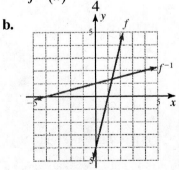

13. $f(x) = y = 4x^2$

Interchange x and y and solve for y:

$$x = 4y^2$$

$$\frac{x}{4} = y^2$$

$$y = \pm\frac{\sqrt{x}}{2}$$

$$f^{-1}(x) = \pm\frac{\sqrt{x}}{2}$$

The inverse is not a function.

14. $f(x) = y = 5x^2$

Interchange x and y and solve for y:

$$x = 5y^2$$

$$\frac{x}{5} = y^2$$

$$y = \pm\frac{\sqrt{5x}}{5}$$

$$f^{-1}(x) = \pm\frac{\sqrt{5x}}{5}$$

The inverse is not a function.

15. $y = 3^x$

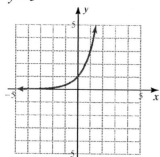

The inverse is a function.

16. $y = 4^x$

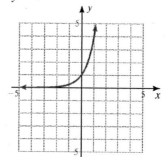

The inverse is a function.

17. **a.** $d = 28 - 16 = 12$

b. $d = f(w) = w - 16$

Interchange w and d and solve for d:

$$w = d - 16$$

$$w + 16 = d$$

$$f^{-1}(w) = w + 16$$

c. $10 = w - 16$

$$w = 26 \text{ in}$$

18. **a.** $f(x) = 2^{x/2}$

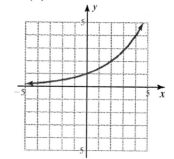

b. $f(x) = 2^{-x/2}$

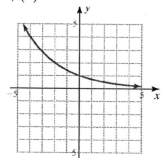

19. **a.** $g(x) = \left|\frac{1}{2}\right|^{x/2}$

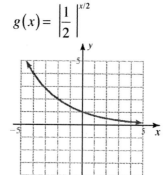

457

b.

$$g(x) = -\frac{1}{2}^{-x/2}$$

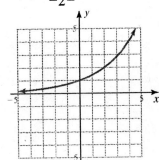

20. **a.** $G = 1000e^{\approx 1.4t}$

b. $G = 1000e^{\approx 1.4(0)} = 1000$ g

c. $G = 1000e^{\approx 1.4(2)} = 1000e^{\approx 2.8}$

   ≈ 61 g

21. $f(x) = \log_5 x$

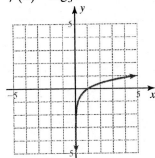

22. **a.** $243 = 3^5$

logarithmic form: $\log_3 243 = 5$

b. $\frac{1}{8} = 2^{|3}$

logarithmic form: $\log_2 \left|\frac{1}{8}\right| = |3$

23. **a.** $\log_2 32 = 5$

exponential form: $2^5 = 32$

b. $\log_3 \frac{1}{81} = -4$

exponential form: $3^{-4} = \frac{1}{81}$

24. **a.** $\log_4 x = -2$

   $x = 4^{-2} = \frac{1}{4^2} = \frac{1}{16}$

b. $\log_x 16 = 2$

   $x^2 = 16$

   $x = 4$

25. **a.** $\log_2 16 = x$

   $2^x = 16$

   $2^x = 2^4$

   $x = 4$

b. $\log_3 \frac{1}{27} = x$

   $3^x = \frac{1}{27} = \frac{1}{3^3}$

   $3^x = 3^{-3}$

   $x = -3$

26. **a.** $\log_b MN = \log_b M + \log_b N$

b. $\log_b M - \log_b N = \log_b \frac{M}{N}$

c. $\log_b M^r = r \log_b M$

27. **a.** $\log 975 = 2.9890$

b. $\log 837 = 2.9227$

28. **a.** $\log 0.00759 = -2.1198$

b. $\log 0.000648 = -3.1884$

29. **a.** inv $\log 2.8215 = 662.9793$

    b. inv $\log(-3.3904) = 0.0004$

30. **a.** $\ln 2850 = 7.9551$

    b. $\ln 0.345 = -1.0642$

31. **a.** inv $\ln 2.0855 = 8.0486$

    b. inv $\ln 2.7183 = 15.1545$

32. **a.** $\log_3 10 = \dfrac{\log 10}{\log 3} \approx 2.0959$

    b. $\log_3 100 = \dfrac{\log 100}{\log 3} \approx 4.1918$

33. **a.** $f(x) = e^{(1/2)x}$

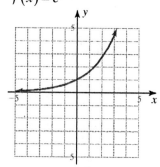

    b. $g(x) = -e^{(1/2)x}$

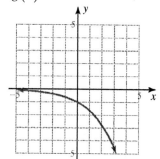

34. **a.** $f(x) = \ln(x+1)$

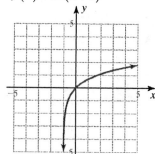

    b. $g(x) = \ln x + 1$

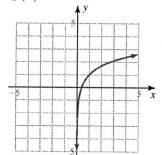

35. $\text{pH} = \approx \log\left(4 \approx 10^{-6}\right)$

$\quad = \approx \left(\log 4 + \log 10^{-6}\right)$

$\quad = \approx \left(\log 4 \approx 6\right)$

$\quad = 6 \approx \log 4$

$\quad \approx 6 \approx 0.602$

$\quad \approx 5.398$

36. **a.** $2^{2x-1} = 32$

$\qquad 2^{2x-1} = 2^5$

$\qquad 2x - 1 = 5$

$\qquad\quad 2x = 6$

$\qquad\quad\; x = 3$

    b. $3^{x+1} = 9^{x-1}$

$\qquad 3^{x+1} = \left(3^2\right)^{x-1}$

$\qquad 3^{x+1} = 3^{2x-2}$

$\qquad x + 1 = 2x - 2$

$\qquad\quad\; 3 = x$

37. a.
$$2^x = 3$$
$$\log 2^x = \log 3$$
$$x \log 2 = \log 3$$
$$x = \frac{\log 3}{\log 2} \approx 1.5850$$

b.
$$5^{2x} = 2.5$$
$$\log 5^{2x} = \log 2.5$$
$$2x \log 5 = \log 2.5$$
$$x = \frac{\log 2.5}{2 \log 5} \approx 0.2847$$

38. a.
$$e^{5.6x} = 2$$
$$\ln e^{5.6x} = \ln 2$$
$$5.6x = \ln 2$$
$$x = \frac{\ln 2}{5.6} \approx 0.1238 \;\cdot$$

b.
$$e^{\approx 0.33x} = 2$$
$$\ln e^{\approx 0.33x} = \ln 2$$
$$\approx 0.33x = \ln 2$$
$$x = \frac{\ln 2}{\approx 0.33} \approx \approx 2.1004$$

39. a.
$$\log x + \log(x - 10) = 1$$
$$\log\left[x(x - 10)\right] = 1$$
$$x^2 - 10x = 10^1$$
$$x^2 - 10x - 10 = 0$$
$$x = \frac{-(-10) \pm \sqrt{(-10)^2 - 4 \cdot 1 \cdot (-10)}}{2 \cdot 1}$$
$$= \frac{10 \pm \sqrt{100 + 40}}{2} = \frac{10 \pm \sqrt{140}}{2} = 5 \pm \sqrt{35}$$

Discard: $x = 5 - \sqrt{35}$ since $\log\left(5 - \sqrt{35}\right)$
$\qquad\qquad\qquad$ is not defined.

The solution is $5 + \sqrt{35}$.

b.
$$\log_3(x + 1) - \log_3(x - 1) = 1$$
$$\log_3\left|\frac{x + 1}{x - 1}\right| = 1$$
$$\frac{x + 1}{x - 1} = 3^1$$
$$x + 1 = 3(x - 1)$$
$$x + 1 = 3x - 3$$
$$-2x = -4$$
$$x = 2$$

Check $x = 2$:
$$\log_3(2 + 1) - \log_3(2 - 1)$$
$$= \log_3 3 - \log_3 1 = 1 - 0 = 1$$

40. a.
$$2P = Pe^{0.05t}$$
$$2 = e^{0.05t}$$
$$\ln 2 = 0.05t$$
$$t = \frac{\ln 2}{0.05} \approx \frac{0.69315}{0.05} \approx 13.863 \text{ yr}$$

b.
$$2P = Pe^{0.085t}$$
$$2 = e^{0.085t}$$
$$\ln 2 = 0.085t$$
$$t = \frac{\ln 2}{0.085} \approx \frac{0.69315}{0.085} \approx 8.2 \text{ yr}$$

41. a.
$$1804 = 1000e^{k(2)}$$
$$1.804 = e^{2k}$$
$$\ln 1.804 = 2k$$
$$k = \frac{\ln 1.804}{2} \approx 0.2950$$

b.
$$1804 = 1000e^{k(4)}$$
$$1.804 = e^{4k}$$
$$\ln 1.804 = 4k$$
$$k = \frac{\ln 1.804}{4} \approx 0.1475$$

42.

a.
$$A(t) = A_0 e^{\approx 0.5t}$$
$$\frac{1}{2} A_0 = A_0 e^{\approx 0.5t}$$
$$\frac{1}{2} = e^{\approx 0.5t}$$
$$\ln \approx \frac{1}{2} \approx = \ln e^{\approx 0.5t}$$
$$\ln 0.5 = \approx 0.5t$$
$$t = \frac{\ln 0.5}{\approx 0.5} \approx 1.3863 \text{ yr}$$

b.
$$A(t) = A_0 e^{\approx 0.02t}$$
$$\frac{1}{2} A_0 = A_0 e^{\approx 0.02t}$$
$$\frac{1}{2} = e^{\approx 0.02t}$$
$$\ln \approx \frac{1}{2} \approx = \ln e^{\approx 0.02t}$$
$$\ln 0.5 = \approx 0.02t$$
$$t = \frac{\ln 0.5}{\approx 0.02} \approx 34.6575 \text{ yr}$$

Cumulative Review Chapters 1–10

1.
$$\lfloor (3x^2 \rfloor 2) + (8x+3)\rfloor \rfloor \lfloor (x \rfloor 2) + (2x^2 \rfloor 6)\rfloor$$
$$= \lfloor 3x^2 + 8x + 1\rfloor \rfloor \lfloor 2x^2 + x \rfloor 8 \rfloor$$
$$= 3x^2 + 8x + 1 \rfloor 2x^2 \rfloor x + 8$$
$$= x^2 + 7x + 9$$

2.
$$\left(4x^4 y^{-2}\right)^2 = 4^2 \left(x^4\right)^2 \left(y^{-2}\right)^2 = 16x^8 y^{-4} = \frac{16x^8}{y^4}$$

3.
$$0.02P + 0.04(1700 - P) = 65$$
$$0.02P + 68 - 0.04P = 65$$
$$-0.02P = -3$$
$$P = 150$$

4.
$$|x - 4| = |x - 8|$$
$$x - 4 = x - 8 \text{ or } x - 4 = -(x - 8)$$
$$-4 = -8 \quad \text{or } x - 4 = -x + 8$$
$$-4 = -8 \quad \text{or} \quad 2x = 12$$
$$-4 = -8 \quad \text{or} \quad x = 6$$
The solution is 6.

5.
$$\{x \mid x < \geq 4 \text{ or } x \geq 4\}$$

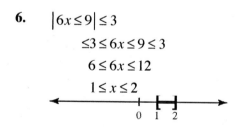

6.
$$|6x \leq 9| \leq 3$$
$$\leq 3 \leq 6x \leq 9 \leq 3$$
$$6 \leq 6x \leq 12$$
$$1 \leq x \leq 2$$

7. $H = 2.85h + 72.69, \ H = 135.39$

$135.39 = 2.85h + 72.69$

$62.7 = 2.85h$

$22 = h$

8. Let L = the length

Let W = the width

$L = W + 20$

$2L + 2W = 160$

$2(W + 20) + 2W = 160$

$2W + 40 + 2W = 160$

$4W = 120$

$W = 30$ ft

$L = 30 + 20 = 50$ ft

The dimensions are 30 ft by 50 ft.

9. $d = \sqrt{(-2-5)^2 + (1-2)^2} = \sqrt{(-7)^2 + (-1)^2}$

$= \sqrt{49 + 1} = \sqrt{50} = 5\sqrt{2}$

10. The slope of the perpendicular line is $-\frac{9}{4}$.

$\dfrac{y-(-4)}{-1-3} = -\dfrac{9}{4}$

$\dfrac{y+4}{-4} = -\dfrac{9}{4}$

$y + 4 = 9$

$y = 5$

11. $8x + 2y = 2$

$2y = -8x + 2$

$y = -4x + 1$

The slope of the parallel line is –4.

$y - 6 = -4(x - (-6))$

$y - 6 = -4(x + 6)$

$y - 6 = -4x - 24$

$4x + y = -18$

12. $|x + 3| > 2$

$x + 3 > 2 \quad$ or $x + 3 < -2$

$x > -1$ or $\quad x < -5$

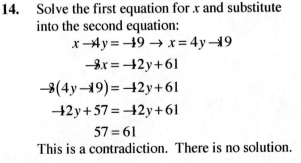

13. $P = \dfrac{k}{V}$

$k = PV$

$k = 1850 \cdot 5 = 9250$

14. Solve the first equation for x and substitute into the second equation:

$x - 4y = -19 \rightarrow x = 4y - 19$

$-3x = -12y + 61$

$-3(4y - 19) = -12y + 61$

$-12y + 57 = -12y + 61$

$57 = 61$

This is a contradiction. There is no solution.

15. Rewrite each equation:

$2x = 5y - 28 \;\rightarrow\; 2x - 5y = -28$

$2y = 5x + 28 \;\rightarrow\; -5x + 2y = 28$

To eliminate y, multiply both sides of the first equation by 2, both sides of the second equation by 5, add the results, and solve:

$$4x - 10y = -56$$
$$\underline{-25x + 10y = 140}$$
$$-21x \qquad\;\; = 84$$
$$x = -4$$
$$2y = 5(-4) + 28$$
$$2y = -20 + 28$$
$$2y = 8$$
$$y = 4$$

Solution: $(-4, 4)$.

16. $3x + y + z = -17$

$\quad\; x + 2y - z = -15$

$\quad 3x + y - z = -21$

Add the first equation to the second equation to eliminate z:

$$3x + y + z = -17$$
$$\underline{x + 2y - z = -15}$$
$$4x + 3y \quad\;\; = -32$$

Add the first equation to the third equation to eliminate z:

$$3x + y + z = -17$$
$$\underline{3x + y - z = -21}$$
$$6x + 2y \quad\;\; = -38$$

Multiply the first result by –2, the second result by 3, add these results, and solve:

$$-8x - 6y = \;\;\; 64$$
$$\underline{18x + 6y = -114}$$
$$10x \qquad\; = -50$$
$$x = -5$$

Substitute and solve for the other variables:

$4(-5) + 3y = -32 \qquad 3(-5) + (-4) + z = -17$

$\;\; -20 + 3y = -32 \qquad\quad -15 - 4 + z = -17$

$\qquad\quad 3y = -12 \qquad\qquad\quad -19 + z = -17$

$\qquad\qquad y = -4 \qquad\qquad\qquad\quad z = 2$

The solution is $(-5, -4, 2)$.

17. $\begin{vmatrix} 5 & -4 \\ -5 & 1 \end{vmatrix} = (5)(1) - (-5)(-4)$

$$= 5 - 20 = -15$$

18. Let x = the height of the building

Let y = the height of the flagpole

$$x + y = 252$$
$$x = 8y$$

Substitute the second equation into the first equation and solve:

$$8y + y = 252$$
$$9y = 252$$
$$y = 28$$
$$x = 8(28) = 224$$

The height of the building is 224 ft.

19.
$$P(x) = x^2 + 5x + 1$$
$$P(-4) = (-4)^2 + 5(-4) + 1$$
$$= 16 - 20 + 1 = -3$$

20.
$$(3h+5)(6h-7) = 18h^2 - 21h + 30h - 35$$
$$= 18h^2 + 9h - 35$$

21.
$$48x^4y + 20x^3y^2 - 12x^2y^3$$
$$= 4x^2y(12x^2 + 5xy - 3y^2)$$
$$= 4x^2y(3x - y)(4x + 3y)$$

22.
$$27n^3 + 8 = (3n)^3 + 2^3$$
$$= (3n + 2)(9n^2 - 6n + 4)$$

23.
$$x^3 + 4x^2 - x - 4 = 0$$
$$x^2(x+4) - (x+4) = 0$$
$$(x+4)(x^2-1) = 0$$
$$(x+4)(x+1)(x-1) = 0$$
$$x+4 = 0 \quad \text{or } x+1 = 0 \quad \text{or } x-1 = 0$$
$$x = -4 \text{ or} \quad x = -1 \text{ or} \quad x = 1$$

24.
$$\begin{array}{r} 3x^2 + 7x + 2 \\ x+5 \overline{)\, 3x^3 + 22x^2 + 37x + 10} \\ \underline{-(3x^3 + 15x^2)} \\ 7x^2 + 37x \\ \underline{-(7x^2 + 35x)} \\ 2x + 10 \\ \underline{-(2x + 10)} \\ 0 \end{array}$$

The factorization is:
$$(x+5)(3x^2 + 7x + 2) = (x+5)(3x+1)(x+2)$$

25.
$$\frac{x+7}{x \cdot 7} \div (x^2 + 14x + 49)$$
$$= \frac{x+7}{x \cdot 7} \cdot \frac{1}{x^2 + 14x + 49}$$
$$= \frac{\cancel{x+7}}{x \cdot 7} \cdot \frac{1}{\cancel{(x+7)}(x+7)} = \frac{1}{x^2 \cdot 49}$$

26.
$$\frac{x-3}{x^2 - 5x + 6} - \frac{x-2}{x^2 - 4}$$
$$= \frac{\cancel{x-3}}{\cancel{(x-3)}(x-2)} - \frac{x-2}{(x+2)(x-2)}$$
$$= \frac{x+2}{(x+2)(x-2)} - \frac{x-2}{(x+2)(x-2)}$$
$$= \frac{x+2-x+2}{(x+2)(x-2)} = \frac{4}{(x+2)(x-2)}$$

464

27. Let x = the first even integer

Let $x + 2$ = the consecutive even integer

$$\frac{1}{x} + \frac{1}{x+2} = \frac{7}{24}$$

$$24x(x+2)\frac{-1}{-x} + \frac{1}{x+2} = 24x(x+2)\frac{-7}{-24}=$$

$$24(x+2) + 24x = 7\,x(x+2)$$

$$24x + 48 + 24x = 7x^2 + 14x$$

$$0 = 7x^2 - 34x - 48$$

$$(7x+8)(x-6) = 0$$

$$7x + 8 = 0 \quad \text{or } x - 6 = 0$$

$$7x = -8 \quad \text{or} \quad x = 6$$

$$x = -\frac{8}{7} \text{ or } \quad x = 6$$

The integers are 6 and 8.

28. $(27)^{-4/3} = \left(\sqrt[3]{27}\right)^{-4} = 3^{-4} = \frac{1}{3^4} = \frac{1}{81}$

29. $\dfrac{\sqrt[5]{7}}{\sqrt[5]{16d^3}} = \dfrac{\sqrt[5]{7} \cdot \sqrt[5]{2d^2}}{\sqrt[5]{16d^3} \cdot \sqrt[5]{2d^2}} = \dfrac{\sqrt[5]{14d^2}}{\sqrt[5]{32d^5}} = \dfrac{\sqrt[5]{14d^2}}{2d}$

30. $\sqrt{18} + \sqrt{50} = 3\sqrt{2} + 5\sqrt{2} = 8\sqrt{2}$

31. $\dfrac{4 + \sqrt{8}}{2} = \dfrac{4 + 2\sqrt{2}}{2} = 2 + \sqrt{2}$

32.

$$\sqrt{x+7} = x + 5$$

$$\left(\sqrt{x+7}\right)^2 = (x+5)^2$$

$$x + 7 = x^2 + 10x + 25$$

$$0 = x^2 + 9x + 18$$

$$0 = (x+6)(x+3)$$

$$x + 6 = 0 \quad \text{or } x + 3 = 0$$

$$x = -6 \text{ or} \quad x = -3$$

Check: $\sqrt{\neq 6 + 7} = \sqrt{1} = 1 \neq \neq 6 + 5$

Discard: $x = \neq 6$

$$\sqrt{\neq 3 + 7} = \sqrt{4} = 2 = \neq 3 + 5$$

The solution is $x = \neq 3$.

33. $(-5 + 9i)(-3 - 5i) = 15 + 25i - 27i - 45i^2$

$$= 15 - 2i + 45 = 60 - 2i$$

34.

$$\frac{x^2}{72} + \frac{x}{9} = \frac{1}{8}$$

$$x^2 + 8x = 9$$

$$x^2 + 8x \cdot 9 = 0, \quad a = 1, \, b = 8, \, c = \cdot 9$$

$$x = \frac{\cdot 8 \pm \sqrt{8^2 \cdot 4 \cdot 1 \cdot (\cdot 9)}}{2 \cdot 1}$$

$$= \frac{\cdot 8 \pm \sqrt{64 + 36}}{2} = \frac{\cdot 8 \pm \sqrt{100}}{2}$$

$$= \frac{\cdot 8 \pm 10}{2} = 1 \text{ or } \cdot 9$$

The solutions are –9 or 1.

Chapter 10 Functions—Inverse, Exponential, and Logarithmic

35. $x^{1/2} - 3x^{1/4} + 2 = 0$

Substitute $u = x^{1/4}$

$u^2 - 3u + 2 = 0$

$(u-2)(u-1) = 0$

$u - 2 = 0$ or $u - 1 = 0$

$u = 2$ or $u = 1$

$x^{1/4} = 2$ or $x^{1/4} = 1$

$x = 2^4$ or $x = 1^4$

$x = 16$ or $x = 1$

The solutions are 16 and 1.

36. $(x+5)(x-2)(x-7) \leq 0$

$(x+5)(x-2)(x-7) = 0$ is the related equation. $x = -5$, $x = 2$, $x = 7$ are the critical numbers.

Separate into regions using –5, 2, and 7 as boundaries.

Select test points: –6, 0, 3, and 8.

When $x = -6$, $(-6+5)(-6-2)(-6-7) = -1(-8)(-13) = -104 \leq 0$ True.

When $x = 0$, $(0+5)(0-2)(0-7) = 5(-2)(-7) = 70 \not\leq 0$ False.

When $x = 3$, $(3+5)(3-2)(3-7) = 8(1)(-4) = -32 \leq 0$ True.

When $x = 8$, $(8+5)(8-2)(8-7) = 13(6)(1) = 78 \not\leq 0$ False.

The solution set is: $\{x | x \leq -5 \cup 2 \leq x \leq 7\}$ or $(-\infty, -5] \cup [2, 7]$.

37. $\dfrac{-2x+10}{x-1} \leq 0$

$x = 5$, $x = 1$ are the critical numbers.

Separate into regions using 1 and 5 as boundaries.

Select test points: 0, 2, and 6.

When $x = 0$, $\dfrac{-2(0)+10}{0-1} = \dfrac{10}{-1} = -10 \leq 0$ True.

When $x = 2$, $\dfrac{-2(2)+10}{2-1} = \dfrac{6}{1} = 6 \not\leq 0$ False.

When $x = 6$, $\dfrac{-2(6)+10}{6-1} = \dfrac{-2}{5} = -\dfrac{2}{5} \leq 0$ True.

The solution set is: $\{x | x < 1 \cup x \geq 5\}$ or $(-\infty, 1) \cup [5, +\infty)$.

38. $y = -(x-3)^2 + 2$

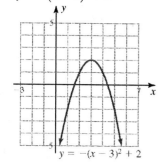

39.
$$x^2 + y^2 - 6x - 10y + 30 = 0$$
$$\left(x^2 - 6x\right) + \left(y^2 - 10y\right) = -30$$
$$\left(x^2 - 6x + 9\right) + \left(y^2 - 10y + 25\right) = -30 + 9 + 25$$
$$(x-3)^2 + (y-5)^2 = 4$$
$h = 3, \; k = 5, \; r = 2$
Center: $(3, 5); \; r = 2$

40. $\dfrac{y^2}{16} - \dfrac{x^2}{4} = 1$

Center: $(0, 0)$

y^2 term is positive, so $a = 4, b = 2$

Vertices: $(0, 4), (0, -4)$

Auxilliary rectangle passes through

$x = \pm 2$ and $y = \pm 4$

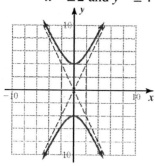

41. $x = y^2 - 4$

Only one variable is squared, thus the conic is a parabola.

42. Find x when $R = C$:
$$x^2 - 55 = 23x + 85$$
$$x^2 - 23x - 140 = 0$$
$$(x - 28)(x + 5) = 0$$
$$x - 28 = 0 \quad \text{or} \quad x + 5 = 0$$
$$x = 28 \quad \text{or} \quad x = -5$$

28 units must be manufactured and sold to break even.

43. Domain of $y = \sqrt{x+9}$

$x + 9 - 0$

$x - -9$

Domain: $\{x \,|\, x - -9\}$

44. $f(x)=x^4; \quad g(x)=4-x$

$(f \circ g)(x) = f(g(x)) = f(4-x) = (4-x)^4$

45. $f(x) = y = 4x - 6$

Interchange x and y and solve for y:

$$x = 4y - 6$$
$$x + 6 = 4y$$
$$\frac{x+6}{4} = y$$
$$f^{-1}(x) = \frac{x+6}{4}$$

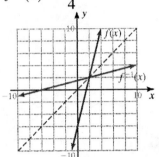

46. $f(x) = 2^x$

Interchange x and y and solve for y:

$x = 2^y$
The inverse is a function.

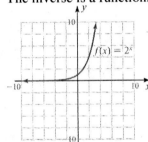

Yes, the inverse is a function.

47.
$$\log_b \sqrt{\frac{21}{83}} = \log_b \frac{-21-}{-83-}^{1/2}$$
$$= \frac{1}{2} \log_b \frac{21}{83}$$
$$= \frac{1}{2} \left(\log_b 21 - \log_b 83 \right)$$
$$= \frac{1}{2} \log_b 21 - \frac{1}{2} \log_b 83$$

48.
$$e^{9.6x} = 7$$
$$\ln e^{9.6x} = \ln 7$$
$$9.6x = \ln 7$$
$$x = \frac{\ln 7}{9.6} \approx \frac{1.94591}{9.6} \approx 0.2027$$

49.
$$\log_3 5 = \frac{\log 5}{\log 3} \approx 1.4650$$

50.
$$8392 = 1000 e^{k(7)}$$
$$8.392 = e^{7k}$$
$$\ln 8.392 = 7k$$
$$k = \frac{\ln 8.392}{7} \approx 0.3039$$

Appendix A: Sequences and Series

A.1 Sequences and Series

Problems A.1

1. $a_1 = 4, a_2 = 8, a_3 = 12, a_4 = 16$

$a_2 = 4 \cdot 2 = 8$

$a_4 = 4 \cdot 4 = 16$

$a_{10} = 4 \cdot 10 = 40$

$a_n = 4n$

2. $a_1 = \cdot 4 = (\cdot 1)^1 \cdot 4 \cdot 1, a_2 = 8 = (\cdot 1)^2 \cdot 4 \cdot 2,$

$a_3 = \cdot 12 = (\cdot 1)^3 \cdot 4 \cdot 3, a_4 = 16 = (\cdot 1)^4 \cdot 4 \cdot 4$

$a_{10} = (\cdot 1)^{10} \cdot 4 \cdot 10 = 40$

$a_n = (\cdot 1)^n 4n$

3. **a.** $6 \cdot 2 = 6 \cdot 2^1 = 12,$

$6 \cdot 2 \cdot 2 = 6 \cdot 2^2 = 24,$

$6 \cdot 2 \cdot 2 \cdot 2 = 6 \cdot 2^3 = 48, \ldots$

b. $6 \cdot 2^4 = 96$

c. $6 \cdot 2^n$

4. $a_n = \frac{1}{3}(n^2 - 1)$

$a_1 = \frac{1}{3}(1^2 - 1) = \frac{1}{3}(0) = 0$

$a_2 = \frac{1}{3}(2^2 - 1) = \frac{1}{3}(3) = 1$

$a_3 = \frac{1}{3}(3^2 - 1) = \frac{1}{3}(8) = \frac{8}{3}$

$a_{10} = \frac{1}{3}(10^2 - 1) = \frac{1}{3}(99) = 33$

5. $a(n) = n(n \cdot 1)$

$a(1) = 1(1 \cdot 1) = 1 \cdot 0 = 0$

$a(2) = 2(2 \cdot 1) = 2 \cdot 1 = 2$

$a(3) = 3(3 \cdot 1) = 3 \cdot 2 = 6$

$a(4) = 4(4 \cdot 1) = 4 \cdot 3 = 12$

$a(5) = 5(5 \cdot 1) = 5 \cdot 4 = 20$

sequence: 0, 2, 6, 12, 20, . . .

6. **a.** $S_4 = -2 + 2 + (-6) + 6 = 0$

b. $S_7 = -2 + 2 + (-6) + 6 + (-10) + 10 + (-14)$

$= -14$

7. **a.** $\sum_{n=1}^{3} n^3 = 1^3 + 2^3 + 3^3 = 1 + 8 + 27 = 36$

b. $\cdot \sum_{k=0}^{3} (3k + 1) = (3 \cdot 0 + 1) + (3 \cdot 1 + 1) + (3 \cdot 2 + 1) + (3 \cdot 3 + 1) = 1 + 4 + 7 + 10 = 22$

8. **a.** $3 + 6 + 9 + 12 + 15 = \sum_{n=1}^{5} 3n$

b. $\frac{1}{3} + \frac{1}{4} + \frac{1}{5} + \ldots + \frac{1}{20} = \sum_{n=3}^{20} \frac{1}{n}$

9. **a.** 9th planet: $2 \cdot 192 + 4 = 388$

b. 10th planet: $2 \cdot 384 + 4 = 772$

Exercises A.1

1. $a_1 = 1, a_2 = 4, a_3 = 7, a_4 = 10$
$a_{10} = 3 \cdot 10 - 2 = 28$
$a_n = 3n - 2$

3. $a_1 = 5, a_2 = 8, a_3 = 11, a_4 = 14$
$a_{10} = 3 \cdot 10 + 2 = 32$
$a_n = 3n + 2$

5. $a_1 = 20, a_2 = 25, a_3 = 30, a_4 = 35$
$a_{10} = 5 \cdot 10 + 15 = 65$
$a_n = 5n + 15$

7. $a_1 = 50, a_2 = 45, a_3 = 40, a_4 = 35$
$a_{10} = 55 \cdot 5 \cdot 10 = 5$
$a_n = 55 \cdot 5n$

9. $a_1 = \dfrac{1}{2}, a_2 = \dfrac{1}{3}, a_3 = \dfrac{1}{4}, a_4 = \dfrac{1}{5}$
$a_{10} = \dfrac{1}{10+1} = \dfrac{1}{11}$
$a_n = \dfrac{1}{n+1}$

11. $a_1 = 1, a_2 = -1, a_3 = 1, a_4 = -1$
$a_{10} = (-1)^{10-1} = (-1)^9 = -1$
$a_n = (-1)^{n-1}$

13. $a_1 = x, a_2 = x^2, a_3 = x^3, a_4 = x^4$
$a_{10} = x^{10}$
$a_n = x^n$

15. $a_1 = x, a_2 = -x^3, a_3 = x^5, a_4 = -x^7$
$a_{10} = (-1)^{10-1} x^{2(10)-1} = (-1)^9 x^{19} = -x^{19}$
$a_n = (-1)^{n-1} x^{2n-1}$

17. $a_1 = x, a_2 = -x, a_3 = x, a_4 = -x$
$a_{10} = (-1)^{10-1} x = (-1)^9 x = -x$
$a_n = (-1)^{n-1} x$

19. $a_1 = x, a_2 = \dfrac{x^2}{2}, a_3 = \dfrac{x^3}{3}, a_4 = \dfrac{x^4}{4}$
$a_{10} = \dfrac{x^{10}}{10}$
$a_n = \dfrac{x^n}{n}$

21. $a_1 = \dfrac{x}{2}, a_2 = \left| \dfrac{x^2}{4} \right|, a_3 = \dfrac{x^3}{8}, a_4 = \left| \dfrac{x^4}{16} \right|$
$a_{10} = (|1)^{10\,|1} \left| \dfrac{x}{2} \right|^{10} = \left| \dfrac{x^{10}}{1024} \right|$
$a_n = (|1)^{n\,|1} \left| \dfrac{x}{2} \right|^n$

23. $a_n = 2n \cdot 3$
$a_1 = 2 \cdot 1 \cdot 3 = 2 \cdot 3 = \cdot 1$
$a_2 = 2 \cdot 2 \cdot 3 = 4 \cdot 3 = 1$
$a_3 = 2 \cdot 3 \cdot 3 = 6 \cdot 3 = 3$

25.

$$a_n = \frac{n(n-2)}{2}$$

$$a_1 = \frac{1(1-2)}{2} = \frac{1(-1)}{2} = -\frac{1}{2}$$

$$a_2 = \frac{2(2-2)}{2} = \frac{2(0)}{2} = 0$$

$$a_3 = \frac{3(3-2)}{2} = \frac{3(1)}{2} = \frac{3}{2}$$

27.

$$a(n) = 1 - \frac{1}{n}$$

$$a(1) = 1 - \frac{1}{1} = 1 - 1 = 0$$

$$a(2) = 1 - \frac{1}{2} = \frac{1}{2}$$

$$a(3) = 1 - \frac{1}{3} = \frac{2}{3}$$

29.

$$a_n = n^2$$

$$a_1 = 1^2 = 1$$

$$a_2 = 2^2 = 4$$

$$a_3 = 3^2 = 9$$

31.

$$a(n) = \frac{n}{2n+1}$$

$$a(1) = \frac{1}{2\cdot 1 + 1} = \frac{1}{3}$$

$$a(2) = \frac{2}{2\cdot 2 + 1} = \frac{2}{5}$$

$$a(3) = \frac{3}{2\cdot 3 + 1} = \frac{3}{7}$$

33.

$$a_n = (-1)^n$$

$$a_1 = (-1)^1 = -1$$

$$a_2 = (-1)^2 = 1$$

$$a_3 = (-1)^3 = -1$$

35.

$$a(n) = (\cdot 1)^n 2^{\cdot n}$$

$$a(1) = (\cdot 1)^1 2^{\cdot 1} = \cdot 1 \cdot \frac{1}{2} = \cdot \frac{1}{2}$$

$$a(2) = (\cdot 1)^2 2^{\cdot 2} = 1 \cdot \frac{1}{2^2} = \frac{1}{4}$$

$$a(3) = (\cdot 1)^3 2^{\cdot 3} = \cdot 1 \cdot \frac{1}{2^3} = \cdot \frac{1}{8}$$

37.

$$\sum_{k=1}^{6} k^2 = 1^2 + 2^2 + 3^2 + 4^2 + 5^2 + 6^2 = 1 + 4 + 9 + 16 + 25 + 36 = 91$$

39.

$$\sum_{k=1}^{4} k^3 = 1^3 + 2^3 + 3^3 + 4^3 = 1 + 8 + 27 + 64 = 100$$

41.

$$\sum_{i=1}^{7} 3 = 3 + 3 + 3 + 3 + 3 + 3 + 3 = 21$$

43.

$$\cdot \sum_{j=1}^{4} \frac{1}{2j} = \frac{1}{2\cdot 1} + \frac{1}{2\cdot 2} + \frac{1}{2\cdot 3} + \frac{1}{2\cdot 4} = \frac{1}{2} + \frac{1}{4} + \frac{1}{6} + \frac{1}{8} = \frac{12 + 6 + 4 + 3}{24} = \frac{25}{24}$$

45.

$$\sum_{k=1}^{7} \frac{k+1}{k} = \frac{1+1}{1} + \frac{2+1}{2} + \frac{3+1}{3} + \frac{4+1}{4} + \frac{5+1}{5} + \frac{6+1}{6} + \frac{7+1}{7} = \frac{2}{1} + \frac{3}{2} + \frac{4}{3} + \frac{5}{4} + \frac{6}{5} + \frac{7}{6} + \frac{8}{7}$$

$$= \frac{840 + 630 + 560 + 525 + 504 + 490 + 480}{420} = \frac{4029}{420} = \frac{1343}{140}$$

47. $\displaystyle\sum_{k=1}^{5}(-1)^{k+1}=1+(-1)+1+(-1)+1=1$

49. $1+2+3+\ldots+200=\displaystyle\sum_{n=1}^{200} n$

51. $1+\dfrac{1}{2}+\dfrac{1}{3}+\dfrac{1}{4}+\ldots+\dfrac{1}{50}=\displaystyle\sum_{n=1}^{50}\dfrac{1}{n}$

53. $1-2+3-4+5-6+\ldots-50=\displaystyle\sum_{n=1}^{50}(-1)^{n-1}\, n$

55. $1+6+11+16+21=\displaystyle\sum_{n=1}^{5}(5n-4)$

57. The depreciation is \$40 less each year. So, $a_n=1420-40n$
   a. $a_8=1420-40\cdot 8=1420-320=\$1100$
   b. $a_{10}=1420-40\cdot 10=1420-400$
       $=\$1020$

59. **a.** 1st bounce: $\frac{9}{10}\cdot 10=9$ ft
   b. 3rd bounce: $\frac{9}{10}\cdot\frac{9}{10}\cdot\frac{9}{10}\cdot 10=\frac{729}{100}$ ft
   c. nth bounce: $\dfrac{9^n}{10^{n-1}}$

61. **a.** 2 hr: $100\cdot 2\cdot 2=400$
   b. 4 hr: $100\cdot 2\cdot 2\cdot 2\cdot 2=1600$
   c. n hr: $100\cdot 2^n$

63. Sales on nth day: $100\cdot 2^{n-1}$
   Sales on Saturday (6th day):
      $100\cdot 2^{6-1}=100\cdot 2^5=100\cdot 32=\$3200$

65. $7+(n-1)(3)=7+3n-3=4+3n$

67. $\dfrac{1}{2}n(16+32n-16)=\dfrac{1}{2}n(32n)=16n^2$

69. $5n^2+n-328=(5n+41)(n-8)$

71. $a_8=a_6+a_7=8+13=21$

73. $a_{10}=a_8+a_9=21+34=55$

75. $a_{12}=a_{10}+a_{11}=55+89=144$

77. Sample answer: A finite sequence has a specific number of terms – it stops somewhere. An infinite sequence goes on forever.

79. Sample answer: If the general term is known, then only one sequence can result. Thus the sequence is unique.

81. $a_n=\dfrac{1}{2}(n^2-1)$
$a_1=\dfrac{1}{2}(1^2-1)=\dfrac{1}{2}(0)=0$
$a_2=\dfrac{1}{2}(2^2-1)=\dfrac{1}{2}(3)=\dfrac{3}{2}$
$a_3=\dfrac{1}{2}(3^2-1)=\dfrac{1}{2}(8)=4$
$a_{10}=\dfrac{1}{2}(10^2-1)=\dfrac{1}{2}(99)=\dfrac{99}{2}$

83. $a_1=-3=(-1)^1\cdot 3\cdot 1,\ a_2=6=(-1)^2\cdot 3\cdot 2,$
$a_3=-9=(-1)^3\cdot 3\cdot 3,\ a_4=12=(-1)^4\cdot 3\cdot 4$
$a_{10}=(-1)^{10}\cdot 3\cdot 10=30$
$a_n=(-1)^n\cdot 3n$

85. a.
$$6+12+18+24=\sum_{n=1}^{4}6n$$

b.
$$2\,\Sigma 4+8\,\Sigma 16+32\,\Sigma 64=\sum_{n=1}^{6}(\Sigma 1)^{n\Sigma}\,2^{n}$$

A.2   Arithmetic Sequences and Series

Problems A.2

1. a. The common difference is $9-5=4$

b. The common difference is $6-12=-6$

2. a. The first term is $a_1=3$.

b. The common difference is $d=8-3=5$

c. $a_n=a_1+(n\cdot 1)d$
$a_8=3+(8\cdot 1)5=3+7\cdot 5=3+35=38$

d. $a_n=3+(n-1)5=3+5n-5=5n-2$

3. a. $a_1=16,\ d=32,\ n=20$
$a_{20}=16+(20\cdot 1)32=16+19\cdot 32$
$\phantom{a_{20}}=16+608=624$
$S_{20}=\dfrac{20}{2}(16+624)=10\cdot 640=6400\text{ ft}$

b. $a_1=16,\ d=32,\ n=x$
$a_x=16+(x\cdot 1)32=16+32x\cdot 32$
$=32x\cdot 16$
$S_x=\dfrac{x}{2}(16+32x\cdot 16)=\dfrac{x}{2}\cdot 32x=16x^2\text{ ft}$

4. a. $S_8=104,\ a_8=20$
$S_8=\dfrac{8}{2}(a_1+a_8)$
$104=4(a_1+20)$
$104=4a_1+80$
$24=4a_1$
$6=a_1$

b. $a_1=6,\ a_8=20,\ n=8$
$a_8=a_1+(n-1)d$
$20=6+(8-1)d$
$20=6+7d$
$14=7d$
$2=d$

5. $a_1=4000,\ n=8,\ d=-200$
$a_n=a_1+(n-1)d$
$a_8=4000+(8-1)(-200)$
$=4000+7(-200)$
$=2600$
$S_8=\dfrac{8}{2}(a_1+a_8)$
$=4(4000+2600)$
$=4(6600)$
$=26,400$
Remaining value:
$$\$50,000-\$26,400=\$23,600$$

6. $a_1=3,\ d=10,\ S_n=304$
$S_n=\dfrac{n}{2}(a_1+a_n),\quad a_n=a_1+(n\cdot 1)d$
$S_n=\dfrac{n}{2}(a_1+a_1+(n\cdot 1)d)=\dfrac{n}{2}(2a_1+(n\cdot 1)d)$
$304=\dfrac{n}{2}(2\cdot 3+(n\cdot 1)10)$
$608=n(6+10n\cdot 10)$
$608=n(10n\cdot 4)$
$0=10n^2\cdot 4n\cdot 608$
$0=5n^2\cdot 2n\cdot 304$
$0=(5n+38)(n\cdot 8)$
$n=\cdot\frac{38}{5}$ or $n=8$ days $\left(\text{Delete }\cdot\frac{38}{5}\right)$

Exercises A.2

1.   5, 8, 11, 14, . . .

    a. $a_1 = 5$

    b. $d = 8 - 5 = 3$

    c. $a_n = 5 + (n-1)3 = 5 + 3n - 3 = 3n + 2$

3.   11, 6, 1, −4, . . .

    a. $a_1 = 11$

    b. $d = 6 - 11 = -5$

    c. $a_n = 11 + (n-1)(-5) = 11 - 5n + 5$

         $= 16 - 5n$

5.   3, −1, −5, −9, . . .

    a. $a_1 = 3$

    b. $d = -1 - 3 = -4$

    c. $a_n = 3 + (n-1)(-4) = 3 - 4n + 4$

         $= 7 - 4n$

7. $\dfrac{1}{2}, \dfrac{1}{4}, 0, -\dfrac{1}{4}, \ldots$

    a. $a_1 = \dfrac{1}{2}$

    b. $d = \dfrac{1}{4} - \dfrac{1}{2} = -\dfrac{1}{4}$

    c. $a_n = \dfrac{1}{2} + (n-1)\left(-\dfrac{1}{4}\right) = \dfrac{1}{2} - \dfrac{1}{4}n + \dfrac{1}{4}$

         $= \dfrac{3}{4} - \dfrac{1}{4}n = \dfrac{3-n}{4}$

9. $-\dfrac{5}{6}, -\dfrac{1}{3}, \dfrac{1}{6}, \dfrac{2}{3}, \ldots$

    a. $a_1 = -\dfrac{5}{6}$

    b. $d = -\dfrac{1}{3} - \left(-\dfrac{5}{6}\right) = \dfrac{1}{2}$

    c. $a_n = -\dfrac{5}{6} + (n-1)\left(\dfrac{1}{2}\right) = -\dfrac{5}{6} + \dfrac{1}{2}n - \dfrac{1}{2}$

         $= \dfrac{1}{2}n - \dfrac{4}{3} = \dfrac{3n-8}{6}$

11. $a_1 = 7, d = 6, n = 15$

    $a_{15} = 7 + (15-1)6 = 7 + 14 \cdot 6 = 7 + 84 = 91$

    $S_{15} = \dfrac{15}{2}(7 + 91) = \dfrac{15}{2} \cdot 98 = 735$

13.   4, 10, 16, 22, . . .

    $a_1 = 4$

    $d = 10 - 4 = 6$

    $a_8 = 4 + (8 - 1)6 = 4 + 7 \cdot 6 = 4 + 42 = 46$

    $S_8 = \dfrac{8}{2}(4 + 46) = 4(50) = 200$

15. $a_1 = 3, a_6 = 8, n = 6$

    $8 = 3 + (6 - 1)d$

    $8 = 3 + 5d$

    $5 = 5d$

    $1 = d$

    $S_6 = \dfrac{6}{2}(3 + 8) = 3 \cdot 11 = 33$

17. $a_1 = 6, \ S_{14} = -280$

$$-280 = \frac{14}{2}(6 + a_{14})$$
$$-280 = 42 + 7a_{14}$$
$$-322 = 7a_{14}$$
$$-46 = a_{14}$$
$$-46 = 6 + (14 - 1)d$$
$$-52 = 13d$$
$$-4 = d$$

19. $d = 40, \ S_{40} = 40$

$$S_{40} = \frac{n}{2}(a_1 + a_1 + (n \cdot 1)d)$$
$$40 = 20(2a_1 + 39 \cdot 40)$$
$$2 = 2a_1 + 1560$$
$$2a_1 = \cdot 1558$$
$$a_1 = \cdot 779$$
$$a_{40} = \cdot 779 + (40 \cdot 1)40 = \cdot 779 + 39 \cdot 40$$
$$= \cdot 779 + 1560 = 781$$

21. $a_1 = 1380, \ n = 20, \ d = -40$

$$a_n = a_1 + (n - 1)d$$
$$a_{20} = 1380 + (20 - 1)(-40)$$
$$= 1380 + 19(-40)$$
$$= 620$$
$$S_{20} = \frac{20}{2}(a_1 + a_8)$$
$$= 10(1380 + 620)$$
$$= 10(2000)$$
$$= 20,000$$

Property's value:
$$\$30,000 - \$20,000 = \$10,000$$

23.   Find S_{10}:

$$a_1 = 1, \ d = 4, \ n = 10$$
$$S_n = \tfrac{n}{2}(a_1 + a_n), \ \ a_n = a_1 + (n \cdot 1)d$$
$$S_n = \tfrac{n}{2}(a_1 + a_1 + (n \cdot 1)d) = \tfrac{n}{2}(2a_1 + (n \cdot 1)d)$$
$$S_{10} = \frac{10}{2}(2 \cdot 1 + (10 \cdot 1)4) = 5(2 + 9 \cdot 4)$$
$$= 5(2 + 36) = 5(38) = 190$$

25. $1, 2, 3, 4, \ldots, n, \ldots$

$$S_n = \frac{n}{2}(1 + n) = \frac{n(n+1)}{2}$$

27. $2, 4, 6, 8, \ldots, 2n, \ldots$

$$S_n = \frac{n}{2}(2 + 2n) = \frac{n(2)(n+1)}{2}$$
$$= n(n+1) = n^2 + n$$

29. r^{n-1} for $r = 2, \ n = 6$

$$2^{6-1} = 2^5 = 32$$

31. $\dfrac{1 - r^n}{1 - r}$ for $r = 2, \ n = 6$

$$\frac{1 - 2^6}{1 - 2} = \frac{1 - 64}{-1} = \frac{-63}{-1} = 63$$

33. $\dfrac{10}{1 - r}$ for $r = \dfrac{1}{2}$

$$\frac{10}{1 - \frac{1}{2}} = \frac{10}{\frac{1}{2}} = 20$$

35. $S_5 = 35,000 - 5,000 = 30,000$

$a_1 = 10,000 \quad n = 5, \quad \text{Find } d.$

$S_n = \dfrac{n}{2}\big(2a_1 + (n-1)d\big)$

$30,000 = \dfrac{5}{2}\big(2(10,000) + (5-1)d\big)$

$60,000 = 5(20,000 + 4d)$

$12,000 = 20,000 + 4d$

$-8,000 = 4d$

$-2,000 = d$

The depreciation for each of the remaining four years is $8000, $6000, $4000, and $2000.

37. **a.** $C = 12,000, \quad S = 6000, \quad N = 5$

$b_t = 12,000 \;\bigg|\; t \;\left|\dfrac{12,000 \;\big|\; 6000}{5}\right|$

$= 12,000 \;\bigg|\; t \left|\dfrac{6000}{5}\right| = 12,000 \;\big|\; 1200t$

b. $b_0 = 12,000 \cdot\; 1200 \cdot 0 = \$12,000$

$b_1 = 12,000 \cdot\; 1200 \cdot 1 = \$10,800$

$b_2 = 12,000 \cdot\; 1200 \cdot 2 = \$9600$

$b_3 = 12,000 \cdot\; 1200 \cdot 3 = \$8400$

$b_4 = 12,000 \cdot\; 1200 \cdot 4 = \$7200$

$b_5 = 12,000 \cdot\; 1200 \cdot 5 = \$6000$

39. Sample answer: An arithmetic series is the sum of the terms of an arithmetic sequence.

41. Answers will vary.

43. $5, 9, 13, 17, \ldots$

   a. $a_1 = 5$

   b. $d = 9 \cdot\; 5 = 4$

   c. $a_{10} = 5 + (10 \cdot\; 1)4 = 5 + 9 \cdot 4 = 41$

   d. $a_n = 5 + (n \cdot\; 1)4 = 5 + 4n \cdot\; 4 = 4n + 1$

45. $S_8 = 136, \quad a_8 = 24, \quad n = 8$

   a. $136 = \dfrac{8}{2}(a_1 + 24)$

   $136 = 4a_1 + 96$

   $40 = 4a_1$

   $a_1 = 10$

   b. $24 = 10 + (8 - 1)d$

   $14 = 7d$

   $d = 2$

47. $a_1 = 30, \quad d = 10, \quad S_n = 750, \quad \text{Find } n$

$S_n = \dfrac{n}{2}\big(2a_1 + (n \cdot\; 1)d\big)$

$750 = \dfrac{n}{2}\big(2 \cdot 30 + (n \cdot\; 1)10\big)$

$1500 = n(60 + 10n \cdot\; 10)$

$1500 = n(10n + 50)$

$0 = 10n^2 + 50n \cdot\; 1500$

$0 = n^2 + 5n \cdot\; 150$

$0 = (n + 15)(n \cdot\; 10)$

$n = \cdot\; 15 \text{ or } n = 10 \qquad (\text{Delete } \cdot\; 15)$

It takes 10 months to accumulate $750.

A.3   Geometric Sequences and Series

Problems A.3

1. **a.**   The first term is $a_1 = 3$

b.   The common ratio is $r = \dfrac{1}{3}$.

c.   Use the formula $a_n = a_1 r^{n-1}$ for $n = 6$:

$$a_6 = 3\left|\frac{1}{3}\right|^{6-1} = 3\left|\frac{1}{3}\right|^5 = 3\left|\frac{1}{243}\right| = \frac{1}{81}$$

d. $a_n = 3 \cdot \left|\frac{1}{3}\right|^{n-1} = 3 \cdot 3^{1-n} = 3^{2-n} = \dfrac{1}{3^{n-2}}$

2. **a.** $a_1 = 4,\ r = \dfrac{-8}{4} = -2$

$a_n = a_1 r^{n-1} = 4(-2)^{n-1}$
$= 2^2(-1)^{n-1}2^{n-1} = (-1)^{n-1}2^{n+1}$

b. $S_n = \dfrac{a_1(1-r^n)}{1-r} = \dfrac{4\left(1-(-2)^n\right)}{1-(-2)}$
$= \dfrac{4}{3}\left(1-(-2)^n\right)$

3. $a_3 = a_1 r^2 = 2r^2$

$S_3 = \dfrac{a_1(1-r^3)}{1-r} = \dfrac{2(1-r^3)}{1-r}$
$= \dfrac{2(1-r)(1+r+r^2)}{1-r} = 2(1+r+r^2)$
$62 = 2(1+r+r^2)$
$31 = 1+r+r^2$
$0 = r^2 + r - 30$
$0 = (r+6)(r-5)$
$r = -6$ or $r = 5$
If $r = -6,\ a_3 = 2(-6)^2 = 2 \cdot 36 = 72$
If $r = 5,\ a_3 = 2(5)^2 = 2 \cdot 25 = 50$

4. $2 - 1 + \dfrac{1}{2} - \dfrac{1}{4} + \dots$

$a_1 = 2,\ r = \dfrac{-1}{2} = -\dfrac{1}{2}$

Since $|r| < 1,\ S = \dfrac{a_1}{1-r}$

$S = \dfrac{2}{1-\left(-\frac{1}{2}\right)} = \dfrac{2}{\frac{3}{2}} = \dfrac{4}{3}$

5. $\displaystyle\sum_{n=1}^{\infty} (0.99)^n = (0.99)^1 + (0.99)^2 + \dots$

$r = \dfrac{(0.99)^2}{(0.99)^1} = 0.99 < 1$

$S = \dfrac{0.99}{1 - 0.99} = \dfrac{0.99}{0.01} = 99$

6. $0.373737\dots = \dfrac{37}{100} + \dfrac{37}{(100)^2} + \dfrac{37}{(100)^3} + \dots$

$a_1 = \dfrac{37}{100},\ r = \dfrac{1}{100}$

$S = \dfrac{\frac{37}{100}}{1 - \frac{1}{100}} = \dfrac{\frac{37}{100}}{\frac{99}{100}} = \dfrac{37}{99}$

7.   Job A: 50, 100, 150, . . . (arithmetic)

$a_1 = 50, \ d = 50, \ n = 7$

$$S_7 = \frac{7}{2}\big(2\cdot50+(7-1)50\big) = \frac{7}{2}(100+300)$$

$$= \frac{7}{2}(400) = \$1400$$

Job B: 0.50, 1.00, 2.00, . . . (geometric)

$a_1 = 0.50, \ r = 2, \ n = 7$

$$S_7 = \frac{0.50\big(1-2^7\big)}{1-2} = \frac{0.50(1-128)}{-1}$$

$$= \frac{0.50(-127)}{-1} = \$63.50$$

8. $a_1 = 1, \ r = 2$

a. $a_{60} = 1\cdot2^{60-1} = 2^{59}$

b. $S_{60} = \dfrac{1\big(1-2^{60}\big)}{1-2} = \dfrac{1-2^{60}}{-1} = 2^{60}-1$

Exercises A.3

1.   3, 6, 12, 24, . . .

a. $a_1 = 3$

b. $r = \dfrac{6}{3} = 2$

c. $a_n = a_1 r^{n-1} = 3\big(2^{n-1}\big)$

d. $S_n = \dfrac{a_1\big(1-r^n\big)}{1-r} = \dfrac{3\big(1-2^n\big)}{1-2} = \dfrac{3\big(1-2^n\big)}{-1}$

$$= 3\big(2^n-1\big)$$

3.   8, 24, 72, 216, . . .

a. $a_1 = 8$

b. $r = \dfrac{24}{8} = 3$

c. $a_n = a_1 r^{n-1} = 8\big(3^{n-1}\big)$

d. $S_n = \dfrac{a_1\big(1-r^n\big)}{1-r} = \dfrac{8\big(1-3^n\big)}{1-3} = \dfrac{8\big(1-3^n\big)}{-2}$

$$= 4\big(3^n-1\big)$$

5. $16, \ -4, 1, \ -\dfrac{1}{4}, \ldots$

a. $a_1 = 16$

b. $r = \dfrac{-4}{16} = -\dfrac{1}{4}$

c. $a_n = a_1 r^{n-1} = 16\left(-\dfrac{1}{4}\right)^{n-1}$

$$= 4^2\big(-4^{-1}\big)^{n-1} = (-4)^2(-4)^{-n+1}$$

$$= (-4)^{3-n}$$

d. $S_n = \dfrac{a_1\big(1-r^n\big)}{1-r} = \dfrac{16\big(1-\left(-\frac{1}{4}\right)^n\big)}{1-\left(-\frac{1}{4}\right)}$

$$= \dfrac{16\big(1-\left(-\frac{1}{4}\right)^n\big)}{\frac{5}{4}} = \dfrac{64}{5}\big(1-\left(-\tfrac{1}{4}\right)^n\big)$$

7. $-\dfrac{3}{5}, \dfrac{3}{2}, \ -\dfrac{15}{4}, \dfrac{75}{8}, \ldots$

a. $a_1 = -\dfrac{3}{5}$

b. $r = \dfrac{\frac{3}{2}}{-\frac{3}{5}} = -\dfrac{3}{2}\cdot\dfrac{5}{3} = -\dfrac{5}{2}$

c. $a_n = a_1 r^{n-1} = -\dfrac{3}{5}\left(-\dfrac{5}{2}\right)^{n-1}$

d. $S_n = \dfrac{a_1\big(1-r^n\big)}{1-r} = \dfrac{-\frac{3}{5}\big(1-\left(-\frac{5}{2}\right)^n\big)}{1-\left(-\frac{5}{2}\right)}$

$$= \dfrac{-\frac{3}{5}\big(1-\left(-\frac{5}{2}\right)^n\big)}{\frac{7}{2}} = -\dfrac{6}{35}\big(1-\left(-\tfrac{5}{2}\right)^n\big)$$

9.
$$-\frac{3}{4},\ \frac{1}{4},\ -\frac{1}{12},\ \frac{1}{36},\ \ldots$$

a. $a_1 = -\frac{3}{4}$

b. $r = \dfrac{\frac{1}{4}}{-\frac{3}{4}} = -\frac{1}{4}\cdot\frac{4}{3} = -\frac{1}{3}$

c. $a_n = a_1 r^{n-1} = -\frac{3}{4}\left(-\frac{1}{3}\right)^{n-1} = -\frac{1}{4}\cdot 3\cdot 3^{1-n}$

$\qquad = -\frac{1}{4}\cdot 3^{2-n}$

d. $S_n = \dfrac{a_1\left(1-r^n\right)}{1-r} = \dfrac{-\frac{3}{4}\left(1-\left(-\frac{1}{3}\right)^n\right)}{1-\frac{1}{3}}$

$\qquad = \dfrac{-\frac{3}{4}\left(1-\left(-\frac{1}{3}\right)^n\right)}{\frac{2}{3}} = -\frac{9}{8}\left(1-\left(-\frac{1}{3}\right)^n\right)$

11.
$$a_1 = 1,\ S_3 = \frac{7}{4}$$

$a_3 = a_1 r^2 = 1r^2$

$S_3 = \dfrac{a_1\left(1-r^3\right)}{1-r} = \dfrac{1\left(1-r^3\right)}{1-r}$

$\qquad = \dfrac{(1-r)\left(1+r+r^2\right)}{1-r} = 1+r+r^2$

$\dfrac{7}{4} = 1+r+r^2$

$7 = 4r^2+4r+4$

$0 = 4r^2+4r-3$

$0 = (2r+3)(2r-1)$

$r = -\dfrac{3}{2}$ or $r = \dfrac{1}{2}$

If $r = -\dfrac{3}{2}$, $a_3 = \left(-\dfrac{3}{2}\right)^2 = \dfrac{9}{4}$

If $r = \dfrac{1}{2}$, $a_3 = \left(\dfrac{1}{2}\right)^2 = \dfrac{1}{4}$

13. $a_1 = 3,\ S_3 = 21$

$a_3 = a_1 r^2 = 3r^2$

$S_3 = \dfrac{a_1\left(1-r^3\right)}{1-r} = \dfrac{3\left(1-r^3\right)}{1-r}$

$\quad = \dfrac{3(1-r)\left(1+r+r^2\right)}{1-r} = 3\left(1+r+r^2\right)$

$21 = 3\left(1+r+r^2\right)$

$7 = 1+r+r^2$

$0 = r^2+r-6$

$0 = (r+3)(r-2)$

$r = -3$ or $r = 2$

If $r = -3$, $a_3 = 3(-3)^2 = 3\cdot 9 = 27$

If $r = 2$, $a_3 = 3(2)^2 = 3\cdot 4 = 12$

15. $r = 2,\ S_8 = 1785$

$a_8 = a_1 r^7 = a_1\cdot 2^7 = 128a_1$

$S_8 = \dfrac{a_1\left(1-r^8\right)}{1-r}$

$1785 = \dfrac{a_1\left(1-2^8\right)}{1-2} = \dfrac{a_1\left(1-256\right)}{-1} = 255a_1$

$a_1 = 7$

$a_8 = 128\cdot 7 = 896$

17.

$$a_1 = -4, \ a_n = 108, \ S_n = 80$$

$$a_n = a_1 r^{n-1} = -4r^{n-1}$$

$$108 = -4r^{n-1}$$

$$-27 = r^{n-1}$$

$$-27 = \frac{r^n}{r}$$

$$-27r = r^n$$

$$S_n = \frac{a_1(1-r^n)}{1-r}$$

$$80 = \frac{-4(1-r^n)}{1-r}$$

$$-20 = \frac{1-r^n}{1-r}$$

$$-20(1-r) = 1-(-27r)$$

$$20r - 20 = 1 + 27r$$

$$-7r = 21$$

$$r = -3$$

$$-27(-3) = (-3)^n$$

$$(-3)^4 = (-3)^n$$

$$n = 4$$

19.

$$a_1 = \frac{16}{125}, \ r = \frac{5}{2}, \ a_n = \frac{25}{2}$$

$$a_n = a_1 r^{n \cdot 1} = \frac{16}{125} r^{n \cdot 1}$$

$$\frac{25}{2} = \frac{16}{125} \cdot \frac{5}{2}^{n \cdot 1}$$

$$\frac{25}{2} \cdot \frac{125}{16} = \cdot \frac{5}{2}^{n \cdot 1}$$

$$\frac{5^2}{2} \cdot \frac{5^3}{2^4} = \cdot \frac{5}{2}^{n \cdot 1}$$

$$\cdot \frac{5}{2}^{5} = \cdot \frac{5}{2}^{n \cdot 1}$$

$$5 = n \cdot 1$$

$$n = 6$$

$$S_6 = \frac{\frac{16}{125}\left(1 \cdot \left(\frac{5}{2}\right)^6\right)}{1 \cdot \frac{5}{2}} = \frac{\frac{16}{125}\left(1 \cdot \frac{15625}{64}\right)}{\cdot \frac{3}{2}}$$

$$= \frac{\frac{16}{125}\left(\cdot \frac{15561}{64}\right)}{\cdot \frac{3}{2}}$$

$$= \frac{16}{125} \cdot \frac{15561}{64} \cdot \cdot \frac{2}{3} = \frac{5187}{250}$$

21.

$$6 + 3 + 1\frac{1}{2} + \dots$$

$$r = \frac{3}{6} = \frac{1}{2}$$

$$S = \frac{a_1}{1-r} = \frac{6}{1-\frac{1}{2}} = \frac{6}{\frac{1}{2}} = 12$$

23.

$$(\ |6) + (\ |3) + \left|\ 1\frac{3}{2}\ \right| + \dots$$

$$r = \frac{|3}{|6} = \frac{1}{2}$$

$$S = \frac{a_1}{1\ |r} = \frac{|6}{1\ |\frac{1}{2}} = \frac{|6}{\frac{1}{2}} = |12$$

25.

$$\sum_{n=1}^{\Sigma} (\Sigma 1)^{n\Sigma} 2^{2\Sigma n} = 2\Sigma 1 + \frac{1}{2}\Sigma\frac{1}{4} + \dots$$

$$r = \frac{\Sigma 1}{2} = \Sigma\frac{1}{2}$$

$$S = \frac{a_1}{1\Sigma r} = \frac{2}{1\Sigma\left(\Sigma\frac{1}{2}\right)} = \frac{2}{\frac{3}{2}} = \frac{4}{3}$$

27.

$$4 - 8 + 16 - 32 + \dots$$

$$r = \frac{-8}{4} = -2 \quad |r| > 1$$

Sum does not exist.

29.

$$\frac{1}{10} + \frac{1}{5} + \frac{2}{5} + \dots$$

$$r = \frac{\frac{1}{5}}{\frac{1}{10}} = \frac{10}{5} = 2 \quad |r| > 1$$

Sum does not exist.

31.

$$0.555\dots = \frac{5}{10} + \frac{5}{(10)^2} + \frac{5}{(10)^3} + \dots$$

$$a_1 = \frac{5}{10}, \ r = \frac{1}{10} \qquad S = \frac{\frac{5}{10}}{1-\frac{1}{10}} = \frac{\frac{5}{10}}{\frac{9}{10}} = \frac{5}{9}$$

33.
$$0.181818\ldots = \frac{18}{100} + \frac{18}{(100)^2} + \frac{18}{(100)^3} + \ldots$$

$$a_1 = \frac{18}{100}, \quad r = \frac{1}{100}$$

$$S = \frac{\frac{18}{100}}{1 - \frac{1}{100}} = \frac{\frac{18}{100}}{\frac{99}{100}} = \frac{18}{99} = \frac{2}{11}$$

35.
$$4.050505\ldots = 4 + \frac{5}{100} + \frac{5}{(100)^2} + \frac{5}{(100)^3} + \ldots$$

$$a_1 = \frac{5}{100}, \quad r = \frac{1}{100}$$

$$S = 4 + \frac{\frac{5}{100}}{1 - \frac{1}{100}} = 4 + \frac{\frac{5}{100}}{\frac{99}{100}} = 4 + \frac{5}{99} = \frac{401}{99}$$

37.
$$2.3161616\ldots = \frac{1}{10}(23.161616)$$

$$= \frac{1}{10}\left[23 + \frac{16}{100} + \frac{16}{(100)^2} + \frac{16}{(100)^3} + \ldots\right]$$

$$a_1 = \frac{16}{100}, \quad r = \frac{1}{100}$$

$$S = \frac{1}{10}\left[23 + \frac{\frac{16}{100}}{1 - \frac{1}{100}}\right] = \frac{1}{10}\left[23 + \frac{\frac{16}{100}}{\frac{99}{100}}\right]$$

$$= \frac{1}{10}\left[23 + \frac{16}{99}\right] = \frac{1}{10}\left[\frac{2293}{99}\right] = \frac{2293}{990}$$

39. $0.140140140\ldots$

$$= \frac{140}{1000} + \frac{140}{(1000)^2} + \frac{140}{(1000)^3} + \ldots$$

$$a_1 = \frac{140}{1000}, \quad r = \frac{1}{1000}$$

$$S = \frac{\frac{140}{1000}}{1 - \frac{1}{1000}} = \frac{\frac{140}{1000}}{\frac{999}{1000}} = \frac{140}{999}$$

41. a. $P_0, \ 1.04P_0, \ (1.04)^2 P_0, \ (1.04)^3 P_0, \ldots$

b. $P_n = (1.04)^n P_0$

$$P_5 = (1.04)^5(20{,}000)$$
$$= 1.21665(20{,}000) = 24{,}333$$

43. $62.5, 50, 40, 32, \ldots$

$$a_1 = 62.5, \quad r = 0.8, \quad n = 4$$

$$S_4 = \frac{62.5(1 - 0.8^4)}{1 - 0.8} = \frac{62.5(1 - 0.4096)}{0.2}$$

$$= \frac{62.5(0.5904)}{0.2} = 184.5 \text{ cm}$$

45. $a_1 = 10{,}000, \quad r = 1.5, \quad n = 5$

$$S_5 = \frac{10{,}000(1 - 1.5^5)}{1 - 1.5} = \frac{10{,}000(1 - 7.59375)}{-0.5}$$

$$= \frac{10{,}000(-6.59375)}{-0.5} = \$131{,}875$$

47.   arithmetic seq: $a,\ a+d,\ a+2d$

geometric seq: $a+1,\ a+d+3,\ a+2d+13$

$a+a+d+a+2d=9$

$\qquad 3a+3d=9$

$\qquad a+d=3$

$\qquad\qquad a=3-d$

geometric seq:

$\quad 3-d+1,\ 3-d+d+3,\ 3-d+2d+13$

or $4-d,\ 6,\ d+16$

$r=\dfrac{6}{4-d}=\dfrac{d+16}{6}$

$(d+16)(4-d)=36$

$-d^2-12d+64=36$

$d^2+12d-28=0$

$(d+14)(d-2)=0$

$d=-14$ or $d=2$

If $d=-14,\ a=3-(-14)=17$

If $d=2,\ a=3-2=1$

The sequence is $1, 3, 5$ or $17, 3, -11$.

49.   sequence: $1, -2, 4, -8, \ldots$

$a_1=1,\ r=-2,\ n=18$

$S_{18}=\dfrac{1\left(1-(-2)^{18}\right)}{1-(-2)}=\dfrac{1(1-262{,}144)}{3}$

$\qquad =\dfrac{-262{,}143}{3}=-87{,}381$ cents

Roberto lost \$873.81.

51.   downward sequence: $8, 6, 4.5, \ldots$

$a_1=8,\ r=\dfrac{3}{4}$

$S_{down}=\dfrac{8}{1-\frac{3}{4}}=\dfrac{8}{\frac{1}{4}}=32$

upward sequence: $6, 4.5, 3.375, \ldots$

$a_1=6,\ r=\dfrac{3}{4}$

$S_{up}=\dfrac{6}{1-\frac{3}{4}}=\dfrac{6}{\frac{1}{4}}=24$

Total distance $=32+24=56$ ft

53. $(a+b)^2=(a+b)(a+b)=a^2+2ab+b^2$

55. $(x+y)^3=(x+y)(x+y)(x+y)$

$\qquad =\left(x^2+2xy+y^2\right)(x+y)$

$\qquad =x^3+x^2y+2x^2y+2xy^2+xy^2+y^3$

$\qquad =x^3+3x^2y+3xy^2+y^3$

57. $(y-2z)^3=(y-2z)(y-2z)(y-2z)$

$\qquad =\left(y^2-4yz+4z^2\right)(y-2z)$

$\qquad =y^3-2y^2z-4y^2z+8yz^2+4yz^2-8z^3$

$\qquad =y^3-6y^2z+12yz^2-8z^3$

59. $\left|\dfrac{1}{x}\ |\ \dfrac{1}{y}\right|^3=\left|\dfrac{1}{x}\ |\ \dfrac{1}{y}\right|\left|\dfrac{1}{x}\ |\ \dfrac{1}{y}\right|\left|\dfrac{1}{x}\ |\ \dfrac{1}{y}\right|$

$\qquad =\left|\dfrac{1}{x^2}\ |\ \dfrac{2}{xy}+\dfrac{1}{y^2}\right|\left|\dfrac{1}{x}\ |\ \dfrac{1}{y}\right|$

$\qquad =\dfrac{1}{x^3}\ |\ \dfrac{1}{x^2y}\ |\ \dfrac{2}{x^2y}+\dfrac{2}{xy^2}+\dfrac{1}{xy^2}\ |\ \dfrac{1}{y^3}$

$\qquad =\dfrac{1}{x^3}\ |\ \dfrac{3}{x^2y}+\dfrac{3}{xy^2}\ |\ \dfrac{1}{y^3}$

61.
$$1 + \frac{2}{5} + \frac{4}{25} + \frac{8}{125} + \ldots$$

Geometric: $r = \frac{2}{5}$

$$S = \frac{1}{1 - \frac{2}{5}} = \frac{1}{\frac{3}{5}} = \frac{5}{3}$$

63.
$$1 + \frac{2}{5} - \frac{1}{5} - \frac{4}{5} - \ldots$$

Arithmetic: $d = -\frac{3}{5}$

65.
$$2 + \frac{7}{4} + \frac{49}{32} + \frac{343}{256} + \ldots$$

Geometric: $r = \frac{7}{8}$

$$S = \frac{2}{1 - \frac{7}{8}} = \frac{2}{\frac{1}{8}} = 16$$

67. $5 + 4 + 2 - 1 - \ldots$

Neither

69.
$$\frac{1}{2} + \frac{2}{3} + \frac{3}{4} + \frac{4}{5} + \ldots$$

Neither

71.
$$4 - 2 + 1 - \frac{1}{2} + \ldots$$

Geometric: $r = -\frac{1}{2}$

$$S = \frac{4}{1 - \left(-\frac{1}{2}\right)} = \frac{4}{\frac{3}{2}} = \frac{8}{3}$$

73. Sample answer: An arithmetic sequence has the same number added to each term to get the next term. A geometric sequence has the same number multiplies by each term to get the next term.

75. Sample answer: An infinite geometric series finds the sum of all the terms in the geometric sequence, while an infinite geometric sequence simply lists the elements in the sequence.

77.
$$2, 1, \frac{1}{2}, \frac{1}{4}, \ldots$$

a. $a_1 = 2$

b. $r = \frac{1}{2}$

c. $a_6 = a_1 r^{6-1} = 2\left|\frac{1}{2}\right|^5 = 2\left|\frac{1}{32}\right| = \frac{1}{16}$

d. $a_n = a_1 r^{n-1} = 2\left|\frac{1}{2}\right|^{n-1} = 2(2)^{1-n}$

$$= 2^{2-n} = \frac{1}{2^{n-2}}$$

79. $a_1 = 2, \ S_3 = 42$

$a_3 = a_1 r^2 = 2r^2$

$$S_3 = \frac{a_1(1 - r^3)}{1 - r} = \frac{2(1 - r^3)}{1 - r}$$

$$= \frac{2(1 - r)(1 + r + r^2)}{1 - r} = 2(1 + r + r^2)$$

$42 = 2(1 + r + r^2)$

$21 = 1 + r + r^2$

$0 = r^2 + r - 20$

$0 = (r + 5)(r - 4)$

$r = -5$ or $r = 4$

If $r = -5, \ a_3 = 2(-5)^2 = 2 \cdot 25 = 50$

If $r = 4, \ a_3 = 2(4)^2 = 2 \cdot 16 = 32$

81. $\sum_{n=1}^{\Sigma}(\Sigma 1)^{n+1}(1.01)^{n}$

$$= 1.01\Sigma(1.01)^{2}+(1.01)^{3}\Sigma\ldots$$

$r=\Sigma 1.01 \quad |r|>1$

Sum does not exist.

83. **a.** 100, 200, 400, 800, 1600

b. $a_{1}=100, \; r=2, \; n=6$

$a_{6}=100(2)^{6}=100(64)=\$6400$

c. $a_{n}=100\cdot 2^{n}$

d. $100\cdot 2^{n}>100,000$

$2^{n}>1000$

$2^{10}=1024>1000$
It would take 10 years to have over
$100,000.

A.4 The Binomial Expansion

Problems A.4

1. **a.** $8!=8\cdot 7\cdot 6\cdot 5\cdot 4\cdot 3\cdot 2\cdot 1=40,320$

  b. $\dfrac{9!}{7!}=\dfrac{9\cdot 8\cdot \not7\cdot \not6\cdot \not5\cdot \not4\cdot \not3\cdot \not2\cdot \not1}{\not7\cdot \not6\cdot \not5\cdot \not4\cdot \not3\cdot \not2\cdot \not1}=72$

2. $\begin{matrix}\cdot 7\cdot \\ : 5:\end{matrix}=\dfrac{7!}{5!(7-5)!}=\dfrac{7!}{5!2!}=\dfrac{7\cdot 6\cdot \not5!}{\not5!2\cdot 1}=\dfrac{42}{2}=21$

3. $\begin{vmatrix}5\\5\end{vmatrix}=1, \; \begin{vmatrix}5\\4\end{vmatrix}=5, \; \begin{vmatrix}5\\3\end{vmatrix}=10, \; \begin{vmatrix}5\\2\end{vmatrix}=10,$

$\begin{vmatrix}5\\1\end{vmatrix}=5, \; \begin{vmatrix}5\\0\end{vmatrix}=1$

4. $(2x\cdot y)^{4}=\begin{matrix}\cdot 4\cdot \\ : 4:\end{matrix}(2x)^{4}+\begin{matrix}\cdot 4\cdot \\ : 3:\end{matrix}(2x)^{3}(\cdot y)^{1}+\begin{matrix}\cdot 4\cdot \\ : 2:\end{matrix}(2x)^{2}(\cdot y)^{2}+\begin{matrix}\cdot 4\cdot \\ : 1:\end{matrix}(2x)^{1}(\cdot y)^{3}+\begin{matrix}\cdot 4\cdot \\ : 0:\end{matrix}(2x)^{0}(\cdot y)^{4}$

$=1\cdot 16x^{4}+4\cdot 8x^{3}(\cdot y)+6\cdot 4x^{2}y^{2}+4\cdot 2x(\cdot y^{3})+1\cdot 1\cdot y^{4}$

$=16x^{4}\cdot 32x^{3}y+24x^{2}y^{2}\cdot 8xy^{3}+y^{4}$

5. $n=7, \; r=3, \; x=x, \; y=\cdot 2$

$\begin{matrix}\cdot 7\cdot \\ : 3:\end{matrix}=\dfrac{7!}{3!(7-3)!}=\dfrac{7!}{3!4!}=\dfrac{7\cdot \not6\cdot 5\cdot \not4!}{\not3\cdot \not2\cdot 1\cdot \not4!}=35$

$\begin{matrix}\cdot 7\cdot \\ : 3:\end{matrix}x^{3}(\cdot 2)^{4}=35x^{3}\cdot 16=560x^{3}$

coefficient is 560

6. $n=7, \; r=3$

$\begin{matrix}\cdot 7\cdot \\ : 3:\end{matrix}=\dfrac{7!}{3!(7-3)!}=\dfrac{7!}{3!4!}=\dfrac{7\cdot \not6\cdot 5\cdot \not4!}{\not3\cdot \not2\cdot 1\cdot \not4!}=35$

7. $n=10, \; r=5, \; p=\tfrac{1}{2}$

$\begin{matrix}\cdot 10\cdot \\ : 5:\end{matrix}=\dfrac{10!}{5!(10-5)!}=\dfrac{10!}{5!5!}=\dfrac{10\cdot 9\cdot 8\cdot 7\cdot \not6\cdot \not5!}{5\cdot 4\cdot \not3\cdot \not2\cdot 1\cdot \not5!}=\dfrac{5040}{20}=252$

$\begin{matrix}\cdot 10\cdot \\ : 5:\end{matrix}\begin{matrix}\cdot 1\cdot \\ : 2:\end{matrix}^{5}\cdot 1\cdot \begin{matrix}1\cdot \\ 2:\end{matrix}^{5}=\begin{matrix}\cdot 10\cdot \\ : 5:\end{matrix}\begin{matrix}\cdot 1\cdot \\ : 2:\end{matrix}^{5}\cdot \begin{matrix}1\cdot \\ 2:\end{matrix}^{5}=\begin{matrix}\cdot 10\cdot \\ : 5:\end{matrix}\begin{matrix}\cdot 1\cdot \\ : 2:\end{matrix}^{10}=252\cdot \dfrac{1}{1024}\cdot =\dfrac{252}{1024}=\dfrac{63}{256}$

Exercises A.4

1. $3! = 3 \cdot 2 \cdot 1 = 6$

3. $10! = 10 \cdot 9 \cdot 8 \cdot 7 \cdot 6 \cdot 5 \cdot 4 \cdot 3 \cdot 2 \cdot 1 = 3{,}628{,}800$

5. $\dfrac{6!}{2!} = \dfrac{6 \cdot 5 \cdot 4 \cdot 3 \cdot \cancel{2} \cdot \cancel{1}}{\cancel{2} \cdot \cancel{1}} = 360$

7. $\dfrac{9!}{6!} = \dfrac{9 \cdot 8 \cdot 7 \cdot \cancel{6} \cdot \cancel{5} \cdot \cancel{4} \cdot \cancel{3} \cdot \cancel{2} \cdot \cancel{1}}{\cancel{6} \cdot \cancel{5} \cdot \cancel{4} \cdot \cancel{3} \cdot \cancel{2} \cdot \cancel{1}} = 504$

9. $\binom{6}{2} = \dfrac{6!}{2!(6-2)!} = \dfrac{6!}{2!4!} = \dfrac{6 \cdot 5 \cdot \cancel{4!}}{2 \cdot 1 \cdot \cancel{4!}} = 15$

11. $\binom{11}{1} = \dfrac{11!}{1!(11-1)!} = \dfrac{11!}{1!10!} = \dfrac{11 \cdot \cancel{10!}}{1 \cdot \cancel{10!}} = 11$

13. $\binom{4}{0} = \dfrac{4!}{0!(4-0)!} = \dfrac{\cancel{4!}}{0!\cancel{4!}} = 1$

15.
$$(a+3b)^4 = \binom{4}{4}(a)^4 + \binom{4}{3}(a)^3(3b)^1 + \binom{4}{2}(a)^2(3b)^2 + \binom{4}{1}(a)^1(3b)^3 + \binom{4}{0}(a)^0(3b)^4$$
$$= 1 \cdot a^4 + 4 \cdot a^3(3b) + 6 \cdot a^2 \cdot 9b^2 + 4 \cdot a \cdot 27b^3 + 1 \cdot 1 \cdot 81b^4$$
$$= a^4 + 12a^3b + 54a^2b^2 + 108ab^3 + 81b^4$$

17.
$$(x+4)^4 = \binom{4}{4}(x)^4 + \binom{4}{3}(x)^3(4)^1 + \binom{4}{2}(x)^2(4)^2 + \binom{4}{1}(x)^1(4)^3 + \binom{4}{0}(x)^0(4)^4$$
$$= 1 \cdot x^4 + 4 \cdot x^3(4) + 6 \cdot x^2 \cdot 16 + 4 \cdot x \cdot 64 + 1 \cdot 1 \cdot 256$$
$$= x^4 + 16x^3 + 96x^2 + 256x + 256$$

19. $(2x - y)^5$
$$= \binom{5}{5}(2x)^5 + \binom{5}{4}(2x)^4(-y)^1 + \binom{5}{3}(2x)^3(-y)^2 + \binom{5}{2}(2x)^2(-y)^3 + \binom{5}{1}(2x)^1(-y)^4 + \binom{5}{0}(-y)^5$$
$$= 1 \cdot 32x^5 + 5 \cdot 16x^4(-y) + 10 \cdot 8x^3y^2 + 10 \cdot 4x^2(-y^3) + 5 \cdot 2xy^4 + 1 \cdot 1 \cdot (-y^5)$$
$$= 32x^5 - 80x^4y + 80x^3y^2 - 40x^2y^3 + 10xy^4 - y^5$$

21. $(2x+3y)^5$
$$= \binom{5}{5}(2x)^5 + \binom{5}{4}(2x)^4(3y)^1 + \binom{5}{3}(2x)^3(3y)^2 + \binom{5}{2}(2x)^2(3y)^3 + \binom{5}{1}(2x)^1(3y)^4 + \binom{5}{0}(3y)^5$$
$$= 1 \cdot 32x^5 + 5 \cdot 16x^4 \cdot 3y + 10 \cdot 8x^3 \cdot 9y^2 + 10 \cdot 4x^2 \cdot 27y^3 + 5 \cdot 2x \cdot 81y^4 + 1 \cdot 243y^5$$
$$= 32x^5 + 240x^4y + 720x^3y^2 + 1080x^2y^3 + 810xy^4 + 243y^5$$

23.
$$\left(\frac{1}{x}\cdot\frac{y}{2}\right)^4 = \binom{4}{4}\left(\frac{1}{x}\right)^4 + \binom{4}{3}\left(\frac{1}{x}\right)^3\left(\frac{y}{2}\right)^1 + \binom{4}{2}\left(\frac{1}{x}\right)^2\left(\frac{y}{2}\right)^2 + \binom{4}{1}\left(\frac{1}{x}\right)^1\left(\frac{y}{2}\right)^3 + \binom{4}{0}\left(\frac{y}{2}\right)^4$$

$$= 1\cdot\frac{1}{x^4} + 4\cdot\frac{1}{x^3}\cdot\frac{y}{2} + 6\cdot\frac{1}{x^2}\cdot\frac{y^2}{4} + 4\cdot\frac{1}{x}\cdot\frac{y^3}{8} + 1\cdot\frac{y^4}{16}$$

$$= \frac{1}{x^4}\cdot\frac{2y}{x^3} + \frac{3y^2}{2x^2}\cdot\frac{y^3}{2x} + \frac{y^4}{16}$$

25.
$$(x+1)^6 = \binom{6}{6}x^6 + \binom{6}{5}x^5\cdot1^1 + \binom{6}{4}x^4\cdot1^2 + \binom{6}{3}x^3\cdot1^3 + \binom{6}{2}x^2\cdot1^4 + \binom{6}{1}x\cdot1^5 + \binom{6}{0}x^0\cdot1^6$$

$$= 1\cdot x^6 + 6\cdot x^5\cdot1 + 15\cdot x^4\cdot1 + 20\cdot x^3\cdot1 + 15\cdot x^2\cdot1 + 6\cdot x\cdot1 + 1\cdot1\cdot1$$

$$= x^6 + 6x^5 + 15x^4 + 20x^3 + 15x^2 + 6x + 1$$

27. $n=6$, $r=3$, $x=x$, $y=-3$

$$\binom{6}{3} = \frac{6!}{3!(6-3)!} = \frac{6!}{3!3!} = \frac{6\cdot5\cdot4\cdot3!}{3\cdot2\cdot1\cdot3!} = 20$$

$$\binom{6}{3}x^3(-3)^3 = 20x^3\cdot(-27) = -540x^3$$

coefficient is -540

29. $n=7$, $r=2$, $x=x$, $y=2y$

$$\binom{7}{2} = \frac{7!}{2!(7-2)!} = \frac{7!}{2!5!} = \frac{7\cdot6\cdot5!}{2\cdot1\cdot5!} = 21$$

$$\binom{7}{2}x^2(2y)^5 = 21x^2\cdot(32y^5) = 672x^2y^5$$

coefficient is 672

31. $n=8$, $r=4$, $x=2x$, $y=-1$

$$\binom{8}{4} = \frac{8!}{4!(8-4)!} = \frac{8!}{4!4!} = \frac{8\cdot7\cdot6\cdot5\cdot4!}{4\cdot3\cdot2\cdot1\cdot4!}$$
$$= 70$$

$$\binom{8}{4}(2x)^4(-1)^4 = 70\cdot16x^4\cdot1 = 1120x^4$$

coefficient is 1120

33. $n=5$, $r=2$, $x=\dfrac{a}{2}$, $y=-1$

$$\binom{5}{2} = \frac{5!}{2!(5-2)!} = \frac{5!}{2!3!} = \frac{5\cdot4\cdot3!}{2\cdot1\cdot3!} = 10$$

$$\binom{5}{2}\left(\frac{a}{2}\right)^2(-1)^3 = 10\cdot\frac{a^2}{4}\cdot(-1) = -\frac{5}{2}a^2$$

coefficient is $-\dfrac{5}{2}$

35. $n=6$, $r=3$

$$\binom{6}{3} = \frac{6!}{3!(6-3)!} = \frac{6!}{3!3!} = \frac{6\cdot5\cdot4\cdot3!}{3\cdot2\cdot1\cdot3!} = 20$$

37. $n=9$, $r=3$

$$\binom{9}{3} = \frac{9!}{3!(9-3)!} = \frac{9!}{3!6!} = \frac{9\cdot8\cdot7\cdot6!}{3\cdot2\cdot1\cdot6!} = 84$$

39. $n=6$, $r=2$, $p=\frac{1}{2}$

$$\binom{6}{2} = \frac{6!}{2!(6-2)!} = \frac{6!}{2!4!} = \frac{6\cdot5\cdot4!}{2\cdot1\cdot4!} = \frac{30}{2} = 15$$

$$\binom{6}{2}\left(\frac{1}{2}\right)^2\cdot1\cdot\left(\frac{1}{2}\right)^4 = \binom{6}{2}\left(\frac{1}{2}\right)^2\left(\frac{1}{2}\right)^4 = \binom{6}{2}\left(\frac{1}{2}\right)^6 = 15\cdot\frac{1}{64} = \frac{15}{64}$$

41. $n=9,\ r=3,\ p=\tfrac{1}{2}$

$$\binom{9}{3} = \frac{9!}{3!(9-3)!} = \frac{9!}{3!6!} = \frac{9\cdot8\cdot7\cdot\cancel{6!}}{3\cdot2\cdot1\cdot\cancel{6!}} = \frac{504}{6} = 84$$

$$\binom{9}{3}\left(\frac{1}{2}\right)^3\left(1-\frac{1}{2}\right)^6 = \binom{9}{3}\left(\frac{1}{2}\right)^3\left(\frac{1}{2}\right)^6 = \binom{9}{3}\left(\frac{1}{2}\right)^9 = 84\left(\frac{1}{512}\right) = \frac{84}{512} = \frac{21}{128}$$

43. $n=4,\ r=2,\ p=\tfrac{1}{6}$

$$\binom{4}{2} = \frac{4!}{2!(4-2)!} = \frac{4!}{2!2!} = \frac{4\cdot3\cdot\cancel{2!}}{2\cdot1\cdot\cancel{2!}} = \frac{12}{2} = 6$$

$$\binom{4}{2}\left(\frac{1}{6}\right)^2\left(1-\frac{1}{6}\right)^2 = \binom{4}{2}\left(\frac{1}{6}\right)^2\left(\frac{5}{6}\right)^2 = 6\left(\frac{1}{36}\right)\left(\frac{25}{36}\right) = \frac{150}{1296} = \frac{25}{216}$$

45.
$$\binom{n}{k} = \frac{n!}{k!(n-k)!} = \frac{n!}{(n-k)!k!} = \frac{n!}{(n-k)!\left(n-(n-k)\right)!} = \binom{n}{n-k}$$

47. Sample answer: $n!$ is the product of all the positive integers from 1 to n.

49. $6! = 6\cdot5\cdot4\cdot3\cdot2\cdot1 = 720$

51.
$$\binom{6}{2} = \frac{6!}{2!(6-2)!} = \frac{6!}{2!4!} = \frac{6\cdot5\cdot\cancel{4!}}{2\cdot1\cdot\cancel{4!}} = 15$$

53.
$$(2a-b)^4 = \binom{4}{4}(2a)^4 + \binom{4}{3}(2a)^3(-b)^1 + \binom{4}{2}(2a)^2(-b)^2 + \binom{4}{1}(2a)^1(-b)^3 + \binom{4}{0}(2a)^0(-b)^4$$

$$= 1\cdot16a^4 + 4\cdot8a^3(-b) + 6\cdot4a^2b^2 + 4\cdot2a(-b^3) + 1\cdot1\cdot b^4$$

$$= 16a^4 - 32a^3b + 24a^2b^2 - 8ab^3 + b^4$$

55. $n=6,\ r=2,\ x=2a,\ y=-3b$

$$\binom{6}{2} = \frac{6!}{2!(6-2)!} = \frac{6!}{2!4!} = \frac{6\cdot5\cdot\cancel{4!}}{2\cdot1\cdot\cancel{4!}} = 15$$

$$\binom{6}{2}(2a)^2(-3b)^4 = 15\cdot4a^2\cdot81b^4 = 4860a^2b^4$$

coefficient is 4860

57. $n=5,\ r=4$

$$\binom{5}{4} = \frac{5!}{4!(5-4)!} = \frac{5!}{4!1!} = \frac{5\cdot\cancel{4!}}{1\cdot\cancel{4!}} = 5$$

Review Exercises

1.   **a.** $a_1=1,\ a_2=3,\ a_3=5,\ a_4=7$
   $a_6 = 2-6-1 = 11$
   $a_n = 2n-1$

b. $a_1=3,\ a_2=6,\ a_3=9,\ a_4=12$
   $a_6 = 3\cdot6 = 18$
   $a_n = 3n$

Appendix A: Sequences and Series

2.   **a.** $4 \cdot 2 \cdot 2 \cdot 2 \cdot 2 = 64$ **b.** $a_n = 4 \cdot 2^n = 2^2 \cdot 2^n = 2^{n+2}$

3.   **a.**
$$a_n = \frac{1}{3}\left(n^2 + 1\right)$$
$$a_1 = \frac{1}{3}\left(1^2 + 1\right) = \frac{1}{3}(2) = \frac{2}{3}$$
$$a_2 = \frac{1}{3}\left(2^2 + 1\right) = \frac{1}{3}(5) = \frac{5}{3}$$
$$a_3 = \frac{1}{3}\left(3^2 + 1\right) = \frac{1}{3}(10) = \frac{10}{3}$$
$$a_{10} = \frac{1}{3}\left(10^2 + 1\right) = \frac{1}{3}(101) = \frac{101}{3}$$
$$S_3 = \frac{2}{3} + \frac{5}{3} + \frac{10}{3} = \frac{17}{3}$$

b.
$$a_n = \frac{1}{4}\left(n^2 + n\right)$$
$$a_1 = \frac{1}{4}\left(1^2 + 1\right) = \frac{1}{4}(2) = \frac{2}{4} = \frac{1}{2}$$
$$a_2 = \frac{1}{4}\left(2^2 + 2\right) = \frac{1}{4}(6) = \frac{6}{4} = \frac{3}{2}$$
$$a_3 = \frac{1}{4}\left(3^2 + 3\right) = \frac{1}{4}(12) = \frac{12}{4} = 3$$
$$a_{10} = \frac{1}{4}\left(10^2 + 10\right) = \frac{1}{4}(110) = \frac{110}{4} = \frac{55}{2}$$
$$S_3 = \frac{1}{2} + \frac{3}{2} + 3 = 5$$

4.   **a.**
$$a(n) = n^2 + 2$$
$$a(1) = 1^2 + 2 = 1 + 2 = 3$$
$$a(2) = 2^2 + 2 = 4 + 2 = 6$$
$$a(3) = 3^2 + 2 = 9 + 2 = 11$$
sequence: 3, 6, 11, . . .
$$\sum_{n=1}^{4}\left(n^2 + 2\right) = 3 + 6 + 11 + 18 = 38$$

b.
$$a(n) = 2n^2 \, \Sigma 1$$
$$a(1) = 2\,\boxed{1}^2 \, \Sigma 1 = 2\,\Sigma 1 = 1$$
$$a(2) = 2\,\boxed{2}^2 \, \Sigma 1 = 8\,\Sigma 1 = 7$$
$$a(3) = 2\,\boxed{3}^2 \, \Sigma 1 = 18\,\Sigma 1 = 17$$
sequence: 1, 7, 17, . . .
$$\sum_{n=1}^{3}\left(2n^2 \, \Sigma 1\right) = 1 + 7 + 17 = 25$$

5.   **a.**   In 1400, value is $1000.
In 2000, value is:
$$1000 \cdot 2 \cdot 2 \cdot 2 \cdot 2 \cdot 2 \cdot 2 = \$64,000$$

b.   In 1600, value is $1000.
In 2000, value is:
$$1000 \cdot 2 \cdot 2 \cdot 2 \cdot 2 = \$16,000$$

6.   **a.**   3, 6, 9, 12, . . .
$$d = 6 \cdot \, 3 = 3$$
$$a_{10} = 3 + (10 \cdot \, 1)3 = 3 + 9 \cdot 3 = 30$$

b.   4, 8, 12, 16, . . .
$$d = 8 \cdot \, 4 = 4$$
$$a_{10} = 4 + (10 \cdot \, 1)4 = 4 + 9 \cdot 4 = 40$$

7.   **a.**   3, 6, 9, 12, . . .
$$d = 6 - 3 = 3$$
$$a_n = 3 + (n-1)3 = 3 + 3n - 3 = 3n$$

b.   4, 8, 12, 16, . . .
$$d = 8 - 4 = 4$$
$$a_n = 4 + (n-1)4 = 4 + 4n - 4 = 4n$$

8.   **a.** $16, 48, 80, 112, \ldots$
$$d = 48 \cdot \, 16 = 32$$
$$a_6 = 16 + (6 \cdot \, 1)32 = 16 + 5 \cdot 32 = 176 \text{ ft}$$

b. $16, 48, 80, 112, \ldots$
$$d = 48 \cdot \, 16 = 32$$
$$a_8 = 16 + (8 \cdot \, 1)32 = 16 + 7 \cdot 32 = 240 \text{ ft}$$

9. **a.** $S_6 = 114, \ a_6 = 24, \ n = 6$

$114 = \dfrac{6}{2}(a_1 + 24)$

$114 = 3a_1 + 72$

$42 = 3a_1$

$a_1 = 14$

$24 = 14 + (6 - 1)d$

$10 = 5d$

$d = 2$

b. $S_6 = 99, \ a_6 = 24, \ n = 6$

$99 = \dfrac{6}{2}(a_1 + 24)$

$99 = 3a_1 + 72$

$27 = 3a_1$

$a_1 = 9$

$24 = 9 + (6 - 1)d$

$15 = 5d$

$d = 3$

10. **a.** $S_6 = 54, \ a_6 = 14, \ n = 6$

$54 = \dfrac{6}{2}(a_1 + 14)$

$54 = 3a_1 + 42$

$12 = 3a_1$

$a_1 = 4$

$14 = 4 + (6 - 1)d$

$10 = 5d$

$d = 2$

b. $S_6 = 54, \ a_6 = 16.5, \ n = 6$

$54 = \dfrac{6}{2}(a_1 + 16.5)$

$54 = 3a_1 + 49.5$

$4.5 = 3a_1$

$a_1 = 1.5$

$16.5 = 1.5 + (6 - 1)d$

$15 = 5d$

$d = 3$

11. **a.** $a_1 = 8000, \ n = 8, \ d = -200$

$a_n = a_1 + (n - 1)d$

$a_8 = 8000 + (8 - 1)(-200)$

$\quad = 8000 + 7(-200)$

$\quad = 6600$

$S_8 = \dfrac{8}{2}(a_1 + a_8)$

$\quad = 4(8000 + 6600)$

$\quad = 4(14,600)$

$\quad = 58,400$

Remaining value:

$\quad \$80,000 - \$58,400 = \$21,600$

b. $a_1 = 8000, \ n = 10, \ d = -200$

$a_n = a_1 + (n - 1)d$

$a_8 = 8000 + (10 - 1)(-200)$

$\quad = 8000 + 9(-200)$

$\quad = 6200$

$S_8 = \dfrac{10}{2}(a_1 + a_8)$

$\quad = 5(8000 + 6200)$

$\quad = 5(14,200)$

$\quad = 71,000$

Remaining value:

$\quad \$80,000 - \$71,000 = \$9000$

12. a. $a_1 = 5$, $d = 4$, $S_n = 230$

$S_n = \frac{n}{2}(a_1 + a_n)$, $a_n = a_1 + (n - 1)d$

$S_n = \frac{n}{2}\big(a_1 + a_1 + (n - 1)d\big) = \frac{n}{2}\big(2a_1 + (n - 1)d\big)$

$230 = \frac{n}{2}\big(2 \cdot 5 + (n - 1)4\big)$

$460 = n(10 + 4n - 4)$

$460 = n(4n + 6)$

$0 = 4n^2 + 6n - 460$

$0 = 2n^2 + 3n - 230$

$0 = (2n + 23)(n - 10)$

$n = -\frac{23}{2}$ or $n = 10$ weeks $\left(\text{Delete } -\frac{23}{2}\right)$

b. $a_1 = 5$, $d = 4$, $S_n = 324$

$S_n = \frac{n}{2}(a_1 + a_n)$, $a_n = a_1 + (n - 1)d$

$S_n = \frac{n}{2}\big(a_1 + a_1 + (n - 1)d\big) = \frac{n}{2}\big(2a_1 + (n - 1)d\big)$

$324 = \frac{n}{2}\big(2 \cdot 5 + (n - 1)4\big)$

$648 = n(10 + 4n - 4)$

$648 = n(4n + 6)$

$0 = 4n^2 + 6n - 648$

$0 = 2n^2 + 3n - 324$

$0 = (2n + 27)(n - 12)$

$n = -\frac{27}{2}$ or $n = 12$ weeks $\left(\text{Delete } -\frac{27}{2}\right)$

13. a. $3, 6, 12, 24, \ldots$

$a_1 = 3$, $r = \frac{6}{3} = 2$, $n = 6$

$a_6 = 3(2)^{6-1} = 3 \cdot 2^5 = 3 \cdot 32 = 96$

b. $4, 2, 1, \dfrac{1}{2}, \ldots$

$a_1 = 4$, $r = \frac{2}{4} = \frac{1}{2}$, $n = 6$

$a_6 = 4 \cdot \left(\frac{1}{2}\right)^{6-1} = 4 \cdot \left(\frac{1}{2}\right)^5 = 4 \cdot \frac{1}{32} = \frac{1}{8}$

14. a. $1, 4, 16, 64, \ldots$

$a_1 = 1$, $r = \frac{4}{1} = 4$

$a_n = 1(4)^{n-1} = 4^{n-1}$

b. $4, -2, 1, -\dfrac{1}{2}, \ldots$

$a_1 = 4$, $r = \frac{-2}{4} = -\frac{1}{2}$

$a_n = 4\left(-\frac{1}{2}\right)^{n-1} = 2^2 \left(-1\right)^{n-1} 2^{1-n}$

$= (-1)^{n-1} 2^{3-n} = \frac{(-1)^{n-1}}{2^{n-3}}$

15. a. $2, -4, 8, -16, \ldots$

$a_1 = 2$, $r = \frac{-4}{2} = -2$

$a_n = 2(-2)^{n-1} = 2^1(-1)^{n-1} 2^{n-1} = (-1)^{n-1} 2^n$

$S_n = \frac{2\big(1 - (-2)^n\big)}{1 - (-2)} = \frac{2}{3}\big(1 - (-2)^n\big)$

b. $1, -\dfrac{1}{2}, \dfrac{1}{4}, -\dfrac{1}{8}, \ldots$

$a_1 = 1$, $r = \frac{-\frac{1}{2}}{1} = -\frac{1}{2}$

$a_n = 1\left(-\frac{1}{2}\right)^{n-1} = \left(-\frac{1}{2}\right)^{n-1}$

$S_n = \frac{1\left(1 - \left(-\frac{1}{2}\right)^n\right)}{1 - \left(-\frac{1}{2}\right)} = \frac{\left(1 - \left(-\frac{1}{2}\right)^n\right)}{\frac{3}{2}}$

$= \frac{2}{3}\left(1 - \left(-\frac{1}{2}\right)^n\right)$

16. **a.** $32 - 16 + 8 - 4 + \ldots$

$a_1 = 32, \ r = \dfrac{-16}{32} = -\dfrac{1}{2}$

Since $|r| < 1, \ S = \dfrac{a_1}{1-r}$

$S = \dfrac{32}{1-\left(-\frac{1}{2}\right)} = \dfrac{32}{\frac{3}{2}} = \dfrac{64}{3}$

b. $18 - 6 + 2 - \dfrac{2}{3} + \ldots$

$a_1 = 18, \ r = \dfrac{-6}{18} = -\dfrac{1}{3}$

Since $|r| < 1, \ S = \dfrac{a_1}{1-r}$

$S = \dfrac{18}{1-\left(-\frac{1}{3}\right)} = \dfrac{18}{\frac{4}{3}} = \dfrac{54}{4} = \dfrac{27}{2}$

17. **a.** $1 + 1.005 + (1.005)^2 + (1.005)^3 + \ldots$

$a_1 = 1, \ r = \dfrac{1.005}{1} = 1.005$

Since $|r| > 1$, the sum does not exist.

b. $1 - 1.001 + (1.001)^2 - (1.001)^3 + \ldots$

$a_1 = 1, \ r = \dfrac{-1.001}{1} = -1.001$

Since $|r| > 1$, the sum does not exist.

18. **a.** $0.313131\ldots = \dfrac{31}{100} + \dfrac{31}{(100)^2} + \dfrac{31}{(100)^3} + \ldots$

$a_1 = \dfrac{31}{100}, \ r = \dfrac{1}{100}$

$S = \dfrac{\frac{31}{100}}{1 - \frac{1}{100}} = \dfrac{\frac{31}{100}}{\frac{99}{100}} = \dfrac{31}{99}$

b. $0.324324324\ldots$

$= \dfrac{324}{1000} + \dfrac{324}{(1000)^2} + \dfrac{324}{(1000)^3} + \ldots$

$a_1 = \dfrac{324}{1000}, \ r = \dfrac{1}{1000}$

$S = \dfrac{\frac{324}{1000}}{1 - \frac{1}{1000}} = \dfrac{\frac{324}{1000}}{\frac{999}{1000}} = \dfrac{324}{999} = \dfrac{12}{37}$

19. **a.** downward sequence: $8, 4, 2, \ldots$

$a_1 = 8, \ r = \dfrac{1}{2}$

$S_{down} = \dfrac{8}{1 - \frac{1}{2}} = \dfrac{8}{\frac{1}{2}} = 16$

upward sequence: $4, 2, 1, \ldots$

$a_1 = 4, \ r = \dfrac{1}{2}$

$S_{up} = \dfrac{4}{1 - \frac{1}{2}} = \dfrac{4}{\frac{1}{2}} = 8$

Total distance $= 16 + 8 = 24$ ft

b. downward sequence: $6, 3, 1.5, \ldots$

$a_1 = 6, \ r = \dfrac{1}{2}$

$S_{down} = \dfrac{6}{1 - \frac{1}{2}} = \dfrac{6}{\frac{1}{2}} = 12$

upward sequence: $3, 1.5, 0.75, \ldots$

$a_1 = 3, \ r = \dfrac{1}{2}$

$S_{up} = \dfrac{3}{1 - \frac{1}{2}} = \dfrac{3}{\frac{1}{2}} = 6$

Total distance $= 12 + 6 = 18$ ft

20. **a.** $8! = 8 \cdot 7 \cdot 6 \cdot 5 \cdot 4 \cdot 3 \cdot 2 \cdot 1 = 40,320$

b. $\dfrac{8!}{4!} = \dfrac{8 \cdot 7 \cdot 6 \cdot 5 \cdot \cancel{4} \cdot \cancel{3} \cdot \cancel{2} \cdot \cancel{1}}{\cancel{4} \cdot \cancel{3} \cdot \cancel{2} \cdot \cancel{1}} = 1680$

21. **a.** $\begin{array}{c} .9. \\ : 3: \end{array} = \dfrac{9!}{3!(9-3)!} = \dfrac{9!}{3!6!} = \dfrac{9 \cdot 8 \cdot 7 \cdot \cancel{6!}}{3 \cdot 2 \cdot 1 \cdot \cancel{6!}} = 84$

b. $\begin{array}{c} -8- \\ =8= \end{array} = \dfrac{8!}{8!(8-8)!} = \dfrac{8!}{8!0!} = 1$

22. a.

$$(a \cdot 3b)^4 = \binom{4}{4}(a)^4 + \binom{4}{3}(a)^3(\cdot 3b)^1 + \binom{4}{2}(a)^2(\cdot 3b)^2 + \binom{4}{1}(a)^1(\cdot 3b)^3 + \binom{4}{0}(a)^0(\cdot 3b)^4$$

$$= 1 \cdot a^4 + 4 \cdot a^3(\cdot 3b) + 6 \cdot a^2 \cdot 9b^2 + 4 \cdot a \cdot (\cdot 27b^3) + 1 \cdot 1 \cdot 81b^4$$

$$= a^4 \cdot 12a^3b + 54a^2b^2 \cdot 108ab^3 + 81b^4$$

b.

$$(3a+2b)^4 = \binom{4}{4}(3a)^4 + \binom{4}{3}(3a)^3(2b)^1 + \binom{4}{2}(3a)^2(2b)^2 + \binom{4}{1}(3a)^1(2b)^3 + \binom{4}{0}(3a)^0(2b)^4$$

$$= 1 \cdot 81a^4 + 4 \cdot 27a^3(2b) + 6 \cdot 9a^2 \cdot 4b^2 + 4 \cdot 3a \cdot (8b^3) + 1 \cdot 1 \cdot 16b^4$$

$$= 81a^4 + 216a^3b + 216a^2b^2 + 96ab^3 + 16b^4$$

23. a. $n = 8, \; r = 4, \; x = x, \; y = 2$

$$\binom{8}{4} = \frac{8!}{4!(8 \cdot 4)!} = \frac{8!}{4!4!} = \frac{8 \cdot 7 \cdot 6 \cdot 5 \cdot 4!}{4 \cdot 3 \cdot 2 \cdot 1 \cdot 4!}$$

$$= 70$$

$$\binom{8}{4} x^4(2)^4 = 70x^4 \cdot 16 = 1120x^3$$

coefficient is 1120

b. $n = 7, \; r = 4, \; x = 3x, \; y = \cdot 1$

$$\binom{7}{4} = \frac{7!}{4!(7 \cdot 4)!} = \frac{7!}{4!3!} = \frac{7 \cdot 6 \cdot 5 \cdot 4!}{3 \cdot 2 \cdot 1 \cdot 4!} = 35$$

$$\binom{7}{4}(3x)^4(\cdot 1)^3 = 35 \cdot 81x^4 \cdot (\cdot 1) = \cdot 2835x^3$$

coefficient is $\cdot 2835$

24. a. $n = 7, \; r = 4$

$$\binom{7}{4} = \frac{7!}{4!(7 \cdot 4)!} = \frac{7!}{4!3!} = \frac{7 \cdot 6 \cdot 5 \cdot 4!}{3 \cdot 2 \cdot 1 \cdot 4!} = 35$$

b. $n = 7, \; r = 3$

$$\binom{7}{3} = \frac{7!}{3!(7 \cdot 3)!} = \frac{7!}{3!4!} = \frac{7 \cdot 6 \cdot 5 \cdot 4!}{3 \cdot 2 \cdot 1 \cdot 4!} = 35$$

25. a. $n = 7, \; r = 4, \; p = \frac{1}{2}$

$$\binom{7}{4} = \frac{7!}{4!(7 \cdot 4)!} = \frac{7!}{4!3!} = \frac{7 \cdot 6 \cdot 5 \cdot 4!}{3 \cdot 2 \cdot 1 \cdot 4!} = \frac{210}{6} = 35$$

$$\binom{7}{4}\left(\frac{1}{2}\right)^4 \cdot 1 \cdot \left(\frac{1}{2}\right)^3 = \binom{7}{4}\left(\frac{1}{2}\right)^4\left(\frac{1}{2}\right)^3 = \binom{7}{4}\left(\frac{1}{2}\right)^7 = 35 \cdot \frac{1}{128} = \frac{35}{128}$$

b. $n = 7, \; r = 3, \; p = \frac{1}{2}$

$$\binom{7}{3} = \frac{7!}{3!(7 \cdot 3)!} = \frac{7!}{3!4!} = \frac{7 \cdot 6 \cdot 5 \cdot 4!}{3 \cdot 2 \cdot 1 \cdot 4!} = \frac{210}{6} = 35$$

$$\binom{7}{3}\left(\frac{1}{2}\right)^3 \cdot 1 \cdot \left(\frac{1}{2}\right)^4 = \binom{7}{3}\left(\frac{1}{2}\right)^3\left(\frac{1}{2}\right)^4 = \binom{7}{3}\left(\frac{1}{2}\right)^7 = 35 \cdot \frac{1}{128} = \frac{35}{128}$$

Notes

Notes

Notes

Notes